Engineering Ethics and Design for Product Safety

About the Author

Kenneth L. d'Entremont, Ph.D., P.E., is an associate professor (lecturer) in the Department of Mechanical Engineering and its Ergonomics & Safety Program at the University of Utah. He also holds an adjunct associate professor position at the Department of Family & Preventive Medicine's Rocky Mountain Center for Occupational and Environmental Health (RMCOEH). Ken is a registered Professional Engineer (P.E.) in two states and worked for years in industry before becoming faculty. His industry experience includes over a decade managing the product-safety efforts for a large, international designer and manufacturer of innovative and hazardous products.

Engineering Ethics and Design for Product Safety

Kenneth L. d'Entremont, Ph.D., P.E.

New York Chicago San Francisco
Athens London Madrid
Mexico City Milan New Delhi
Singapore Sydney Toronto

Library of Congress Control Number: 2020943727

McGraw Hill books are available at special quantity discounts to use as premiums and sales promotions or for use in corporate training programs. To contact a representative, please visit the Contact Us page at www.mhprofessional.com.

Engineering Ethics and Design for Product Safety

1 2 3 4 5 6 7 8 9 LCR 25 24 23 22 21 20

ISBN 978-1-260-46053-7
MHID 1-260-46053-3

This book is printed on acid-free paper.

Sponsoring Editor
Robert Argentieri

Editorial Supervisor
Stephen M. Smith

Production Supervisor
Lynn M. Messina

Acquisitions Coordinator
Elizabeth M. Houde

Project Manager
Jyoti Shaw, MPS Limited

Copy Editor
Girish Sharma, MPS Limited

Proofreader
Sanjeev Mishra, MPS Limited

Indexer
Maria van Beuren

Art Director, Cover
Jeff Weeks

Composition
MPS Limited

This book is dedicated to product-safety engineers and professionals from the past, in the present, and of the future.

Keep your values positive because your values become your destiny.

—MOHANDAS GANDHI[1]

[1] Also sometimes attributed to a Chinese or Buddhist proverb; wist.info.

Contents

Preface

I have found that a story leaves a deeper impression
when it is impossible to tell which side the author is on.
—LEO TOLSTOY[2]

Nothing can be all things to all people. This book is certainly no exception to that dictum. It is my hope that this book represents but a single step in a journey—to be made along with readers and other authors—to advance the understanding, study, and practical application of engineering ethics and product-safety engineering to better serve society.

Some of the material contained herein was written by others and has been available before. Other material is the author's original work. Of the existing material, much of it has been dispersed across time and various fields of study. These product-safety engineering elements have never been aggregated into a cohesive, easily referenced, or readily applicable collection for students of engineering or practicing engineers. These elements, or small "jig-saw puzzle pieces," have existed without benefit of a picture showing the completed project. This book creates an ad hoc product-safety engineering "toolbox" along with a set of assembly instructions. Many product-safety engineers in practice have, no doubt, already assembled their own toolboxes. It is hoped that my original material fills in some of the gaps that still exist in the overall jig-saw puzzle that is product-safety engineering and that my presentation of all material helps readers see the manner in which the material contained in the book may be applied to real-world engineering-design problems.

It is not the intent of this book to impart great wisdom upon seasoned product-safety engineers, although each of us should always remain open to learning new things at any point of life—and sometimes from unlikely sources. Instead, it is hoped that this book will help younger and less-experienced product-safety engineers, and other professionals with an interest in product safety, get a "jump start" in understanding this field. Many accomplished practitioners have had to cobble together product-safety knowledge bases of their own. There is no reason to continue this difficult, time-consuming learning path to gain information and understanding about such an important topic. We should all learn from one another. This book serves as my initial attempt to make a modest contribution to this field.

[2] In a letter to a friend; wist.info.

This book is fundamentally a design-engineering book which focuses upon the fields of engineering ethics and product-safety engineering. These are two important engineering-design areas to engineer and consumer alike. The first of the two areas, engineering ethics, is far too broad to treat in general. The discussion is, therefore, limited to the application of engineering ethics within the design-and-manufacturing corporation. The second area, product-safety engineering, is not well represented in the literature and, for all practical purposes, cannot be addressed without the first area. The term *Design for Product Safety* (DfPS) is coined and used to represent a concerted corporate effort to *design-in* safety to consumer products from the very start of product conception and design. Proper execution of DfPS should minimize the need for post-production "fixes" to product defects leading to product-safety recalls and other problems. This integrated approach to design engineering should deliver superior results for both company and consumer.

Unfortunately, there is no one-size-fits-all solution to product-safety engineering. There will be some product-safety engineers and professionals facing actual, important problems in their daily roles that are not directly addressed in this book. Product-safety engineers and professionals face unique issues, especially across the numerous positions, organizations, industries, and geographic regions involved. I ask that such readers be inventive and patient. Perhaps this book can still be of *indirect* help to their situations from parallels which may be drawn from an issue that *is* covered in this book. Lines had to be drawn when completing this book that necessarily limited its scope. Also, please consider that I view this book as but one installment in a continuing body of work.

It is my plan to continue the expansion of issues, products, and geographical regions considered—as well as the depths to which they are investigated and the perspectives from which they may be viewed—as future editions of this book are released.[3] I look forward to the *insightful* and *constructive* feedback[4] from readers that will ultimately lead to such improvements to this continuing work. I anticipate keeping, and making available, a list of inevitable clarifications and corrections that may help readers.[5] In addition, I hope to continue learning and, consequently, growing wiser in the interim. Some of my conclusions may indeed "evolve" over this period and through subsequent editions of this book.

Since this is my first comprehensive work on product-safety engineering, the focus is admittedly on potentially high-risk products and on engineered products which are mass-produced and innovative, or otherwise unique, in nature.[6] The modern consumer demands new levels of performance as well as features and characteristics which have never before existed in products. Therefore, some of these ground-breaking and challenging products do not have years of user experiences upon which to base engineering-design decisions during a fast-paced and short modern product-design program. Such products are frequently the most demanding for a design-and-manufacturing company and its engineers since they are generally working without design guidance from

---

[3] I am hopeful that my editor and publisher are in agreement.

[4] I can be reached at *DesignForProductSafety@gmail.com*.

[5] At *www.designforproductsafety.com*.

[6] It is further assumed that the companies, unless otherwise stated, are competent design-and-manufacturing firms which are *not* looking to simply get into a market in order to sell low-cost, low-quality merchandise in large quantities and then depart the market or quickly dissolve the corporation.

established knowledge bases for well-established products—let alone product standards and regulations. Innovative products may well be a decade ahead of any published industry standards for the product. Competitors will likely enter the new market segment and that market segment may, consequently, evolve and not stabilize for several years. Regulators could similarly struggle with such products and may attempt to provide *political* solutions to *technical* product-safety problems[7] simply because they are under public pressure to "do something." Due to such consumer demands and because of the current wave of start-up businesses in innovative product areas, many lower-risk products—however prevalent—with years of user experience and well-developed design practices and standards are simply *not* the focus of the *first* edition of this book.

This book will completely please no one—nor could it or should it. Product-safety opinions and conclusions are heavily guided by the value systems of individuals and groups. There will never exist a functional consumer product that will be sufficiently safe for all parties involved in judging it.

In a conscious effort to prevent self-censorship in the writing of this book, I have not been involved in expert-witness work for several years—nor do I seek such work. Because of my history with one former employer that was a large designer and manufacturer of powersports products—an experience recently documented[8]—I empathize all the more with the plight of design and product-safety engineers at design-and-manufacturing companies struggling to do conscientious work and deliver safe products to consumers.

Product designers and manufacturers may not like certain parts of my book, just as regulators and consumer advocates may not like other portions of the book. No single party involved in product safety has a monopoly on virtue. It is my wish that this book be considered *intellectually honest* rather than merely *superficially pleasant*. The users of consumer products are best served when a spirit of respect and cooperation exists between some traditionally adversarial parties involved in product safety. Despite my blatant and acknowledged Utopian perspective—since it flies in the face of business and political realities—it is the perspective espoused in this book. It is also my conviction that this is the perspective that best serves society.

Each party involved in consumer-product safety can probably be doing at least a *slightly* better job of advancing product safety than they presently are. Having worked as product-safety manager for an international design-and-manufacturing company, as an engineering consultant, as a researcher, and now as engineering faculty, I am well aware of the difficulties in producing "reasonably safe" products and the ethical quandary in which practicing, ethical engineers at aggressive, for-profit enterprises may find themselves.

Reasonable people can indeed reach different conclusions from the same evidence or "facts," but the welfare of the public is the most important goal. All parties must always remember this. It is when this synergistic relationship is *not* remembered, or valued, that the product-safety "machine" breaks down. This product-safety machine is a vulnerable mechanism that can easily be rendered inoperative through misguided

---

[7]Companies may be equally culpable for going along with regulators and for, sometimes, even proposing such "solutions" themselves.

[8]Jeans, David. "The Polaris RZR, an Off-Road Thrill That Can Go Up in Flames." *New York Times*, September 6, 2019. (https://www.nytimes.com/2019/09/06/business/polaris-rzr-fires.html)

actions from the parties involved. These parties should congratulate one another when the product-safety machine works and delivers safe products to consumers. Therefore, my sense of engineering ethics urges me to pursue product-safety engineering in this *idealistic* fashion. I do not believe my position to be *naïve* since I fully understand the obstacles to free-and-open communication and collaboration between companies, regulators, advocates, and consumers. I also understand that not all suggestions from well-meaning parties for improving the safety of products stand up to rigorous examination. Reasonable people can also disagree on what may pose an acceptable risk. However, the ultimate goal of safer products for the public is reason enough for the various parties to continue working together.

It is my hope that practicing engineers follow the directions indicated by their respective ethical compasses once technologies and numbers can take them no further. It is also my hope that engineering *management*—including the executive—ultimately becomes engineering *leadership* and creates a healthy environment to both support and reward ethical decision-making by engineers, and other professionals, who design-in the safety characteristics of the products used by consumers.

I support all those who seek to advance the safety of products used by consumers. Although many have often disagreed in the past and will continue to disagree into the future, so long as parties are acting for the greater good of consumers and society, these parties should continue to respect and listen to one another. We should all also permit people and their conclusions to grow and mature over time. One should not necessarily be eternally "wed" to a position on a topic for decades without the opportunity to change. As observed by Sorensen, "Consistently wise decisions can only be made by those whose wisdom is constantly challenged."[9]

Any opinions and conclusions—as well as any mistakes—in this book are solely mine. Opinions and conclusions expressed in the book are not necessarily those of any person, group of people, organization, or institution. There *will* be differences of conclusion between some readers and me. This is not a bad thing, however. Dialogue between the parties involved in delivering safe products to consumers is necessary.

The shortcomings in this book are admittedly many; yet, it is my hope that the positives significantly outweigh the negatives. There is certainly room for improvement in the second edition of this book. The examples used here are few in number relative to the numbers and types of consumer products available today. Perhaps, the discussion appears too heavily biased toward off-road vehicles (ORVs) and not enough toward simple consumer products such as electrical appliances or toward medical devices, for example. I am both blessed and prejudiced by my background in the powersports industry. This may color my perception, but this product area provides an example-rich environment having parallels which may be extended to other innovative products also presenting significant hazards and risks. I also do not believe that there would be great interest in a book addressing the engineering design and safety of decades-old products posing only nominal risks to their uses. The goal of identifying hazards and reducing risks in consumer goods is, however, of great interest to many, I hope, and the ultimate purpose of this work.

---

[9] Sorensen, Theodore C. 1963. *Decision-Making in the White House: The Olive Branch or the Arrows.* New York, NY: Columbia University Press.

I do not doubt that there will be at least a modicum of truth in whatever criticisms are made of my book. As stated before, this book could be neither as wide nor as deep as all readers might justifiably like it to be. And, despite my best efforts, there may still linger predilections in my bases and conclusions—and, as already admitted, I still learn more each day about this field. Let us learn and work together to improve the safety of products used by consumers each day.

Some engineering students, whom I have had the honor of teaching, have had the same justifiable question at the start of my product-safety engineering and ethics course. Namely, *how safe does my product need to be?* This is a great question because it demonstrates that they recognize the need for their engineering designs to perform at some level of safety. It is also a great question to ask because it is so difficult to answer.

Thus, this book.

Sincerely,
KLdE
Salt Lake City, Utah
August 2020

Acknowledgments

Professional life as an engineer has its ups, its downs, and its detours. I am grateful to have been blessed over the years with wonderful friends, mentors, coworkers, acquaintances, competitors, and even those not wishing me well. Each has contributed to either my professional or my personal growth in some way. I truly thank each of them—without reservation.

I am also blessed to have found an academic position in an environment permitting me to develop the product-safety engineering curriculum serving as one basis for this book. I would like to particularly thank the following people at the University of Utah Department of Mechanical Engineering for their support (in alphabetical order): Professors Timothy A. Ameel, Donald S. Bloswick, Bruce K. Gale, and Andrew S. Merryweather. I also wish to express my gratitude to Prof. Kurt T. Hegmann, M.D., Department of Family & Preventive Medicine and Director of the Rocky Mountain Center for Occupational and Environmental Health (RMCOEH), for his support. The opportunity to teach in both of these departments has been tremendously rewarding. I have been able to continue learning from wonderfully talented faculty and staff members—as well as students—in both departments. I learn and take away much from my classroom time. I consider this opportunity to be a true gift.

My family members know who they are because I share my love for, and appreciation of, them regularly. Regardless, I again say thank you.

However, there are also those whom I must explicitly thank for their friendship and support over the years—even though some are truly family members. I owe much to Prof. Kenneth M. Ragsdell for the unmerited confidence and faith he placed in one particular forgettable undergraduate mechanical-engineering student at the University of Arizona decades ago. It was his support, guidance, and mentorship that has made a large and lasting impact upon me for the better—both as an engineer and as a person. It was from him that I learned much about engineering design, design optimization, and life.

It is to Mr. Barry Toone, Esq. that I also owe tremendous thanks. He started off as my attorney, but has since become my friend and my brother. To elaborate further would likely just further embarrass both of us. Yet, without his being there for me, this book may never have materialized.

I am also fortunate to have worked with Prof. Ralph Barnett for several years. During both my employment by him and the time since, I had occasion to enjoy his personality as well as his work. Each of these experiences has been rewarding. My appreciation also extends to Ms. Chris Smith at Triodyne Inc. for her assistance over the years and with this book. My gratitude also goes to Triodyne for permission to reproduce figures within this book.

I would like to thank Ms. Yvonne Halpaus and QNET LLC for their help over the years. They have helped me better understand CE Marking, exporting to Europe, and European directives and regulations.

Several people helped me get this book to print and I would like to thank them for their contributions. These include Professors Roger C. Jensen at Montana Technological University and Richard F. Sesek at Auburn University for their reviews of my publisher proposal. Also to be thanked are Mr. Ken Ross, Esq. for guest lecturing my course and reviewing some material in the book, Mr. Jon Bready and Collision Safety Engineering, L.C. for their continued support through guest lecturing in my engineering course and for permission to use figures from their work, Mr. Markus Fiebig and the JOB Group for providing information on and photographs of their product, and Jeffrey P. Rosenbluth and Tetradapt for permission to use photographs of the TetraSki. There, too, are others and I apologize for not being able to name each one.

Finally, I would like to thank the wonderful professionals at McGraw Hill for their support of and assistance with this book. This includes Mr. Robert Argentieri, Editorial Director, and Ms. Elizabeth Houde, Editorial Coordinator. I wish to also thank Mr. Jeff Weeks, Senior Art Director, for taking time to consider the content of the book and then design a great cover. I am indebted to MPS Limited and their entire staff, but especially to Ms. Jyoti Shaw, for finding my errors, helping me correct them, and getting the book to print. They all were pleasures with whom to work.

I look forward to working with the McGraw Hill staff again on the second edition!

Engineering Ethics and Design for Product Safety

CHAPTER 0

Notice

Executing all of the steps, methods, procedures, and analyses presented in this book is neither *necessary* nor *sufficient* for the design and manufacture of safe products.

- A safe product can result from an incomplete, sloppy, and even clueless engineering-design process. A good engineer can triumph over poor procedures in some cases.
- An unsafe product can result from the most disciplined and organized product design and manufacturing processes even with highly talented engineering personnel.
- It is the *final product* itself that must ultimately be evaluated as being safe or otherwise—and that only in conjunction with the product users and environment.

The focus of this book is product safety—not product liability.

- The goal of this book is to help engineers design and manufacture safe(r) products without blind obedience to compliance and legal implications.
  - Consequently, engineers should be focused on the design and manufacture of reasonably safe, and progressively safer, products rather than on product liability.
  - Engineers should work *not* to liability-proof a company, but should instead work to risk-reduce a product to the extent feasible when all aspects of product use are evaluated.
  - Nothing can prevent a design-and-manufacturing company from ever being sued.

Any example in this book is simply that—merely a tool to demonstrate a concept or a tool.

- Any data, numbers, and/or determinations for the fictional and incompletely specified products discussed are for illustrative purposes only.
- The designs of products discussed in this book lack much of the detail that is necessary to know when assessing and mitigating risks to users.
- These discussions of fictional products do not reflect the results of the necessary in-depth analyses, tests, and other efforts required to properly engineer actual, real-world products.

- Although several manufacturers or industries may be mentioned by reference in this book, this should in no way be taken to mean that either a manufacturer or an industry is guilty of anything or worse than other manufacturers or industries not appearing in this book.

The author is not an attorney. Nothing in the book shall be considered legal advice.

The author has no financial interest in any party, topic, or outcome discussed or referenced in this book.

CHAPTER 1
Introduction

Your beliefs don't make you a better person,
your behavior does.

—SUKHRAJ DHILLON[1]

1.1 Priorities versus Values

Product safety should never be a *priority*.

Product safety should always be a *value*.

Priorities come and go.

Values remain.

1.2 The Crisis Cycle

The company-versus-regulator battlefield is littered with the corpses of product-safety issues and recalls that brought much unwanted attention to numerous design-and-manufacturing companies and their products. Some recalls have been large in magnitude—many units were recalled and repaired or replaced at manufacturer expense (Burrows 2018). Other recalls have been repeats or expansions of prior recalls [Schafer 2018; "Broken: Deadly Dressers (S1, E3)" 2019], recalls of similar products but by other manufacturers (Frankel 2018), and even "fixes" to problems introduced through previous recall repairs (Hurd 2020). Some have strained prior agreements between manufacturer and regulator (Frankel 2019).

For the moment, imagine a company with engineers to design, analyze, test, manufacture, and distribute its own products. Also, imagine that this company has recently undergone a spate of product safety–related recalls. One could safely wager that product safety quickly became the top *priority* of that company as recalls, consumer complaints, and adverse publicity rapidly mounted. Once that crisis was over, however, one might also wager that other longstanding concerns eventually became the new priorities of the company. That company's new business priorities might now return to best-in-class product performance, enhanced user experience (UX), or simply lower manufacturing costs. Each of these objectives is a worthy or necessary pursuit for a profitable business. Yet, what emerges after the crisis subsides—whether it be new procedures, new policies, a "customer-focused" mentality, or an environment where the

[1] https://quotesgram.com/quotes-about-being-a-better-person/

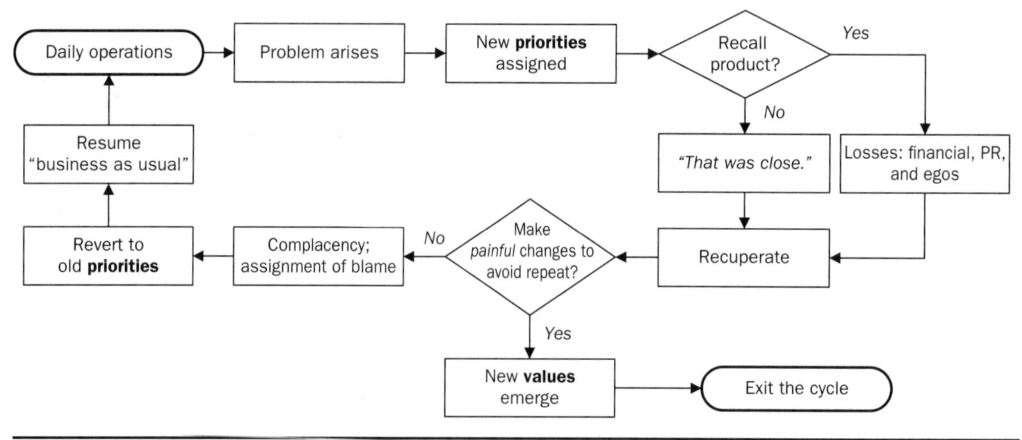

FIGURE 1.1 The crisis cycle.

simple truth may be discussed openly and without fear of reprisals—reveals the true *values* of that organization for better or for worse.

After an extended, post-crisis period of time, some companies will return to the same "business as usual" operating mode that landed the company in trouble with consumers, regulators, media outlets, and attorneys in the first place. Priorities can, and sometimes do, change once the external pressures on an organization arising from a crisis situation are removed.

Unless executives and senior management consciously and actively create the environment and work climate needed for an organization that truly values product safety, a *crisis cycle* of problem, action/reaction/panic, recall, recuperation, and continued complacency may well be repeated. This crisis cycle is shown in Figure 1.1.

To start with, a design-and-manufacturing company is conducting its daily operations as it has always done. This normal activity includes program planning, project management, product design, analysis, testing, manufacturing, logistics, sales, marketing, accounting, and executive decision-making. All is going well at this time. Then, a serious product safety–related problem with a product of theirs raises its ugly head. Product safety suddenly becomes a top corporate priority. A flurry of new company activities commences in order to figure out just what is going on with the product which is now in the hands of consumers. There can be a mixture of motions and emotions among company personnel including action, reaction, and panic. Ultimately, there should soon be a decision made to either recall the product or not recall the product.[2] If the decision is to not to recall the product, then those involved will likely breathe a collective sigh of relief. If, on the other hand, it is determined that a product-safety recall is needed, then a recall is initiated and, sometimes, handed off to a group dedicated to completing all of the required recall actions. These actions may include communicating with regulators, the design and testing of a remedy, a press release, and distributing the repair, for example. The results of product-safety recalls include financial losses, a tarnished corporate image, and bruised management egos. Regardless of the

[2] There are legitimate instances when a product-safety recall is not necessary.

decision made, there will generally be a short period of recuperation for those involved in evaluating the product problem after the product-recall decision has been made and executed, if necessary.

Following this brief respite from the trauma of the recall-decision process, the company *will* make a decision about its future. However, the company may not be aware that such a decision was even actually made. This is unfortunate both for the company and for the consumer. This decision takes place after the recall decision has been made and the personnel involved resume their pre-crisis activities. Executive management will ultimately decide whether or not they want to make the fundamental organizational changes that may be necessary to prevent the recurrence of problems with its products.

Make no mistake. Making significant changes within a large organization takes time, commitment, and a tolerance for pain. The company will be at either of two places in the near future. For some companies, the pain of reform outweighs its perceived benefits and, after all, things were going well before the product problem. Management may simply assume that the prior trend of success through luck will continue. That company will simply revert to complacency and the old business priorities of sales, marketing, and growth of market share. The return of the company to "business as usual"—after possibly assigning blame to individuals for the problem—will follow if the old priorities are simply readopted by company personnel. No changes are made; no lessons are learned.

On the other hand, an enlightened executive should, after the recall or potential recall, realize that the same problem may be just as likely to reappear as it is to go away unless significant changes are made to the business. Management commitment to change for the better, regardless of pain, will ultimately result in changes to corporate values to the benefit of all parties at stake. This change will not happen quickly, nor will it happen without a continued top-down commitment—in more than mere "lip service"—by management. New corporate values can emerge if this harder path is taken. Many truly worthy pursuits take much longer and require more commitment than simply taking the easy way out through "business as usual." Therefore, management must take a long-term approach to solving any such systemic product-safety problems.

The rank-and-file employees who keep the company running have been well trained to smell insincerity. After all, many of these employees have long been fed a steady diet of words without meaning by management. No employees are going to change their existing behaviors unless they believe that management really means what they are saying and that employees will not be punished for believing what they may be told. But, unless executive management is truly dedicated to establishing new values throughout the organization to better serve both the company and its consumers, the crisis cycle may continue *ad infinitum*. A quote, apparently by Rita Mae Brown,[3] comes to mind. Namely, "Insanity is doing the same thing over and over again and expecting different results" (Sterbenz 2013). Make painful long-term changes, and the crisis cycle can be broken as new corporate values and their resulting behavioral changes emerge.

Ultimately, it is executives and other upper management that are to be either *credited* with intentionally establishing positive corporate and engineering environments where product safety can be pursued without risk to career through retaliation or *blamed* for

[3] The quote is often misattributed to other people, including Albert Einstein, who denied having said it.

permitting a negative environment to fester where product-safety issues cannot be openly voiced and acted upon by those having legitimate concerns. Negative work environments can be generated and nurtured either through *explicit action* or through *benign neglect*. Either way, since the extant corporate work culture is management's choice, the resulting climate—as well as its results—remain management's credit or blame *alone*.

1.3 Product-Safety Engineering

There will be engineers whose tasks will be helping the company design, manufacture, and distribute safe products. These same people performing this task may help the company escape, or even avoid, the crisis cycle just discussed. This task is *product-safety engineering*. It may be performed by a dedicated employee or a group of employees, or it may be done by design engineers in the course of their regular engineering-synthesis activities. It is not so important precisely *how* product-safety engineering is performed. It is much more important *how well* it is performed. If successful, these engineers will be performing *design for product safety* (DfPS).

In the author's mind, the two terms "product-safety engineering" and "design for product safety" are essentially synonymous. However, the use of DfPS will give an added emphasis to performing the product-safety engineering "up front," or during the engineering-*design* phase of the product-design process (PDP). In this way, safety-positive characteristics can be *designed into* the product in a proactive fashion. Some companies may only seem concerned with product safety once a product-safety problem has erupted after the sale of a product. It should be the desire of engineering and management alike to *prevent* such problems and their potential severe-to-fatal injuries to consumers.

The field of product-safety engineering is practiced in the real world, but is not well known to the general public and perhaps also not well acknowledged within engineering itself. As a result, this engineering discipline is not well represented in the literature. There are exceptions. However, one solitary author appears to have written much of this literature (Hammer 1972, 1980, 1993). In addition, in these works, Hammer *comingles* product safety and system safety. Indeed, this comingling may become unavoidable when addressing significant product-safety engineering issues. There exists significant overlap between engineering for product safety and engineering for system safety. Many system-safety methods are useful, and sometimes even necessary, when engineering some products for safe consumer use. However, this comingling also leaves product safety without an identity of its own. In addition, system-safety engineering sometimes piles unnecessary methodological expectations from system safety onto the back of product-safety engineering.

Even the terms "product" and "system" are themselves ambiguous. Just about any product can be considered a system. A system—even when large, complex, and costly—can also be considered to be a product. For the most part, the book will use "product" to mean a consumer product and will reserve the term "system" to a large, complex product such as an airliner or an advanced weapons system. These aerospace and military programs provided much of the impetus toward the latter half of the twentieth century to develop and formalize what is now known as "system safety" (Ericson 2006). The sizes and complexities of aerospace and military systems and the consequences of their failures have forced these industries to establish methods for pursuing safety now known as *system safety*. Because of its genesis, system safety sometimes brings with it the need to treat every engineering-design program as if it was a crewed, interplanetary NASA-spacecraft program. With this system-safety treatment comes the requirement for many deep and detailed analyses that require dedicated teams of engineers, scientists, and analysts along with years of uninterrupted scheduling to accomplish. Much design-engineering work simply does not need—or cannot practically permit—this extent of design analysis due to project type, human resources, financial resources, and scheduling realities in a competitive marketplace.

For example, if a company is designing a new and improved, yet simple, paperclip, it might be unreasonable to expect that company to perform a full-bore, system-safety "deep dive" for that new product including failure modes, effects, and criticality analysis (FMECA), fault-tree analysis (FTA), event-tree analysis (ETA) and bow-tie analysis. However, if even competent engineers are working on such rudimentary products, it is likely that portions of many system-safety analyses are being conducted, at least on a mental level, by those engineers during the PDP. Unfortunately, these considerations of hazards, risks, failure modes, and accident scenarios are often not be written down and preserved in document form. Some might presume that such engineers were not acting responsibly, but this is not always the case.

In the instance of a *legacy* product—which is a product that a company has been producing for many years and the original designers may no longer be alive or with the company—there may be no records of detailed safety analyses that were possibly conducted. Yet, over time, the failures, failure modes, and other risk along with their countermeasures, may well have been duly considered and adequately addressed by the passing engineers working on the product. This is especially true if a company has been designing and manufacturing a particular type of product for decades and that product has been iteratively refined year-by-year during that period. Those engineering design-and-analysis results were just, unfortunately, not documented. There may be no real benefit to conducting such analyses on well-established products other than creating a written record which itself could be a worthwhile goal.

The product-versus-system discussion is continued in the next section.

1.4 Scope of the Book

There are three emphases in this book. This book is primarily an *engineering-design* book focused on *product safety* accomplished by using *engineering ethics* when engineering-design guidance is insufficient or nonexistent. Each of these three topics can become worlds unto themselves. Therefore, each of these topics is further refined. First, the topic of engineering design is limited to a product-design process (PDP) used by an innovative designer and manufacturer of potentially hazardous mass-produced durable consumer goods in a competitive, fast-paced, and regulated industry. It is presumed that this company already has an integrated and functioning PDP in place to produce well-engineered products. Such a design, analysis, and testing process could differ significantly from this example in the case of a company designing, testing, and delivering a single, massive coal-fired electric-power plant for one specific customer having large teams of highly skilled operating and maintenance personnel.

Second, the focus on product safety is through product-safety engineering and, in particular, the proactive DfPS approach. The integration of product-safety engineers and the product-safety function within an established PDP is presented in this book as the best[4] way to deliver sound products to consumers. Such an approach is beneficial to the company as well. Regulators, consumer advocates, and consumers themselves also appreciate this active-prevention approach to public safety.

Finally, engineering ethics—let alone the complete field of classical Western ethics—is far too broad to cover in any single book. Even engineering ethics is multi-dimensional and includes numerous aspects such as professionalism, safety, health, research, human-subject experimentation, and the environment. Although all aspects of engineering ethics are important, only the consumer-product safety aspect of engineering ethics will be considered in this book. Even this will be further limited to the *application* of those ethics through DfPS—an integrated engineering-*design* approach—during the PDP.

The three distilled-down aspects of this book are listed below:

1. The product-design process
2. Design for product safety (DfPS)
3. Applied engineering ethics

To help understand the three aspects of this book—and to also help explain the relationship between product-safety engineering and system-safety engineering—Figure 1.2 has been developed. It is admittedly a contrived "space," but this figure should be helpful nonetheless. The horizontal axis of the figure represents the relative magnitudes of particular product or system properties. The vertical axis shows the ethics axis and starts with engineering ethics and extends up to western ethics.

Under the product or system property axis are listed several characteristics of products and systems. Primary characteristics of products and systems include size, complexity, cost, lead time, as well as the number of potential fatalities to which a product or system failure or accident could lead. Beneath the origin of the horizontal axis, and to its near left are the smaller, simpler, less-expensive, quicker, and fewer-fatality magnitudes for these properties, respectively. To the far right along this axis are the larger-magnitude values for each property of a product or a system.

Below the magnitudes of properties in Figure 1.2 are two dashed lines. The upper dashed line denotes the usual range of properties conventionally covered in system-safety engineering textbooks (Ericson 2016; Gullo and Dixon 2017). Generally, system-safety engineering is aimed toward the larger, more-complex, more-costly, longer-lead-time, and higher potential-fatality products or systems. There has been a "trickle-down" of systems-safety engineering practices to non-aerospace and non-military products such as automobiles which occupy the center region of the horizontal axis in this figure. Many systems-engineering methods have been successfully applied in, and some even further developed by, the automotive industry over the decades.

There exists an area neither explicitly nor usually covered by system-safety engineering. This is the area just to the right of the horizontal axis' origin. It is this region of the horizontal axis that will be covered under product-safety engineering. This area will also extend to the right into—and consequently overlap with—system-safety engineering under the medium-scale product or system. This is seen in the horizontal

[4] From the author's professional experience with such products.

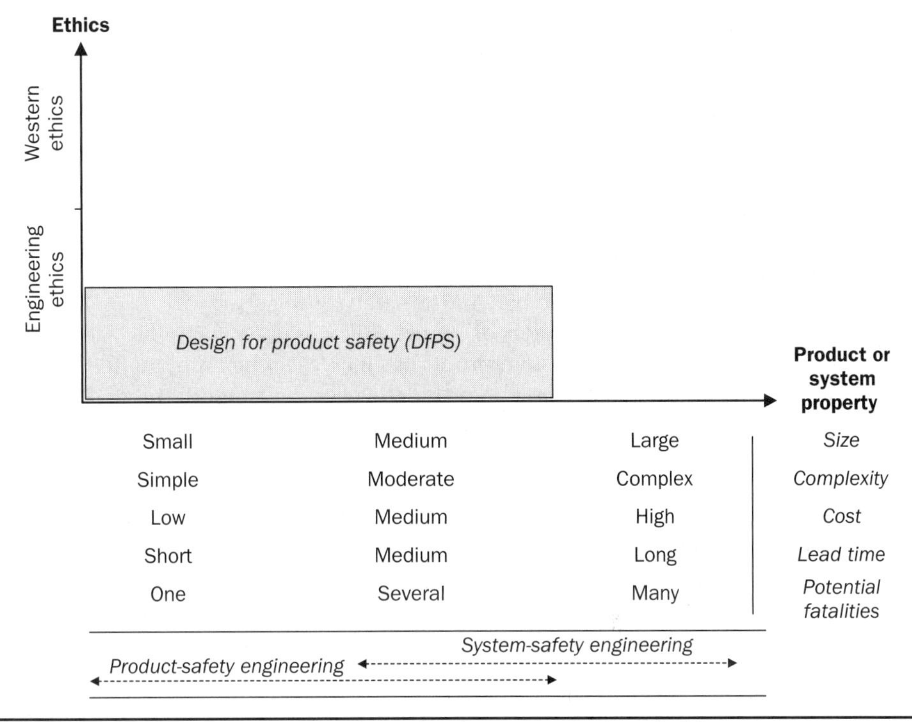

FIGURE 1.2 Design for product safety in product/system-property and ethics space.

coincidence of the two dashed safety-engineering lines. Therefore, some system-safety engineering methods are completely relevant to and useful for product-safety engineering. Yet, as in the case of the new-and-improved paperclip mentioned above, some system-safety engineering methods may be neither necessary nor even productive if performed formally so long as skilled engineers are involved in the design. The engineering-design team may well recognize and address the inherent hazards, risks, failure modes, accidents, and their resulting outcomes of such simple products even if records and documentation are not maintained, but it is good to generate and retain documents. Such documentation will be helpful in answering questions in the future about a product.

From Figure 1.2, one might get the impression that these low-to-high relationships between product or system properties always exist. This is, of course, not always true. There are exceptions. For example, the modern smartphone is technologically complex, long in lead time, and not small in cost; however, a smartphone is a small product or system and any severe injuries or fatalities involving the device would likely be limited to the immediate user. Yet, it is certainly conceivable that a smartphone-battery failure such as a thermal-runaway event, resulting in a fire, could result in a catastrophic accident whose outcome could bring down an airliner and result in many fatalities. Thus, these property relationships are not hard-and-fast rules, but they do help to conceptually differentiate product safety from system safety in many cases.

At this time, a box is placed in Figure 1.2 upon the horizontal axis to partially occupy the horizontal expanse of product-safety engineering. This box is titled Design for Product Safety and denotes that it is an engineering-*design* approach to product safety. It is proactive, not reactive, in nature.

The DfPS box extends vertically to only partially span the vertical axis of Figure 1.2. The height of this box is limited to show that Western ethics are not covered by this book. Even the already-limited topic of engineering ethics is not completely included. Engineering-ethical areas such as health, the environment, and sustainability are excluded from the discussion in this book. This should not be viewed so much as an exclusion of worthwhile material. It should, instead, be seen as a focus upon the *application* of engineering ethics during the *design* of *products* to help guide design engineers in the decision-making process to benefit the safety of society.

Before leaving the discussion of Figure 1.2, it is constructive to state that not all magnitudes for product or system properties increase when moving to the right of the horizontal axis. Take the case of the product or system property being the *likelihood* of an accident taking one's life. Fortunately for the general population of the United States as a *whole*, the larger, more-complex, and more-deadly systems may have a lower likelihood of killing people than some more-simple products. Unfortunately for the *individual* U.S. citizen, that person is more likely to be killed in a motor-vehicle accident than by an airliner crash ("Facts + Statistics: Mortality Risk" n.d.). The lifetime odds of dying in a motor-vehicle accident involving any type of vehicle is one in 103 (1:103), whereas the same odds from an air-or-space transport accident is one in 10,764[5] (1:10,764). As a reference, the odds of being killed by lightning strike is reported as one in 218,106 (1:218,106). At the other end of the likelihood scale are the lifetime odds of dying from any cause, which is, of course, one in one (1:1). Everyone is eventually going to die.

1.5 Perspective of the Book

The perspective taken by the book is that of a large designer, manufacturer, and distributor of mass-produced durable consumer products for worldwide distribution. This company's products are innovative and also potentially risky to users especially if the consumer is not vigilant when using the products. This company operates in a competitive, fast-paced engineering-design environment and is also subject to regulatory oversight regarding product safety.

In addition, when addressing the safety of a product, the engineering synthesis, or design, and analysis will be of a comprehensive nature and include the *system* of the product rather than merely the product by itself. Factors such as users and use environment will also be taken into account since these non-product effects can significantly affect the resulting safety level of a product user during real-world use. This triad is illustrated in Figure 1.3 as the product-user-environment (PUE) model. Each of the three elements of the model is visible as are the overlapping areas of the three elements. These overlaps indicate the interactions between the three elements. In addition, a given product might be perfectly safe to use in one environment by a specific user. However, a significant change of environment or a major action by a user could drastically change the product-use experience and result in accident, injury, or death to that user.

[5] Sadly, the lifetime odds of dying from opioid-only—either legal or illegal—exposure is higher than either of these two odds at one in 96 (1:96).

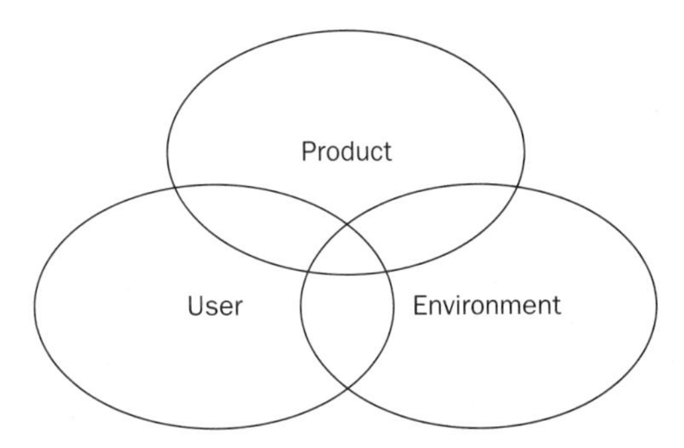

Figure 1.3 The product-user-environment (PUE) model.

Accordingly, the book will address product-safety engineering thoughtfully by not evaluating products in a vacuum devoid of external effects, by not responding with knee-jerk reactions potentially leading to unintended consequences as a result, and by avoiding one-size-fits-all solutions to engineering problems by regarding each individual engineering-design task on its own merits.

This book will not focus on industries and applications such as process-related industries or companies designing and manufacturing large, single industrial facilities. Examples of the former include refineries and chemical factories; examples of the latter include electrical-power plants and steel mills. The concepts, as illustrated through examples in the book, can still provide useful safety insight into these other applications, however.

1.6 Approach of the Book

Unlike many engineering-analysis textbooks, but similar to many engineering-design textbooks (Pugh 1991; Dym and Little 2004; Ulrich and Eppinger 2012), there will be little mathematics used. Figures and tables will instead be used in place of formulae to illustrate concepts and methods. The reader not being a mathematician will not hinder the grasping of the book's straightforward concepts and methods. The simple mechanical examples, such as the two-slice toaster and the hand-powered mechanical winch, used within are such that engineers of any discipline—and even those without engineering training—shall be able to readily understand the material and the important aspects of product-safety engineering and DfPS.

Chapters do not have exercises which can be easily assigned as homework problems. However for course instructors, students, and those wishing to simply self-study the book, materials on key concepts and discussion topics, and questions for chapters are available at the author's website.[6] These materials may either be used in an educational setting to stimulate student discussion or assigned as writing[7] or analysis exercises. Because of the potentially large range of acceptable answers, instructors will

[6] www.designforproductsafety.com

[7] It is the author's opinion that writing skills, along with critical thinking, are among the most important abilities for a product-safety engineer to possess.

need to work out their own solutions and then use their own judgments when grading the assignments.

1.7 Parts of the Book

This book will help the design engineer address the concept of product safety in her/his products and take the necessary steps to *design-in* safety to products for consumers and users, rather than try to add it in at a later design or manufacturing stage. Beginning in Part 1, *Concepts*, Chapters 2 through 6, the book will present safety, engineering ethics, product-safety concepts, hazard and risk, accident and outcome, and an integrated, functional PDP.

Having laid the conceptual groundwork on product safety in Part 1, the book proceeds to Part 2, *Application*. Chapters 7 through 11 will cover the role of the product-safety engineer within the PDP, sources of product-safety design guidance, product-safety facilitators (such as warning signs and manuals), engineering methods that may be used to enhance product safety during the PDP, and how products—once sold—must be monitored to look for and respond, as appropriate, to product-safety defects through safety recalls.

Throughout this book, the term "safety" will mean "product safety" unless otherwise stated in order to facilitate the reading of the material. Similarly, "safety engineer" and "safety engineering" will be interpreted to mean "product-safety engineer" and "product-safety engineering," respectively. The term "engineer" will generally mean "design engineer" and "design" will signify "engineering design" as opposed to industrial design or fashion design. An organization involved in the design, manufacturing, and distribution of a product may be referred to as an "organization," "corporation," or "company."

Finally, a list of acronyms and abbreviations and a glossary of terms used in the book are included at the end. Also included at the end are two appendices. The first is an appendix containing a simple list of hazards and some of their associated risks to help the reader understand the difference between the two often-confused terms. The second appendix presents an example, a hand-operated mechanical winch, used in the book to illustrate product-safety engineering methods.

References

"Broken: Deadly Dressers (S1, E3)." 2019. Netflix.Com. 2019. https://www.netflix.com/title/81002391.

Burrows, Dan. 2018. "10 Biggest Product Recalls of All Time." Kiplinger.com. 2018. https://www.kiplinger.com/slideshow/investing/T052-S000-10-biggest-product-recalls-of-all-time/index.html.

Dym, Clive L., and Patrick Little. 2004. *Engineering Design: A Project-Base Introduction*. Second edition. New York, NY: John Wiley & Sons.

Ericson, Clifton A. 2006. "A Short History of System Safety." *Journal of System Safety*. 2006. https://system-safety.org/ejss/past/novdec2006ejss/clifs.php.

———. 2016. *Hazard Analysis Techniques for System Safety*. Second edition. Hoboken, NJ: John Wiley & Sons.

"Facts + Statistics: Mortality Risk." n.d. Iii.Org. Accessed November 10, 2019. https://www.iii.org/fact-statistic/facts-statistics-mortality-risk.

Frankel, Todd C. 2018. "Product Recalls under Trump Fall to Lowest Level in 16 Years, but New Signs Emerge of a Tougher Regulator." *Washington Post*. January 13, 2020. https://www.washingtonpost.com/business/2020/01/13/product-recalls-under-trump-fall-lowest-level-16-years-new-signs-emerge-tougher-regulator/.

________. 2019. "Britax Avoided One Recall for Its BOB Stroller. But Its Crash Fix Leads to a Recall Now." *Washington Post*, April 25, 2019. https://www.washingtonpost.com/business/economy/britax-avoided-one-recall-for-its-bob-stroller-but-its-crash-fix-leads-to-a-recall-now/2019/07/25/8fea37e0-af19-11e9-8e77-03b30bc29f64_story.html?noredirect=on&utm_term=.0e96de25bde6.

Gullo, Louis J., and Jack Dixon. 2017. *Design for Safety*. First edition. Hoboken, NJ: John Wiley & Sons. https://doi.org/10.1002/9781118974339.

Hammer, Willie. 1972. *Handbook of System and Product Safety*. First edition. Englewood Cliffs, NJ: Prentice Hall.

________. 1980. *Product Safety Management and Engineering*. First edition. Englewood Cliffs, NJ: Prentice Hall.

________. 1993. *Product Safety Management and Engineering*. Second edition. Northridge, IL: ASSE.

Hurd, Byron. 2020. "GM Alerting Truck and Sedan Owners to a Do-Over on Brake Recall." Autoblog.Com. 2020. https://www.autoblog.com/2020/01/29/gm-brake-recall-do-over/.

Pugh, Stuart. 1991. *Total Design: Integrated Methods for Successful Product Engineering*. Reading, MA: Addison-Wesley.

Schafer, Lee. 2018. "Rolling Recalls Test Polaris' Commitment to Running Lean." *StarTribune*, April 14, 2018. http://www.startribune.com/rolling-recalls-test-polaris-commitment-to-running-lean/479724953/.

Sterbenz, Christina. 2013. "12 Famous Quotes That Always Get Misattributed." Businessinsider.Com. https://www.businessinsider.com/misattributed-quotes-2013-10.

Ulrich, Karl T., and Steven D. Eppinger. 2012. *Product Design and Development*. Fifth edition. New York, NY: McGraw-Hill.

PART 1

Concepts

CHAPTER 2

Product Safety

What is ~~truth~~ safety?

—PONTIUS PILATE[1]

2.1 Introduction

Being able to define a concept is crucial to being able to understand it, discuss it, and advance it. This is true with all forms of safety, including product safety. There is often misunderstanding of these fields by the public and even by engineers. There is also often confusion between safety and other properties.

And so, the discussion of what safety is—and is not—begins.

2.2 What Is Safety?

Although precise definitions for many words that will be used in this book are covered in Chapter 4, an early and meaningful discussion can still be had using imprecise definitions for some words and concepts. In professional and organizational settings, the function of safety is often incorrectly lumped under the same umbrella as the function of health. There are similarities between the two disciplines, undoubtedly. However, there are significant differences between the two as well.

Both safety and health are entirely desirable and necessary attributes to have within any society, environment, workplace, system, or product. These two attributes lead to higher expectancies for duration of life, quality of life, and sustainability of the environments in which they flourish. Yet, there is a temporal element that is often overlooked when safety and health are addressed in a singular manner. *Health* addresses the hazards that can injure or kill someone over time. *Safety* addresses those hazards that can injure or kill someone immediately.[2]

Prolonged exposure at or above some level to a hazard, such as lead (Pb), is necessary to pose a health concern to a person. Conversely, unskilled or incompetent people operating table saws under poor conditions can immediately injure themselves in practically no time. This is an example of a safety concern.

---

[1] John 18:38—with apologies to Pontius Pilate.

[2] This book will focus on human injury, including death, and not necessarily consider environmental impact.

Example 2.1: Health versus Safety 1

In recent years, many citizens spread across the world have become increasingly aware of the environment and humankind's impact upon its health. One area receiving much attention is the proliferation of single-use plastics by society. The stability of modern plastics means that they are tough and durable, but do not biodegrade within any reasonable time. Much of the world's plastic waste winds up in, among other places, floating on the oceans. One particular single-use plastic item has received much attention of late: the plastic drinking straw. This type of straw is used in large quantities each day by consumers in developed countries around the world. It is doubtful that many of these consumers have malicious environmental intent. Instead, the utility and convenience of the plastic drinking straw makes it an attractive method of drinking beverages, especially in a fast- or convenience-food environment.

Recently, there have been drinking-straw options either newly developed or reintroduced from the past. Some people have sworn off of the use of plastic straws. Some locales have outlawed the item. Other companies and establishments have stopped offering such straws. Paper drinking straws are now being offered by some businesses to their customers. They are not as durable as their plastic counterparts since they could become soft and "mushy" after prolonged soaking in a drink, but are only intended to be used once. Such paper straws are considered to be more "Earth friendly" since they will biodegrade in what is considered an acceptable time, but are still single-use in nature and can, consequently, contribute to landfill pollution. Overall, many consider the paper straw to be an improvement in environmental health over the plastic straw.

Other options to single-use plastic straws include multi-use straws made from metal, such as stainless steel, or from polymers, such silicone. These options are durable and are intended to be reused. Therefore, they should also contribute to the environmental health of the planet when compared to single-use plastic straws and, perhaps, even paper straws.

Various assorted drinking-straw options are shown in Figure 2.1. The item labeled as "1" is the ubiquitous single-use plastic drinking straw that has raised so much public ire. Such straws are offered by many companies across the country and this discussion is focused on none in particular. The item labeled as "2" is just one example of a single-use paper straw.

Although the multi-use nature of the metal and polymer alternatives to plastic straws can help the environmental health of the planet, if they are not kept sufficiently clean, these options could themselves pose a personal-health hazard to users. Bacteria can grow and present illness risks to those using unsanitary multi-use straws. Items "3A" through "5B" show three sets of multi-use options and the cleaning equipment provided with the purchase of the straws. The most elementary of these straws is item "3A," a straight metal tube. Also sold with "3A," in this case, was item "3B," an angled straw of similar construction. This item has a bend, or angle, closer to the drinking end of the straw than to its drink end. Item "3C" is the cleaning brush for "3A."

Although the multi-use aspect of the metal straw has its advantages, it has led to at least one product-safety recall ("Laceration Injuries to Children Prompt Starbucks to Recall Stainless Steel Beverage Straws to Provide New Warnings" 2016). Injuries sustained by children were cuts to their mouths. The remedy appears to have been "Consumers should not allow children to handle or use stainless steel straws." It could be supposed that item "3D" in Figure 2.1, a silicone straw tip, was meant to provide

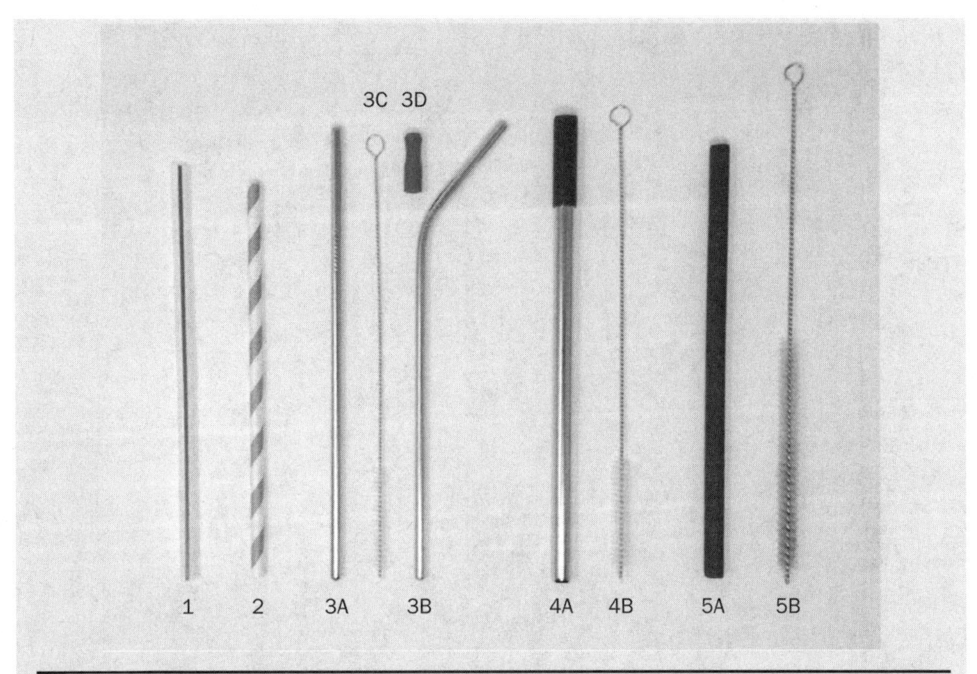

Figure 2.1 Drinking straws.

protection from mouth lacerations; however, the tips included with the straws were not attached to the straws within the package. It could be suspected that the interface between the straw and the straw tip could itself harbor bacteria if not sufficiently cleaned between uses. Therefore, the straw safety tips could present health problems to users.

Presuming that the silicone tip to the metal straw does, indeed, protect a user from a sharp edge at the end of a straw,[3] the metal, multi-use straw still presents an injury risk to users through its mechanical strength. The straw shown as "3A" is a tiny, hollow steel column. It is essentially a hollow nail and can be driven into lumber with a hammer just as a nail can be.[4] This hazard of the strong, metal straw has resulted in at least one fatality in the United Kingdom (Vigdor 2019) in which the victim suffered a traumatic brain injury after the straw pierced her eye following a fall. Whether the straw involved in the fatality was straight, as with "3A," or angled, as with "3B," has not been revealed to the author's knowledge. Although the straight metal drinking straw might pose a higher impalement risk than the angled straw, when the angled straw fails, it can also present an impalement hazard. The angled straw will not be as strong under a purely compressive-loading situation. However, once the angled straw fails as a column in its "buckling" mode, it may look as the straw shown in Figure 2.2. Although not as sharp as the straight, metal straw, this angled, metal straw could also prove injurious to fall upon.

[3] All metal straws seen, to date, by the author had rolled edges on each end and presented little serious-cut risk to straw users.

[4] The author regularly demonstrates this phenomenon to students and visitors.

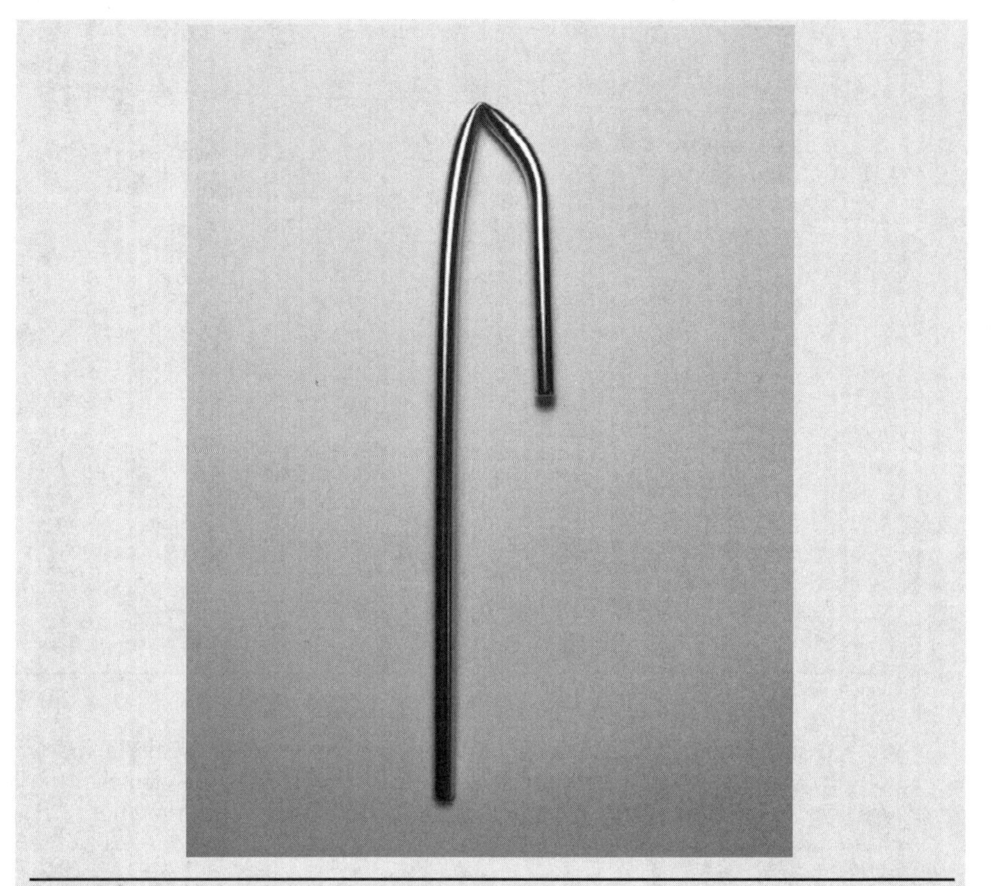

FIGURE 2.2 Buckled angled drinking straw.

Items "4A" and "4B" in Figure 2.1 are a metal drinking straw and its cleaning brush, respectively. Item "4A" is larger in diameter than item "3A" although both are straight, stainless-steel multi-use straws. Due to its larger diameter, "4A" would be stronger than "3A," but may not be significantly more likely to injure than "3A." However, unlike "3A," item "4A" had a plastic end already installed by the manufacturer when purchased.

Items "5A" and "5B" are a silicone drinking straw and its corresponding cleaning brush, respectively. This straw, "5A," is highly flexible and should not injure anyone if fallen upon. Therefore, the silicone drinking straw would be safer, if fallen upon, than either of the two metal drinking straws. Yet the health concern remains for all multi-use straws when not properly cleaned before use(s).

Since speed and convenience of use is a major "driver" in the continued use of plastic, single-use straws, it remains to be seen if multi-use straws will gain significant acceptance. It is quite inconvenient to thoroughly wash a multi-use straw between each use which would also require that the brush be carried by users before cleaning when dry and after cleaning when wet. So, it also remains to be determined whether future multi-use straw users will benefit when both health and safety concerns, such as pandemics, are considered.

In the real world, the functions of health and the safety are sometimes combined into one operating or engineering group at a product design-and-manufacturing company. It is here that striking differences appear. This is because safety and health are "different animals."

Namely, the health function may simply become one of *compliance*. In fact, the health function often becomes the Compliance group or department. This is due to the many sometimes-strict health regulations and directives regarding the incorporation of hazard substances such as Pb, mercury (Hg), cadmium (Cd), hexavalent chromium [Cr(VI)], and others. There also exist environmental regulations regarding the permeation of gasoline vapor from fuel systems and the emissions of hydrocarbons (HC), carbon monoxide (CO), and nitrogen oxide (NO_x) from spark-ignition (SI) internal-combustion (IC) engine exhaust gases. In addition, compression-ignition (CI) engines, such as diesels, must meet particulate-matter (PM) requirements. In many modern and internationally competitive businesses, once compliance with such requirements has been achieved, the work is done. Production soon commences.

Of course, this is not always the case. There is nothing preventing a responsible organization using gasoline IC engines in its products from further decreasing HC, CO, and NO_x emissions. Perhaps, some organizations do. This is a good thing and kudos is due to those organizations for their levels of corporate social responsibility (CSR). However, due to the aggressiveness of many new regulations and the business need to get products to market quickly and affordably, there is sometimes little margin between the test results of the engine-emissions compliance tests and the specified pass/fail criteria.[5] There may also be little time elapsing between the passing of these tests and the start of production.

There is nothing necessarily simple or easy about the compliance function in an organization. For example, the patchwork of different regulations from different regulatory agencies with differing jurisdictions turns assuring compliance with *all* applicable requirements into an exercise fraught with peril. Overlooking a single word placed somewhere in an obscure sentence can be difference between product compliance and non-compliance, product salability and non-salability, and career success and failure. Strict compliance with all pertinent regulations is mandatory and penalties exist for selling non-compliant products. The compliance process is further complicated as regulations constantly change. Also changing are suppliers and sub-suppliers thereby potentially compromising a company's supply-chain integrity. Only domestic compliance has been mentioned thus far but, as international markets are eyed by sales executives as a way to boost product sales, compliance personnel must quickly understand the regulatory framework and requirements in markets abroad.

Despite hardships of compliance work, its silver lining is that there are clear criteria for what is sufficient. If the requirement for maximum Pb content is not more than 0.10 percent by weight, and the product being designed and developed is 0.05 percent, then the product *is* objectively and completely compliant. No further work is necessary aside from proper documentation and records retention.

---

[5]Perhaps kudos is also due to these environmental and other regulators for providing manufacturers with aggressive, yet attainable, criteria.

Example 2.2: Health versus Safety 2

At the time of this writing,[6] the nation and the world are experiencing the public-health crisis of COVID-19, also known as the "coronavirus disease" ["Coronavirus Disease 2019 (COVID-19) Situation Report—29" 2020]. Before the need to shut down gatherings for "social distancing" efforts, the author was co-instructing a course requiring weekly visits to manufacturing and other industrial sites. The hosts of many of these visits required students and instructors alike to watch the sponsor's health and safety video content before being escorted around the various physical facilities. Many of these sites have stairways which, consequently, present hazards from slips, trips, and falls. One prudent countermeasure to these hazards is the mandatory use of hand rails when climbing stairs. This makes good sense from the perspective of industrial safety. However, many citizens have been taking precautions to help prevent the spread of COVID-19 by avoiding close contact with strangers, by washing their hands frequently, and by not touching objects that are also touched by others. These precautions by the public make tremendous sense from the perspective of public health—but runs contrary to public-safety interests.

Industrial, OSHA-type "compliance safety" would mandate that hand rails *must always* be used. Personal self-preservation might, on the other hand, countermand the hand-rail decree in the interests of avoiding infecting oneself with an unusually serious strain of influenza. One possible personal-reasoning process follows from these simple observations: "I had the flu two years ago and it was a very bad experience; yet, I have never fallen down a flight of stairs as far back as I can remember."[7] Ergo, one may decide to keep one's hands in one's pockets rather than exposing oneself to the germs of strangers despite fall risks.

The above reasoning and actions are, truly, a calculated risk taken by the stair climber. The risk of injury from a fall does persist but, in this case, the risk of infection and sickness is considered—either accurately or inaccurately—to be the larger of the two risks. Thus far and fortunately, neither hazard has harmed anyone involved with the author's course, so neither risk-control strategy has demonstrated itself to be superior to the other.

What this simple example does illustrate is that *safety* and *health* can become competing objectives especially when any *compliance* requirement is blindly adopted without second thought. Although holding a hand rail is generally a good countermeasure to a falling, tripping, or loss-of-balance hazard, it is not always without other risks—in this case the spread of a virus. One measure that could be considered to improve the safety of compliance with the organizational mandate on hand-rail use could be the installation of effective hand-sanitization stations at both the tops and bottoms of stairways. Such hand-sanitization measures must, of course, *actually be effective* in reducing the probability of disease propagation ["Interim Guidance for Preventing the Spread of Coronavirus Disease 2019 (COVID-19) in Homes and Residential Communities" 2020], otherwise this step and its associated costs are for naught.

---

[6] Summer 2020.

[7] This is the author's current reasoning, at least.

2.3 When Is a Product Safe Enough?

Unfortunately, the requirements for product-safety engineering (PSEg) are rarely, if ever, so clear-cut as they are in the world of compliance. Even if an engineer is not designing innovative, unprecedented products, there seldom is such objective guidance regarding safety. Assume that an engineer is designing a product which incorporates a rope to transfer force and there exists a standard for that product. Suppose that the standard requires a rope with a tensile strength of at least 1000 lbf (4448 N). The conscientious design engineer had laboratory tensile testing performed on the rope provided by the supplier to ensure that the rope is, indeed, capable of withstanding a load of 1000 lbf. A sample of several rope samples provides an average tensile strength of 1047 lbf (4657 N) with little variance in the results. Thus, rope meets the standard's strength criterion. However, is the engineer done with respect to product safety? Not really. After the product has been used for a while and rope begins to wear and decay, the engineer knows that the rope may no longer meet the standards strength requirement. It is highly unlikely that the standard will provide any guidance to the design engineer on this matter. In addition, many will argue, and perhaps rightfully so, that engineers *should always* design-in a factor of safety that is larger than unity, for example 1.5 or 2.0. Consequently, many could argue that the specification for the rope strength should have been 1500 lbf (6672 N) or 2000 lbf (8896 N). Again, the standard will likely provide little-to-no guidance on this matter. The design engineer understands the potential consequences of a rope failure, but is simultaneously tasked by corporate management to make an inexpensive product. Whatever decision the engineer makes, there will likely be someone who will disagree with it.

The age-old dilemma facing product-design engineers is *When is our product safe enough*? For reasons that will be explained throughout this book, no simple engineering or technical answer can be given to the engineer for this valid and crucial question. In a regulatory or legal sense, there are only five people in the United States who are able to make this determination. They are the commissioners on the U.S. Consumer Product Safety Commission.[8] Furthermore, when they do make such decisions, it is not uncommon for there to be a split decision where some commissioners believe that a product is safe while others believe that it is not safe. Given this, what hope do conscientious design or product-safety engineers have? Despite this reality, it is hoped that this book will help such engineers answer this question for themselves through an understanding of product-safety concepts and methods. It is further hoped that the engineer's ultimate determination will be an ethical *conclusion*, based on values, that can withstand scrutiny over time and not merely a convenient *opinion* that satisfies current business priorities. Furthermore, that conclusion should be fully considered—if not backed—by engineering management and corporate executives if it is justified and ethical.

2.4 What Safety Is *Not*

Before diving too deeply into safety, it is sensible to differentiate safety from some concepts that are sometimes grouped, and sometimes confused, with safety. This will help eliminate some potential distractions from the discussion and assist on focusing upon safety specifically.

[8] www.cpsc.gov

Many people believe that a safe product and a reliable product are one and the same. In an ideal world, this might be so. Yet the world is a real one and, sadly, safety and reliability are not synonymous. They are "related but distinct" (Sommerville 2004, 32) properties of a product or system. A safe product performs as *intended* without harming users or bystanders. A reliable product performs as *specified* with little interruption or no failure. One engineering model puts forth the concept of *dependability* (Sommerville 2004). The four key properties, or characteristics, of product dependability are shown in Table 2.1. The dependability model by Sommerville is taken largely from software engineering, but is applicable to many engineering disciplines. Some minor changes have been made to amend the dependability model for PSEg use. The dependability of a product or system "reflects the user's degree of trust in that system."

The four properties of dependability are availability, reliability, safety, and security. Availability is the property of a product being able to function as intended whenever called upon. Reliability is the property of a product functioning as intended without failure. Safety is the product property of functioning as intended without accident or personal harm. Security is a property of a product where it is immune to malicious acts such as sabotage.

Many people might characterize availability as an aspect of reliability. Some may see security as an aspect of safety. This safety component of security can be especially true when considering and developing standards for cybersecurity and the Internet of Things (IoT) ["Status Report on the Internet of Things (IoT) and Consumer Product Safety" 2019].

This book will cover product safety as being the pursuit of products which, through an integrated and disciplined product-development process (PDP), possess positive design characteristics that reduce safety risks to users from accidental harm. Health risks and malicious-intent, security risks will not be covered in this book. A partial list of properties that safety is *not* is shown in Table 2.2.

Dependability		
1	Availability	Functions as intended whenever needed
2	Reliability	Functions as intended without failure
3	Safety	Functions as intended without accident or personal injury
4	Security	Functions as intended in spite of malicious actions

Table 2.1 Elements of Product Dependability

Safety Is Not Necessarily	
1	Health
2	Law
3	Dependability
4	Availability
5	Reliability
6	Security
7	Compliance
8	Quality

Table 2.2 Properties that Safety Is Not

Case Study 2.1: Reliability versus Safety

On September 14, 1993, Lufthansa Flight 2904, en route from Frankfurt, Germany, was approaching Warsaw's Okęcie International Airport for landing. The local weather conditions in Warsaw were poor with a passing cold front producing heavy rain and high-speed winds. Not only was the wind large in magnitude, but the wind had been changing in direction. Therefore, the weather was challenging; however, good pilots do successfully land aircraft under such conditions with regularity.

Upon aircraft touchdown, passengers reportedly began clapping their hands in approval of the piloting skills exhibited by the pilot and the copilot. But soon afterward, rather than experiencing deceleration, one passenger said that the aircraft "started to gain speed on the runway." Regardless of whether or not the plane did, in fact, accelerate, the plane was not slowing down significantly and as would be expected by seasoned air travelers.

Thirty-four seconds after touch down—and after running off of the runway at 72 knots (83 mi/h; 133 km/h) and traveling 90 m (295 ft)—the left wing of the airliner hit an embankment. The plane continued over the embankment and hit an antenna before coming to rest.

The impact resulted in the rupture of aircraft fuel tank(s). The fuel spilled to the left side of the aircraft. The jet fuel was soon ignited, producing a large fire initially located on the left wing. However, the fire quickly breached the passenger compartment, which later filled the cabin with dense smoke. Despite emergency responders arriving on the scene quickly, it was not possible to suppress the fire within the passenger compartment until a large hole had burned through the roof of the airplane.

Although some crew members received incapacitating injuries preventing them from helping with the ensuing evacuation, other crew members were able to help passengers get out of the burning aircraft. Of the 64 passengers and six crew members, one passenger and one crew member (the pilot) lost their lives. Forty-nine passengers and two crew members received injuries considered to be major. The remainder received only minor injuries or none at all.

The airliner manufacturer had provided the pilots of the aircraft with a software system to assist, through automation, with the braking of the aircraft upon touchdown. This braking system controlled three components involved with slowing the aircraft once it was on the ground: (1) the main-wing spoilers, (2) the engine thrust reversers, and (3) the wheel service brakes.

The investigation of this airline accident revealed that, despite the accident, the airliner and its automated braking system worked as designed ("Report on the Accident to Airbus A320-211 Aircraft in Warsaw" 1994). This report was later annotated by a second author (Ladkin 1999). It was discovered that the automated-braking system worked perfectly because it did exactly what it was instructed to do by its software programmers.

Built into the system were safeguards to either protect the aircraft from damage and its passengers from injury or prevent premature wear of service-brake components. For instance, the *full* application of wing spoilers is not usually done during commercial flight. Spoilers are the control surfaces seen by airline passengers out of the side windows when looking at the main wing of an airliner. They are on the top, hinged surfaces of each main wing and are positioned in front of the landing flaps of the wing. Spoilers perform two functions: they increase the airplane's aerodynamic resistance, or drag, by increasing the frontal area of the aircraft and they decrease the lift of the wing by "spoiling" the airflow.[9] Each of these functions of spoilers is helpful in slowing the aircraft *once* it is on the ground. This increase in drag is helpful in slowing while the aircraft is still traveling fast[10] and the spoiling of the lift of the wing helps the aircraft settle onto the ground and stop flying once ground contact has been established.

Similar to the case of spoilers, it would generally be unwise to deploy the engine thrust reversers of an aircraft while in normal commercial flight. Thrust reversers are doors on the turbofan bypass-air engine nacelles that, when deployed, direct some of that engine's thrust forward toward the front of the aircraft. This, of course, slows the aircraft. Passengers sometimes experience a pilot increasing power to the engines immediately upon touchdown. This is typically done only *after* the thrust reversers have been deployed to slow down the plane.

The service brakes on an airplane are quite similar to those on an automobile. They provide resistance to wheel-and-tire rotation. They can slow down a plane or hold a stationary plane in place. They provide no thrust or forward torque to assist with the aircraft's forward motion. Due to the characteristics of the brake system's design with respect to a loaded airliner, the service brakes are generally not applied by the pilot or copilot until the aircraft has slowed down significantly from its initial touchdown speed through the application of thrust reversers and spoilers. Applying the service brakes too early—at too high of a ground speed—can lead to either premature brake-component wear or even brake-system fires and other resulting failures.

As mentioned, the aircraft's automated braking system took these factors into consideration. Spoilers and thrust reversers are not deployed until the aircraft concludes that it is on the ground; the service brakes are not applied until the airplane has slowed to below a given speed. Generally speaking, the aircrew would activate the aircraft's automated braking system just prior to landing and the computerized system would activate the spoiler, thrust-reverser, and service-brake systems as the plane contacted the ground and slowed down. However, and unfortunately, once activated, the aircrew members on Flight 2904 were unable to override this computerized aircraft-braking system.[11]

The report on the accident showed that the automated-braking system on the airliner functioned as it was *designed* to function. This system simply did not perform as it was *intended* to function. The braking system was completely *reliable*; it simply was not *safe*. The specifications, or criteria, that were given to the braking system by its software programmers—although well intentioned—did not accommodate all of the eventualities that would be faced by the airliner during use. The landing conditions faced by Flight 2904 were among the unaccounted-for situations.

The software logic of this system wanted to be "sure" that the aircraft was on the ground before deploying the spoilers and then the thrust reversers. Likewise, the automated-braking system wanted the aircraft to be sufficiently slowed before applying the

---

[9] Spoilers are also used for roll control of certain aircraft during normal flight, but the *full* application of *both* spoilers is rare while a commercial airplane is still airborne.

[10] Spoilers are ineffective at low airspeeds.

[11] This situation of computerized flight control has strong parallels to a design problem with another airliner that has led to two recent crashes (Cassidy 2019).

service brakes in order to protect those friction components from excessive wear and consequential heating. These are, of course, prudent countermeasures to the accidental or premature activations of these systems.

Before the automated-braking system deployed the spoilers, **either** *both* landing-gear struts needed to be loaded to at least 6300 kg (13,800 lbf) **or** *both* sets of main landing-gear wheels needed to be rotating at no less than 72 knots (133 km/h or 83 mi/h). Before deploying the thrust reversers, the automated-braking system had to sense that *both* main landing-gear struts were loaded. In order to apply the service brakes, the automated-braking system required that *both* sets of main landing-gear wheels were rotating slower than 0.8 times V_0, a pre-programmed speed value.[12] A flow chart of the automated-braking system control logic for the three aircraft components is shown in Figure 2.3.

Although it might be tempting to stop here in the accident investigation, these characteristics of the aircrafts systems were aggravated by dynamic weather conditions and avoidable human errors. The pilots were told by the airport's control tower to accommodate a crosswind during final approach and landing, but that crosswind was, in fact, a tailwind. The tailwind led to two problems. First, since aircraft always try to land with a headwind to minimize landing speed, the touchdown speed of Flight 2904 was 20 knots (37 km/h or 23 mi/h) greater than that recommended for its current weight at 172 knots (319 km/h or 197 mi/h). Second, the pilots landed the plane, as customary in a crosswind landing, with the anticipated upwind wing lower than the downwind wing. This is so that the aircraft will touch down first in the upwind main landing gear. The crosswind will then push the downwind side of the airplane down onto the runway. Since there was no crosswind, the airliner remained on one main landing gear, the starboard or right, for 9 seconds until the port, or left, landing gear touched down. It was only then that the spoilers and engine thrust reversers were deployed.

The weather conditions not only created the tailwind, but also produced heavy rain that left a layer of water on the runway surface of several millimeters (approximately 0.1 in.) in places. The post-accident investigation revealed that three of the four tires on the airliner's main landing gear showed significantly wear which decreased the depths of the water-dispersion grooves on those tires. Therefore, the combination of high landing speed, standing rainwater on the runway, and worn tires all contributed to the landing-gear tires "hydroplaning" upon touchdown. Because the tires of the aircraft were riding on a film of water and not in direct contact with the runway surface, the wheels on the main landing gear did not rotate.

Once the main landing-gear wheels had started rotating and had showed the aircraft as having slowed down sufficiently,[13] the automated-braking system then permitted the application of service braking—even though the pilots had been trying to apply the brakes since initial touchdown. This consumed another 4 seconds after the deployment of spoilers and thrust reversers. By the time that pilots had gained service-brake authority, there was too little runway left to bring the airplane to stop.

---

[12] The author was unable to find the explicit value for V_0. However, from calculation, it appears to be 90 km/h (56 mi/h).

[13] An *indicated* rotational speed of 72 knots (133 km/h or 83 mi/h).

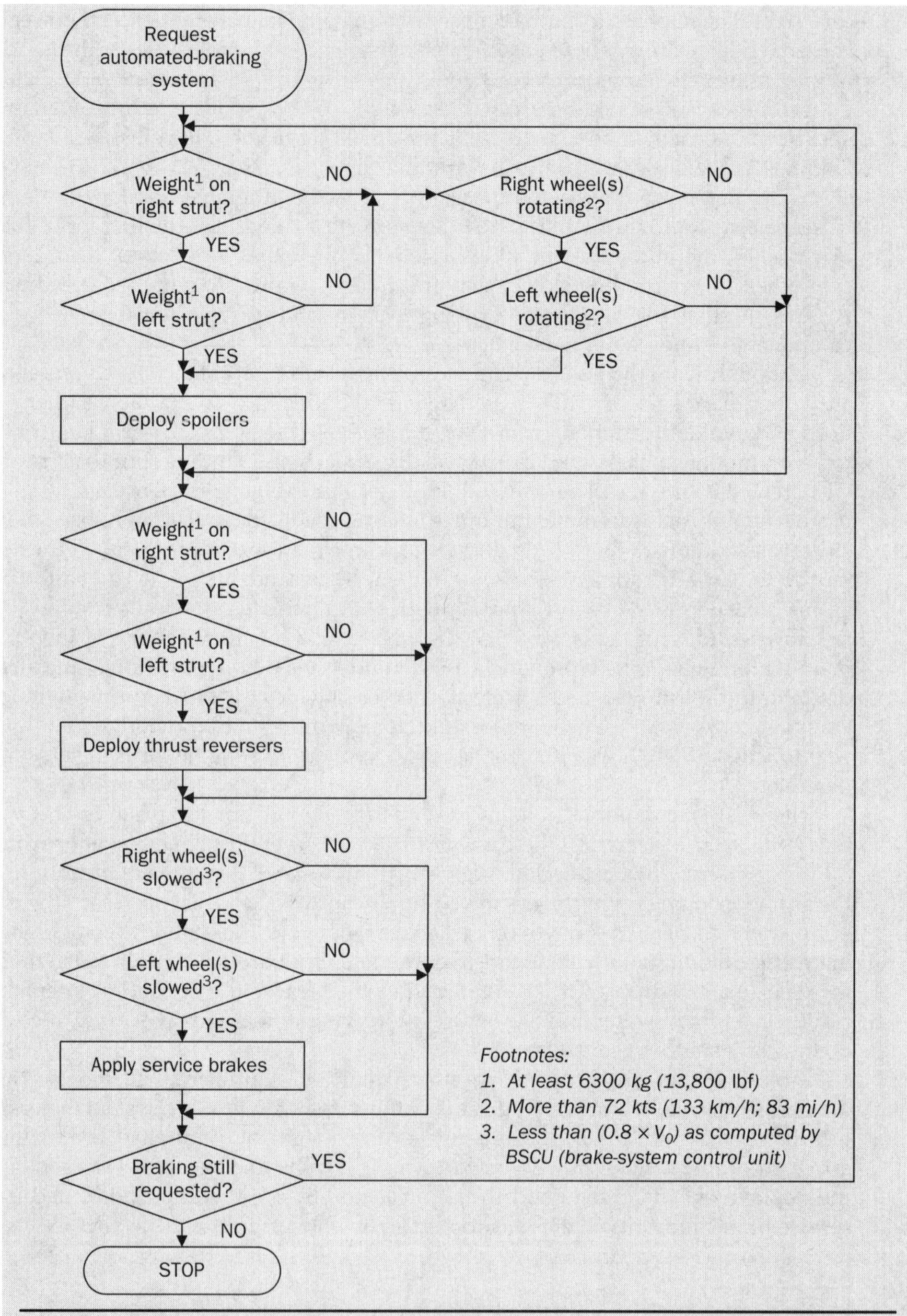

Figure 2.3 Automated-braking system control logic.

The pilot "managed only to deviate the aircraft to the right" of the runway. The plane then left the runway and impacted the embankment.

Another factor which could have contributed to this accident, according to the report, was that there was pertinent information not given to the aircrew in the aircraft's flight manual. Although the manual did say that higher touchdown speeds can extend aircraft rollout, or stopping distance, "by about 25%," the manual did not provide a method of determining that distance precisely. This, unfortunately, is not the only case of an airliner flight manual containing insufficient information (Gollom, Shprintsen, and Zalac 2019).

Whenever possible, engineers have a duty—to society and to themselves—to help prevent injuries from taking place in the first place. If engineers are unable to reasonably prevent injuries, then it is simply being responsible to use such prior accidents as learning opportunities. This aircraft crash was obviously a tragedy which cost two humans their lives. Therefore, it should be studied in order to see what went wrong and how its repetition could be prevented.

In this case, there is no indication that anyone involved with the design, analysis, or testing of the automated-braking system acted in bad faith. It is likely that much time and effort was spent in the implementation of the braking system by conscientious people. This automated system is also exactly this type of innovation that can lead to breakthroughs in product and system safety. The automated-braking system was indeed a good idea. It was probably even a *great* system for most conditions that it encountered; however, it was not a *perfect* system. Among the flaws with the entire automated-braking system were the specifications given to the system's software programmers about what constituted an aircraft being "on the ground." This led to faulty software logic and to the events of Flight 2904. This should be considered a failure in *requirements engineering* (Sommerville and Sawyer 1997) in system design. Such failures are often small and simple by themselves—a seemingly insignificant number or condition—but may bring with them severe consequences when implemented within the larger system.

Even with a simple traffic accident involving one automobile, a larger set of factors must be considered by investigators to extract the most information from the event. This airliner accident, however, was a highly complex event with numerous factors and agents involved which affected the accident's outcomes. Consider the PUE model shown back in Figure 1.2. It shows the three major factors affecting the safety of a product. It is a fruitful exercise to think about what went wrong and how similar accidents, injuries, and deaths may be prevented in the future.

This example airline accident amply demonstrates the difference between *reliability* and *safety*. This accident has also been chronicled by Sommerville in a video lecture (Sommerville 2014a) and the corresponding presentation-slide deck (Sommerville 2014b). This tragic event, in which two people lost their lives, was caused by an aircraft's auto-braking system which performed as designed, but not as intended. This automated-brake system was completely reliable and yet proved to be unsafe.

Just as safety is not reliability, safety is not compliance. It, quite simply, *cannot* be. For example, consumers and engineers alike want a safe version of a new type of product. Since this is a ground-breaking product that has never before been thought of in detail, there exist no standards, let alone regulations, with which the new product must comply. Therefore, compliance cannot be attained. However, there will be *expectations* to meet from both consumer and designer/manufacturer. Thus, the product's characteristics—along with its users and environment—will provide some level of safety for product users. It is hoped that this level of product safety will meet the manufacturer and consumer expectations and be considered a "safe" product irrespective of the compliance vacuum.

Example 2.3: Compliance versus Safety

To further demonstrate that compliance and safety differ in a significant way, consider the following example. Suppose that the manufacturer of a product is taken to court under the allegation that the product was, among other things, "dangerous." This accusation lacks specificity, but the discovery portion of a lawsuit permits the injured party to request documents from the designer and manufacturer of the product. Also suppose that the manufacturer subsequently produced two emails from the engineering-testing group. One of these emails concerned a regulatory-compliance test; the other, a product-safety test. The first email, about compliance, says something like the following:

> The product barely passed allowable Lead (Pb) content level of 0.1% (by weight) and also just squeaked by the Cadmium (Cd) allowable limit of 0.01% (by weight).

The second email, about product safety, says something like the following:

> The product just barely passed our company vertical-strength test by supporting the required load of 100 lbf (445 N) for the required 60 seconds, but it completely collapsed at 64 seconds.

In the case of compliance, passing is a binary event. Something either passes or fails; it is a 1 or a 0. In this case, the product passed tests for both Pb and Cd levels. It would be difficult for an injured party to assail the adequacy of these maximum concentration levels required for compliance when such levels have been set by national or international bodies or governments. In this case, the maximum-concentration levels were taken from the European Commission's *Restrictions on Hazardous Substances* (RoHS) Directive ("Directive 2011/65/EU on the Restriction of the Use of Certain Hazardous Substances in Electrical and Electronic Equipment" 2011).

When looking at product safety, results are often more gray than black-and-white. Although the test was passed by the product, the test was an internal test and was not developed and scrutinized by other companies, the industry, a standards-development body, or regulators. The sufficiency of the pass/fail criteria—both the load and duration—can be easily attacked by an opposing attorney. In addition, the reporting test engineer seemed to express some concern about the adequacy of the product in the email by saying that it "just barely passed" the test.

Although the first email is not a great example of communication, it does unconditionally show that the product is compliant with maximum limits of two hazardous substances. The second email, on the other hand, is rather cringeworthy. It passed the internal metric for strength—just barely—but the engineer expressed concern about the product, nonetheless. The intangible, visceral reaction that many readers may have had to the second email indicates that there is a different reaction to safety matters than to health-compliance matters.

This example also shows poor communications skills and practices exhibited by the test engineers. They were submitting opinion in addition to fact. This material would be of grave concern to a manufacturer's legal counsel, but of lesser concern to an engineer simply committed to producing safer products.

Quality is also different from safety. At least, quality is usually not the same as safety when considered by a design-and-manufacturing company. Consumers would generally disagree and opine that a "dangerous" product was of low quality. To an engineer or a company, a product or component of a product was of high quality if it met the specifications assigned to it regarding performance, construction, material, dimension, and durability. These products would be sampled and their quantities would be measured statistically. The result of this quality-assurance (QA) or quality-control (QC) effort would be tabulated and provide arithmetic means, variances, ranges, and perhaps more. There are many dedicated, highly competent people in QA/QC who help in the production of safe, reliable, dependable, compliant, and legal products. Yet, a low-quality product that fails may be a completely safe product—so long as it does not fail in a particular way at a particular time. A product that is broken is unlikely to harm someone since it often possesses no energy and will generally be left alone by people.

Similarly, legality has little to do with safety. Society would like to believe that laws make sense, but citizens have learned over time that many laws make no sense whatsoever. This will be discussed later in this chapter in Section 2.7, *Value Systems.* Practices that are legal may be unsafe, or at least considered unsafe in another location. For example, some states allow commercial truck drivers to operate with two trailers; other states limit a truck driver to one trailer. It was only recently that all 50 states permitted drivers to attach electronic devices to the windshields of their vehicles although restrictions still apply. Some states permit motorcyclists to "lane split," or filter through stopped traffic; others do not. Some states and counties permit people to ride in the back of pickup trucks while others do not. It is highly unlikely that the actual levels of safety for these activities truly vary by much from place to place.

Just as some legal actions may be unsafe, some safe activities may be completely illegal. If there is assuredly no other vehicle within 1 mile (1.6 km), it is completely safe to not stop at stop signs and to drive on the wrong side of the road. Both of these actions are most definitely illegal, however. Additional aspects of the law are covered later in this chapter.

2.5 Product-Safety Management versus Product-Safety Engineering

Although it might upset some people, it must be clearly stated that—as they are traditionally practiced and described—product-safety management (PSMt) and PSEg *may be* quite different things. Many professionals involved with PSMt are concerned primarily with product specifications, supply-chain management, and compliance matters since many products in the United States are both designed and manufactured abroad. The departments involved in PSMt may be purchasing, compliance, or legal. Make no mistake, PSMt is an important job and those doing it should be recognized and applauded for their good work. Those in PSMt also perform a vital role in managing product-safety recalls and remediation or withdrawal of the product from the market. Many such products are also used by "at risk" populations including children, so that their risks may be higher than to adult populations. These products include toys, clothing, home goods, and furniture. Important potential hazards and risks of many of these products are toxicity resulting in poisoning, flammability leading to burns, and detachable parts culminating in choking. Many such characteristics are strictly regulated and, consequently, are tested in many ways for compliance to these third-party or governmental criteria. However, some individuals in PSMt are removed from the actual product-design and development processes that may affect the overall safety characteristics of the products which they have suppliers ship to them.

Product-safety engineers are, in contrast, involved with the design, analysis, development, testing, and manufacturing of the product down to a minute level. Rather than simply being given specifications—although this sometimes happens—product-safety engineers work with product-design teams in an integrated PDP to develop the product and then determine what product-safety characteristics and levels are needed for users. They may also need to devise test procedures to verify levels of product performance or safety. If the product is of a new type, then knowledge about what was safe in the past may no longer valid. The product-safety engineer (PSEr) may be involved with the planning and scoping of the project if the new product is expected to pose new hazards or higher risk levels that previous products have. Product-safety engineers must work to *design-in* positive product-safety characteristics to their products from the very beginning of the engineering-design programs. Having said all of this, there will at times be poor decisions made or product-performance problems that have escaped the rigorous design, analysis, prototyping, and testing protocols that the better engineering companies have established. Accordingly, there will be times when product-safety recalls will be necessary. The PSEr should be involved in both identifying and repairing those issues that become product-safety defects and prompt a recall.

From the above descriptions, PSMt personnel and product-safety engineers share some duties. The term PSMt should also not be confused with a product-safety manager (PSMr), who may be an engineer in charge of the product-safety *engineering* effort and not directly involved with the purchasing department's product-safety *management* efforts. Product-safety managers, or directors, may have several product-safety engineers working for them and be within the engineering department.

2.6 Product Safety and Product-Safety Engineering

Literally, PSEg is—or should be—the engineering, or designing, of safety *into* a product from the beginning, rather than trying to improve the safety of a product after it has been manufactured, shipped, and delivered—which is generally a losing proposition for all parties involved. Although the concept of safety has been introduced and definitions will follow in another chapter, there are still numerous obstacles that the engineer who simply wishes to design safe products must overcome. Engineers have spent hour upon hour studying physical phenomena and the mathematical equations, which permit their applications. In typical textbook engineering, there exist two types of answers: those that are *right* and those that are *wrong*. It is that simple—and objective.

For example, Newton's second law of motion (Newton 1687) can be reduced down and written as $\bar{f} = m\bar{a}$. So long as the mass of the body remains constant and the speed of light is not approached, this equation is considered to be true—or to be a fact.

Similarly, the law of conservation of energy (Planck 1910, 42), when interpreted very loosely, states that energy is neither created nor lost. Therefore, energy is conserved. Theories or answers violating such physical laws are considered to be incorrect and flawed.

Consider the example of an aircraft, such as the one shown in Figure 2.4. If the upward-acting lift force is greater than the downward-acting weight force, then the plane will get off of the ground. If the forward-acting thrust force is greater than the rearward-acting aerodynamic-drag force, then the plane will accelerate forward. It does not matter whether one knows these physical realities or whether one agrees with them. These are objective relationships that are indisputable. These are also the relationships that engineers enjoy and study.

There is terrific intellectual challenge, purity, and even beauty in pursuing and applying such scientific truths. Such truths have served society well by enabling the construction of public-works systems, roads, bridges, dams, automobiles, aircraft, rockets, electronics, communications systems, and inexpensive or strong new materials. These same truths are now also being used to conserve energy and to protect or reclaim the environment. Many people have been attracted to study and practice engineering and the sciences for these very reasons. Unfortunately, PSEg has no such "laws" to dictate what is safe and what is unsafe.

PSEg involves many factors and functions beyond the typical engineering curriculum of even the most reputable of colleges and universities. PSEg, specifically, involves many concepts, parties, and interests. The waters that a PSEr may need to navigate include engineering (design, analysis, prototyping, testing, and manufacturing), science (new technologies), health, business, law, information technology (IT), intellectual property (IP), logistics, purchasing, regulatory affairs, and also ethics—both professional and personal at times.

As just mentioned, many, if not most, PSEg criteria cannot be represented by simply looking to a formula such as Newton's second law of motion. Reasonable people can, and indeed do, disagree on what is sufficiently *safe*. This state of discord is what makes product safety so difficult to discuss—let alone resolve. Therefore, understanding the role of *value systems* is critical to successfully practicing PSEg.

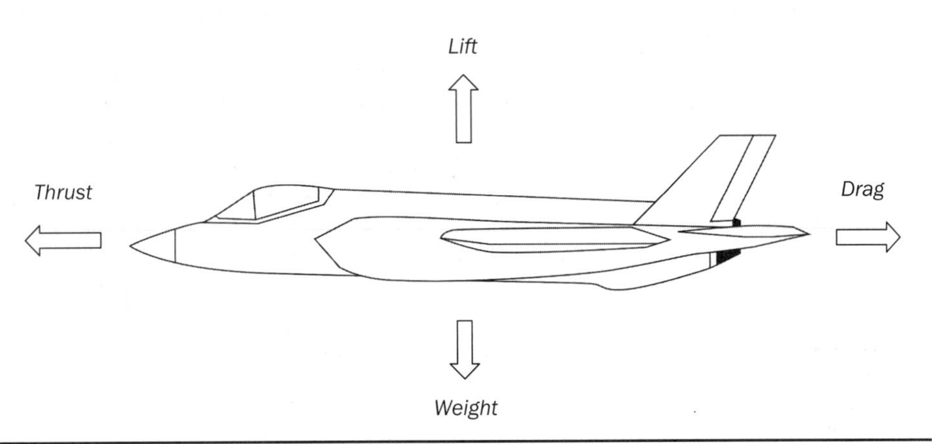

Figure 2.4 Forces acting upon an aircraft in flight.

PSEg is truly a "different animal" than other fields of safety. For example, engineers working in the field of *industrial* safety have many pass/fail criteria already well established for them through extensive workplace regulations and other requirements at Federal and state levels. Assuring industrial safety often becomes a pure *compliance* effort,[14] without regard for any side effects that may arise out of blind compliance. PSEg is anything but a compliance exercise. Although there do exist some product-safety requirements, many of these sources, such as voluntary industry standards, offer minimal true guidance to the design engineer during the product-design process who is making numerous detailed decisions affecting ultimate product safety. This is especially true when that engineer is creating an innovative product for which no standard has even been conceived, such as described above. Even when product-specific safety and performance criteria *do* exist in standards, the designer and manufacturer (read, design engineer) might be wise to treat these as *necessary* conditions rather than *sufficient* conditions. Therefore, industrial-safety knowledge and experience are not necessarily transferable to PSEg despite both fields containing the word *safety* in their names.

To illustrate this lack of objective truth, the author goes through the following exercise[15] with students on the first day of class for the "Product-Safety Engineering and Engineering Ethics" class he teaches. This is to demonstrate to the class the difficulty of PSEg and how it differs from other forms of engineering.

First, students are presented with the first fundamental canon of the NSPE (National Society of Professional Engineers) Code of Ethics (NSPE 2018) which states that:

> Engineers, in the fulfillment of their professional duties, shall:
> 1. Hold paramount the safety, health, and welfare of the public.

Other engineering societies and groups have similar language in their codes of ethics. The point is that there is little disagreement with this sentiment among engineers—professional or otherwise. This makes sense because engineers, especially those working on products or projects with the ability to inflict great harm upon a person or groups of people, are entrusted to make ethical decisions thereby protecting public safety, health, and welfare.[16]

Students are then asked if any one of them disagrees with this canon. No student has yet voiced or otherwise demonstrated a disagreement with this obligation to society. Thus, engineering students willingly consent to their future duty to defend the public's safety through their professional actions.

Students are then asked to think of or name one single general fact. It may be any fact such as $1 + 1 = 2$, the sun is hot, or the classroom lecture is too long. No students have had difficulty with this first question to date. The more-vocal students readily volunteer answers to the class, some of which are quite entertaining.

The follow-up question to the general-fact question is to name one single *safety* fact. There generally follows dead silence in the classroom. After some time has elapsed, some answers are offered. These answers vary, but regular responses include "read the manual" and "expect the unexpected." Reading the manual is indeed a worthwhile

---

[14] It is rightfully acknowledged that many dedicated and conscientious industrial-safety professionals do indeed go *beyond* mere compliance in the performance of their duties.

[15] This exercise has been borrowed from Ralph Barnett.

[16] This welfare also includes *economic* welfare.

activity provided that the manual is a good one, but not all manuals are good manuals. The slogan "expect the unexpected" is impossible and not actionable as is the slogan "Safety First." The catch phrase "expect the unexpected" is equivalent of telling someone to know the unknowable. In addition, it implies that injury accidents are unpredictable. A short search of any video-sharing website will rapidly uncover scores of accident-video compilations. Looking at both the activities—or stunts—being performed as well as the people attempting them—and the possibly inebriated crowds usually cheering them on—will demonstrate the expected outcomes from these endeavors. These unsafe actions lead to the anticipated, predictable results. While it is good practice to look for events leading to accidents and injury outcomes which have never before happened, it can be a useless waste of time and effort to suddenly expect that tomorrow morning the sun will rise in the west and that gravity will operate in the opposite direction. Therefore, the sample admonitions to read the manual and to expect the unexpected are not *laws* of product safety.

Another counter example to safety fact is that there exist other long-held and widely accepted safety adage "always wear a helmet" in motorized recreational activities. This is generally sound advice; however, there is at least one exception. Users of personal watercraft (PWCs) do not generally wear helmets unless they are competing with other PWCs and operators and the collision threat is significant. It is believed—although not proven—that falling off from a PWC at speed with a helmet can lead to a twisted or broken neck if the water catches the helmet in a precise manner. So, at the moment, the helmet is not suggested for general PWC use by that industry.

Finally, it may be widely held that all moving vehicles must have brakes in order to slow that vehicle. There is at least one exception to this rule—the vélodrome bicycle. These are also known as "track" bicycles since that are rather impractical for use off of the track and may even be considered unsafe for general-purpose use for several reasons. These bicycles have single-speed, fixed-drive, chain-and-sprocket drive systems. The pedals will always move when the bicycle is in motion and cannot "freewheel" as do conventional bicycles. This characteristic can lead to "pedal strikes" if such a bicycle is used on an unbanked surface. Many velodrome tracks are banked to more than 40 degrees in turns to accommodate such bicycle characteristics. Another property of the track bicycle is the absence of service brakes. Any deceleration, or braking action, is accomplished kinetically by the rider applying "back torque" to the pedals which transfer this torque to the bicycle's rear wheel. Since velodrome racing heats can field 20 to 30, or more, riders on the track at one time, the *differences* in bicycle speeds should be kept to a minimum or massive pile-ups will occur.

It is hoped that readers will see that for every potential product-safety *law*, there may be numerous exceptions. Although the focus of the book is generally on probable events, a true physical or mathematical law will be valid for all cases. There should be no exceptions for a law. No product-safety axiom presented thus far has met this criterion for becoming a law. Therefore, there exist no PSEg "laws" to guide the product-safety or design engineer to a true, unique, and unassailable result.

In closing, the author summarizes the recently self-admitted quandary in which these future product-design engineers just found themselves. Namely:

1. Nothing is more important to these students than protecting the safety of the public
2. These students now realize that they know nothing about safety

Of course, much of this is the construction of a straw man who will be knocked down. Engineers do know something about safety. For example, reducing the probability of the severity of an incident will lead to greater safety, generally. However, the exercise does show the students that they are now embarking upon a different kind of engineering journey than most of those they have taken in the past with such subjects as statics, dynamics, thermodynamics, aerodynamics, hydrology, materials, circuit design, and control theory.

Because there are no engineering laws for the safety of products, students will have to approach PSEg differently. Having no objective metrics for what is "safe," engineers must approach the design of safe products in a different manner than the areas of engineering. The next section will be the start of the journey.

2.7 Value Systems

The prior section spoke of the lack of objective metrics in the field of product safety. That is, there are no universally acceptable objective criteria for assessing the safety of a new product. Lacking an authority, how does one make a determination of whether or not a product is safe? Before attempting to answer that question, it is important to understand how people make such decisions. People—either in groups or individually—make such decisions based on their *value systems*. A value system is a set of principles and ideals accepted by a group or individual that influences thoughts and resulting actions. Value systems play a major part in forming how people think and how people behave. As a result, value systems affect people's beliefs and judgments. These judgments include the assessment of whether some product is safe or unsafe.

Societies possess value systems that manifest themselves through customs, beliefs, and behaviors both demonstrated and expected by its citizens. The international traveler is confronted by many such customs, beliefs, and expectations. Cultures express their expectations through their stances on such practices as punctuality versus tardiness for meetings, finishing all of one's food versus leaving food on one's plate when dining, tipping versus not tipping after dinner, and many more. None of these is either right or wrong. Each is perfectly logical given their respective societies and histories.

Any misstep of the social values mentioned above will probably result in no harm beyond social ostracization. However, running afoul of a society's value system embodied in its legal system can lead to punishment including incarceration and capital punishment. In fact, any country's value system is evident in their laws to a great degree, especially if the country is democratic in structure. Legislation certainly does not keep up with rapidly changing social norms. Yet, a society's laws generally do represent the consensus of the population. In no way is every person's opinion taken into account. There is never unanimity. Not everyone chooses to vote in and not everyone agrees with the results of elections. But, through plebiscites and properly elected officials, the will of the people can speak to guide the law of the land. In the case of a culture dominated by an oppressive dictatorship, it is unclear to whether the imposed value system conforms to society, or the inverse.

The above looked at people in large groups, or societies. If people are taken individually, that is where value systems become the most difficult to address. An individual person's value system will be affected by many things including age, historic period, geographic location, socio-economic status, political leanings, faith/religion, ethnic group, education level, life experiences, financial interests, and even prejudices to name but a few. In addition, value systems need not be logical and consistent. This vast range of diverse value systems leads to problems for the engineer who is tasked with designing, for instance, a new gasoline-powered chain saw.

The members of the engineering team will often have their own collective opinion regarding the safety of the new chain saw through a consensus value system as well. If there is anything atypical about the design or operational characteristics of the new product after competitive benchmarking has been performed, it may be a wise step to seek out the voice of the customer (VoC) and listen to their feedback. A set of focus groups, for example, could be used to help out the design team. However, as has been discussed already, since value systems vary and since value systems affect a decision such as safe or unsafe, the results from these focus groups will vary. These results may vary by little or they may vary by quite a bit. If choosing from the same population, say potential purchasers who will likely be experienced chain-saw operators, there is likely to be little variation between groups—unless there is some drastically new feature or characteristic to the proposed chain saw. However, should two focus groups contain drastically different populations, the results may well differ significantly. If by chance, one group contained lifetime firearm-rights organization members and the other group contained lifetime environmental activists, then one should expect to see the focus-group results diverge from one another on many questions asked. Furthermore, it would not be surprising to see the design-and-manufacturing company disagree with regulators on the issue of product safety should accidents and injuries result from the product at a later time.

A situation such as the above may leave a traditionally educated engineer at a loss. There exist no product-safety laws which can be universally applied. In addition, as was suggested from the results of the fictional focus groups, there is likely to be no universal agreement within society as to what is *safe enough* for product designs. Engineers, who themselves are people, do not wish to produce any unsafe products. Even if not for altruistic reasons alone, these engineers may themselves use the product and also have family and friends who may cross paths with the product once it is released. Therefore, such engineers are in a quandary.

It has been astutely observed (Barnett and Glowiak 2004) that:

> If technologists design systems by appealing only to first principles,[17] they will not be able to satisfy the safety requirements of society. Formalization of safety expectations requires codes, standards, and statutes. Indeed, the most important design decisions cannot be made by science or technology—they ultimately rest on doctrine advanced by value systems.

These authors correctly opine that looking no further than the fundamental principles of science, engineering, and mathematics will never adequately address the safety needs of the public and that manifestations of collective value systems, such as codes, standards, and laws, help to outline societal expectations of product safety for design engineers. They proceed to state that the most-important design decisions must incorporate the consideration of values. It might be too strong to say that *the* most-important

---

[17] "First principles" are those objective scientific and mathematical laws discussed earlier.

decisions rest upon values, but it is clear that *many* important, and ultimately significant, engineering-design decisions rest upon them.

This conclusion forces engineers to confront a dirty little secret of engineering: *engineers hate studying people.* In fact, some engineers went into engineering in order to avoid working with people.[18] Human behavior is neither deterministic nor invariant. Human behavior varies widely both between and within people. No clean, neat mathematical equations exist for modeling humans. Some fields of engineering work diligently to produce models for human actions. These engineering fields include highway traffic-flow management, ergonomics, and human factors. However, dealing with the human element remains troubling for many engineers due to subjectivity, imprecision, and variability.

2.8 History and Safety

When attempting to understand the role of value systems in product safety, it is helpful to look at history. What was once considered safe may no longer considered so. What is considered safe today may not be thought of as safe tomorrow.

As a right of passage in American high schools, many students were forced to read a book which was highly influential in its time—the early 1900s. That book was written by its author to be social commentary about unfair labor practices and the exploitation of disadvantaged workers. Instead, the public perceived the book to be about the horrors of the American food-supply chain. Upton Sinclair (1906a) wrote about the despicable labor practices of the Chicago meat-packing industry, but coincidentally exposed the deplorable sanitary and health conditions of that industry. The author later said that "I aimed at the public's heart, and by accident I hit it in the stomach" (Sinclair 1906b). Upon the publication of *The Jungle*, the Federal government soon responded with the Food and Drug Act (*The Food and Drug Act* 1906) and the Federal Meat Inspection Act (*Federal Meat Inspection Act* 1906). The Food and Drug Act made producers of drugs label them accordingly if they contained addictive substances including alcohol, cannabis, morphine, or opium; they could, however, still be sold for a time. Over three decades later, the nation passed the Food, Drug, and Cosmetics Act (*The Food, Drug, and Cosmetics Act* 1938). This 1938 act regulated foods, drugs, cosmetics, tobacco, medical devices, and devices which emitted radiation.

The "bar" on what society considers safe and healthy generally only ratchets upwards toward greater levels of each. An era of much citizen advocacy, which included a product-safety awakening, took place in the United States in the 1960s. Perhaps much of this activity was spurred onward by a solitary book from a Washington, DC attorney named Ralph Nader (Nader 1965). Although this book was about numerous issues involving the automobile industry and the environment, its title *Unsafe at Any Speed: The Designed-In Dangers of the American Automobile* took square aim at one particular vehicle: the Chevrolet Corvair.[19] Nader speaks of design decisions made both by engineers and

[18] The impression that being an engineer means not having to deal with people is a false one. Today's practicing engineer spends considerable time in meetings, writing emails, and working in collaborative teams. Successful engineers must be able to work with people, but perhaps this was not known to all students when entering engineering school. Communication with people is among the most valuable skills for a successful engineering career today.

[19] Those with an interest in safety history are encouraged to read Nader's book.

by accountants at General Motors, the car's manufacturer. The Corvair was to be an inexpensive car with an air-cooled rear engine introduced in the early 1960s. Although some European cars had been so constructed, Detroit automakers were unfamiliar with the design of such cars. Of particular interest to Nader were the Corvair's resulting handling and crashworthiness characteristics which he and many others found to be quite lacking. These two characteristics also appear to have been affected through cost-cutting measures. There was much about this car to criticize, including its aesthetics. Although Chevrolet responded to critics with significant changes to the vehicle, the damage from adverse publicity had done much damage to sales and the production of the car was halted by the late 1960s. The uproar raised by or facilitated through this book help led to the creations of the U.S. Department of Transportation's National Highway Traffic Safety Administration (NHTSA)[20] and the Environmental Protection Agency (EPA).[21]

This automotive example is just one example of how criteria for acceptable levels of safety change over time. The automobile industry had long made cars with little in the way of crashworthiness properties. Many automobile interiors had aggressive, sharp components such as radio knobs and door handles as well as unforgiving, hard surfaces throughout. New technologies have also permitted accidents to be both avoided and survived. Traction-control (TCS) and anti-lock brake (ABS) systems and technologies have helped vehicles avoid accidents. Technologies such as two-point seat belts (lap belts) followed by three-point seat belts (lap-and-shoulder belts) and airbags, now throughout a vehicle interior, have aided people in surviving the accidents that still occur.[22] As new technologies are invented and existing technologies are refined to enhance the safety of product users, society will undoubtedly be the beneficiary.

2.9 Product Liability

The topic of product liability—or "products" liability in a legal setting—is only raised in order to address why it will not be covered in depth. An earlier example on emails produced during legal discovery was but a minor, temporary excursion into the legal implications of product safety. The focus of the book is PSEg. Since it is *engineering*, its pursuit should reward reason and logic. If society's system of laws was rational, product liability would be a worthwhile pursuit for a design engineer to both study and follow. However, the system of laws is a *legal* system, not a *justice* system to paraphrase U.S. Supreme Court Justice Oliver Wendell Holmes, Jr.[23]

The above is more than a distinction without a difference. The engineer, and scientist for that matter, is after the technical equivalent of justice. Justice being the pure and true way that things *should* turn out in an ideal world free of artificial influences. The legal system, on the other hand, is a contrived world of human-made rules and procedures. In a legal system, a given input may produce different outputs depending upon

---

[20] www.nhtsa.gov

[21] www.epa.gov

[22] This demonstrates the difference between an accident and its outcome, or result. This relationship will be explored in Chapter 5.

[23] Some people believe that Justice Holmes once said to an inexperienced attorney in his courtroom that "This is a court of law, young man, not a court of justice." The author has been unable to find a reliable reference for this quote, however.

where, and when, such a legal experiment might be conducted. In an ideal world where justice rules, a given input to a system would always produce the same output—and that output would always be the most virtuous one.

Laws are inconsistent within a country. In the United States, for instance, laws vary at least from state to state. What a design engineer considers to be safe should not vary according to user locale. Furthermore, engineers should not spend their time trying to anticipate the machinations of a product-liability trial. There are far too many legal nuances and capricious decisions to motions made by judges to accurately anticipate the resulting outcomes. The author is aware of instances involving high-profile automotive product-liability litigation where jurors could not be told that victims of the accident were intoxicated and not wearing seat belts. With occurrences such as these, engineers must better devote their time to the design of safe products and let product-liability litigation take its own irrational and convoluted path. This is a book about designing safe products, not a book about liability-proofing a product. Professional-ethical considerations make designing safe products a higher engineering value than designing liability-proof products. These two company goals, however, need not be mutually exclusive at all times.

There are many sound reasons for having an attorney skilled in products liability speak to an organization's engineers about legal theory, such as negligence and strict liability, and certain aspects of best engineering practices within an organization. Preventable mistakes should be avoided, especially if those mistakes could result in a trial court mistaking organizational chaos or incompetence for malice. However, organizations should not expend effort on trying to make a bad group look good. Instead, that effort should be spent transforming that group into a better one to withstand the scrutiny that will one day come. This transformed group will also design superior products.

An engineer will never "win" in a court of law. The best that one should hope for is to not lose. If a design engineer does A, then plaintiff's counsel will say that the engineer should have done B. If the engineer did B, then plaintiff's counsel will say that the engineer should have done A. If the engineer did both A and B, then plaintiff's counsel will say that the engineer should have done C. It never ends. When all is said and done, sound, informed, and rational decisions—based on ethical *conclusions* rather than baseless *opinions*—are the engineer's best defense in a court of law.

Many practicing engineers will find themselves providing either deposition or trial testimony at some point in their careers. This is a cost of doing business. Advice on providing sworn testimony, in depositions and at trial, that the author received as an inexperienced consulting engineer was the following three points:

1. Tell the truth.
2. Do not speculate.
3. If something is unknown, simply admit it.

If these points for a fact witness to remember do not serve an engineer of any type well, that engineer should consider changing either the type of work or the employer. Engineering ethics will be discussed in later chapters. For engineers wanting to learn more about the law and products liability and their intersections with engineering practice, there are numerous sources available including Gayton (2017) and Ross (n.d.).

2.10 Conclusion

Safety is often confused with characteristics such as health, compliance, legality, security, and reliability. They are not the same things. They may be, but they also may not be. It is hoped that the reader now has a good grasp on what safety and product safety mean—and what they do *not* mean. It is also hoped that the reader seen that there will rarely be a functional product that will be "safe enough" to everyone. The engineer should operate and make decisions consistent with ethical conduct, such as that set forth by a suitable engineering code of ethics. This chapter forms a foundation on which other book chapters will build.

Value systems have been shown to be a vital part of how people decide upon what is acceptable and what is not. Value systems change over time, from place to place and from person to person. A study of history shows that what was once acceptable behavior and risk may no longer be greeted with open arms.

PSMt is more of a business function than PSEg is. Both of these topics will be compared and contrasted later in the book.

Product liability and PSEg need not be mutually exclusive. It is this author's hope that they are instead complimentary to one another. This is not always the case, however. When there is a divergence in actions and goals between engineering and business, the design engineer should follow the path leading to greater product safety for consumers of the product.[24]

References

Barnett, Ralph L., and Suzanne A. Glowiak. 2004. "Standards—Impact and Impotence." *Triodyne Safety Brief* 27 (2): 4. http://www.triodyne.com/SAFETY~1/sb_V27N2.pdf.

Cassidy, John. 2019. "More Questions Than Answers about Boeing, the 737 Max, and the F.A.A." *The New Yorker*, May 16, 2019. https://www.newyorker.com/news/our-columnists/more-questions-than-answers-about-boeing-the-737-max-and-the-faa.

"Coronavirus Disease 2019 (COVID-19) Situation Report—29." 2020. Geneva, CH: World Health Organization (WHO). www.who.int › docs › situation-reports › 20200219-sitrep-30-covid-19%0A.

"Directive 2011/65/EU on the Restriction of the Use of Certain Hazardous Substances in Electrical and Electronic Equipment." 2011. Brussels, BE.

Federal Meat Inspection Act. 1906. https://en.wikipedia.org/wiki/Federal_Meat_Inspection_Act.

Gayton, Cynthia M. 2017. *Legal Aspects of Engineering, Design, & Innovation*. Tenth edition. Dubuque, IA: Kendall Hunt. https://he.kendallhunt.com/product/legal-aspects-engineering-design-innovation.

Gollom, Mark, Alex Shprintsen, and Frederic Zalac. 2019. "737 Max Flight Manual..." Cbc.Ca. 2019. https://www.cbc.ca/news/business/boeing-737-manual-mcas-system-plane-crash-1.5065842.

"Interim Guidance for Preventing the Spread of Coronavirus Disease 2019 (COVID-19) in Homes and Residential Communities." 2020. Cdc.Gov. 2020. https://www.cdc.gov/coronavirus/2019-ncov/hcp/guidance-prevent-spread.html.

[24] As stated earlier in the book, the author is not an attorney, nor does he pretend to be one.

"Laceration Injuries to Children Prompt Starbucks to Recall Stainless Steel Beverage Straws to Provide New Warnings." 2016. Cpsc.Gov. 2016. https://www.cpsc.gov/Recalls/2016/starbucks-to-recall-stainless-steel-beverage-straws-to-provide-new-warnings.

Ladkin, Peter B. 1999. "Annotations to 'Report on the Accident to Airbus A320-211 Aircraft in Warsaw.'" Bielefeld, DE. 1999. http://www.rvs.uni-bielefeld.de/publications/Incidents/DOCS/ComAndRep/Warsaw/warsaw-report.html.

Nader, Ralph. 1965. *Unsafe at Any Speed: The Designed-In Dangers of the American Automobile*. First edition. New York, NY: Grossman.

Newton, Isaac. 1687. *The Principia (Philosophiæ Naturalis Principia Mathematica).*

NSPE. 2018. "Code of Ethics for Engineers." Alexandria, VA: NSPE. https://www.nspe.org/sites/default/files/resources/pdfs/Ethics/CodeofEthics/NSPECodeofEthicsforEngineers.pdf.

Planck, Max. 1910. *Treatise of Thermodynamics*. Third edition. Mineola, NY: Dover Publications, Inc.

"Report on the Accident to Airbus A320-211 Aircraft in Warsaw." 1994. Warsaw, PL. http://www.rvs.uni-bielefeld.de/publications/Incidents/DOCS/ComAndRep/Warsaw/warsaw-report.html.

Ross, Kenneth. n.d. "Product Liability Prevention." Accessed January 24, 2020. www.productliabilityprevention.com.

Sinclair, Upton. 1906a. *The Jungle*. New York, NY: Barnes & Noble, Inc.

———. 1906b. "What Life Means to Me." *Cosmopolitan Magazine*, October 1906. http://dlib.nyu.edu/undercover/what-life-means-me.

Sommerville, Ian. 2004. "Critical Systems." https://ifs.host.cs.st-andrews.ac.uk/Books/SE7/Presentations/PDF/ch3.pdf.

———. 2014a. *Warsaw Airbus Accident*. UK: YouTube.com. https://www.youtube.com/watch?v=wzoxek74RTs.

———. 2014b. "Warsaw Airbus Accident." Slideshare.net. https://www.google.com/url?sa=t&rct=j&q=&esrc=s&source=web&cd=1&cad=rja&uact=8&ved=2ahUKEwjvtZ-Fu6viAhVlmK0KHUJ_B3oQFjAAegQIAxAB&url=https%3A%2F%2Fwww.slideshare.net%2Fsommerville-videos%2Fwarsaw-airbus-accident&usg=AOvVaw371UGr6M1tUbobLflWIAJy.

Sommerville, Ian, and Pete Sawyer. 1997. *Requirements Engineering: A Good Practical Guide*. New York, NY: John Wiley & Sons.

"Status Report on the Internet of Things (IoT) and Consumer Product Safety." 2019. Bethesda, MD. Source: U.S. CPSC.

The Food, Drug, and Cosmetics Act. 1938.

The Food and Drug Act. 1906.

Vigdor, Neil. 2019. "Fatal Accident with Metal Straw Highlights a Risk." 2019. *New York Times*. https://www.nytimes.com/2019/07/11/world/europe/metal-straws-death.html.

CHAPTER 3
Engineering Ethics

In sport, winning is almost universally accepted as being right, and losing wrong. It's such an absolutist binary, what results is an enormous middle space, where right and wrong depend upon whom you ask, and who's listening to the answer.

—MATTHEW BEAUDIN[1]

3.1 Introduction

The opening quote was penned recently about the sport of men's professional road cycling. Although the book's discussions are not about sports, there are many parallels which may be drawn between sport and business. The participants in neither endeavor want to lose. Since many business leaders are former high-school or college athletes,[2] some *executive* meetings may even take on the atmosphere of a fraternity house. This devolution may be especially rapid and severe if the products designed and manufactured by the company are used by risk-taking and "macho" types of users—and the company's engineers feel obliged to push product performance boundaries.

Business becomes the sport of "beating the competition," "moving the ball," and posting "good numbers at the end of the quarter." At times, there may be incredible pressure to sacrifice personally for the good of the team and, hence, be a good "team player." There are occasionally official employee meetings which sometimes take on the trappings of a high-school pep rally. It is in such environments—which also bring with them financial incentives for delivering strong business performance—that some people may be torn or become confused between duty to an employer and duty to society and *themselves*.

Unfortunately, people do not always choose wisely. Regarding sports, the men's professional-cycling peloton has been ravaged by several doping scandals, crippled by the resulting investigations, and decimated by the suspensions of riders or banning of cycling-team managers, the most-recent large investigation about a decade ago ("U.S. Postal Service Pro Cycling Team Investigation" 2012).[3] Similarly, some international

[1] *The Spirit and the Letter: The Fine Line between Competing and Cheating*, VeloNews, May 2014, 20.

[2] Of varying proficiencies.

[3] Unfortunately, it appears that doping remains a phantom which cycling continues to battle (Ballinger 2020).

corporations have been, will be, or will again be subjected to fines and litigation following actions leading to mishaps including fatalities (Parloff 2018; DePass 2018; Cassidy 2019). Not only can product design-and-manufacturing companies lose reputation and even market share, but stockholders may lose money and consumers could be injured or killed when poor or unethical decisions are made by corporations. One consequence may be loss of revenue for the company which, by itself, should not be a disastrous result for a profitable company. Other costs for such actions, however, may be paid by innocent employees through job losses, and their resulting personal upheavals, through the "right sizing" that may follow from business "margin pressures." The last consequence of poor decision-making, personal injury or death, often lead to litigation and its associated costs for the company. People involved may even face prison time (Vlasic 2017a; Vlasic 2017b) for severe or repeated unethical actions if these acts are legally prohibited.

This chapter will, it is hopefully said, provide some ethical guidance to engineers—and other professionals involved in product-safety engineering and anywhere in the product-development process (PDP)—as ethical issues arise in their professional careers. There will also be guidance given to organizational leaders about their responsibilities in providing safe products to consumers.

3.2 Priorities versus Values

The opening of Chapter 1 provided a quick glimpse of the difference between priorities and values. When one flies aboard a commercial airliner, a flight attendant will likely announce to passengers that "Your safety is our first *priority*...." Although certainly well intended, their focus is probably better stated as "Your safety is among our *values*." Since airlines have indeed spent much time and money on passenger safety—required or otherwise, airlines do demonstrate compliance with safety regulations and, when necessary, flight crews have placed passenger interests ahead of their own as demonstrated by the airliner-crash example in Chapter 2 and elsewhere in the media (Cohen 2010). So, perhaps, many airlines do indeed *value* the safety of their passengers.

However, to this day, airlines continue to seat passengers who are non-ambulatory in seating positions that compromise the safety of ambulatory passengers in the cabin should a true emergency arise and evacuation becomes necessary. Although this appears to be in compliance with U.S. Federal Aviation Administration (FAA) regulations, if airlines were indeed interested in passenger safety as a value, they would explicitly address the seating situation—by making tough decisions which would, no doubt, have repercussions for the air carriers.

What about *safety*—namely *product safety*? Many factories have "ethics statements" about workplace safety posted in meeting rooms and hallways or on their websites. Many industrial workplaces place posters bearing sayings such as "Safety First" everywhere in hopes that such "sloganeering" will help improve industrial safety by making workers safer. Many people believe that this is true. Some believe that such sloganeering only makes people feel good. Furthermore, one television personality contends that such banners, posters, and placards may actually lead workers into complacency (Rowe 2015). Yet, any employer not posting such slogans would likely be criticized should an industrial injury occur. Critics would say that not hanging such banners in a workplace would be an indication of a poor corporate safety culture. Yet there is so much more to a "culture of safety" than simply hanging signs in the workplace or placing an ethics or "safety" statement on the corporate website.

When it comes to *product* safety, such platitudes are generally lacking for the reasons mentioned in the previous chapter. Exhortations such as "follow the instructions" and "read the manual" are trite and could be perceived as passing on product-safety responsibility to users. There are few product-safety "rules," if any, that can be implemented in a general and meaningful way.

One purpose of this book is to demonstrate that product safety is a much-different phenomenon than is industrial safety. The results of poor industrial-safety practices are often evident immediately. For example, severe injuries and even fatalities at the worker's factory or warehouse serve to quickly and graphically illustrate lapses in industrial-safety coverage. The effects of poor product-safety practices and decisions may be borne out by accidents and injury outcomes that are delayed by months or years; they may even take place halfway around the world. Unlike the industrial worker, poor decisions made by product-development personnel may never affect them personally, in fact, these design personnel may never know about such accidents. Such people, including design engineers and their management and executives, must be aware of the repercussions of their decisions, work together to produce safe products, and be provided with accurate feedback from real-world consumer usage unfiltered by a corporation's legal department.

It is easy for an organization to state that product safety is their priority. It is much harder and takes longer to demonstrate product safety as a true value. A value is something that people work toward because they consider it the proper thing to do. A corporate priority is a dynamic-in-time ranked goal that reflects present business expediencies. Priorities in competitive businesses frequently mirror the economic climates facing the company. Any business should, and in fact must, acknowledge these realities in order to succeed, provide products to users, and provide jobs for employees. Consequently, such corporate priorities have been moves of production to low-cost countries (LCCs), the drive to reduce administrative waste and manufacturing costs,[4] creating products with class-leading performance, and the like. There exist people in many organizations who exhibit greater enthusiasm than either intelligence or ethics. Sometimes, these people rise rapidly within a company to find themselves in management and executive positions whose rewards come from meeting specific performance targets, manufacturing output, sales goals, or other business metrics without ever questioning the virtue of them.[5] If product-design engineering discovers a safety problem with a new product that just started rolling off of the assembly lines that could stop a superior from meeting a goal, how will that superior respond? Her/his compensation and career trajectory could be seriously affected by not making the quarter. If product safety is truly a company *value*, there will be no question, or "push back," about this matter from management. The decision will be a "no brainer" and that decision will be to make sure that there is no product-safety issue with the new product prior to its shipping and retail release. Managers and executives will be openly supportive of the decision and understanding of its consequences. In a just world, such an employee would be lauded by the corporation.

---

[4] Such cost and waste-reduction efforts are known simply by the adjective *Lean*.

[5] There certainly exist many business decisions and goals without product-safety implications. The decision to offer a new bicycle in red, but not in blue, is one which should result in no injuries and should compromise no engineer's ethics even if that final decision is objectionable for other reasons, e.g., aesthetics.

If the same company only has product safety as a *priority*—and just as one of many priorities—it is unclear how things will play out. Sure, product safety is a priority, but so is beating last quarter's earnings. Hence, there may be several now-competing priorities. Which priority is the highest? It will depend upon the company and, by company, that means managers and executives. It will be a different matter, however, if the product has already been sold when a product-safety problem is discovered. The matter of product-safety recalls which may result in such a case is covered in a Chapter 11.

If company executives elect to do nothing and, by good fortune, no user ever gets injured, then the matter of *value* versus *priority* becomes only an academic question. If, unfortunately, users are harmed resulting in regulatory and media pressures coming to bear on the company, then one can surely wager that product safety will quickly become that company's number-one priority—at least for the time being. Once the panic and pressure have subsided, it is up to company executives to determine if product safety truly rises to value status or remains as a transitory priority. Only the passage of time will tell what executives decide and how the corporation ultimately acts. This will be evident through whom management rewards and whom management punishes—and whether management chooses to repeat the crisis cycle[6] again.

From this, it appears that priorities follow real-world pressures; these pressures wax and wane over time. Values, on the contrary, are much more permanent. Values reflect individual or collective ideals and principles. In the realm of product-safety engineering, the difference between a company priority and a company value can potentially lead to dire consequences both for the company and for the consumer. However, even if engineers find themselves working for companies not making ethical decisions, this is *no* excuse for those engineers to sink to—and remain at—the ethical levels of their employers. This will be discussed later in the chapter.

3.3 Systems of Professional Ethics

Several professions, including engineering, have systems of professional ethics or codes of ethical conduct. Such ethical systems exist for practitioners of medicine, science, finance, real estate, and the law. Practitioners are expected to obey these codes in their respective practices. In general, these codes of ethics help society. The focus on this section will be on the ethics of physicians and attorneys—perhaps among the highest-profile professions—along with scientists.

The Hippocratic Oath [c. 400 BCE ("Hippocratic Oath" 2019)] originally served as the physician's promise to society and to patients. The public knows this as "First, do no harm." Although this phrase does not appear in all versions of the oath and the actual oath is much longer than this simple phrase, its intention is clear for physician interactions with the public. Today, some medical schools may ask their graduates to follow the Hippocratic Oath or some other oath, while some institutions require nothing at all (Shmerling 2015). Looking at the American Medical Association's (AMA) history of its Code of Ethics ("History of the Code" 2017) shows that the first edition of their Code appeared in 1847. At least six more editions followed until, in 2008 following member feedback, the Code was determined to have become "unwieldy." The AMA code survives today in eleven chapters ("Code of Medical Ethics Overview" 2019) complete

[6] Refer to the Figure 1.1.

with its own concordance ("AMA Code of Ethics Concordance" 2017). The crux of the chapters of AMA medical-ethics code is distilled down, with losses, to nine principles of medical ethics ("AMA Code of Medical Ethics" 2016).

It would be unrealistic to think that the complexities of the *practice* of medicine and costs of the *business* of medicine would not clash and render the resulting Code of Ethics suspect in some ways. For example, the AMA states in its Preamble that the code was "developed *primarily* for the benefit of the patient."[7] There are some "soft" or vague phrasings in some of the principles that follow the preamble but, in all fairness, principle VIII does indeed state that "A physician shall, while caring for a patient, regard responsibility to the patient as *paramount*."[8] This responsibility of physicians is similar to the ethical responsibilities that the engineering profession places upon its practicing members.

Next, consider the legal profession. It must be explicitly noted that the ethical code for attorneys is *not* designed to protect society. That model code ("ABA Model Code of Professional Responsibility" 1980), adopted by many states, is constructed to protect the attorney's client regardless of that representation's impact upon society. Although possibly beneficial to society in some ways, the ABA code should not be used as a benchmark for engineering-ethical behavior.[9]

The reality of today's legal climate is that some engineers will find themselves required to provide testimony in legal actions, some of them product-liability lawsuits, at some point in their professional careers or personal lives. These engineers must be reminded that, even when a corporate attorney for a designer-and-manufacturer company is with the engineer during either deposition or trial testimony, that attorney represents the interests of the employer or its executive staff and *not* necessarily those of the engineer.

Before leaving professional ethics in general, scientists also have ethical responsibilities. They sometimes address issues that many engineers never need to even contemplate: for example, the creation of new forms of life as well as contamination and pollution by humankind during interplanetary exploration. Two authors (Resnik and Elliott 2016) have found that scientists find themselves faced with ethical dilemmas of three kinds: (1) dilemmas related to problem selection, (2) dilemmas related to publication and data sharing, and (3) dilemmas related to engaging society.

Engineers generally do not face the first two of the above three issues. Engineers at a design-and-manufacturing company function as employees hired to perform the prescribed tasks. Therefore, few engineers face the quandaries of problem selection mentioned in point 1 above. Those decisions are made *for* them by management.[10]

Since many design and manufacturing engineers work in for-profit companies within competitive industries, the sharing of data, information, and best practices will

[7] Emphasis added by the author.

[8] Emphasis again added by the author.

[9] Having said this, the author knows many attorneys who are among the finest of people.

[10] Ethical engineers can and should voice their objections to management on projects that the engineers consider unethical to them, e.g., weapons systems. Management will determine the results from these voicings of concern. Even if the engineers cannot stop such projects, engineers can sometimes be reassigned to other projects. This may placate the issue at a personal level; however, misguided projects may still pose a risk to society even if job re-assignment takes place.

likely rarely happen due to the real-world pressures of competition and the advantages of maintaining intellectual property. This addresses the second point above.

However, the third dilemma for scientists is shared, somewhat, by engineers. That is the obligation to "engage" the public on the "implications of research" (Resnik and Elliott 2016, 38). This could be accomplished through publications, expert testimony, non-governmental organizations (NGOs), and even whistleblowing. Some scientists may be rightfully concerned that their objectivities could become compromised through such public engagement and the resulting notoriety. Unlike engineers who work directly on products that will be delivered to consumers, scientists sometimes work on abstract phenomena far removed from its potential, but far-from-certain, incorporation into a technology or a product at a much-later date. Science often wishes to remain value-neutral by assigning neither good nor bad to a research effort or to a discovery. In many cases, the ethical principles of objectivity, openness, and fairness seem to suffice in scientific research.

Science is a tremendously broad topic ranging from microbiology and its study of microscopic organisms to physics to its study of the cosmos. Resnik and Elliott have encouraged scientific scholarly societies to create codes of conduct for "socially responsible [scientific] practice."

There are some codes of ethics by scientific groups, for example the American Chemical Society (Asc.org 2016). In addition, research funded by the U.S. National Institutes of Health (NIH) require that researchers follow the Responsible Code for Research (RCR) (Steneck 2007). In his review of the RCR, Kalichman sees it as being "aspirational," ill-defined, but still valuable as general ethics training (Kalichman 2013). This RCR subject matter goes far beyond the engineering-ethics areas that will be discussed in this book, including human experimentation.

3.4 Codes of Engineering Ethics

Being a practicing engineer carries with it certain obligations to society. These obligations—and penalties for failure—have evolved with time. Perhaps the first code for "engineering" ethics was that of Hammurabi. In 1758 BCE, the King of Babylon decreed (Martin and Schinzinger 2005):

> If a builder has built a house for a man and has not made his work sound, and the house which he has built has fallen down and so caused the death of the householder, the builder shall be put to death....

Over the ages, things have gradually become more civilized. However, the code of Hammurabi makes it abundantly clear that "engineers" have a duty to perform competent work on behalf of the public regardless of whether the motivation is ethics or fear of death.

Modern codes for the practice of engineering champion the same obligations to the public but without the ancient, harsh penalties for engineering failures. Although there are several codes of ethics for engineers, this book will use the National Society of Professional Engineers (NSPE) Code of Ethics for Engineers (NSPE 2018) as the example for engineers to follow. Several other engineering societies have codes with the identical or similar wording ("ASME Code of Ethics of Engineers" 2012), ("ASCE

Code of Ethics" 2017), ("AIChE Code of Ethics" 2014) ("IEEE Code of Ethics" n.d.).[11] It has been pointed out that such engineering-ethics codes have not always looked out for the good of the public before the good of the employer (Davis 2001), but these codes have evolved to where they are today and this is beneficial to society.

The NSPE code of ethics does not focus on a single discipline of engineering and nothing appears to be lost with the continued use of this code. Continuing from Chapter 2 of this book, the NSPE code of ethics for engineers states that:

> **Preamble**
>
> Engineering is an important and learned profession. As members of this profession, engineers are expected to exhibit the highest standards of honesty and integrity. Engineering has a direct and vital impact on the quality of life for all people. Accordingly, the series provided by engineers require honesty, impartiality, fairness, and equity, and must be dedicated to the protection of the public health, safety, and welfare. Engineers must perform under a standard of professional behavior that requires adherence to the highest principles of ethical standards.
>
> 1. Fundamental Canons
>
> Engineers, in the fulfillment of their professional duties, shall:
>
> 1. Hold paramount the safety, health, and welfare of the public.

Engineering students or engineers reading this should stop, reread, and reflect on their professional responsibilities.

Engineering truly does have an impact on society's quality of life. It is immaterial whether that slice of affected society is just one person or many people. A large societal impact is certainly true for civil engineers providing public clean drinking-water systems; for electrical/computer engineers designing, building, and maintaining communications devices and systems; and for aerospace engineers designing and constructing airliners which carry hundreds of people at a time at high speed and high elevation from point to point. Mechanical engineers often design products to provide a function, sometimes required and sometimes desirable. In the case of consumer products, these products are usually in close proximity to an individual or a small group of users who are depending on those product functions to improve their quality of life in some fashion. From these examples, one sees the upside of good engineering work and its furthering of the public's safety, health, and welfare. The opposite side of the same coin is that poor engineering can result in injury and even death to users of those affected products and systems. With the practice of engineering comes great responsibilities regardless of whether or not those responsibilities are recognized by the engineer. These responsibilities are sometimes difficult to see or grasp since engineers may be quite isolated from the ultimate effects of their work. This situation is further aggravated by the structure of the engineering profession itself where an engineer quickly moves from one project to another.

In American society, the engineer is not as highly regarded as in some other cultures. It is perhaps unfortunate that engineering does not carry with it the social importance

---

[11] The February 2006 version ("IEEE Code of Ethics" 2006) did not contain this language, but instead said "to accept responsibility in making decisions consistent with the safety, health and welfare of the public, and to disclose promptly factors that might endanger the public or the environment." It was later updated to coincide with the other engineering codes of ethics.

some people ascribe to physicians and attorneys.[12] Consequently, some engineers may forget the true importance and full impact of their professional roles upon society—both in its advancement and in its protection. Although not frequently acknowledged, a few *indirect* actions by an engineer may influence the welfare of as many lives as a lifetime of *direct* client interaction in medicine.

This situation is exacerbated by the standard practice, or business model, of engineering in today's corporate world. The typical design-and-manufacturing engineer is but an employee—a mere *hireling*—who works on pre-selected projects semi-anonymously for a prescribed compensation from a business entity. This employee–employer relationship has also been noted by others (Fleddermann 1999, 20). Because of the above reasons, engineers do not generally consider themselves to be engineering *professionals*. Most engineers, aside from consulting engineers, rarely have "clients" other than their permanent employers who can exercise greater control of the engineer than a transient client can. As a result, many engineers see engineering as a *career*, not as a *profession*.

Many physicians and attorneys are clinical or business practitioners; they interact intimately with their patients or clients as their professional services are rendered. These interactions often provide tangible and rapid, if not immediate, feedback to those professionals. The average design engineer and other engineers considered in this book do not have such clients, per se. The engineers may also not see any feedback from their actions for years if involved in a large, complex engineering-design program. They also have an employer who makes many critical decisions for the engineer during the course of a project.

Relatively few engineers in the design-and-manufacturing corporation are either professional engineers or belong to a state professional-engineering society. The engineer has neither the client contact nor the sense of professional participation that physicians and attorneys, for example, enjoy—or have thrust upon them. The engineer is essentially an employee and not a member of a recognized profession in the same way as is a physician or an attorney. Few engineers ever actually vow, explicitly, to uphold the NSPE code of ethics, or a similar code, during their careers. In fact, ethics is rarely ever spoken of during the course of undergraduate education or the practice of engineering design for consumer products. However, the author's experiences indicate that many, if not most, engineers readily understand, or do see once it has been demonstrated, the ethical implications of their actions. This is fortunate for society. Yet, at the same time, practicing design engineers may easily become conflicted about whether their societal obligation or their business obligation is the greater responsibility.

So, *to whom does the engineer owe allegiance—the employer or society?* This discussion continues.

3.5 Ethics and Engineers

Countless tomes have been written on the subject of human ethics and morality. No attempt will be undertaken here to address what may touch upon a fundamental

[12] The author is not arguing a case for "engineer worship."

essence of human existence and behavior, namely "What is right and what is wrong?" History and libraries are filled with the names of great Western philosophers who have attempted to resolve this and other questions about existence and acceptable behavior. These names include Socrates, Plato, Aristotle, Confucius, Hegel, Kant, Mill, Russell, Schopenhauer, and too many others to continue naming.

The treatment of ethics that will be presented in this book is one born out of practicality and is, admittedly, superficial in depth when compared to the great philosophers. Among the eternal questions not being raised are those such as whether it is acceptable to kill one person who is about to kill one hundred people since such situations rarely arise in modern engineering practice. Many classical ethicists will, no doubt, be dismayed at the brevity and shallowness of this discussion since so much is left out the vast and intellectually rewarding expanse of the field of human ethics. However, to do otherwise would be a betrayal of the intended readers of this book—engineering students, practicing engineers, and other professionals.

Engineers are generally, by training and by work environment, practical people who strive to efficiently and economically solve problems. In addition, they are employed by business people who *expect* these engineers to provide immediate and tangible results. Unlike scientists who are trained and encouraged to think and to hypothesize, engineers are paid to *deliver*. The engineer/corporation business relationship is just that simple. There is little incentive within the practice of engineering to exceed that which is required to meet the task at hand—on schedule and on budget—before moving onto the next project. This efficiency in itself can become a problem; product-safety engineering often requires that more than the bare minimum be done. Consequently, ethics and engineering ethics will be presented in a brief and practical manner. Perhaps, if this chapter is successful, many readers will go on to further study some of the names and references provided in this chapter.

It is not surprising that the field of engineering ethics is a distinct field of ethics. Engineering ethics can be approached either from the perspective of the ethicist or from the perspective of the engineer. This treatment will be from the latter. Before going too far, the separation of classical intellectuals from technical intellectuals is a pertinent division to recognize early in this pragmatic discussion of engineering ethics. While it might be common sense to conclude that the technically minded, in general, and engineers, in particular, are not like other people, the general distinction between the classical intellectual and the mathematics or physical-science intellectual has been well described by one author (Snow 1959). There, he explains, exist two cultures separated by a "gulf of mutual incomprehension—sometimes (particularly among the young) hostility and dislike, but most of all lack of understanding." This author will not attempt to span this gulf and, thereby, create greater mutual understanding. Instead, this author simply recognizes this incomprehension, or even suspicion, and addresses engineering ethics from the perspective of the engineer.

There have been and continue to be excellent textbooks and websites on engineering ethics (Harris, Pritchard, and Rabins 1995; Martin and Schinzinger 2005; van de Poel and Royakkers 2011; Heywood 2017; Roeser n.d.). Some such books are broad in expanse covering many ethical areas of engineering practice. Others are more narrowly focused and dive more deeply into a narrower set of ethical topics. Some books are philosophical in their natures while others are more applied in their approaches. Due to the nature of this book, the engineering-ethics discussion here will not include social experimentation, workplace rights, environmental ethics, global policies, sustainability issues, internet ethics, "big data," human health, product stewardship, and others. Making advances in each of these areas is a worthwhile and meaningful pursuit, but simply not the emphasis of the book. This book will focus on the engineering ethics necessary for engineering teams at a design-and-manufacturing company mass producing safe products for consumers to use. This is an admittedly short-term perspective to some people. For instance, product-life cycle issues such as end-of-life recycling are not considered, neither are worker conditions throughout the entire supply chain. One of the results of these exclusions is seen in Figure 1.2. The box for "Design for product safety" does not extend throughout the "Engineering ethics" region since not all topics within engineering ethics are discussed. Consequently, this box also does not reach the "Western Ethics" region of the figure's vertical axis.

In order to proceed, a working definition of *ethics* is needed. One discussion of ethics ("Ethic" n.d.) is given as:

> Morals often describe one's particular values concerning what is right and what is wrong…
>
> While ethics can refer broadly to moral principles, one often sees it applied to questions of correct behavior within a relatively narrow area of activity…
>
> In addition, morals usually connotes an element of subjective preference, while ethics tends to suggest aspect of universal fairness and the question of whether or not an action is responsible.

This discussion forms a delineation between ethics and morals. The field of morals is more concerned with right and wrong than the field of ethics which covers responsibilities. Ethics, per this discussion, will be applied to a narrow area of activity—in this case, engineering-design activities and product-safety considerations. Furthermore, ethics will be used to evaluate responsible behavior within the confines of engineering design and product-safety engineering. Ethics will be discussed; morals will not.

For the purposes of this book, the term "engineering ethics" will mean acting in accordance with an accepted engineering code of ethics, such as that of the NSPE. Thus, engineers are ethically obligated to act in ways that further the safety, health, and welfare of society.

Although the above obviously narrows down the treatment of ethics, whatever may be lost in the process is necessary in order to provide practicing design engineers with a pragmatic and actionable definition of the term. This book will avoid discussion of *morals* and what is *right* and *wrong*.[13] Doing otherwise could be a spirituality-based exercise and would certainly be beyond the scope of this book which is intended to be a practical initiation of the role that ethics plays with the product-safety evaluations made within the engineering-design process.

Ethics and values, and consequently value systems, are obviously intertwined. No true attempt to unweave them will be undertaken here. However, whether one receives

---

[13] Some discussion of morals appears in quotations from other authors, however.

values through *nature* or develops them through *nurture*, they affect how one thinks and should, therefore, affect how one acts. In this book, it will be presumed—unless otherwise stated—that product-design and product-safety engineers have value systems which esteem human safety as well as consider the treatment of others as they themselves would wish to be treated. This is what will be meant when the term *ethical engineer*[14] is used. Indeed, it is hoped that the ethical engineer is simply an ethical person who remains unchanged once the mantle of engineer is put on. It is for this reason that the term "ethical engineer" can be defined prior to defining the term "engineering ethics" which will be done later in the chapter.

3.6 Applied Engineering Ethics

What is it like to be an engineer? This simple, yet potentially profound, question may appear foreign or incongruous to many readers. After all, readers may ask *what is unique about engineers?* The engineer stereotype is of a nerdy social misfit who seldom goes out and is rarely invited to attend social gatherings.[15] Why bother worrying about them or their ethics? Popular media—television and cinema—are filled with storylines where doctors and lawyers are constantly facing grave ethical quandaries. This makes sense since their episodic clients usually face either serious illnesses or long-term incarcerations, respectively. But, in many cases, the engineer's ultimate customers also face hazards, risks, and possible permanent and debilitating injury. These potential design-failure consequences are just not as readily evident to the casual observer as deaths by cancer or lethal injection.

The question of *what it is like to be an engineer?* does not go as far as Nagel's classic *what is it like to be a bat?* question (Nagel 1974). However, some of the same existential themes can be discussed. Nagel speaks of the "conscious experience" of the bat and the experiences that bats must "experience." Nagel does realize that he can only visualize what it would be like for *him* to behave as a bat—not what it is like for a *bat* to behave as a bat. Engineers are not as far removed from the general population as are bats, so there is value in looking at the "engineer experience."[16]

Of course, not all engineering experiences will be alike. One set of authors outlines the differences between engineers, their identities, and their perceived national responsibilities in three different countries. When answering the questions: "Who is an engineer? Or, what makes one an engineer?" the authors conclude that engineers in France, Germany, and Japan have differing perspectives due to history, war, and culture, respectively (Downey, Lucena, and Mitcham 2007).

One of the above authors later joins with another to discuss that, even within a single country, that nation's engineers struggle to find their place in society (Han and Downey 2014). The South Korean government continually reiterates the importance of technological excellence to the nation's identity, yet engineers cannot find the professional respect and rewards that they seek.

---

[14] The term *ethical professional* could also be used for those individuals who are not actually engineers.

[15] This is of course untrue. My apologies to all engineers for even reiterating this stereotype.

[16] This is, of course, an extreme generalization. Each engineer is truly an individual with unique perspectives and experiences that will ultimately bear upon that "experience."

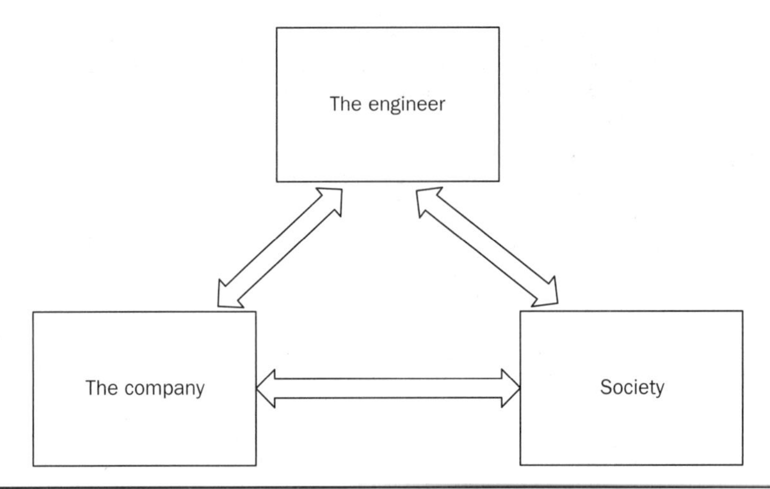

FIGURE 3.1 The engineering-ethics model.

Despite the above differences, there will be similarities among many of the engineering experiences that are *experienced* even by large groups of engineers. This is because many engineers of all disciplines do similar tasks. They apply science and mathematics to solve real-world problems. The engineering experience of those employed by design-and-manufacturing companies will further converge toward similarity.[17]

In this book, Nagel's question—and that of the author—will be treated as rhetorical ones. The question of what it is like to be an engineer is far too broad to answer with any meaningful specificity for such a large group of sentient beings each having a vast range of life experiences and perspectives. However, for the purposes of this book, it is possible—and even necessary—to pose a possibly more-fundamental question than the original one.

Namely, *to whom does the engineer owe allegiance?* The answer to this question is important since it *should* affect the decisions made by an engineer involved with engineering design and product safety. These decisions may well also affect the risk posed to users of the resulting engineered products. In the end, engineers themselves will formulate their final answers—which will be evident through their lifetime of professional actions—to the engineer-existence question.

In the engineer-business relationship, there exist *three* entities. They are: (1) the engineer, (2) the design-and-manufacturing company, and (3) consumers or society as a whole. This engineering-ethics model is shown in Figure 3.1. The three parties are shown within the figure.

It is argued that each entity in the model from Figure 3.1 has an ethical obligation to the others. This discussion will not focus so much on company-to-society and society-to-company ethical obligations which are less clear than the other obligations. For example, the society-to-company obligation could be that so long as a company pays its taxes, society, through its government, provides workforce and an infrastructure

[17] There will certainly remain uniqueness to each professional engineering journey undertaken.

for the necessary business activities. Many companies also support society, as a whole, through tax revenue, support of benevolent organizations and other causes however self-serving when publicized. Some of these same companies strive to reduce or repair their environmental footprints as well.

Looking at this ethical model from the perspective of the practicing design or manufacturing engineer, there are several "arrows," or directed ethical obligations, which will be discussed. In particular, four *directed* arrows are recognized. *First*, engineers have a responsibility to their employing companies to work hard and complete assignments on time and on budget to the best of their abilities. They apply mathematics and science, learned from years of study and experience, to solve problems that are of value to their employers. In return, engineers are paid a salary. This, again, is the employee obligation to employer and constitutes a straightforward *business* arrangement.

Second, engineers also have a responsibility to society per the NSPE canon to "hold paramount the safety, health, and welfare of the public." As stated earlier, this is a responsibility that most engineers willingly take on once an engineering code of ethics is offered to them, but most engineers have never been explicitly sworn to abide by such a professional code. Many sincere engineers simply recognize that people could be hurt by a poorly designed product. The engineers, then, voluntarily act to control such risks. Such engineers actually do no more after explicitly adopting these principles than they would have done beforehand due to their generally good character. However, also as mentioned earlier, many engineers to not deal with the *direct* ethical implications of their work on a regular basis. Engineers are often removed from their "client" by both distance and time. Indeed, engineers working for a supplier or a sub-supplier of an original-equipment manufacturer (OEM) may have little comprehension of the final risks posed by their work once embodied into the product ultimately delivered to the consumer through the designer-manufacturer-distributor-retailer chain.

Third, the engineer-company ethical responsibility arrow points both ways. Yes, engineers to indeed have an ethical responsibility to do a good job and to look after the employers' interests. However, the company also has an ethical responsibility to its engineers to provide them with the resources—in both finances and time—and support needed for them to do the engineering job that engineers believe is necessary. This includes the ability, and even autonomy, to *sufficiently* design, develop, analyze, and test products before releasing them to the public.

Fourth and finally, engineers are people with personal responsibilities in the forms of family, friends, loved ones, and self. Yes, engineers have ethical responsibilities to themselves. These include professional, financial, quality-of-life concerns for self and family, and the ability to live with the decisions that they have made professionally. The engineer is an employee, usually hired on an "at-will" employment basis that can be terminated at any time by either side. Engineers are expendable and may be "jettisoned" at the whim of the company. Few engineers in the United States enjoy any protection in the form of labor unions, although there are exceptions ("Professional Aerospace Union" n.d.).[18] These arrows of ethical responsibility are shown and numbered in Figure 3.2.

These arrows have each been discussed. However, the author constructs a concise set of rules following from these four arrows. These four *Rules of Applied Engineering*

[18] The need for engineers to unionize is not being argued here. What is being recognized is that engineers generally lack professional protection, as do practitioners of some other professions and trades.

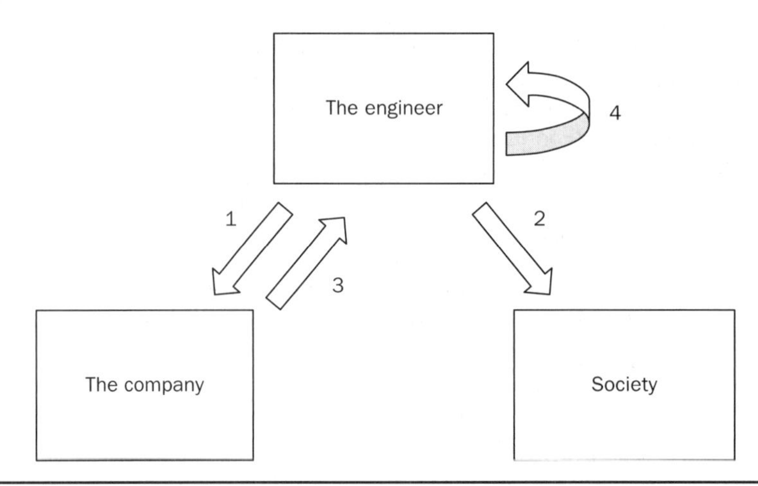

Figure 3.2 The applied engineering-ethics model.

Ethics are summarized in Table 3.1. *Rule One* is to work hard and earn the salary that the employer pays the engineer for her/his work. *Rule Two* is to do the right thing. Be able to sleep at night and be able to look oneself in the mirror without wincing. *Rule Three* is to make corporate executives and managers earn their salaries—just as the engineer earns hers/his—by weighing in on important product-safety decisions. This forces management to "put skin in the game." The engineer, alone, should not be made responsible—or *blameworthy*—in large decisions, especially if severe fallout will result in from a poor decision. Corporate management *must* be made to help solve difficult decisions, particularly if the safety of product users will be involved. Consequently, executives and managers *will* share the blame or success of these decisions. Blame *should not* fall upon the engineer alone. *Rule Four*, do NOT go to jail, embodies the fourth arrow of Figure 3.2, the engineer's obligation to self. This is not an excuse for self-serving engineer behavior, but is instead a self-preservation action when all else fails. The engineer must protect her/his own personal and family interests. Resignation or even whistleblowing may be necessary to protect oneself from bad corporate situations. In earlier decades, this exhortation would have been nothing but a joke; however, examples have already been adduced in this book showing that imprisonment for engineering misdeeds is a possibility. Corporations are full of ambitious and unethical personalities that may see fit to blame one or more engineers for corporate-wide failures. An engineer should not become a victim. Although this final point could be perceived by some readers as being

Rules	
1	Work hard
2	Do the right thing
3	Make managers and executives earn their salaries
4	Do NOT go to jail!

Table 3.1 Rules of Applied Engineering Ethics

overly dramatic, this presentation of that point suits the author's purposes of stressing the importance of this ethical responsibility. The engineer simply cannot depend upon the corporation to protect her/his interests. This table, when presented by the author during course lectures, usually facilitates a lively discussion of engineering ethics with students.

So, for purposes of this book, the term "applied engineering ethics" will mean actions in accordance with the rules of Table 3.1. This is a logical and practical extension of an engineering code of ethics. The NSPE code of ethics looks out for the safety, health, and welfare of the public. This is incorporated in these rules as number two. The practicing engineer should always fulfill the obligation to employer and, when necessary, the public. The code is silent on the employer's responsibility to the engineer. This reciprocity is vital in order for engineers to fulfill their obligations to the public. Without employer participation, engineers would be working in a vacuum and be isolated from the public. Similarly, the code remains silent on the engineer's obligation to self and to self-preservation.

If the engineers cannot resolve their ethical values with employer actions—even after repeated attempts—they may be forced to choose between career—not *profession*—and societal good through demonstrating disobedience, or even "blowing the whistle," over unethical behavior by their employers. Again, the engineer's career may come to a cataclysmic end, but the profession of engineering—as well the public good—will be well served through such a selfless action.

Before proceeding further on a severe action such as whistleblowing, it must be stated that *any* action must only be done for ethical reasons. A disobedient or whistleblowing engineer should never expect to be rewarded, nor should whistleblowing ever be done as an act of revenge. In fact, the economic and personal penalties are likely to far outweigh any readily perceived benefits of such actions. Furthermore, there are few, if any, legal protections for ethical disobedience and whistleblowing unless large amounts of money can be recovered by attorneys. Engineering codes of ethics are silent on this matter as well. There appears to be no "upside" for the engineer undertaking such behavior other than its intrinsic virtue.

There are several ways in which an engineer can demonstrate severe disagreement with employer actions. Harris, Pritchard, and Rabins list three forms of protest, or "organizational disobedience," in which engineers can express their disagreement (Harris, Pritchard, and Rabins 2005, 176). These are: (1) disobedience by contrary action, (2) disobedience by nonparticipation, and (3) disobedience by protest.

The first of these behaviors involves the engineer acting in a manner opposite that which is decided by management. This will be seen as acting against the business interests of the organization. For example, rather than working as has been decided or as directed, engineers instead continue working as their ethical convictions direct them. In doing so, the engineer is working for the good of society rather than the good of the company.

The second way to express disagreement with company direction is to not be a party to any of the actions to which the engineer object. Rather than acting against the interests of the employer, the engineer simply does nothing—or tries not to assist the objectionable effort. Such a situation could arise in the case of work on a weapons system. Although many employees may choose to decline employment by defense contractors, it is possible that a civilian project, with which an engineer is ethically content, could morph into a weapons project if military sales are pursued by the company. Those engineers who object to working on such products may choose to stop working on them. Of course, the employer will likely see the nonparticipation as an adverse action since it is in opposition to perceived company interests. To management, any difference between contrary action and nonparticipation will be viewed as a distinction without a difference. The engineer should expect this response by employers to such actions.

The final way for an engineer to express strong ethical disagreement with organization behavior is through protest. Protest can take one of several forms. Protest can be silent. The engineer can simply quit—either having or not having another job at the time.[19] Or, engineer protest can be public and open. An open form of protest could be an organized protest demonstration, but would more likely take on the form of "whistleblowing."

Whistleblowing is going public with information that would show unethical behavior or decision making by an employer. Make no mistake, whistleblowing is a serious step to take. The decision to become a whistleblower must not be made without much thought and reflection. Whistleblowing also involves overcoming the loyalty to an organization that many employees have toward their employers. Duska recognizes this but astutely observes (Duska 1991) that:

> I don't owe loyalty to just anyone that I encounter. Rather I owe loyalty to persons with whom I have a special relationship... It is important to realize that in any relationship which demands loyalty the relationship works both ways and involves mutual enrichment…
>
> Since loyalty is a kind of devotion, one can confuse devotion to one's job (or ends of ones' work) with devotion to a company.

This statement affirms the rules arrows in the applied engineering-ethics model and the rules of applied engineering ethics put forth here; it also alludes to the sense of professional responsibility by the engineer. Regardless, the whistleblowing engineer must strongly feel that such actions are necessary due to ethical conviction.

Whistleblowing should also not be done without competent legal counsel. Some employment agreements might make whistleblowing a violation to which the employer is entitled to compensation. Much like the David-versus-Goliath analogy,[20] the engineer is David and the company is Goliath when it comes to the legal and public-relations arenas. The engineer is a sole individual who has bills to pay and a family to support. Corporations have nearly bottomless resources to use in countering and neutralizing the whistleblower. Unlike in the scriptures, David usually does not win against Goliath. A large price will be paid by the whistleblowing engineer. Will it be worth it? Only that engineer can answer that question.

Anyone contemplating whistleblowing might want to read two articles on the topic (Hoke 1995; Sharp 1995) even though these articles concern academia and science rather that the corporate life that this book has been discussing. This notwithstanding, the first of the articles states that "Despite the extreme personal difficulties almost every whistleblower reports having endured, many say that, if faced with the same choices today, they would again speak out." This conclusion is worth remembering.

An engineer may ask if whistleblowing is necessary. One set of authors has provided a set of criteria to consider if a social responsibility, such as whistleblowing, has

---

[19] This point is illustrated in the newspaper article referenced by the footnote in the Preface to this book.
[20] 1 Samuel 17.

become necessary (Simon, Powers, and Gunnemann 1972). These authors state that "... there are some situations in which a combination of circumstances thrusts upon us an obligation to respond." If whistleblowing is being contemplated, consider four factors. These factors are: (1) need, (2) proximity, (3) capability, and (4) last resort. Each criterion is examined in turn.

First, Duska summarizes the other authors' in this step by "There must be a clear and present harm to society that can be avoided by whistle-blowing." This criterion would certainly be met if, by engineer inaction, hundreds of people would literally perish in the immediate future. Such a case would be if a person knew of an obvious, at-hand problem with a commercial airliner about to take off with a full load of passengers. However, if a minor-to-modest problem could perhaps arise sometime in the indeterminate future, then that engineer would not have a strong case to present. Yet, it is finally up to the engineer to decide if ethics compels the engineer to go public regardless. That engineer will doubtlessly find willing reporters for interviews to fill local-media airtime, but it is doubtful that such a situation alone would garner the national attention, and public outrage, the engineer hopes could change the actions of a large corporation. As a result, this engineer would probably end up as casualty of having, and having acting upon, engineering ethics without a strong case to make. In such a case, this engineer's reputation will be damaged.

Second, the engineer must be close to activities on which the engineer reports. The engineer must have factual knowledge, not merely suspicion or speculation, about an unethical course of action. Documentation can always help the engineer, but that engineer must also beware of company-property considerations and employment agreements. Again, always consult with legal counsel if contemplating such an act. The engineer must also be completely objective and aware of if another person could see the same information and reasonably reach a conclusion vastly different from the whistleblower's about the ultimate effects of not blowing the whistle.

Third, the pragmatic aspect of whether or not the engineer's actions could reasonably lead to the avoidance of the harm. Duska states the "You are not obligated to risk your career and the financial security of your family if you can't see the case through to completion or you don't feel that you have access to the proper channels to ensure that the situation is resolved." This book's author is less prone to advise those engineers having ethical crises about what they should do or not do. An engineer certainly has no *obligation* to report employer misdeeds since the engineering codes of ethics remain silent on the matter. On the other hand—and as stated throughout this book—an engineer's value systems must guide the engineer to do what is right, regardless of reward or risk. This author takes no swipe at any engineer following the path of "no obligation," yet that engineer is reminded of the professional obligation to pursue the safety, health, and welfare of the public. Engineers finding themselves in such situations should ask themselves if remaining in such an environment is healthy for the engineer. Engineers might also ask themselves, "Will suspect behavior by my group affect me? Can I expect my questionable *professional* behavior to *not* affect my *personal* behavior?"

It is unfortunate that engineering books and papers do not speak about the importance of professional and personal reputations. Some, perhaps many, professionals consider their reputations to be among their most valuable assets. Once a reputation is lost, it is nearly impossible to restore. By remaining at an unethical company, an engineer may be considered guilty, by mere association, of the unethical behavior taking place at that organization—even if that engineer acted with undeniable engineering ethics. Even if you change jobs, you may not change reputations. Remaining at an unethical company may preclude employment by other companies in the future.

Fourth and finally, whistleblowing should only be undertaken if all else fails. Engineers should not be quick to rush off and talk to the media. In some instances where there are alternative means to address the problem within the company, talking to the media can be counter productive in several ways. For one, an engineer's career may be ruined for no reason whatsoever and, second, the solution to the problem could be delayed from all of the publicity that is generated.

Engineers do have a responsibility to do a good job by working hard and also by working smart. Nothing changes this. Engineers are hired to do a job and the business relationship with employers requires that the job be done. This is more of a business arrangement than an ethical one, but conscientious engineers will do their best out of loyalty and a sense of duty to the company. Most engineers exemplify this strong work ethic.

When it comes to engineers' societal obligations, the waters can become somewhat muddied if concrete terms are not used. That is why the term "applied engineering ethics" has been chosen for this section. Abstract classical-ethical concepts such as the "greater good" and "intrinsic virtue" do nothing to advance what is a *practical* subject to *practical* people in *practical* environments. Engineers and their employers have no time to waste—and society does not want to injured by inferior products. These are simple and pragmatic relationships.

3.7 Ethics in Engineering Education

The engineer experience typically begins with undergraduate students being educated in a mathematics- and science-heavy curriculum with only limited room for non-technical electives such as philosophy, political science, languages, literature, art, and so on. A good engineering program can be difficult for an undergraduate student to complete in the span of 4 years. This is because so much effort is required to earn the bachelor of science (BS) in an engineering field at a good college or university.

Engineering curricula at colleges and universities accredited in the United States by the Engineering Accrediting Commission of ABET (Accrediting Board for Engineering and Technology), abide by the requirements in ABET's manual ("Criteria for Accrediting Engineering Programs" 2019). ABET's general ethical requirements for all bachelor's-level engineering-degree programs are contained within Part I's criterion 3, Student outcomes. That criterion and outcome 4 read as follows:

> The program must have documented student outcomes that support the program educational objectives. Attainment of these outcomes prepares graduates to enter the professional practice of engineering...
>
> 4. an ability to recognize ethical and professional responsibilities in engineering situations and make informed judgments, which must consider the impact of engineering solutions in global, economic, environmental, and societal contexts

ABET, and those institutions of higher learning, rightfully recognize the need to produce engineers with a sense of ethical responsibilities to the world and to the environment. However, when delving into the details of specific engineering programs, only three programs—civil, construction, and cybersecurity—explicitly contain mention of ethics under "Curriculum." The curriculum requirements for civil and construction engineering programs mention ethics in conjunction with professional licensure. The cybersecurity ethics requirement is to be considered "as appropriate." Despite these sparse explicit curriculum requirements for engineering programs as a whole, in fairness to ABET, it has been the author's experience that ABET's evaluators of engineering programs do indeed, and invariably, stress the importance of students learning and demonstrating ethical responsibilities in all engineering disciplines without limitation to professional registration. Of course, ethical behavior by practicing engineers should never be conditional upon licensure as a professional engineer (PE).

During the course of many BS programs, students are either advised to or required to take the fundamentals of engineering (FE) examination[21] ("FE Exam" n.d.). The most-recent FE-exam pass rates for the various engineering disciplines were reported to range from 59% to 80%. This one of the first steps to becoming a PE and is often listed on a young engineer's curriculum vitae. Once the BS has been completed, most new graduates immediately start a new job or career as an employee of a business—but usually nothing resembling a traditional profession such as medicine or law.

The FE examination specifications ["Fundamentals of Engineering (FE) MECHANICAL CBT Exam Specifications," n.d.] show that the 6-hour closed-book exam contains questions on 15 topics. Topic 4 is "Ethics and Professional Practice." This topic has five areas in total. These five areas are: (1) codes of ethics, (2) agreements and contracts, (3) ethical and legal considerations, (4) professional liability, and (5) public health, safety, and welfare. Three of these areas pertain to the ethics and safety discussion at hand. Those areas are 1, 3, and 5. So, although FE-exam coverage of ethics and safety is but a small portion of the overall test, the subject is covered by ABET-accredited programs, to a degree, before engineering students enter engineering practice.

New engineering hires are eager to start work on real-world problems where the answer is not in the back of the book. In fact, some real-world engineering problems are difficult to pose let alone solve. It is not surprising that many new engineers find themselves overwhelmed by real-world engineering duties when they begin engineering employment. Most of these engineers probably want to serve society in some manner—however small that may be. But at the same time, they are also wanting to advance their careers, get a promotion, and earn more money. Many are also starting families. None of these is wrong. However, in order to achieve these career goals, satisfying your superiors is a necessity. Not only is engineering competence usually required, but so is meeting company goals which may be either passing priorities or long-standing values. Failing to accomplish these goals is a sure path to unemployment or to relegation to the company's obscure product-line group, for example, the three-slice toaster division in a remote location, never to be heard from again.

It would be wrong for the public to assume that the typical engineer dwells upon the matter of ethics daily as part of a professional life. It would be equally wrong for engineers to presume that they do not have to. There should be reflection by engineers

[21] Formerly known as the engineer-in-training (EIT) examination.

upon the consequences of design-and-manufacturing decisions made by them since the safety of the public might depend upon their choices.

The PE exam in America may be taken after the engineer has accumulated several, often 4, years of experience. Licenses are issued at the state-government level and the years of experience necessary may vary according to the state of residence. One admittedly older standard reference for engineers studying to take the PE exam (Lindeburg 2001),[22] has only five pages dedicated to engineering ethics and the book is over 2.5 in. (6.4 cm) thick. Furthermore, engineering ethics is Chapter 71 out of 72 chapters, thus engineering ethics could be perceived as an afterthought or an "Oh, by the way...." This should not be taken as a slight to that author. Admittedly, there is little material available beyond the NSPE code of ethics. That author sets the engineer's *highest* obligation as that to society and the public. This priority is followed by six others: the law, the engineering profession, the engineer's client, the engineer's firm, other involved engineers, and the engineer personally. Although the order of the last six priorities could easily be argued, the primacy of the first cannot. That book's chapter also covers topics including dealing with clients, employers, suppliers, other engineers, and the public. When speaking of clients and employers, Lindeman states:

> Engineers must not be bound by what the client wants in instances where such desires would be unsuccessful, dishonest, unethical, unhealthy, or unsafe.

Thus, those engineers taking the PE exam are bound to see some material similar to this depending upon the study materials used in preparation for the test. But again, relatively few engineers take, and pass, both the FE and the PE exams. It is also worth noting that the legal ability to affix the "PE" stamp to a document is much more valuable to some engineers than to others. The PE is especially valuable to civil engineers working on public-works types of projects including highways, bridges, commercial buildings, and irrigation plans. It is much less useful to a mechanical or electrical engineer and may never be required[23] unless, for example, that engineer offers services directly to the public. Being a PE does quite little practically for an engineer involved with designing and manufacturing consumer products. One article estimates that only 20 percent of *all* engineers are licensed as PEs (Gearon 2014) and many of this group will likely be civil engineers for the reasons cited above. Consequently, few mechanical, electrical, industrial, and other engineers working on consumer products will be registered PEs.

The need for ethical instruction has been recognized by others (Stephan 2001). But it is difficult to discuss engineering ethics in either a curriculum or a study book without being limited to saying such things as "behave acceptably" and "be a good person." Outside of the United States, some have studied the instruction of engineering ethics to engineering students and have even connected it with the task of engineering design (van Gorp and van de Poel 2001). These authors note that among the difficulties facing

[22] This is the reference book used by the author in studying for his PE examination.
[23] The author has never, in two decades of professional-engineering practice, used his P.E. stamp.

practicing design engineers are ill-defined problems and lack of definitive solutions. This connection of ethics with design engineering, design requirements, safety requirements, and requisite trade-offs advances pedagogy in the right direction—toward the real engineering world.

Some universities offer science-and-society types of courses, often through liberal-arts departments. Numerous engineering departments have, over the years, offered courses to their students that have included as reading books such as these (Florman 1994; Petroski 1982; Pirsig 1974). Although old, these books are good at pointing out and then discussing less-obvious aspects of engineering—or at least some that are not always discussed in the classroom.

One author, McGinn, has listed three possible ways to get engineering-related ethical course material to undergraduate engineering students (McGinn 2018). He also critiqued their efficacies. These three approaches are:

1. Requiring a traditional philosophy department ethics course
2. Adding ethics into an existing engineering course through a non-technical ethics lecturer
3. Integrating engineering ethics within an engineering course taught by an instructor with experience and expertise in both ethics and engineering instruction

McGinn correctly surmises that the first two approaches listed above are "unlikely to be fruitful." He sees the third option as "more promising," however.

The *first* approach will likely become bogged down, when viewed by a practical undergraduate engineering student, when classic ethical concepts, theories, and their proponents are explained. Again, this is all worthwhile material to many students, but some undergraduate students may not *presently* see it as being pertinent to their studies.

The *second* approach has merits but also has challenges. The ethical material could be tailored to engineering and even to current events. Yet, time constraints could limit the ethics portion of the class to only superficial treatment of the subject due to the volume of technical material necessary in modern undergraduate-engineering courses.

It is the third approach that McGinn looks at with the greatest favor. With this approach, students study ethical issues in engineering within a dedicated course. Case studies and other real-world and current-event examples would be considered by students.[24] The instructor for such a class should "[have] expertise in teaching engineering ethics, [have] an abiding interest in engineering education, and [be] familiar with the realities of engineering practice." The ethical evaluations should be conducted with attention to context and consequences.

One example of a current university-level course combining engineering ethics with product-safety engineering, taught by the author, is provided as a case study next.

---

[24] McGinn provides several useful case studies in his book.

Case Study 3.1: Integrating a Course Including Engineering Ethics into an Undergraduate-Engineering University Curriculum

An accreditation authority cited earlier indicates that, even though ethics may not explicitly need to be taught in an undergraduate curriculum for engineering students in the United States, there, nonetheless, must be documented demonstration of students being able to understand engineering-ethical obligations and act accordingly. It is far too common that any formal education of engineering students in the field of ethics will take place through the handful of non-technical electives required for graduation. Such classes are likely to be from an abstract human-ethics philosophy perspective. These classes are, of course, valid for a well-rounded liberal-arts education. However, some curricula for undergraduate-engineering students are simply too full of required technical classes to permit much in the way of student study in literature, arts, languages, and philosophy. In addition, many[25] engineering students are quite pragmatic by nature—and by graduation requirements—and simply do not wish to study philosophical abstractions that they may regard, consequently, as useless to them. Therefore, it is important to provide some exposure of applied engineering ethics to future engineers while still in school, *especially* if the exposure can help them meet ABET-curricular requirements.

The author has been fortunate to have been able to develop and teach a university-level course which combines engineering ethics with product-safety engineering. This course[26] has been offered for over 5 years in an accredited mechanical engineering (ME) department within a large, public university. The course was initially offered as an experimental class, but has since been given a permanent course number within the department and is offered each Fall semester. This course is a technical elective to undergraduate students in the ME department. Graduate students in engineering are also able to take this course for graduate credit.

In a ME department of approximately one thousand undergraduate students, enrollment for this course has recently been running from 35 to 45 senior students and 10 graduate students. Undergraduate enrollment has primarily been from mechanical-engineering students, but students from other engineering programs have also taken the course. Details of this unique course have been published in a paper (d'Entremont and Merryweather 2018).

The content of the course is broken into two parts: the concepts and the application. General engineering-ethics and product-safety considerations are covered in the first part while specific applications of these considerations during the engineering design, development, and testing phase as well as the post-sale evaluation phase of the PDP are covered in the second. Course materials include publicly available documents and reports from companies and governments around the world.[27] Highly qualified guest lecturers[28] supplement the instructor by covering such topics as corporate

[25] But certainly not all.

[26] University of Utah, Mechanical Engineering (ME EN) 5150/6150, "Product-Safety Engineering and Engineering Ethics."

[27] This book is intended to serve as a textbook for that course. Therefore, that course's material and presentation closely follow the chapters of this book.

[28] The author is fortunate to have several highly qualified guest lecturers available to him locally, although some guest lectures are delivered through videoconferencing.

social responsibility (CSR), CE marking, forensic engineering, product-liability law, medical-device safety, safety philosophy and technology, and the engineer's role in product-liability issues.

The course culminates with a semester project to either design and/or analyze the *safety system* of a product which includes the product itself as well as any facilitators.[29] In this project, engineering principles are applied as well as ethical ones when assessing the product-safety characteristics of an item or system.

Students have been highly receptive to both this course and its content. Feedback in the form of ratings and comments from students are collected at the end of each semester. Some representative student comments are included in Table 3.2.

Student responses indicate that this topic was new to them—even as fourth-year engineering students. Guest lecturers also were able to bring a wider breadth of information and insight to students than the instructor alone could. The real-world, industrial experience aspect of the course was also valued by students. Several students commented that the course content was good, that such a course should be required of all engineering students, or that the course "makes you think about new aspects of engineering."

This case study of a product-safety engineering course, including engineering ethics, is just one example—one course by one professor at one university—of what can be done to help educate future engineers about engineering ethics and professional responsibility. In this case, the ethical component was directly related to product-safety engineering decision-making. Connections between engineering ethics and environmental, sustainability, renewable-resource, energy conservation, and industrial engineering, just to name a few, could easily be made—and are being made—by some engineering programs. The number of such courses should only increase as awareness of these connections with the practice of engineering increases.

The course in the above case study is not exactly what McGinn had in mind in the third of his approaches. The case-study course studies only a sub-set of engineering ethics and not the entire spectrum of the topic. Furthermore, the course is not completely dedicated to the study of engineering ethics. Instead, this course employs engineering ethics as a crucial portion of the practice of a sub-set of engineering pursuit; in this case, design and product safety.

Even when a suitable course involving engineering ethics is offered to students in their undergraduate curricula, the reader may wonder if it is even *possible* to teach ethics to young adults—some perhaps beyond ethically formative years. The following discussion examines this valid question.

Discussion: It is Possible to Teach Ethics to Engineering Students or to Practicing Engineers?

It may be fair to reason that, by the time many people reach their late teenage years, they have already had large portions of their value systems established. It may be difficult to re-wire, or re-program, what has already been instilled regarding some values including appropriate ethical social behaviors and attitudes toward, for example, the safety of products. These and other outlooks have been formed through parental

[29] *Facilitators* are covered in Chapter 7.

Question: Comments on Course Effectiveness
• The lectures and information shared were helpful in learning about product safety. The guest lecturers had fascinating industry insight that was extremely useful and beneficial.
• The class presentations and the guest lecturers were excellent sources of information for this course.
• Made you think critically and play the devils advocate on things. Product safety is a course that all should take I think. Not many students think of a lot of things that are covered in this class.
• Great topics and necessary information for engineers
• I appreciated the number of guest lectures that we were given, it gave us a chance to see face to face some industry professionals in their given fields. This gave us a chance to hear real world applications of what we were learning. This was very very helpful....
• Excellent course content!
• The homework and tests felt real, as in real world experience problems which is great. The guest lecturers were great because they brought in real-world experience and knowledge.
• Guest lecturers were very interesting and I enjoyed the added insight they brought to the subject. The real-world application of this course is great. A course similar to this required for all undergraduate students would be great thing to drive home the importance of safety in design, and the societal implications of what we do as engineers.
• Helpful to see full aspects of engineering in industry perspective.
• I actually enjoyed the class more than I thought I would... It was a solid class and I enjoyed it.
Question: Instructor Comments*
• Guest lecturers were great. The homework assignments provided an opportunity to dive into product safety and different forms of safety evaluation methods that were very beneficial.
• I tell everyone to take this class because it makes you think about new aspects of engineering.
• Helped the students to think about the situation and propose a solution without making students feel like there was only one right way
• Real world life experiences shared with the class were very informative and helpful.
• Taught critical aspects of being an engineer.

* Related to course content

TABLE 3.2 Representative Verbatim Student Comments on the Course

upbringing, faiths, social experiences, school experiences, life experiences, work history, interactions with friends and associates, current events, and personal reflection. These experiences and the resulting value systems, however fluid they may prove to be, do become part of what individuals consider themselves to be. It is, therefore, reasonable to ask if it is possible to teach ethics to people with existing value systems. This question has also been asked before (Stephan 2004).

There are at least two answers to the question. The *first* answer is that some engineering students from other countries and cultures have expressed gratitude to the author at the end of his semester course for explaining the expected ethical conduct of the engineer in the United States. This initially came as a great surprise to the author; however, it makes sense after some thought. These particular engineering students had never known society's expectations of engineers in the United States since they were not raised in this country and took up their studies here often in their early adulthood. Many students are also removed from the day-to-day interactions and other social influences that helped establish the value systems of native students. It is unfortunate that this type of information had never been transferred earlier to some international students in the United States. It is likely, however, that relatively few engineering students and practicing engineers do not know what is expected of them by a Western culture. So, this first answer applies to few engineers or future engineers.

The *second* answer to the discussion question addresses the majority of engineering students and engineers who find themselves in design-and-manufacturing companies making products for consumers. This second is answered by asking another question.

Whether or not young adults and adults can be taught ethical behavior presupposes that only the engineers themselves are responsible for the ethical decision-making in the corporation which, in the end, actually produces safe products for consumers. Even the most-ethical of engineers may be unable to take the actions necessary to keep unsafe products out of the market—or fix unsafe products already in the market. Management, from the departmental level up to the chief executive officer, is ultimately responsible for the risks posed to consumers by product. Corporations may simply not want to hear bad news about their products from their engineering staff and may shut down dissent within the engineering staff.[30]

From the author's industrial experiences, practicing engineers actually do know ethics and generally wish to follow the norms of ethical behavior. The question that *should* be asked about engineering ethics is this:

> How can design-and-manufacturing corporations create and maintain an engineering work culture that encourages—and rewards—ethical behavior as a value?

This question should be directed to executive management at companies. Such positive work environments do not germinate and flourish on their own through bottom-up, grassroots, working-engineer efforts. Such environments only happen under the strong and continued direction and engagement of top company leaders.[31] Even once established, such environments are weak and unstable and prone to collapse without continued support from the same top corporate leaders. There should also be reward—or, at least, not punishment—for engineers who act ethically. Strong business priorities, such as revenue demands, will destroy a fledgling value—such as a culture of ethical engineering behavior—without executive attention and support. Such a temporarily espoused company value will quickly relapse back into but a passing priority.

---

[30] The ethical engineers on staff must, of course, raise their valid concerns on product safety to management.

[31] *Leaders*, not *managers*.

Ethical engineers often voice their concerns about products, but the more-timid engineers may be reluctant to express their worries when they have seen engineers before them be neutralized, or even punished, for proper behavior. This does not excuse the unethical behavior of engineers in any way because engineers should always act ethically. It is, however, equally reprehensible to permit the corporations and executives—the very ones making revenue from these possibly compromised products—to have a "get out of jail free" card by passing ethical responsibilities on to their employees. Such corporations have a social responsibility to both hire ethical personnel and to enable and encourage them to act ethically—even when their competitors are not acting in a similar manner. A vacuous, platitudinous engineering-ethics policy posted on the corporate website saying that it is an ethical company, has membership in an ethics-related industrial group, or has signed a letter of support for an effort means nothing if corporate values and resulting actions do not follow those measures.

If some younger engineers, who individually may not know much about ethics, see other engineers being rewarded for ethical behavior, then that would be the *instruction on ethics* for those engineers that the original question pondered. Therefore, *it is possible to teach ethics to engineers*. Such effective instruction on engineering ethics, however, just may not follow a classically trained educator's expectations of a typical teaching model.

To reiterate the side-discussion question, "Is it possible to teach ethics to adult engineers?" Readers may rightfully be asking themselves the same question because it is a proper question to ask. However, this curiosity presupposes that the question sums up—in its entirety—the motivation of this book. It does not. This is only half of the equation.

In order to produce ethically design products for consumers through the design-and-manufacturing company model, there are multiple components within that model each with an effect on the final product produced. Among these components are engineers along with engineering and executive management.

Ralph Barnett has a quote that goes as follows: "Practicing ethics is great as long as you are not the only one doing it."[32] So, perhaps the better question to ask is "Is it possible to teach ethics to engineering and executive management?" Several books on engineering ethics either state or provide evidence that "Ethics is good business" (Schlossberger 1993, 234). However, if this matter was unquestionably settled, there would exist little need to state the obvious. It is hoped that many organizations and executives see ethics as being necessary for proper and sustainable business operations. Yet, it appears—from the stream of product recalls and corporate penalties visible on some websites[33]—that some organizations may be able to continue to make money and meet shareholder expectations without a strong sense of ethical responsibility to society.

3.8 Conclusion

The book started with a discussion of the difference between a priority and a value. This chapter extended that conversation. A temporary panic may shuffle the business priorities of a company facing a crisis as shown in the crisis cycle of Figure 1.1. Will things

[32] From numerous conversations with Barnett.

[33] For example, www.cpsc.gov.

return to the way they were before the problem, or will this change be permanent? What remains after that panic has subsided will reveal the true values of the organization.

Engineering societies for many disciplines of engineers have codes of professional ethics. They are nearly identical to one another when it comes to the primary responsibility of engineers. This responsibility is to look out for the safety, health, and welfare of the public. Although few engineers ever take such an oath explicitly and because there is little practical incentive for practicing engineers to join these societies, many engineers only voluntarily and implicitly accept these responsibilities to the public. It is also argued here that engineering is more of a career than a profession in contrast to the practices of medicine and law.

The philosophical aspects of ethics will assure that the topic remains controversial irrespective of any engineering focus. Engineering ethics will also remain a topic difficult to teach and one on which the examination of students is formidable. Engineering ethics, although narrowed from the original field of ethics, is still both broad and deep. Many important aspects of engineering ethics exist aside from the product-safety engineering concerns covered within this book. This chapter focused on an abbreviated form of engineering ethics—applied engineering ethics—whose application will be demonstrated throughout the book.

An engineering-ethics model was provided showing the some of the directed ethical-responsibility arrows existing between the engineer, the employer, and society. Engineering codes of ethics, although basically good, do not adequately represent the multiple parties involved in real-world engineering since only the engineer is obligated to any form of behavior. These codes have been shown to not sufficiently address the realities of engineering practice. The four rules of applied engineering ethics also hold engineers ethically responsible to their employers to be competent engineers who do their best for their employers. But in turn, the employers have an ethical responsibility to support and to participate in the difficult decisions facing their hired engineers. In addition, employers of engineers have a responsibility—both to the engineers and to the public—to create and nurture an environment which rewards ethical conduct. The goal should not be to punish unethical engineers, but instead to see employer-inspired ethical behavior emerge from the engineering staff. In the case that all of the efforts made by engineers fail, then engineers have an obligation to protect themselves in the manner that they see fit, including whistleblowing. Engineers are warned to be cautious in their actions, however.

There is no doubt that much work remains in order to make ethics a regular part of the undergraduate-engineering curriculum. At present, there may be an assumption made that someone along the way will teach it to students or it may be an afterthought "shoe horned" into a technical curriculum. However, some progress has and is being made on this front in both Europe and the United States regarding the instruction of ethics within the engineering curriculum.

Although this book's title includes "engineering ethics," this primary chapter on ethics is not particularly lengthy. Yet, ways in which engineering ethics can and should be applied by practicing engineers and other professional are interwoven throughout the book. Rather than being a standalone, isolated topic, engineering ethics is at the core of many aspects of real-world engineering and is raised where and when appropriate within the product-design process. Although ethical aspects do not affect the calculated strength of a structural beam or column as it is being designed, ethical considerations may affect the determination of whether that beam or column is strong enough for its intended purpose. Also, since crucial design criteria are often lacking for the specific decisions that design engineers must make daily, the importance of ethical behavior and decisions by engineers is vital. Roeser—while speaking of morals rather than ethics—believes that "Rather than delegating moral reflection to 'moral experts,' engineers should cultivate their own moral expertise. They have a key moral responsibility in the design process of risky technologies, as they have the technical expertise and are at the cradle of new developments."

The application of engineering ethics—along with critical thinking—will be specifically discussed again in Chapter 11 regarding product-safety recall considerations. In particular, there is one simple question that members of engineer teams can always ask themselves when facing a possible product-safety recall decision:

> Would I let my loved ones use this product?

The answer to the question itself is important, of course, but so is the amount of time needed to respond. Perhaps further engineering-design work needed if the answer to this simple question is not affirmative *and* immediately forthcoming. The ethics of an engineer may be the final criterion for this decision. Company priorities should never impede ethical actions by engineers.

In engineering practice and in life, some people behave in ways that other people find unacceptable. Perhaps those *mis*behaving people even appear to be rewarded for their actions. In the practice of engineering ethics, it is vital that engineers act in accordance to their own *values*—not organizational *priorities*. Doing otherwise is to betray one's own essence. Additionally, company executives have a responsibility—*if* they truly want ethical behavior from their engineering staff—to both *create* and *nurture* an environment that encourages ethical behavior and rewards ethical decisions. Unfortunately, ethical behavior is not always without cost either for the company or for the engineer.

This chapter on engineering ethics closes using the same article with which it opens. Matthew Beaudin continues in his chapter-opening quote with the following which includes what Jonathan Vaughters[34] said about a professional bicyclist engaging in blood doping:

> There is a simple way to look at all this. Think about your mother…
>
> "If your mom walked in on you…" Vaughters said. "If you're sitting there getting a blood transfusion, is she going to freak out? Probably."
>
> Then again, some kids had more lenient moms than others, didn't they?

References

"ABA Model Code of Professional Responsibility." 1980. Chicago, IL: American Bar Association. https://www.americanbar.org/groups/professional_responsibility/publications/model_rules_of_professional_conduct/.

"AIChE Code of Ethics." 2014. New York, NY: American Institute of Chemical Engineers (AIChE).

[34] Former professional cyclist, current cycling-team manager, and author.

"AMA Code of Ethics Concordance." 2017. Chicago, IL: American Medical Association. https://www.ama-assn.org/sites/ama-assn.org/files/corp/media-browser/public/ethics/ama-code-ethics-concordance.pdf.

"AMA Code of Medical Ethics." 2016. Chicago, IL: American Medical Association. https://www.ama-assn.org/sites/ama-assn.org/files/corp/media-browser/principles-of-medical-ethics.pdf.

Asc.org. 2016. "Global Chemists' Code of Ethics." Washington, DC: American Chemical Society. https://www.acs.org/content/acs/en/global/international/regional/eventsglobal/global-chemists-code-of-ethics.html.

"ASCE Code of Ethics." 2017. *Asce.Org*. Reston, VA: American Society of Civil Engineers (ASCE). https://www.asce.org/code-of-ethics/.

"ASME Code of Ethics of Engineers." 2012. Asme.Org. 2012.

Ballinger, Alex. 2020. "'In Cycling There Is 90 per Cent Doping,' Says Stefan Denifl's Lawyer." Cyclingweekly.Com. 2020. https://www.cyclingweekly.com/news/latest-news/in-cycling-there-is-90-per-cent-doping-says-stefan-denifls-lawyer-448411.

Cassidy, John. 2019. "More Questions than Answers about Boeing, the 737 Max, and the F.A.A." *The New Yorker*, May 16, 2019. https://www.newyorker.com/news/our-columnists/more-questions-than-answers-about-boeing-the-737-max-and-the-faa.

"Code of Medical Ethics Overview." 2019. Chicago, IL: American Medical Association. https://www.ama-assn.org/delivering-care/ethics/code-medical-ethics-overview.

Cohen, Aubrey. 2010. "Hudson Landing Pilot, Flight Attendant Retiring from US Airways." Seattle Post-Intelligencer. 2010. https://blog.seattlepi.com/aerospace/2010/03/03/hudson-landing-pilot-flight-attendant-retiring-from-us-airways/.

"Criteria for Accrediting Engineering Programs." 2019. Baltimore, MD: ABET. https://www.abet.org/accreditation/accreditation-criteria/.

D'Entremont, Kenneth L., and Andrew S. Merryweather. 2018. "Integrating Product-Safety Curriculum to Enhance Design and Reinforce Engineering Ethics." In *2018 ASEE Annual Conference & Exposition*, 16. Salt Lake City, UT: American Society for Engineering Education. https://www.asee.org/public/conferences/106/papers/21752/view.

Davis, Michael. 2001. "Three Myths about Codes of Engineering Ethics." *IEEE Technology and Society Magazine*, 2001. https://ieeexplore.ieee.org/document/952760.

DePass, Dee. 2018. "Polaris Hit with $27.25 Million Penalty for Failing to Report Vehicle Problems." *StarTribune*, April 2, 2018. http://www.startribune.com/polaris-industries-pays-27-25-million-settlement-to-consumer-product-safety-commission/478547013/.

Downey, Gary Lee, Juan C. Lucena, and Carl Mitcham. 2007. "Engineering Ethics and Identity: Emerging Initiatives in Comparitive Perspective." *Science and Engineering Ethics* 13 (4): 463–87. https://doi.org/10.1007/s11948-007-9040-7.

Duska, Ronald. 1991. *Ethical Issues in Engineering*. Edited by Deborah G. Johnson. Upper Saddle River, NJ: Prentice Hall.

"Ethic." n.d. Merrian-Webster.Com. Accessed November 20, 2019. https://www.merriam-webster.com/dictionary/ethics.

"FE Exam." n.d. NCEES.Org. Accessed November 22, 2019. https://ncees.org/engineering/fe/.

Fleddermann, Charles B. 1999. *Engineering Ethics*. Upper Saddle River, NJ: Prentice Hall.

Florman, Samuel C. 1994. *The Existential Pleasures of Engineering*. Second edition. New York, NY: St. Martin's Griffin.

"Fundamentals of Engineering (FE) MECHANICAL CBT Exam Specifications." n.d. Seneca, SC: NCEES. https://ncees.org/engineering/fe/.

Gearon, Christophe J. 2014. "Mandated Master's Degrees Could Change the Engineering Game." *U.S. News and World Report*, March 2014. https://www.usnews.com/education/best-graduate-schools/top-engineering-schools/articles/2014/03/17/mandated-masters-degrees-could-change-the-engineering-game.

Gorp, Anke van, and Ibo van de Poel. 2001. "Ethical Considerations in Engineering Design Process." *IEEE Technology and Society Magazine*, 2001.

Han, Kyonghee, and Gary Lee Downey. 2014. *Engineers for Korea*. Edited by Gary Lee (Virginia Tech) Downey and Kacey (Purdue University) Beddoes. San Rafael, CA: Morgan & Claypool.

Harris, Charles E., Michael S. Pritchard, and Michael J. Rabins. 1995. *Engineering Ethics: Concepts and Cases*. Belmont, CA: Wadsworth.

Harris, Charles E., Michael S. Pritchard, and Michael J. Rabins. 2005. *Engineering Ethics: Concepts & Cases*. Fourth edition. Belmont, CA: Wadsworth.

Heywood, John. 2017. *The Human Side of Engineering*. San Rafael, CA: Morgan & Claypool. https://doi.org/https://doi.org/10.2200/S00748ED1V01Y201612ENG028.

"Hippocratic Oath." 2019. Encylopaedia Brittanica. 2019. https://www.britannica.com/topic/Hippocratic-oath.

"History of the Code." 2017. American Medical Association. https://www.ama-assn.org/sites/ama-assn.org/files/corp/media-browser/public/ethics/ama-code-ethics-history.pdf.

Hoke, Franklin. 1995. "Veteran Whistleblowers Advise Other Would-Be 'Ethical Resisters' To Carefully Weigh Personal Consequences before Taking Action." www.the-Scientist.Com. 1995. https://www.the-scientist.com/profession/veteran-whistleblowers-advise-other-would-be-ethical-resisters-to-carefully-weigh-personal-consequences-before-taking-action-58515.

"IEEE Code of Ethics." n.d. Ieee.Org. Piscataway, NJ: Code of Ethics. Accessed November 19, 2019. https://www.ieee.org/about/corporate/governance/p7-8.html.

———. "IEEE Code of Ethics." 2006. Piscataway, NJ: IEEE.

Kalichman, Michael. 2013. "A Brief History of RCR Education." *Accountability in Research* 20 (5–6): 380–94. https://doi.org/10.1080/08989621.2013.822260.

Lindeburg, Michael R. 2001. *Mechanical Engineering Reference Manual for the PE Exam*. Eleventh. Belmont, CA: Professional Publications, Inc.

Martin, Mike W., and Roland Schinzinger. 2005. *Ethics in Engineering*. New York, NY: McGraw-Hill.

McGinn, Robert. 2018. *The Ethical Engineer: Contemporary Concepts & Cases*. Princeton, NJ: Princeton University Press.

Nagel, Thomas. 1974. "What Is It Like to Be a Bat?" *Duke University Philosophical Review* 83 (4): 435–50. http://www.jstor.org/stable/2183914.

NSPE. 2018. "Code of Ethics for Engineers." Alexandria, VA: NSPE. https://www.nspe.org/sites/default/files/resources/pdfs/Ethics/CodeofEthics/NSPECodeofEthicsforEngineers.pdf.

Parloff, Roger. 2018. "How VW Paid $25 Billion for 'Dieselgate'—and Got Off Easy." Fortune. 2018. http://fortune.com/2018/02/06/volkswagen-vw-emissions-scandal-penalties/.

Petroski, Henry. 1982. *To Engineer Is Human: The Role of Failure in Successful Design*. New York, NY: Vintage Books.

Pirsig, Robert M. 1974. *Zen and the Art of Motorcycle Maintenance: An Inquiry into Values*. New York, NY: Morrow Quill.

Poel, Ibo van de, and Lamber Royakkers. 2011. *Ethics, Technology, and Engineering: An Introduction*. Chichester, UK: Wiley-Blackwell.

"Professional Aerospace Union." n.d. SPEEA.Org. Accessed November 22, 2019. https://www.speea.org/.

Resnik, David B., and Kevin C. Elliott. 2016. "The Ethical Challenges of Socially Responsible Science." *Accountability in Research* 23 (1): 31–46. https://doi.org/10.1080/08989621.2014.1002608.

Roeser, Sabine. n.d. "Engineering Ethics 2.0." Tudelft.Nl. Accessed November 5, 2019. https://www.tudelft.nl/ethics/ethics/for-educators/ethics-20/.

Rowe, Mike. 2015. "Safety Third!" Mikerow.Com. 2015. https://mikerowe.com/2015/01/safety-third/.

Schlossberger, Eugene. 1993. *The Ethical Engineer*. Philadelphia, PA: Temple University Press.

Sharp, Phillip. 1995. "Advice to Whistleblowers." www.the-Scientist.Com. 1995. https://www.the-scientist.com/letter/advice-to-whistleblowers-58338.

Shmerling, Robert H. 2015. "First, Do No Harm." Website. 2015. https://www.health.harvard.edu/blog/first-do-no-harm-201510138421.

Simon, John G., Charles W. Powers, and Jon P. Gunnemann. 1972. *The Ethical Investor: Universities and Corporate Responsibility*. New Haven, CT: Yale University Press.

Snow, C. P. 1959. "The Two Cultures." Cambridge, UK: Cambridge University Press. https://www.google.com/url?sa=t&rct=j&q=&esrc=s&source=web&cd=3&cad=rja&uact=8&ved=2ahUKEwi4uO_J357kAhVJsZ4KHWeBAxcQFjACegQICRAC&url=http%3A%2F%2Fs-f-walker.org.uk%2Fpubsebooks%2F2cultures%2FRede-lecture-2-cultures.pdf&usg=AOvVaw3Jie1uCjb4UNhw4kFTD-wX.

Steneck, Nicholas H. 2007. "Introduction to the Responsible Conduct of Research." Washington, DC: NIH Office of Research Integrity (ORI). https://ori.hhs.gov/ori-introduction-responsible-conduct-research.

Stephan, Karl D. 2001. "Is Engineering Ethics Optional?" *IEEE Technology and Society Magazine*, 2001.

———. 2004. "Can Engineering Ethics Be Taught?" *IEEE Technology and Society Magazine*, 2004.

"U.S. Postal Service Pro Cycling Team Investigation." 2012. Colorado Springs, CO: USADA. http://cyclinginvestigation.usada.org/.

Vlasic, Bill. 2017a. "Volkswagen Engineer Gets Prison in Diesel Cheating Case." *New York Times*, August 25, 2017. https://www.nytimes.com/2017/08/25/business/volkswagen-engineer-prison-diesel-cheating.html.

———. 2017b. "Volkswagen Official Gets 7-Year Term in Diesel-Emissions Cheating." *New York Times*, December 6, 2017. https://www.nytimes.com/2017/12/06/business/oliver-schmidt-volkswagen.html.

CHAPTER 4

Product-Safety Concepts

Concern for man and his fate must always form the chief interest of all technical endeavors .... Never forget this in the midst of our diagrams and equations.[1]

—ALBERT EINSTEIN

4.1 Introduction

Chapter 2 discussed some of the difficulties of discussing safety, in general, and product safety, in particular. Chapter 3 raised the topic of ethics and the subsequent obligations of engineers to society, to their employers, and to themselves. This chapter will cover product-safety concepts that will help practitioners of engineering better understand, and ultimately fulfill their goals of designing and manufacturing safe products for consumer use.

Many of this chapter's concepts will help in considering and weighing design approaches, guarding-device options, device and system safeguarding strategies, and potential user behavior. Other concepts will be the consideration of safety hierarchies, a product's combination of product-safety risk and user activity, user task loading, at-risk product users, and the Haddon matrix.

4.2 Safety Hierarchies

The first safety concept covered will be the safety hierarchy. Some readers may already be aware of the first step of using this tool, namely, remove the hazard. As will be shown, using this simple tool is neither so simple nor so useful.

Some readers may also have noticed that the title of this section is in the plural. The reason is because there is not a single, unique *safety hierarchy*. There are many. The reader could even devise one of her/his own liking. Nothing is stopping anyone from doing just that. As will be shown, there is nothing sacred about any safety hierarchy.

What many today acknowledge, through its general acceptance, as *the* safety hierarchy is more accurately described as the *consensus* safety hierarchy (Barnett and Brickman 1985) and is shown in Figure 4.1. These authors have assembled and listed

[1]Speech at the California Institute of Technology, *New York Times*, February 6, 1931, p. 6.

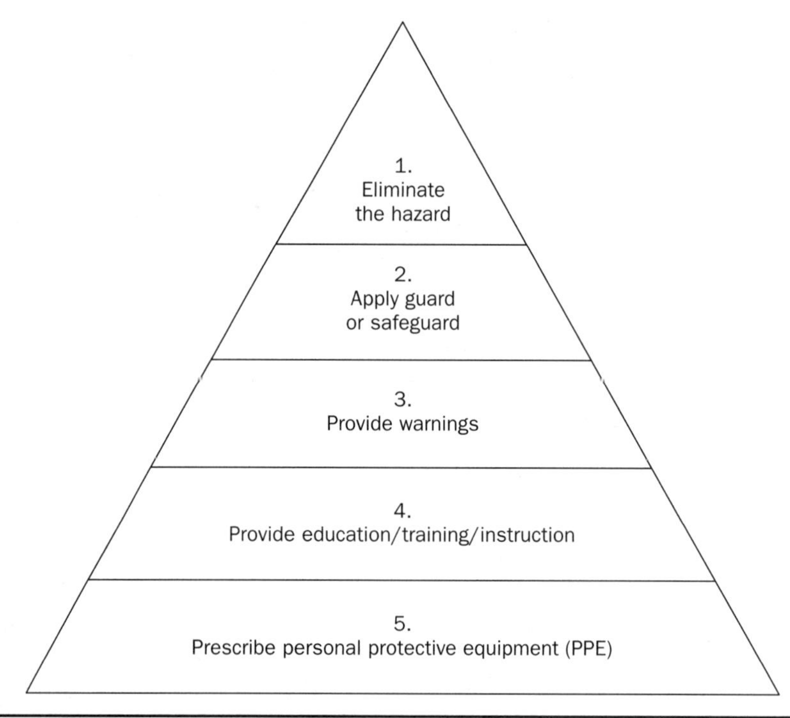

Figure 4.1 The consensus safety hierarchy.

39 different safety hierarchies in their work several decades ago. What the safety community would recognize today as the safety hierarchy may have first been published by the National Safety Council (NSC) in 1955 (NSC 1955). The three steps of their simple hierarchy were:

1. Elimination of the hazard from the machine, method, material, or plant structure.
2. Guarding or otherwise minimizing the hazard at its source if the hazard cannot be eliminated.
3. Guarding the person of the operator through the use of personal protective equipment (PPE) if the hazard cannot be eliminated or guarded at its source.

The preface to the third edition says that section 4 entitled "Removing the Hazard from the Job," from which the above is quoted, was added with that edition.

Although the various safety hierarchies uncovered by the authors are similar to one another, some hierarchies have as few as two levels and others as many as four. Some hierarchies share actions (categories), but may have them in different ranked orders (priorities). All hierarchies evaluated, however, had one form of hazard elimination as their first priority. The headings and categories of safety hierarchies reviewed are summarized in Table 4.1.

The reader is referred to the original work for the full details of that safety-hierarchy collection as well as the identities of their creators.

Priority	Heading	Category
1	Eliminate danger	Eliminate hazard
		Eliminate danger
2	Guard/safeguard	Guard hazard
		Isolate hazard
		Protect against hazard
		Minimize exposure to hazard
		Reduce severity
		Safeguard
		Safeguard danger
		Reduce danger
3	Warn	Warn
4	Train	Educate
		Train
5	Guard person	Guard person

Table 4.1 Summary of Safety-Hierarchy Categories

The consensus safety hierarchy shown in Figure 4.1 rank orders, or "prioritizes," five possible steps that can potentially be taken by a design engineer to improve the safety of a product. These steps are:

1. Eliminate the hazard from the product
2. Apply guarding devices or safeguarding technologies on the product
3. Provide warning labels, signs, or messages to product users
4. Provide education, training, or instruction to product users
5. Prescribe PPE for product users

The consensus safety hierarchy has five steps as compared to the NSC's first safety hierarchy which had three steps. The early NSC version of the safety hierarchy did not incorporate what are steps three and four of the consensus safety hierarchy. Perhaps it was the propagation of warning labels on many products in the latter half of the twentieth century which led to the third step and the formalization of training methods, especially on hazardous machinery and equipment, which led to the fourth. Regardless, the safety hierarchy as it is known today generally has the five steps shown in Figure 4.1. A discussion of guards and safeguards follows shortly.

It should come as no surprise to readers that several approaches to accomplishing increased safety and health have been presented over the years. The consensus safety hierarchy primarily resulted from work among engineers and technologists. In this work, technical people addressed what they knew best—machines and equipment. The accident scenario that they addressed is machinery as the source of a human injury. Put another way, a human would be the "target" of a mechanical "threat."

Another way to address this target-and-threat environment is to focus on the human instead of the machine. This is, in fact, the approach taken by William Haddon, Jr.

#	Strategy
1	Prevent the energy accumulation
2	Reduce the amount of energy accumulation
3	Prevent the energy release
4	Reduce the energy-release rate
5	Separate energy and target in space and time
6	Separate energy and target by barrier
7	Use non-aggressive contact surfaces
8	Strengthen the target—living or non-living
9	Move quickly to detect, evaluate, and counter harm or damage to target (short-term response)
10	Repair and rehabilitate the target (long-term response)

Table 4.2 Summary of Haddon Energy-Damage Reduction Strategies

His background was in public health and, in particular, epidemiology. His work (Haddon, Jr. 1973) framed human injuries as results of energy releases. He focused on treating these injuries as an *epidemic* and applied public-health concepts to develop countermeasures. He developed 10 strategies for avoiding injurious energy releases. These energy-damage reduction *strategies* are listed by rank order in the "strategy" column of Table 4.2.

One should see parallels between Haddon's energy-damage control strategies and the safety hierarchy. Each is a list of strategies whose intentions are to protect people from harm. Haddon's approach is based on energy, humans, and epidemiology; the safety hierarchy, on machinery, engineering design, and technology. This similarity has also been noted by others (Barnett and Switalski 1988). Haddon's considerations of design paths to attenuate the release of hazardous energy towards people have also been recognized, at least in essence, as "inherently safe design measures" ("ANSI 12100:2012 Safety of Machinery—General Principles for Design—Risk Analysis and Risk Reduction" 2012). Through the application of inherently safe design, objects are to be made smaller, lighter, slower, softer, lower in pressure, and lower in voltage. It may, however, not always be possible to follow all such energy-mitigation routes during real-world engineering-design efforts of functional products.

Discussion: Guard versus Safeguard

The words "guard" and "safeguard" are used throughout this book. To some people, they may be the same thing. In fact, even within the study of product-safety engineering, they *may* be identical in some cases, but the two need not always be.

When discussing a guard, what is often meant by engineers, for example, is a metal shield placed around or in front of a mechanical hazard. For example, chain-and-sprocket mechanical-power transmission systems sometimes have a *barrier* guard around them to prevent injury. This physical[2] barrier may prevent people from getting their fingers caught in the in-running nip point formed by the rotating sprocket and the roller chain. Such a person could easily receive a severe injury from placing a finger

[2] In the mechanical sense.

between the sprocket and chain even leading to amputation. Immediate bystanders could also be injured by having either long hair or dangling jewelry become entangled in this mechanical-drive system.

The barrier guard presented above *prevents passage* into the hazardous area. No further human action is needed to prevent injury with a barrier guard. There are other guards which merely *present an obstacle* to someone who is determined to gain entry into the area of the hazard. Examples of this type of guard are light sensors used on industrial machines and on residential garage-door openers. With heavy equipment, an array of light sensors, or a *light curtain*, will shut down the machinery if a person even momentarily obstructs any of the numerous light beams. These types of guards are *barricades*, rather than *barriers*, because some other action is necessary to prevent injury once the barricade has been breached. At many homes, the light sensor will halt the descent of the garage door, and then the garage-door opener system will reverse the door's motion to raise it, if a child or family pet walks through the garage entryway.

Safeguards are countermeasures to prevent injury from hazards and their associated risks. *Guards are but one form of a safeguard*. However, there are many forms of safeguards and safeguarding. Product-safety facilitators, such as product warning signs, manuals, education and training materials, hangtags, and well-executed advertising materials are safeguards against consumer injury.[3] One of the most-important forms of safeguarding is the designing-in of positive safety characteristics to consumer products. The better the job of safeguarding a product through such *Design for Product Safety* (*DfPS*) that the engineering-design team does, the less of a need there will be for additional safeguarding—let alone guarding.

[3] In an industrial-safety setting, administrative controls, to be introduced in the next section, also serve as safeguards.

It is both implicit and apparent that, within a safety hierarchy, the first step is more desirable and effective than the second in preventing injury, the second step is more desirable and effective than the third, and so on. Therefore, according to the consensus safety hierarchy, the use of PPE to protect a product user from harm is the least favorable and least effective way to address a product hazard that cannot be eliminated or otherwise sufficiently controlled.

Even after agreeing upon a specific safety hierarchy to use, the design engineer quickly realizes that this prioritization of safety measures provides insufficient guidance to proceed without second thought on design synthesis and analysis. Although the consensus safety hierarchy is perhaps the most widely known and accepted safety "principle" by many people and professions, it has been properly concluded by Barnett and Brickman that:

> [The consensus safety hierarchy] does not rise to the level of a mathematical theorem or law. This safety hierarchy was born out of consensus, not research, and its general validity can be disproved by numerous counterexamples.

For efficiency, from here on, the term "safety hierarchy" will mean "consensus safety hierarchy" unless otherwise specified.

The following is a representative instance of when the safety hierarchy fails the engineer tasked with designing a safe version of a product which many people would consider to be hazardous.

Example 4.1: The Design of a Gasoline-Powered, Walk-Behind Lawn Mower

Problems facing the designer: On the surface of this design task are at least two obvious and significant hazards to the lawn-mower user. The two largest safety hazards of this product appear to be the selection of gasoline to provide power to the lawn mower through an internal-combustion (IC) engine and the motion of a rotating blade employed to cut the grass.

Assuming that this is to be a competitively (read, low) priced, walk-behind (nonriding) lawn mower, the lawn mower will necessarily be of a conventional design incorporating no novel technologies or design approaches and, consequently, appear similar to the mower shown in Figure 4.2. Simply put, it will look like—and work like—a typical push lawn mower. It will have four wheels. It will have one long handle on which the controls are mounted to the rear of the mower chassis. The chassis will hold the engine assembly while shielding the rotating blade from above and have a discharge chute on the right-hand side to expel cut grass clippings. Assume that there will be no bag to gather the lawn clippings. Also, for the sake of simplicity, assume that there is no self-propulsion drive system on the mower.

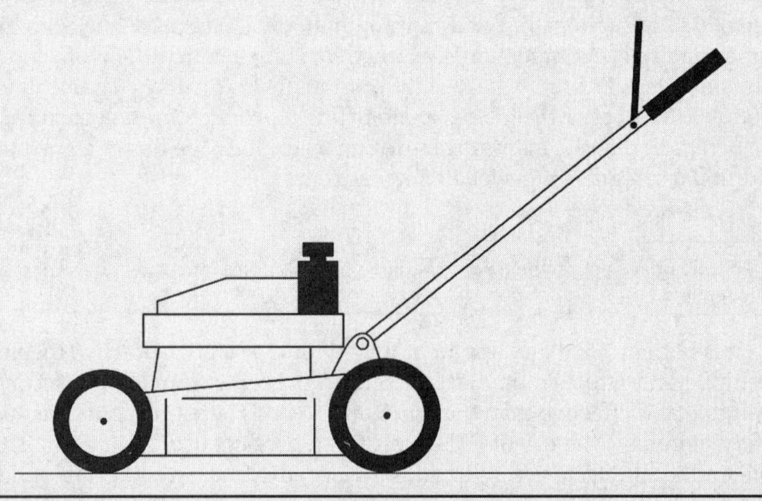

Figure 4.2 Gasoline-powered, walk-behind lawn mower.

Analysis: Taking the fuel hazard first, the presence of gasoline brings with it several harmful characteristics. Gasoline can ignite. Gasoline vapors can displace the oxygen-containing air within confined spaces leading to asphyxiation. Gasoline vapors can explode when ignition sources and combustible concentrations of air-and-fuel vapor exist simultaneously. Gasoline is also poisonous if ingested.

Evaluating at the rotating blade used to mechanically cut the grass on the lawn, a person readily recognizes a sharp blade rotating at high speed. Obviously, hands, fingers, children, and small animals must be kept away from the point of operation which is primarily below the chassis. However, there is exposure to the hazard from the side-discharge chute of the chassis. The person using the lawn mower must also be cognizant of the potential for the rotating mower blade to throw objects beneath the mower through this chute and thereby injure bystanders.

Each of these two hazards—gasoline and the rotating blade—is capable of injuring, if not killing, a user or a bystander. These are indeed two serious hazards with consequences that must be addressed by the design engineer.

Solution by safety hierarchy: In neither case—gasoline nor rotating blade—can the hazard be eliminated without drastically changing the characteristics, the performance, or the utility of the product. Although electrical power in the form of a battery or an electrical cord could be substituted for gasoline power, the public has expressed a desire for, and a certain acceptance of, gasoline-powered lawn mowers. Furthermore, even if a battery is substituted to power an electric motor, the other significant hazard, the rotating blade, still exists and remains vital for the purpose for which the lawn mower is purchased: to cut grass. Furthermore, electrical hazards such as the stored chemical energy of batteries have associated risks of their own such as electrocution, fire, and explosion. Both gasoline and electricity continue to pose hazards to users and cannot be eliminated.

Focusing on only the rotating-blade hazard for the moment and using the safety hierarchy to lead the designer to first try to eliminate the hazard which is the rotating blade. It is apparent that without resorting to an alternate technology to cut the grass—which is not a design option since this is a low-relative-cost product—there is no way to economically eliminate that blade hazard and still perform its intended function—cutting grass. Smart tinkerers and engineers have been designing lawn mowers for centuries and have yet to discover a way to efficiently cut grass not involving one or more sharp blades in motion. It is unlikely that *this* design engineer will be able to devise a significantly different way to cut grass given the budget and schedule no doubt given to this engineer.

Unable to eliminate the moving-blade hazard, the engineer then must guard,[4] or protect, the user from the rotating blade. This has largely already been accomplished—to a degree—by the current state-of-the-art lawn mower chassis. The main deck of the chassis does guard the product user both from much accidental bodily contact with the blade and from objects thrown by rotating blade. It is still possible to get feet and hands under the chassis deck and become seriously injured. It is also still possible to be injured by thrown objects. The blade guard in the form of the chassis deck does not completely eliminate the possibility of injury; it only reduces the probability of injury—but significantly so—from the blade. The second step of the safety hierarchy is now complete—at least partially so. It is unclear whether step 2 of the safety hierarchy requires one-hundred percent (100%) protection of the product user from the hazard. Unlike a totally enclosed power-transmission device, but similar to an industrial punch press, the lawn mower requires an opening through which to insert the "workpiece" (the grass) into the "point of operation" (the rotating blade). Therefore, some access to the rotating blade must always exist. In this case, the access is at the bottom of the mower with some access from the side-discharge opening. The designer, however, can and should work to reduce the risk of users and bystanders to the blade hazard. This exposure risk will just never reach zero, unfortunately.

---

[4] Through many decades of lawn-mower design engineering and product use, the only socially acceptable emergent option is to provide a barrier guard on a gasoline-powered lawn mower. This is not a "clean sheet" design from a societal perspective. In a true clean-sheet design of a product, each design option or characteristic might be weighed, without preconception, on its benefits and its drawbacks.

Since total blade guarding cannot be achieved, the designer proceeds to step 3, and beyond, of the safety hierarchy: Provide warnings to product users. When looking at a modern lawn mower, one may see a plethora of warning signs for numerous hazards including gasoline and the blade. The product user reading the warning signs should be directed to read, understand, and follow the owner's manual and to wear PPE and clothing such as eye and ear protection, long pants, and sturdy, closed-toed shoes.

It is at this point that the safety hierarchy again fails the design engineer. The hierarchy does not indicate *when*—or even *if*—a set of warning labels (step 3), training information in the owner's manual (step 4), and PPE use (step 5) fully compensate for the lack of complete barrier guarding for the lawn mower's rotating blade. Thus, it is evident that the safety hierarchy fails to live up to the faith placed in it by some people who may never have actually attempted to apply it.

Returning to the unresolved hazard of gasoline, the safety hierarchy again fails to deliver meaningful results to the design engineer. Having already acknowledged the need to retain gasoline as the mower's energy source, the design engineer must now guard the lawn-mower user from gasoline—both while the lawn mower is in use and when it is in storage. Current IC lawn mowers keep gasoline in a fuel tank generally attached to the engine. There is a fuel-tank cap which screws securely into place to seal the tank and is usually tethered to the tank to prevent its loss. A fuel line carries gasoline from the tank to the carburetor's float bowl where it is then metered into the engine's combustion chamber through passing engine-intake air. There will no doubt be significant testing of the fuel system in the laboratory and in the field by the mower's designer and manufacturer to verify the final fuel-system design as being functional and reasonable safe to use. The design engineer will work to protect the user against the fuel hazard by minimizing the amount of gasoline lost through "sloshing" liquid and vented gases. Either of these fluid-escape mechanisms can lead to explosion, fire, or asphyxiation. There are also factors beyond the control of the designer such as the care taken by the user when refilling the mower's fuel tank. Will the user spill large amounts of gasoline? Will the user refill the fuel tank when the mower's exhaust system is still very hot? These are beyond the direct control of the engineer.

As with the rotating-blade hazard, the gasoline hazard will be the subject of warning signs and operator instructions in the manual provided with the product. PPE will not likely be a factor regarding the handling of gasoline for this residential consumer equipment. (The use of PPE could be a safeguarding action in certain racing, industrial, or military fuel-handling applications.) There are no sufficiency criteria for warnings in the safety hierarchy regarding fuel evaporation and spillage. The engineer is on her/his own over the safety of numerous design characteristics of the lawn mower.

Final-design utility of safety hierarchy: It has been shown that the safety hierarchy has been unable to guide the engineer in a meaningful way when making safety-critical decisions during lawn-mower design. Fortunately, design engineers of real-world products cannot and do not rely upon the safety hierarchy for this reason and are able to advance product safety regardless.

In defense of the lawn-mower industry, there have been significant improvements made over the years in user safety to the simple, walk-behind lawn mower used in this example. Deadman controls shut off lawn mowers when their handles are released; some lawn mowers have the fuel tank and the potentially hot exhaust system on opposite sides of the engine; lawn mowers can be pull-started from the rear of the mower rather than from the side thereby moving the user farther from the blade; and the undersides of many mowers can be rinsed off from the outside/top of the mower reducing the need to tip or elevate the mower deck to expose the blade, especially with the engine running, for cleaning purposes.

In fairness to consumers, advocates, and regulators, lawn mowers continue to be the subject to safety recalls[5] involving their gasoline systems as well as risks from thrown objects. Therefore, work remains to be done to improve the safety of lawn-mower users, but progress has been demonstrated. It is up to designers and manufacturers of lawn mowers to continue this trend.

[5] U.S. Consumer Product Safety Commission, https://www.cpsc.gov/Recalls

The above example should serve as a design case study where the safety hierarchy proves less than useful once numerous critical and concrete design decisions must be made by the design engineer. A significant hazard can rarely be eliminated—the first priority in the safety hierarchy—if the product is to perform a useful task. Instead, what frequently must be done are the second and third priorities of the safety hierarchy, applying safeguards and providing warning signs, respectively. Equipment safeguarding can often reduce the exposure to the user of a product, but can rarely eliminate a product user's potential risk from a hazard. For example, the design of the lawn-mower chassis in Example 4.1 reduces the probability of injury to the user from the rotating blade, but does not completely protect the user from that hazard since the blade must be accessible from the ground in order to cut the grass. After having applied a guard to equipment, some continued exposure sometimes still lingers, but that exposure probability has been lowered from the pre-guard level. Therefore, warning signs must sometimes be provided according to the safety hierarchy. In the case of many lawn mowers, warnings are provided for numerous hazards in several places on the mower—and in the owner's manual.

The very existence of safety-hierarchy priorities beyond the first confirms that first priority cannot always be accomplished. The existence of third priority confirms that priority two cannot always be accomplished, and so on. Therefore, product-hazard risks cannot always be eliminated and controlled. Guards cannot always completely protect a product user. The willing reader can easily add to this sequence of insufficiencies in the safety hierarchy with little effort.

Ross has also observed that the safety hierarchy often fails to provide the guidance needed by engineers facing real-world design challenges (Ross 2015). Information lacking in the safety hierarchy includes the following:

- When a warning sign is sufficient to protect a user from incomplete guarding of a hazard that cannot be eliminated
- When a warning sign is sufficient to protect a user from a guard that cannot completely protect the hazard which cannot be eliminated
- When education and training instructions are sufficient for warnings that are insufficient
- When PPE must be used since the instructions are insufficient

The above list shows that the safety hierarchy is a concept that sounds good to the layperson, and may even be successfully argued in court of law by a talented attorney representing an injured party, but remains a theoretical concept that fails the engineer once real-world product-design work commences.

In practice, the safety hierarchy includes a set of consideration *necessities* rather than an order of prioritized *sufficiencies*. Due to the limited practical utility of the safety hierarchy, it is better considered as a philosophical concept rather than a useful and practical design tool for a real-world engineer doing product-design work. The same conclusion has also been reached by others (Hall et al. 2010).

4.3 Industrial Hierarchy of Controls

Although this book is not about industrial safety, the safety hierarchy used in the field of industrial safety is presented and briefly discussed. This section is added in order to avoid the confusion some readers might experience when studying the field of safety in general. This industrial-safety hierarchy—called the *Hierarchy of Controls*—is shown in Figure 4.3 and has historically been applied to industrial machinery in manufacturing industries and to workplaces such as factories to cover both safety and health concerns.

The U.S. Centers for Disease Control's (CDC) National Institute for Occupational Safety and Health (NIOSH) uses this hierarchy in their materials (NIOSH 2015). Compared to the consensus safety hierarchy, the hierarchy of controls is shown as an *inverted* pyramid. The hierarchy of controls contains five levels of control numbered from 1 through 5 in ranked order of descending efficacy. Levels 1 and 5 are generally similar to the consensus safety hierarchy. Level 1 is to eliminate the hazard if possible.

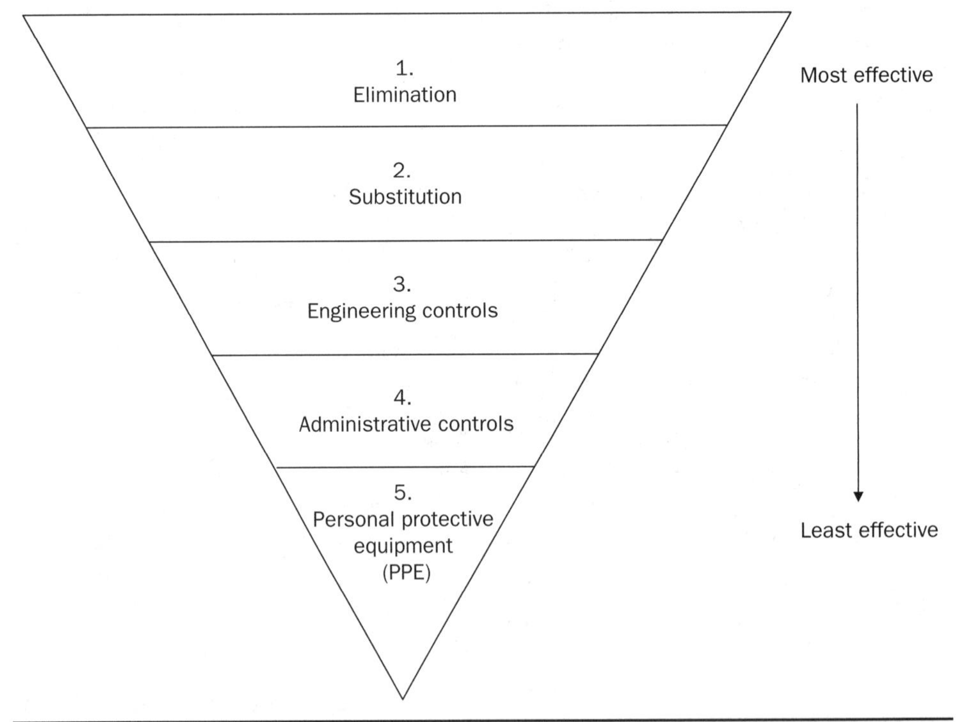

Figure 4.3 The hierarchy of controls.

Level 5 is to use PPE when necessary. In between steps 1 and 5, there exist differences as well as similarities.

The hierarchy of controls' step 2 is the substitution of one hazard for another. Limiting this conversation to safety, rather than health, consider the following example. If a machine of type A has a high level of hazard risk during use but a machine of type B can accomplish the same manufacturing operation at a lower hazard risk than a machine of type A, then a machine of type B should be used according to the hierarchy of controls. This is reasonable and would appear to be common sense to many engineers and readers.

Step 3 in the hierarchy of controls is to employ *engineering* controls to protect workers. Engineering controls are frequently hardware devices in place to protect workers. In the safety context, an engineering control would be the placement of a barrier guard between a worker and an in-running nip point formed by two rotating gears on a machine, for example. An engineering control can physically prevent a worker from reaching into and being injured by machinery. Such a control may only make it harder to reach into a machine and injure oneself. In a perfect world, an engineering control will be inexpensive and offer minimal-to-no interference with product function.

Step 4 of the same hierarchy is to use *administrative* controls to protect workers. Unlike engineering controls which are often physical barriers preventing bodily contact with a hazard, administrative controls cannot prevent an injury from happening without additional actions. Instead of physical or mechanical barriers, administrative controls frequently consist of policies, procedures, checklists, best practices, and other human behavior-based management efforts which must be known, understood, and *followed* by workers. Warning signs on machinery and products would also full under this category.

Because three comes before four, the use of engineering controls is deemed more effective than the use of administrative controls and rightfully so. Therefore, the use of a guard over the power-transmission gears discussed above is more effective in preventing worker injury than a yellow line painted on a factory floor delineating a hazard zone near the gears. This is compatible with the consensus safety hierarchy which places the use of guarding on a machine above and preferable to the provision of a warning marking on the floor to workers.

If one further researches industrial safety, the term "Prevention through Design" (PtD) may be encountered in references such as NIOSH's website (NIOSH 2013). PtD is a national initiative to eliminate occupational risks, redesign workplaces, and enhance the work environment. Although PtD, on its surface, may appear similar to product-safety engineering or Design for Product Safety (DfPS), it is not the same thing. The worthy objective of PtD is to prevent or reduce occupational injury and illness through redesigning how tasks are performed in order to reduce the probability or severity of an industrial accident and its injury outcome. Also involved are groups of stakeholders from various industrial sectors and efforts including researchers and educators. Product-safety engineering, however, should be addressed somewhere within the large, integrated product-design process (PDP), system since the design of machinery can definitely contribute to workplace-injury reduction.

4.4 Classification of Safeguard Devices

Engineers are faced with a multitude of decisions to be made when designing new products, especially when the products are innovative. It can be overwhelming, as well

	Characteristics		
Type	Safety Increased	Safety Unaffected	Safety Decreased
I	•		
II	•	•	
III		•	
IV	•		•
V	•	•	•
VI		•	•
VII			•

Table 4.3 Classifications of Safeguard Devices

as confusing, to sort through all of it in order to make the best decisions. Many of these design decisions are conceptual choices made early in the PDP which is covered later in the book. Among the choices to be made are decisions regarding hard numbers and specifications such as length, weight, design performance criteria (such as load capacity and top speed), and more. A subset of these design choices may be the guards and safeguards which will be integrated into the product that can affect the safety of the product when used by its consumers. It may well be difficult to even sort through some of these potential hazard countermeasures in order to design-in the appropriate levels of safety for users.

The engineer may confront expectations, suggestions, and requirements for the product under design. If the product will be similar to existing products from other companies, the engineer should also have competitive-benchmarking[6] information and other data to help guide the design decisions to be made. Many people making decisions receive much advice—both solicited and unsolicited. There may also exist standards, regulations, and agreements that may impact the engineer's ultimate decisions regarding product-safety characteristics. These are covered in Chapter 8.

Before making decisions impacting user safety, it is good for engineers to understand the fundamental characteristics of each potential safeguard device under consideration for the product being designed. One methodology to do this is called the classification of safeguard devices (Barnett and Barroso 1981a). Using this scheme, a proposed safeguard device—of any type—is simply evaluated to see which of scheme's seven classifications best suits it. Table 4.3 shows the safeguard-device classification types as well as their corresponding product-safety attributes. For each potential safeguard, an evaluation is made regarding whether that safeguard improves safety, leaves safety unaffected, and/or decreases safety. A safeguard may have one effect, two effects, or all three effects, but it will at least have one effect on product safety. Each type of safeguard device is described in turn below.

A *Type-I* device is a safeguard that only improves the safety of the product on which it is used. A guard that completely covers the power transmission system, for example the chain-and-sprocket drive system on a chain saw, shown in Figure 4.4, generally only

[6] The practice of evaluating and testing a competitor's product to identify its strengths and weaknesses in order to improve a company's own product's design and manufacturability.

Figure 4.4 Chain-saw drive sprocket and chain.

has positive impact on product safety. Such a guard would be a barrier to virtually all access to the hazard of the in-running nip point posed by the chain and drive sprocket and not have any side effects.

A *Type-II* safeguarding device will either increase the safety of a product or leave it unaffected. On example of a Type-II device would be the use of color, such as red colorant in plastic or a yellow-and-black striped decal, to identify a product hazard or hazard location to the product user. One such action would be to paint red the part of a product that will be hot if touched by a user during operation and result in a burn. Such a safeguard may work—or it may not. The successful prevention of user injury may well depend upon numerous other factors including product characteristics and attributes, the user,[7] and the use environment. However, the use of color as a safeguard device is unlikely to injure anyone by itself.

A *Type-III* safeguard will leave the product unaffected with respect to safety. In some instances, the addition of a redundant system to a reliable, but little-used, system would fall into this classification. For example, many smoke alarms emit a "beep" when the battery becomes low in voltage. The *addition* of a back-up, or redundant, beeper would likely add little to the overall safety of the smoke alarm. The beeper safeguard systems already in smoke alarms have, for the most part, proven to be quite reliable and, consequently, adding a secondary beeper makes little engineering sense and would add little, if anything, in the way of consumer safety. The addition of such a redundant system would only add cost to each unit without measurable benefit. The severity of a fire accident would not be reduced and the probability of avoiding a housefire accident would not be markedly improved either.

A *Type-IV* safeguard is one which may either increase or decrease product safety. One such example is fitted on most consumer clothes dryers in the form of the loading/

[7]The color-blind population might become an "at-risk" population, discussed in Chapter 9, if a single color, rather than a conspicuous high-contrast and alternating pattern, is used as a safeguard.

unloading door. Although many people take the dryer door for granted, it is in fact an interlocked barrier guard that protects users from the rotating internal-drum hazard of the dryer. Getting caught in the drum could severely injury a person, especially a small child. The clothes-dryer door does an admirable job of preventing such injuries; however, the door is frequently misused as an ON/OFF dryer-control system. The user also expects the dryer's drum to slowly stop rotating when the loading door is opened. This may not always happen. Thus, the potential for harm is somewhat increased through user expectation and dependent behavior. The drum has always stopped before; therefore, the drum will stop each time the dryer door is opened. The user depends upon the dryer drum to stop turning and acts accordingly. This example shows that some product users have developed a pattern of misusing the dryer door and depending upon its unintended control-system properties.[8]

Type-V safeguard devices are those devices which may protect the user, do nothing, or harm the user. Barnett and Barroso offer the automobile seat belt as this type of safeguard device. In many frontal impacts and some side impacts, the standard three-point automobile seat belt does provide protection to automobile users. In rear impacts, however, seat belts generally provide little in the way of occupant protection, especially if the seat back collapses as a result of the rear-impact accident. A seat belt can also contribute to automobile-occupant injury if the seat belt is worn improperly—either too high on the abdomen or with the shoulder-strap portion of the seat belt behind the back.

A *Type-VI* safeguard device is one which may either not affect safety or lead to greater danger for a product user. One example of this type of safeguard device is the locating of an operating control near a product hazard. For example, placing an ON/OFF control on a power tool near a sharp, moving component hazard would not affect user safety if the sharp component in motion is not physically encountered but, if it is encountered by the user during operation, user safety would indeed be decreased. The location of the control increases the likelihood that the tool's hazard will be encountered by a user without an off-setting safety benefit.

The final safeguard-device type, *Type VII,* includes those safeguard devices which only increase danger to users. The Type-VII safeguard device offered by Barnett and Barroso is the crane personnel baskets that can be attached to the cable of a crane in order to move people from one worksite position to another. OSHA regulations permit the use of such devices with restrictions (OSHA 2010), but crane manufacturers, according to the authors, categorize the use of the personnel basket as an unsafe practice. Perhaps the only "indicated" use for a personnel basket is in instances where all other alternatives for getting people to or from a site s are even more hazardous to people or if time is a critical factor. Readers may sometimes see personnel baskets used during dam construction when vertical canyon walls preclude the use of ladders (with their own hazards) and other ascent/descent systems or during emergency situations where rescue workers need immediate access to an otherwise inaccessible location.

It is vital in this discussion of safeguard-device types to focus upon *probabilities* rather than upon *possibilities*. There are few devices that would have only positive safety attributes if one goes to extreme efforts to construct extremely *improbable*—yet

[8]The product-safety concept of consumer dependency leading to product misuse is explored later in this chapter.

entirely *possible*—situations. The power-transmission guard adduced as the example of the Type-I device could, possibly, be dangerous to someone if it were to fall off of a tall machine and strike a person in the head.[9] Such considerations are beyond the scope of this classification discussion and this book. Including all possibilities of potential events would likely render all safeguarding devices as Type V where each device could either enhance, have no effect on, or compromise safety.

The evaluation of safeguard devices is one instance in which it is necessary to apply *engineering judgment*—the common sense of engineering—into the PDP. There are other places where the exercise of engineering judgment, in conjunction with applied engineering ethics, will be critical.

It is hoped that several of these safeguard-device types are never actually encountered by engineering-design teams due to elimination—either consciously or by accident—by the design team before serious consideration even begins. In fact, the authors of this classification methodology state that it is unethical to offer safeguards of types VI and VII since doing so adds cost to a product without delivering benefit. The same might also be argued for type-III safeguarding devices. Nevertheless, all safeguard-device types are included in this discussion for the sake of completeness.

During real-world design efforts, it is vital to view potential safeguard devices in light of the product, the product user(s), and the product-operating environment. It is important to look at engineering design and product safety in such a comprehensive and systematic fashion. Safeguard devices that make sense and work well on one product, or type of product, may prove unsuitable in other cases. There are few "no brainer" decisions in the product-design and product-safety engineering, especially when designing unique and innovative products as some readers presently or ultimately will do. A safeguard-classification example follows below as Example 4.2.

Example 4.2: Classification of Safeguard Devices—ROPS

Many four-wheeled off-road vehicles (ORVs) have a rollover protective structure (ROPS). The purpose of the ROPS is to provide some level of occupant protection in the event of a vehicle tip-over or rollover accident. Of course, the ROPS cannot protect all occupants under all accident conditions, nor does the ROPS act alone. The ROPS is typically an element of an occupant protective system (OPS). The ROPS is just one component of the OPS which also may include seats, headrests, seat belts, nets, doors, handholds (including the steering wheel), and other occupant-compartment features provided to positively affect accident outcome. Occupants would also be wise to wear PPE such as helmets, eye protection, and gloves.

A side view of an ORV, in this case a side-by-side (SSV), is shown in Figure 4.5. Some elements of an OPS have been omitted for greater clarity. The ROPS is readily identified as the structure above the seating area of the ROV.

Imagine the SSV ROPS to be a traditional, or "typical," ROPS. That is to say that the ROPS bars are as shown in Figure 4.6 when viewed from above. These ROPS bars are along the perimeter of the vehicle with a top cross member to reinforce the side members. This is the type of ROPS seen on many ROVs and even on some other vehicle types.

---

[9] Barnett recently wrote that "Every physical entity created by man or nature is a hazard capable of causing harm." (Barnett 2020)

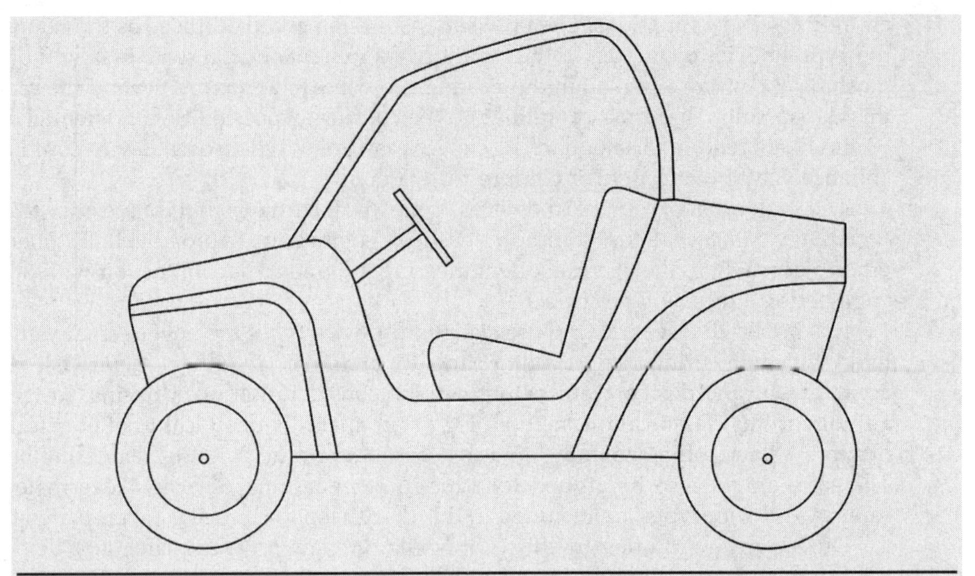

Figure 4.5 Side view of SSV with ROPS.

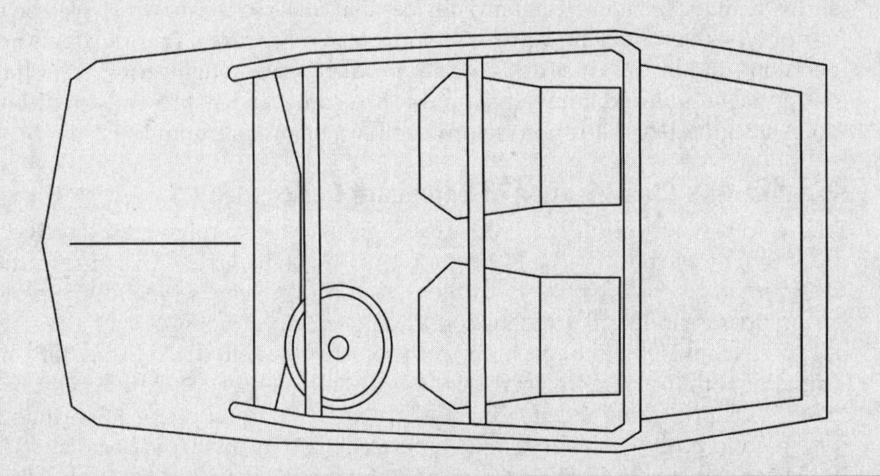

Figure 4.6 Top view of SSV with traditional ROPS.

Classifying the traditional ROPS as a safeguard device may result in the following conclusions:

- ***The ROPS may increase safety*** in the case of a moderate-speed rollover accident when all vehicle occupants are properly restrained and using handholds and any doors or nets
- ***The ROPS may do nothing*** in the case of a low-speed rollover accident in which the ORV simply tips onto its side and stops its rotation before the ROPS may even touch the terrain

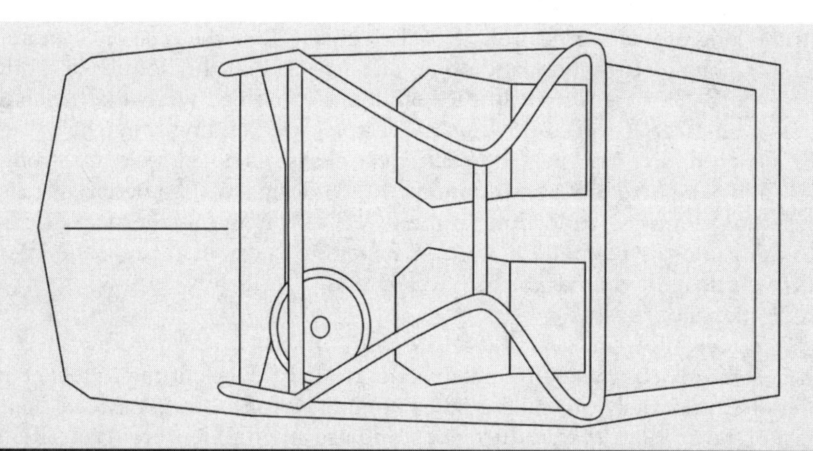

Figure 4.7 Top view of SSV with alternate ROPS.

- ***The ROPS may do harm*** in the case of a rollover accident in which an occupant gets an arm or a hand trapped between the top of the SSV ROPS and the terrain

Therefore, this traditional SSV ROPS would be a *Type-V* safeguard device according to the Barnett and Barroso safeguard-classification criteria.

Also, imagine that there is an alternate ROPS design for a similar SSV. A top view of this alternate design is shown in Figure 4.7, also from above. This ROPS differs from the traditional ROPS design in that the top bars of the ROPS are not placed along the perimeter of the vehicle. Instead, these top bars of the ROPS are shaped in an "hourglass" figure and are narrower in the longitudinal middle (above the occupants) of the ROPS than at the ends where the ROPS attaches back to the SSV.

Applying the safeguard-classification criteria to the alternate ROPS might tempt the design engineer to conclude that the ROPS is now no longer able to injure the occupants' arms or hands through a crush injury. *If this was true,* the device would move from being a Type-V device to a Type-II device. However, although the probability of upper-extremity injury has, perhaps, been reduced—it has not been eliminated. It would still be possible for vehicle occupants to suffer hand and arm injury from the front and rear ROPS upright bars that extend from the SSV chassis up to the ROPS top bars. Given the certain conditions and accident situations, these hourglass-shaped top ROPS bar sections could prove just as risky to SSV occupants as the traditional-design ROPS bars.

Since the alternate ROPS design could—potentially—be less hazardous to ROV occupants, should the alternate design be adopted without further delay? Real-world engineering is rarely so simple. Although there could be some safety advantages to the alternate ROPS design when compared to the traditional ROPS design, it might be optimistic to presume that these advantages would be significant and measurable. Indeed, there may be factors beyond product-safety considerations alone that must be weighed before making such significant SSV ROPS-design changes. It is important to understand how such a design change will impact the overall function of the SSV and what other changes would be necessary to compensate for changes in the ROPS' mechanical properties. For example, with some SSV designs, the ROPS may provide a significant amount of structural strength to the chassis, especially in bending stiffness in the longitudinal vertical plane of the vehicle. Take the case of "bottoming out" the suspension system when landing an airborne SSV at the bottom of a ditch. The two ends of the SSV are forced upward while the combined weight of the chassis, passengers, and cargo is forced downward. This vehicle-bending moment will put a significant compressive load into the ROPS if the chassis is not sufficiently strong on its own. All things being equal, the traditional ROPS design will be better able to handle this resulting compressive loading in the ROPS bars than the alternate ROPS design due to the geometry of the ROPS. Other factors to consider include the intended use of the vehicle and whether the SSV was intended for heavy-duty work or for light-duty recreational purposes.

This example illustrates the care with which critical product-safety engineering decisions must be made. It is much better to analyze the entire product being designed, than to focus solely on one particular component of the larger system and exclude the rest of the product, the product users, and use environment from consideration.

4.5 Classification of Safeguard Systems

Section 4.4 discussed individual safeguard *devices*. This section will discuss how safeguard devices interact, or complement, one another as a *system* of safeguard devices. This permits the engineer to further examine the range of effects that an individual proposed safeguarding device may have on product-safety characteristics. This section helps the engineer gain insight into multiple safeguard devices through their interactions.

The functional hierarchy of safeguarding systems (Barnett and Barroso 1981b) is a scheme for assessing the function of a safeguard within a particular product-safety system. The structure of the functional hierarchy is listed in Table 4.4. It can be seen from this table that there is no theoretical end to the ranking of safeguarding function in a product.

Rather than being actual safeguarding devices, *Zero-Order* safeguard systems are instead characteristics of a product hazard itself. Some products may not possess

	Characteristics	
#	**Rank**	**Properties**
0	Zero order	Intrinsic safety properties of a product
1	First order	Safeguarding devices that improve upon the safety of Zero-Order systems
2	Second order	Safeguarding devices that improve upon the safety of First-Order systems
3	Third order	Safeguarding devices that improve upon the safety of Second-Order systems
...	...	
N	*n*th order	Safeguarding devices that improve upon the safety of (*n*−1)-Order systems

Table 4.4 Functional Hierarchy of Safeguarding Systems

FIGURE 4.8 Simple kitchen knife.

zero-order safety system characteristics. A simple kitchen knife, with a sharp blade as shown in Figure 4.8, is an example of a product which does exhibit zero-order safety properties. Although the sharpened blade of the knife itself may indeed be capable of inflicting severe personal injury, the other characteristics of the knife render the product acceptably safety. These other knife characteristics include the following:

- The blade and its sharp edge are in the open to the knife's user; they are not hidden from view.
- The knife is of simple design; there are no moving or hidden parts.
- The knife is manually operated; there is no automatic control system which operates the knife.
- The knife, through the user, generally operates at slow speeds and within the perception and reaction times of the users.
- The hazard of knives is well known to most adults.

Knife designs deviating from the simple kitchen knife shown may not possess the same zero-order safety properties as the simple kitchen knife. For example, a rolling circular-bladed kitchen-knife design with a floating handle ("Meet Bolo" n.d.) may contain hazards that are not as easily discernable as with the traditional knife design. In the case of the traditional kitchen knife, there is little more that can apparently be done by design engineers to further safeguard the product while it is in operation cutting food and ingredients. Adding additional safeguarding hardware to the knife system is likely to compromise the function of the knife while in operation and in other phases of its operation including the necessary cleaning of the product.

Revisiting the guard over the chain-sprocket-roller power-transmission system in the chain saw example used in Section 4.4 reveals that the drive system does *not* possess zero-order safety system characteristics. Even the simple one drive-sprocket, one roller-sprocket, and one-chain system contains one in-running nip point[10] as well as potential entanglement hazards for users, their hair, their clothing, and their jewelry. The fitment of a guard over the chain and driven sprocket serves as a *First-Order* safety system. This chain and sprocket constitute a hazard against which it is relatively straightforward to guard. The guard does not inhibit the functioning of the chain saw and often only needs to be removed for maintenance and service purposes. Such a guard may also provide additional benefits to people in the forms of potential noise reduction and shielding of cast-off lubricant and debris.

Continuing to focus on the guard over the chain-and-sprocket system driving the chain saw, the method of securing the guard has not yet been specified. The guard will be relatively easy to circumvent by a user if it can be removed by simply pulling it with a hand. Securing the sprocket-access guard in place with machine screws or bolts requiring tools and conscious user effort in order to remove the guard, makes these fasteners a *Second-Order* safety system since they enhance the efficacy of the sprocket-access guard which is a first-order safety system.

A potential *Third-Order* safety system could be a small cable which tethers the sprocket guard to the chain saw so that the guard is not lost during servicing or adjustment and that it is more likely to be put back into place on the chain saw prior to resuming its use. As with many safeguards, there could be a down side to having a small cable in the vicinity of the chain saw's chain and that might be the entanglement of the cable in the chain.

If the design engineers were to further pursue increasing chain-saw safety, the placement of an interlock controlling the chain-saw ignition system could be studied. If such an interlock was to be incorporated, it would serve as a *Fourth-Order* safety system since it would enhance the protective properties of the guard tether which is of third-order nature. The author is not suggesting that such an interlock be incorporated into the chain saw's design. There are numerous technical and practical challenges to pursuing such a path. To begin with, the interlock would be subject to the harsh environment of vibration, shock, and lubricant exposure. Furthermore, fitment of such an interlock could adversely affect the operational reliability of the chain saw since interlock malfunctions could easily render the chain saw as inoperable. A user experiencing such interlock-reliability issues would be tempted to bypass that interlock altogether which would likely be a simple operation to a mechanically inclined person. Perhaps it is or will be feasible to supply such a safeguard device in the future, but it may not be a simple design task. Much analysis, laboratory testing, and field testing would be required to assure a useful and reliable chain saw—as with any product.

A possible *Fifth-Order* safety system could be the placing of a warning sign inside of the sprocket-and-chain cavity that can only be seen when the guard is removed. Such a warning sign may tell a user not to use the chain saw if this warning sign is visible. The down sides of doing this go back to the harsh environment of the chain saw and its drive system. It is hot, oily, and subject to flying wood particles. Any warning sign in this area would need to be durable in addition to being clean in order to remind a user to replace the guard before using the chain saw.

---

[10] On many chain saws, the "blade" of the chain saw serves as a barrier guard protecting access to the in-running nip point at the non-driven sprocket of the saw blade.

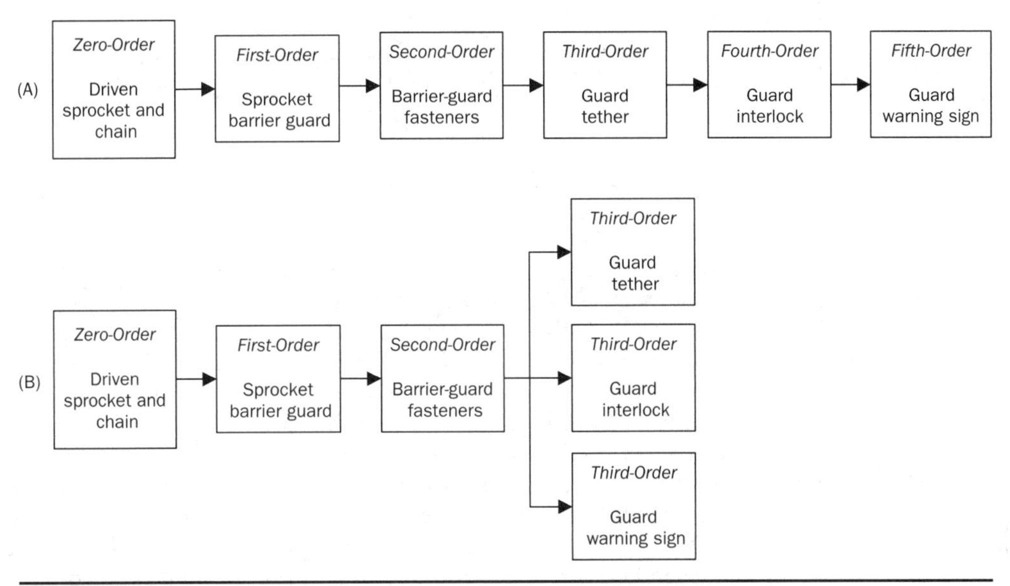

FIGURE 4.9 Series and parallel function in chain-saw safeguarding systems.

Taking a look at the fourth- and fifth-order safety systems above shows that at some point it may no longer be clear which lower-order safety system an additional safeguard device specifically helps. With the guard interlock, it seems reasonable that it would improve the effectiveness of the guard itself, a first-order safeguarding system; so, maybe the guard tether truly is a second-order safety system. Perhaps the fifth-order safety system, the warning sign, does indeed further the safety of the fourth-order guard interlock if the guard is not replaced—or perhaps it helps increase the safety of the guard itself. This and other examples seem to bring to light that some safeguard devices may function in parallel, rather than strictly in series. Figure 4.9 shows that the guard's tether, interlock, and warning sign may all be third-order safety systems that increase the safety effect of the guard and its fasteners. It may simply be enough to know that in order for one safeguard to work, that safeguard requires that another safeguard be present on the product. This parallel-behavior aspect of the functional hierarchy of safeguarding systems was not mentioned by the authors in their original work. Figure 4.9(A) shows the serial nature of the chain-saw safeguarding system as initially discussed in this section while Figure 4.9(B) illustrates the parallel nature of some safeguarding-device functionality just mentioned above.

4.6 Dependency

Many products rarely operate in an isolated environment completely by themselves. It is even rarer for consumer products to do so. Therefore, it is likely that people will be nearby when consumer products are in use. Many products have controls that are implemented in hardware, software, or a combination of both. It is through these controls that humans interact with, to operate, products. It is through regular interaction with products that users become familiar and comfortable with the products and their controls.

#	Type of Safeguard Misuse
1	Misuse as a control system
2	Misuse in kind
3	Misuse in magnitude

Table 4.5 Misuses Due to User Dependency

Many of these same products present hazards, with associated risks, to users. Due to these risks, many products employ safeguarding, through devices and systems of devices, to protect the user. The safeguards employed are sometimes transparent to the user and may be taken for granted in some cases. In other instances, these safeguards may even become a "crutch" for the user of the product. This can lead to the misuse of the product and, ironically, compromise its safety for the user.

This dependency has been summarized by the statement *"Every safety system gives rise to a statistically significant pattern of user dependence"* (Barnett, Litwin, and Barroso 1983). Put more succinctly within the same work, the authors say that *"[u]ser dependence on safety systems is foreseeable."* These quotes point to a pattern of misuse demonstrated by the users of many kinds of products. Although these products and controls were not intended for use in the emergent manner, such product users have come to depend upon the characteristics of the safeguards engineered into the products. The three types of product misuse due to user dependency listed by the authors are shown in Table 4.5.

The *first* type of product-safeguard misuse is that where a product's safety feature or characteristic is used as a *control system*. Earlier in this chapter, the example of the clothes-dryer door being used as an OFF/ON switch—or a control system—was provided. Frequently, and in the same manner as the clothes dryer, the door on a microwave oven is used as a control system because it can be difficult to properly identify and operate the OFF control on a microwave oven which one does not use with regularity. It is the assumption of the user that the safeguards built into the microwave will turn off the oven, through one or more interlocks or sensors, whenever the door is opened. The user of the microwave oven is usually correct in their assumptions and dependency; however, this is not how the designers of microwave ovens intended for the products to be used. It is quite likely that microwave-oven designers had taken extra caution in the designs of their products to take this dependency into account and are as certain as practical that the microwave oven shuts down each time the door is opened. (In the case of clothes dryers, some dryers have a Start button, but lack a Stop button. A clothes-dryer Stop must be executed by turning the time dial to zero time. It truly is much easier for the user to simply open the dryer door to Stop the dryer. In this case of dryer design, such dependent user behavior could even be expected by the design engineers.)

After walking into an elevator, one may see another person hurry to get into the same elevator car even as the elevator door starts to close. Out of instinct and courtesy, one may extend an arm into the path of the elevator door to prevent its closure. That courteous person is depending upon the force-sensing element of the elevator door, or the light-beam sensors found in some elevator doors, to detect the presence of the arm and reverse the direction of the elevator door. This works almost always. Yet, this self-reversing feature of elevator doors is a safety device to prevent people from being struck by closing elevator doors *by accident*. The correct way to stop the door from closing before the late arrival can board the car is to press the door-open button.

Operating the elevator door properly out of this courtesy may truly take some effort. Although elevator buttons and their pictograms appear standardized, the control panels on elevators may be in different places, elevator lighting may be poor, there may be no backlighting of the elevator controls, and the viewing angle of the control panel may make it difficult to quickly locate, identify, and press the door-open button. In addition, the door-open and door-close buttons are often next to one another and their pictograms look similar to one another when one is rushed to act. The well-intentioned person might just as easily press the door-close button instead of the door-open button and appear to be disrespectful of the very person she or he is trying to help if observed by others on the elevator car. It is for these reasons that many people continue to extend an arm to stop an elevator although one may sometimes encounter a force-sensing edge that does not respond to the light-force response found on other doors. Therefore, anyone using this non-recommended operating method should be cautious of such doors.[11]

The *second* type of product-safeguard misuse is that of *in kind*. An example of this type of misuse is the wearing of PPE when participating in recreational activities. One such example is in mountain biking/bicycling (MTB). MTB participants frequently wear helmets, goggles, gloves, knee and shin pads, elbow pads, chest protectors, and sturdy footwear. Sometimes, neck braces are even worn by participants. The purpose of this PPE is to protect the rider. However, one can easily see how wearing all of this protective gear may lead some riders to take on risks that they otherwise would not.

Another misuse in kind example is the tire-pressure management system (TPMS) found on automobiles today. Prior to TPMS, automobile owners needed to regularly check the air pressure in their car's tires. In fact, they still do. However, the advent of the reliable TPMS has made automobile owners dependent upon the low-air pressure pictogram illuminated in their dashboards to tell them to put more air into their tires. The reliability and durability of the modern automobile and its systems, including tires, is truly marvelous. The strides made in automobile safety, however, may have made society more dependent upon automotive safeguarding technologies than ever before. These safeguarding technologies include the TPMS just mentioned as well as anti-lock braking systems (ABS), traction-control systems, and vehicle electronic stability-control systems.

The *third* and final type of product misuse arising from user dependency is that of *in magnitude*. A common instance of misuse in magnitude is "knowing" that the engineers have designed a factor of safety (FoS) into the product being used. Thus, if a ladder has a rated capacity of 225 lbf (102 kg), then one "knows" that the ladder will safely support at least 10 or 20 percent more than that rating—and probably more—before failing. Generally, this is a valid assumption. Yet, the amount of margin built into products is likely not uniform. Any particular product unit may also have been weakened in some way over its lifetime of use, especially a ladder which is frequently handled roughly and exposed to the elements. This treatment over time might have degraded the structural strength of the product thereby cutting into the presumed design margin upon which a user is depending.

Another example is that of an automobile driver being aware that roadway turns are frequently designed to accommodate vehicles traveling faster than the posted speed limit. This knowledge may lead a driver to enter a turn at a supra-legal speed and depend

[11] The author confesses to using the extended-arm method to prevent elevator-door closure but remains prepared to quickly pull back his arm if there is any sign of the door not opening.

upon the highway turn's super-elevation (banking) or run-off space (shoulder) to provide the safety margin upon which the driver is depending. In most cases, the highway-design engineer has been generous and has provided the margin necessary for the dependent behavior demonstrated by the driver above. It may not always be so, however.

Barnett, Litwin, and Barroso posit that there is nothing intrinsically wrong with these dependent behaviors. They do say that these "secondary effects" exist and that designers should consider these effects that can adversely impact product-safety levels. This is one of many considerations that should be taken into account during the engineering-design process.

The field of user error is too broad and deep to cover in this book—at least in the present edition—so the author will not venture too deeply into it. Readers interested in studying this important topic are encouraged to do so. The following references should be useful for those starting an exploration into the field including basic concepts (Barnett and Switalski 1988), human behavior (Embrey 2005; Swain and Guttman 1983), and learning and communication (Rasmussen 1983; Woods 2009).

4.7 Uniformity

When an engineer "delivers" a product to a consumer, there are generally social agreements that are made between the parties involved. In other words, the engineer and her/his employer are providing a product at a fair price with reasonable levels of safety to a consumer. The consumer presumes that this product will perform adequately when it is used as indicated.

Throughout the engineering design process, the engineers have taken steps to countermeasure many hazards and risks contained in the product's final-design configuration. These hazards may include mechanical, thermal, mass, and laceration. Both burns and cuts can range from negligible to severe—and even to fatal. If engineers have taken steps to prevent minor-cut injuries to consumers, it is not unreasonable for those consumers to presume that the engineers have taken such steps to safeguard the consumers from all hazards that are *equal to or more severe* than those minor cuts. For example, if precautions have been taken to prevent minor burns from one product feature to users, it is natural for consumers to assume that any other minor-burn hazards—let alone any major-burn hazards—have also been protected against. This "Principle of Uniform Safety" has been observed (Barnett 1994) and has been stated as: *Similarly perceived dangers should be uniformly treated.* The word "risk" may be substituted for the word "danger" in this quote for the purposes of this section.

One example used by Barnett in this work is the machine with n hinged maintenance-access doors. These doors also serve as barriers (guards) for machine operators. Personnel must gain access to the machine through these doors in order to perform inspections and preventive maintenance (PM). Assume that the machinery and the operation may be started and stopped without loss, except for time, so that these maintenance personnel can perform their duties throughout the work shift. Also assume that on a particular day, a new maintenance person starts work on the machine and observes that each door (so far) has been interlocked, i.e., it has a switch which senses when a barrier-guard door has been opened, sends a signal to the machine control system to shut down the machine, and the machine promptly shuts down. An open interlocked door should also prevent the re-start of the machine before the door has been closed. This new person has, after observing interlocks on $(n - 1)$ doors, now starts misusing the interlocked door as a control system. Consequently, she or he has resorted to—or has become dependent upon—opening the interlocked door, having the machine come to an immediate halt, and then reaching into the machine to perform service.

Is it *reasonable* to her/him to assume that the *n*th door is so interlocked and immediately reach into the machine after opening that door? Is it *safe* for her/him to presume this? The answers to these two questions are probably yes and no, respectively. The current design of the machine—which may not be how the machine was initially designed and delivered—encourages personnel to assume that all of the access doors are interlocked. If many, but not all, of the doors are interlocked, it may well be safer to not interlock any doors at all. The absence of interlocks would prevent the *inductive* reasoning, and its resulting behavior, that may lead to a decreased level of vigilance of product users, in this case maintenance personnel.

Inductive reasoning is based upon observations only made thus far by the observer and may, consequently, be wrong. If every frog that a person has seen in life has been green, then she or he might conclude that all frogs are green. This is, of course, incorrect. Frogs come in many different colors. Just because something has not yet happened does not mean that it cannot happen. Design engineers should, when possible, work to prevent consumer behaviors based on faulty inductive reasoning from using their product designs.[12] Safeguards should be considered by engineers throughout products and systems to provide a *uniform* level of safety for consumers from product hazards thereby preventing consumer expectations that will go unfulfilled and may lead to harm. Although this may not always be possible, it remains a worthy design goal.

4.8 Decreased Vigilance

Many years ago, perhaps up to the early 1960s, product safety was not the high-profile topic that it sometimes is today. Changes in consumer awareness, advocacy, and regulation have brought forward many effective and necessary product improvements. Industrial workers decades ago, for example, frequently had little or no guarding on the heavy metal-working machines they operated. Instead of relying upon guarding, these workers by necessity relied upon their own diligence and vigilance to avoid personal injury.

As mentioned in Section 4.6, user expectations may also lead to decreases in the attention paid by consumers to the use of their products. In addition to this reason, the distractions of daily life in the days of the Internet of Things (IoT), especially mobile telephones and other forms of communication have seemingly lowered the attention levels of some people doing certain tasks, for example, texting while driving an automobile despite ubiquitous laws to the contrary.

With the advent of safeguarding devices and technologies, personal vigilance has declined according to at least one work (Barnett, Litwin, and Barroso 1983). These authors write about the dependency-driven changes in user behavior leading to the misuse of products as mentioned in Section 4.6. In their later work (Barnett, Litwin, and Barroso 1984), they opine that some current groups of users of hazardous products have exchanged their earlier vigilance for reliance upon modern safeguard devices.

[12] Before moving on, the reader is offered an example of *deductive* reasoning, which is the opposite of the inductive reasoning just discussed. Deduction is based on truths rather than on observations. As a result, deductions are always correct. If all mammals are vertebrates, and all dogs are mammals, then it correctly follows that all dogs are vertebrates. This induction and deduction dichotomy will be used elsewhere in the book.

One could also argue that today's crop of automobiles with many electronic accident-avoidance technologies, such as anti-lock brake systems, vehicle-stability control, adaptive cruise control, lane guidance, automatic emergency braking, and blind-spot monitoring, have led to an erosion of basic driving skills and attentiveness among motor-vehicle operators. This does not even include the rise in semi-autonomous and fully autonomous vehicles. Few would argue that the positives in automobile safety do not outweigh the negatives, but some drivers may come to depend upon these technologies as "crutches" and drive more aggressively even when roadway conditions have deteriorated alarmingly. This behavior could be classified as a misuse in kind from the Dependency section of this chapter.

It may be necessary at certain times and with certain products to take such user dependencies into consideration during the PDP. It may not always be possible to design countermeasures to some adaptive user behaviors, especially when a product feature or characteristic works well and has always proved reliable for the user in the past. When practical, however, design engineers should be vigilant of user dependency and, perhaps, consider the fault-adverse and fault-tolerant design approaches presented in Chapter 10 in order to not surprise product users if, or when, product fault or failure might disable a product characteristic upon which a user may be depending. User surprises should be avoided to the extent practicable.

4.9 Compatibility

At times, consumers may bypass, or circumvent, the safeguards which engineers put in or on products to provide protection from hazards. Sometimes this consumer behavior is quite unpredictable and unanticipated. At other times, such behavior could have been expected.

To address this phenomenon, the "Compatibility Hypothesis" was constructed (Barnett and Switalski 1988). It is stated as: *The larger the perceived improvement in utility compared to the perceived risk, the greater will be the motivation to circumvent a [product's] safeguarding system.* So, simply put, the more benefit there is to defeating a protective device, the more likely it is to be circumvented by users. A list of reasons that a consumer may want to bypass a safety device on a product is provided in Table 4.6. The author has made minor changes to the original authors' list.

Examples of such circumventions are numerous. Older readers may remember that, in the early 1970s, automobiles were required to have buzzers that sounded when the driver's seat belt was not fastened. These buzzers would remain on for as long as the car was operated with an unfastened seat belt. Many more people then drove without their seat belts than today and the buzzing became intolerable to them. The solution: buckle the seat belt before getting into the car and then simply sit on top of it while driving. Problem solved!

It is unclear just how predictable this response was to designers and regulators of the day, but this example should be remembered today whenever proposing an onerous burden on consumers in the name of safety.[13] Engineers, regulators, consumer advocates, and others must all remember that consumers are quite clever and adaptable.

[13] Several years ago, when an SSV manufacturer used such a scheme to limit the speed of its vehicles, the aftermarket responded with an inexpensive device attaching to the seat-belt buckle that "fooled" the vehicle speed-control system into believing that the seat belt was fastened when it was not.

#	Reasons
1	To increase [fill in the blank]
2	For ease of access/maintenance
3	Comfort
4	Worn or damaged
5	Aesthetics
6	Not reinstalled after removal for maintenance/repair
7	Machismo / Bravado / Hubris
8	Horseplay
9	Not originally equipped with it
10	Lost during use
11	Cannibalized to repair other products
12	Had undesirable side effects (types IV & V)
13	Removed to improve safety (types VI & VII)
14	Incompatible with other safety systems
15	Sabotage

Table 4.6 Circumvention of Product Safeguards

Consumers are intelligent creatures who respond in a *closed-loop* fashion wherein they adapt—change their behaviors—in order to attain their goals. Some designs and regulations appear to see consumers as *open-loop* creatures who will not modify their actions when their environments change. This latter, open-loop approach to consumer behavior is unrealistic and destined to failure. (The reader wanting to know more about closed-loop and open-loop behavior should consult a source on feedback-control systems.)

The first entry in Table 4.6 can be virtually any reason suiting the consumer. A consumer may want to lighten the product (decrease the weight), increase the speed, or increase the output. The blank may be filled with nearly anything.

Some of the entries in Table 4.6 apply more to industrial equipment than to typical consumer products, in general. The reader should go through each and consider examples, but not every entry will be discussed here. The third entry, comfort, is a vital consideration for any product. If a safeguard is particularly uncomfortable, users will not employ it at all times. If a seat belt, for example, is incorrectly attached to a vehicle, a user—even when convinced of the virtue of a seat belt—may not always wear it if it is too uncomfortable. This is an instance where human factors and anthropometry (Bailey 1996; Bridger 2018; Kuczmarski et al. 2002; Sanders and McCormick 1993) are important aspects to a good design.

Engineering designers—and product regulators—must remember that product users are intelligent and adaptive creatures. They can often find "workarounds" to product-safety measures that they find inconvenient or counterproductive to their purposes. It is important to understand these user purposes as well as the product under consideration and its environment. Only then can truly effective ways to increase product-safety levels for users become realizable. Trial attorneys and consumer advocates should also keep this in mind when striving to improve the safety of consumer products through edict.

4.10 User Activity and Safety Risk

Most of society, along with ethical engineers, wants a product to possess the greatest level of safety that is practical. It is proper that consumers demand safe products just as ethical engineers should want to design and manufacture them. One thing that can happen, however and unfortunately, is that some people will expect levels of safety in one type of product to be the same as a level of safety in another type of product. This is neither always reasonable nor even possible. For example, some consumer may want the same level of safety from a table saw as that demonstrated by a two-slice kitchen toaster. This expectation is unlikely to ever be met in real life, however. These two products are significantly different in several ways.

The toaster, as seen in Figure 4.10, is a product generally used in a highly controlled environment and serves a solitary purpose: to toast bread products. The toaster is also totally enclosed with respect to the electrical hazard except for the opening required to insert the bread, or the workpiece. Regarding the thermal hazard, prepared (sliced) bread (or pastries) are loaded through the top of the toaster without need for further manipulation until extraction of the toasted bread; non-conducting tongs or agile finger may be used to remove the hot items. Although bread can vary in shape, size, and thickness, the toaster is able to handle various slices of bread. The housing of the toaster may sometimes get hot but there is no need to hold the toaster stationary during use, many toaster housings are well insulated, and most adults are well aware of the burn hazard from earlier experiences in life. So long as the toaster owner's residence meets building and electrical codes and the toaster is used in a dry setting, the toaster is a relatively safe product. The most severe injury likely to be seen is a mild burn on a finger.

FIGURE 4.10 Two-slice toaster.

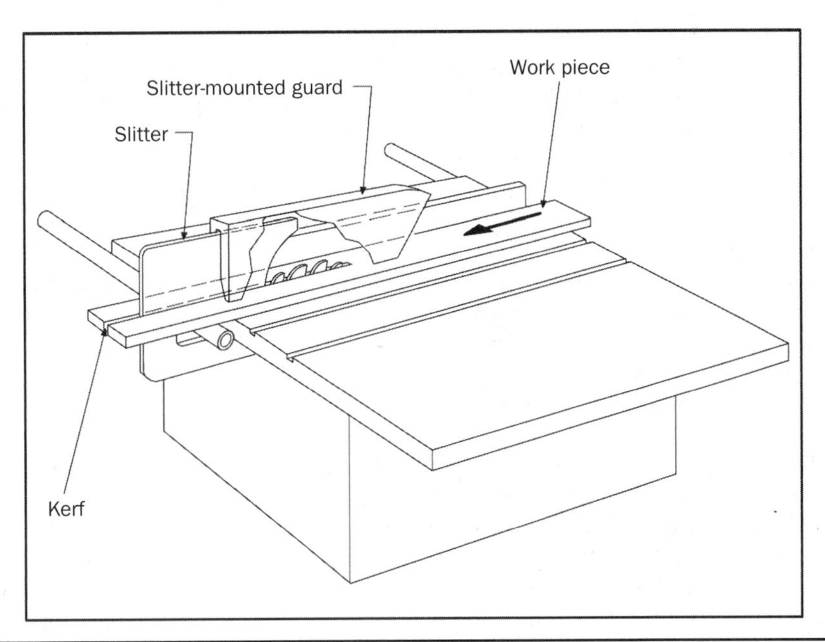

FIGURE 4.11 Table saw. (Used by permission of Triodyne Inc.)

The table saw shown in Figure 4.11 does not have the heat hazard, but does have the electrical hazard that will be controlled by proper design and an adequate residential electrical system. In addition to these hazards, the primary hazard of the table saw is the rotating, semi-exposed blade. Table saws are sold and fitted with an adjustable blade guard, yet some users choose to remove the guard and it remains possible to injure yourself even with the guard still in place. (There are other severe hazards to using a table saw, including workpiece "kick-back," but the focus here will be upon the laceration hazard.) As with the toaster, the table saw's hazard is not completely enclosed but, unlike the toaster, the table-saw operator must manually feed the workpiece (assumed to be wood) into the table-saw point of operation during use. The wood may be large or small, thin or thick, light or heavy, and short or long. The blade, table, and guard must be able to handle these variations. The table-saw operator must also move her or his hands and body in order to handle workpieces. The balance of the operator may also be affected through these movements. When feeding in the wood, operators must move their hands *toward* the blade in many cases. There are push sticks available for in-feeding work, but they are not always used and may not be compatible with quickly completing the desired woodcutting operation.

The table saw is being used in a less-controlled setting and requires greater user interaction with both the product and the workpiece when contrasted to the toaster. Operators must also constantly hold and feed in a workpiece to the hazard. A typical severe injury from table-saw use is the laceration, or even amputation, of fingers.[14]

[14] According to a conversation with U.S. CPSC staff in 2019, there has only been one fatality from table-saw use reported to the CPSC in the United States.

From the above comparison, it appears that the two products differ in significant ways. One way to compare different products is to compare the "user activity" of the product to the risk of using that product. Figure 4.12 shows how user activity and safety risk of particular products may be positioned in a two-dimensional plot. Along the horizontal axis is the degree of user activity of the product. A product requiring little user interaction and judgment will appear at, or near, zero. A product requiring, or inviting, much user decision making and skill will appear at, or near, one or unity.

The vertical axis denotes the amount of product risk which ranges from zero to ∞ (infinity). In a practical sense, these risks would correspond to a simple nuisance to one or more permanent disabilities or deaths, respectively. In between these two extremes, would be cuts, burns, impacts, temporary disabilities, and amputations.

There are no specific numbers assigned to either axis of the safety risk versus user activity plot. This graphic is a conceptual tool only. Yet, this tool remains useful. Through use of this plot, it is possible to contrast two or more products by placing them in User Activity-Safety Risk "space." After doing so, it is logical to ask if it is feasible—or even possible—to provide the same levels of safety for two products located in completely different areas of the plot. Looking at Figure 4.12 and using the toaster as a reference point, the table saw is further to the right and above it. This indicates that the table saw requires more user input—both decision and action—from the user and poses a greater safety risk than the toaster. What this means to the design engineer is that there are significant differences and design challenges to providing toaster-level safety to table-saw users. These challenges may prove insurmountable in practice regardless of acceptable cost to the consumer. What this means is that all parties involved in product safety should also acknowledge these product-characteristic differences and adjust their demands accordingly.

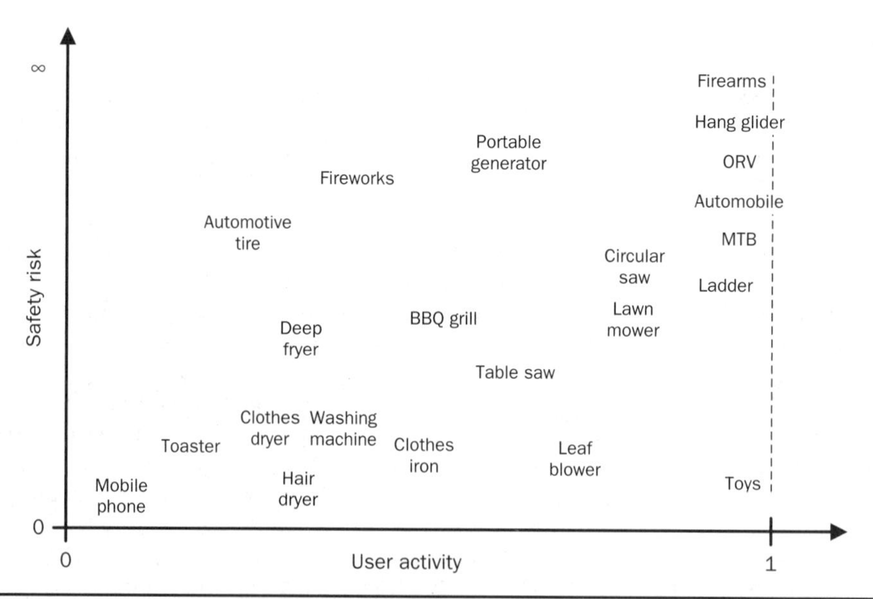

FIGURE 4.12 Safety risk versus user-activity plot of selected products.

From this figure, it would be foolish to expect toaster safety levels from a product such as an MTB. Further complications of direct comparisons between products include:

- Motivation of the user—simple operation versus thrill-seeking behavior
- Controlled versus uncontrolled environments—level versus inclined surfaces, dry versus wet, etc.
- Variable environments—weather, temperature, lighting, etc.
- One person's actions affecting another person's safety—vulnerabilities of bystanders or passengers
- User awareness of hazard, risk, and possible injury
- Amount of user education and skill required or recommended
- Enforced licensure for use
- Avoidability or rate at which a potential injury takes place—slowly (permitting user intervention) or quickly (precluding human perception and reaction)
- Compounding factors—alcohol, drugs, and the social enabling of poor decisions

Readers should intuitively understand that the use of effective safeguards on a product in the Figure 4.12 plot will move that now-safeguarded product to a lower position on the same plot. For example, an effective table-saw blade guard or user-protective system would move the table saw downward on the plot by an amount dependent upon that safeguard's effectiveness. This could prove useful to designers for explaining and demonstrating incremental product-safety improvements made during the engineering-design process.

4.11 User Task Load

This section will further complicate the life of the design engineer. Its purpose is to be certain that apples are always compared to apples and not to oranges. Section 4.10 dealt with the degree of user activity needed for the proper use of a product. It was this user-activity aspect that requires decision making and proper execution of those decisions by the user in order to get the desired performance out of a product safely. Here, the conditions under which those decisions must be made and carried out are evaluated. This aspect of user decision making and ability to respond accordingly under given conditions is known as task loading.

An example of when task loading is relevant is comparing the operation of two different types of motor vehicles: the automobile and the motorcycle. Each is a motorized vehicle which transports one or more people from point A to point B. The operators of these two vehicle types, however, perform their operational tasks on widely differing vehicles and in much different environments. For example, motorcyclists operate their vehicles while wearing PPE (helmets), in the windblast, and largely unprotected from weather conditions and debris while automobile operators function within a well-protected and controlled environment. Some of these use-environment differences affect the user's ability to perform tasks. Other differences may affect how long a vehicle operator may be able to effectively or safely perform those tasks.

In the particular case of motorcycling, years ago, the Motorcycle Safety Foundation (MSF) (McKnight and Heywood 1974) published an extensive report investigating the

#	Sources of Load	Scale Range	Descriptions
1	**Mental Demand**	Low/High	• How much mental and perceptual activity was required (e.g., thinking, deciding, calculating, remembering, looking, searching, etc.)? • Was the task easy or demanding, simple or complex, exacting or forgiving?
2	**Physical Demand**	Low/High	• How much physical activity was required (e.g., pushing, pulling, turning, controlling, activating, etc.)? • Was the task easy or demanding, slow or brisk, slack or strenuous, restful or laborious?
3	**Temporal Demand**	Low/High	• How much time pressure did you feel due to the rate or pace at which the tasks or task elements occurred? • Was the pace slow and leisurely or rapid and frantic?
4	**Own Performance**	Good/Poor	• How successful do you think you were in accomplishing the goals of the task set by the experimenter (or yourself)? • How satisfied were you with your performance in accomplishing these goals?
5	**Effort**	Low/High	• How hard did you have to work (mentally and physically) to accomplish your level of performance?
6	**Frustration Level**	Low/High	• How insecure, discouraged, irritated, stressed, and annoyed versus secure, gratified, content, relaxed, and complacent did you feel during the task?

Table 4.7 NASA TLX Sources of Load, Scales, and Descriptions

behaviors, knowledge, and skills required to safely operate a motorcycle. This was an in-depth task-load study which broke motorcycle operation into many aspects which will not be covered here. Perceptive readers could readily see the clear differences between the operations of passenger cars and motorcycles with respect to driver and rider demands, respectively. Fortunately, there are simpler ways to assess the workload placed on people performing tasks. The NASA Task Load Index (TLX) is one such tool. However, those involved with designing complex products in widely ranging use environments are encouraged to study the MSF work.

The NASA TLX (Hart 1986) is described as "a multi-dimensional rating procedure that provides an overall workload score based on a weighted average of ratings on six subscales: Mental Demands, Physical Demands, Temporal Demands, Own Performance, Effort, and Frustration." Table 4.7 contains the definitions of these TLX rating scales. The TLX may be completed in paper form, but NASA also offers the TLX evaluation process through a smartphone application.[15] The development of the

[15] https://humansystems.arc.nasa.gov/groups/tlx/

A. Toaster Operation				
#	Source of Load	Tally	Magnitude of Load	Product
1	Mental Demand (MD)	1	30	30
2	Physical Demand (PD)	1	15	15
3	Temporal Demand (TD)	3	70	210
4	Performance (P)	3	40	120
5	Effort (E)	5	40	200
6	Frustration (F)	2	30	60
	Check and Sum:	*15*		*635*
	Sum/15:			*42.3*
B. Lawn-Mower Operation				
#	Source of Load	Tally	Magnitude of Load	Product
1	Mental Demand (MD)	2	80	160
2	Physical Demand (PD)	2	95	190
3	Temporal Demand (TD)	1	60	60
4	Performance (P)	3	80	240
5	Effort (E)	5	85	425
6	Frustration (F)	2	60	120
	Check and Sum:	*15*		*1195*
	Sum/15:			*79.7*

TABLE 4.8 Example TLX Results

TLX was described by Hart and Staveland (Hart and Staveland 1988). Years later, the TLX method was revisited by Hart (Hart 2006).

The author administered the NASA TLX to students in one of his product-safety engineering classes. Two tasks were compared for user/operator task loading: using a toaster and using a lawn mower. Each of these examples was discussed earlier in this chapter. This administration was not performed as is recommended by the TLX's developers because it was not done in a controlled experimental setting. It was given as part of an outside-of-classroom homework assignment, but this portion of the homework was submitted anonymously. In addition, the group was not necessarily representative of toaster and lawn-mower users in either intent or prior experience.[16] Therefore, the combined results will not be presented. However, one arbitrary, yet representative, sample response set is analyzed per the TLX method.

The results of the responses from one unidentifiable student for the task loads of using a toaster and a lawn mower were recorded and processed by the TLX procedure in the original reference. The author does not wish to reproduce this procedure which is easily accessible to those with the interest from Internet material. The processed responses are included in Table 4.8. These processed results are then plotted in Figure 4.13.

[16] The author presumed that all students had prior experience with both products which, it turned out, appears to be reasonably valid.

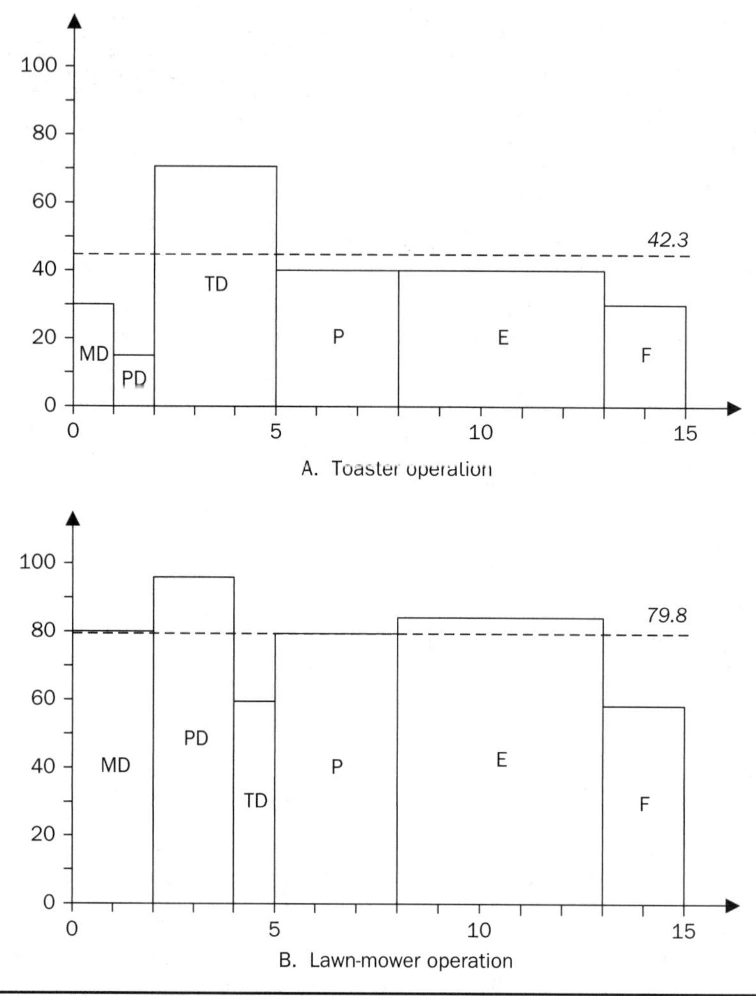

FIGURE 4.13 Graphical representations of example TLX results.

The figure shows the amount of user loading for each source of workload through the *area* of its corresponding box. The horizontal width of a box indicates how many times a source was mentioned as dominant in pair-wise comparisons of sources. The height of a box indicates the effort level attributable to the workload source.

For each task, the overall workload, a weighted average of the boxes, is shown by a dashed horizontal line with its numeric value to the right-hand side of the line. As could be expected, the user workload for mowing a lawn is larger than that for making toast. Lawn mowing was perceived to be 79.9 out of 100 in load magnitude, while toasting bread load-magnitude level was 42.3 out of 100. Thus, mowing the lawn is almost twice as hard, from a user task-loading perspective, as making toast. This is not a surprising result. However, it does confirm the methodology, at least in a coarse sense, despite the lack of experimental controls admitted by the author.

Results of a user task-loading analysis, such as TLX, can be used by engineers and others to study impact of the task loading for users along with the user-activity level of a product during the PDP. The safety of products with low user activity may be less susceptible to the task loading of users than highly user-active products with high levels of user task loading. In the latter case, it could be beneficial to either lower the level of user activity or to lower user task loading. This may not always be possible, however.

4.12 User Qualification and At-Risk Populations

This section of the chapter will discuss user populations which may need to be consciously considered during the PDP—perhaps as early as Phase I or II—for some products. Not all potential product users have the same skills and abilities and not all potential users are equally resilient to injury from product-hazard risks. The treatment of the two populations—which may overlap—is not intended to be complete. This coverage, however, should still provide design and product-safety engineers with sufficient insight to recognize either when *user qualification* or when risk-estimate levels should perhaps be adjusted for *at-risk* populations.

The *first* of the two product-user combinations is user qualification. This is the recommendation that only certain people use a product. This concept is not new, at least, for some products. For example, automobiles are designed only for use by licensed drivers. Therefore, licensure of drivers is a necessary *qualification* for users of motor vehicles. Through the licensing of drivers, at the state level in the United States, several other qualifications are met. In order to get a driver's license that person must be at least X years old; this age varies by state. In addition to eliminating legal motor-vehicle operators below a given age, driver's licensure requires that students demonstrate a certain level of knowledge about safe vehicle operation through a written test and possess a minimum amount of vehicle-operation skill through a driving test.

Some other products have age restrictions on them. Adult ATVs are not intended to be operated by those under 16 years of age; special youth ATVs have their own age restrictions depending upon the ATV type ("ANSI/SVIA 1-2017: American National Standard for Four-Wheel All-Terrain Vehicles" 2017). Other products such as tobacco and video games[17] have their own age restrictions placed upon them. Although there is enforcement of these age restrictions, that enforcement is far from absolute. Youths still acquire some products which were not designed for them. This mismatch of user and product may depend on such characteristics as cognitive ability, physical strength, physical stature, and reflexes. Some members of society feel that such products should only be used by those people old enough to consciously accept all of the risks of using such products while others feel that some such products should be used by no one.

The *second* product-user combination is the at-risk population or, more specifically, those people at greater risk due to injury from product use. The reasons for which one may be classified as "at risk" include lesser or diminished cognitive and physical abilities. Cognitive ability can be affected by age and illness. Physical ability includes strength, coordination, and sensation.

Readers will likely already see that some users falling out of the population of qualified users will fall into an at-risk population. Young and very-young children will not

[17] https://www.esrb.org/ratings/

have the cognitive and reasoning abilities needed to make decisions when severe injury may be the result. They may also lack physical strength and stature to perform some tasks. The elderly can also be cognitively affected through diseases such as dementia even when they may still possess sufficient physical strength. Some people lack the ability to rapidly recover from what would be a mild, temporary injury to many others. This is an additional source of injury risk. Other potential product users may be insensate due to accident or illness. If a person has lost sensation in her/his lower body, then simple products such as heating pads and seat warmers have the potential to injure that person without her/his being aware of it. These are only some of the considerations of which design engineers must sometimes be aware.

There may also be specific *environments* that give rise to at-risk populations. For example, a fully abled person who is crossing a road could become a member of an at-risk population with respect to road use. In fact, there is an at-risk group called Vulnerable Road Users (VRUs) (Constant and Lagarde 2010). VRUs include pedestrians, bicyclists, and motorcyclists. Therefore, not only must the condition and ability of a person be considered when looking at product-injury risk, but the environment may also play a significant role in escalating the vulnerability of a group of people. Both the physical and socio-economic environments may be worthy of consideration when estimating risk. In the pedestrian example, a designated crosswalk with lighting to warn approaching drivers of a crosswalk could lower risk. A driver approaching this crosswalk could be distracted by a text message from a friend and not see the pedestrian despite the flashing lights at the crosswalk.

The above shows the vital role that the PUE model from Figure 1.3 can play in helping provide a broad and thorough analysis of product-safety issues. There are no universally applicable solutions for all product-safety problems. Each problem should be looked at in its entirety in order to design a product appropriate for its task, users, and environment. Such consideration may cause the design or product-safety engineer to modify a traditional risk estimate if an at-risk person could be affected to a greater degree or with greater frequency than the typical user.

4.13 Haddon Matrix

William Haddon, mentioned earlier in this chapter, became the first Administrator of the newly formed National Highway Traffic Safety Administration (NHTSA) at the U.S. Department of Transportation in 1966 (*Washington Post* 2002). Due to his epidemiological background, Haddon approached his new task with his existing perspectives: scientifically analyze accidents and see if they could be prevented and, if not, how injuries could be prevented or reduced. Although this seems to be the only rational way to approach this problem today, Haddon's approach was rather novel at a time when many people thought that accidents were still simply "acts of God" and, therefore, unavoidable.

Haddon broke down the timing of an accident into three phases:

1. Pre-accident
2. During the accident
3. Post-accident

His goal was to identify when certain countermeasures to an accident could be applied to the system that would be involved in an accident. As an example, some actions would be constructive prior to an accident while other actions would only be useful after an accident.

Haddon then broke the factors contributing to the accident into four types of factors:

1. The human
2. The product
3. The environment
   a. Physical environment
   b. Socio-economic environment

The human is sometimes called the host, user, or operator. The product is sometimes called the vehicle or agent of the hazard. The particular nomenclature depends upon the application. This fits into the PUE model shown earlier as Figure 1.3 and demonstrates the interactions between the user, the product, and the environment.

Haddon understood that there were contributing causes to any public-health epidemic. These factors could include such things as poor personal hygiene, a contaminated water supply, rodent infestation, overcrowded living conditions, and poverty. Some of these factors, or conditions leading to an epidemic, were human in nature (poor personal hygiene), some were "agents" of the hazard (contaminated water and rodents), some were related to the physical environment (overcrowded living conditions), and finally some were related to socio-economic conditions (poverty).

Furthermore, Haddon understood that a public-health epidemic called for different actions by public authorities at different phases of an epidemic. Of course, the best way to address an epidemic is to prevent it altogether. Therefore, pre-event actions could include educating the public on the importance and effectiveness of good personal hygiene practices. Once an epidemic is underway, it is important to identify the cause and to isolate the cause and, perhaps, the affected population by quarantine. Following the outbreak and control of the epidemic, medical aid should be provided, the causal factors remedied—the water source cleaned and the rats exterminated. The socio-economic factor, in this case poverty, may prove quite difficult to address—both years ago and today.

Since this epidemic model was by its nature two-dimensional, it was logical to construct a matrix of these elements. This became known as the Haddon matrix and is shown in Table 4.9. The four factors are listed across the top of the table and the three phases of the accident are listed vertically on the left-hand side of the matrix.[18]

If a sample Haddon matrix was completed for the epidemic discussed above, it might look something like Table 4.10.[19]

This Haddon matrix contains examples of what might be done before, during, and after an epidemic for the four different factors. In this case the factors are humans, agent, physical environment, and socio-economic environment. For some accidents, it can be

[18] The Haddon matrix is sometimes shown with only three factors across the top when the two environments are combined into a single environment. Although this combination of environments makes the Haddon matrix easier to teach, the separation of the two environmental factors makes the matrix easier to use.

[19] It is unlikely that two different people or groups would construct *identical* Haddon matrices or other analyses discussed in this book. However, the primary points should be similar to one another.

Phase		Factors			
		User	Product	Physical Environment	Socio-Economic Environment
1	Pre-Accident				
2	During Accident				
3	Post-Accident				

TABLE 4.9 The Haddon Matrix

Phase		Factors			
		Human	Agent	Physical Environment	Socio-Economic Environment
1	Pre-Accident	• Proper personal hygiene	• Supply clean water • Control rodent population	• Provide suitable living conditions	• Widespread employment • Stable economy
2	During Accident	• Provide medical care	• Eliminate contaminated water source • Trap rodents	• Isolate the afflicted/ infected	• Allay fears of the citizenry through communication
3	Post-Accident	• Nurse back to health those physically affected	• Ensure clean water • Reduce rodent population	• Improve physical living conditions	• Improve economic living conditions

TABLE 4.10 The Haddon Matrix for the Example Epidemic

difficult to fill in all of the boxes with meaningful responses. The lack of an appropriate response in the Haddon matrix element box indicates that there is a missing element, or "gap," in providing complete coverage of this event by the parties involved. There may also be dearth of useful countermeasures in either an entire row or an entire column of the Haddon matrix. This, too, would indicate that a larger, systemic gap may exist. Engineers may, at this point, object to the use of such a construct since they have little direct control over the physical environment and no direct control over the physical environment. This is true. That is why the Haddon matrix is a tool often used by a coalition of technical people and policy makers. This tool provides a "high-level" view of a situation to encourage cooperation of various portions of organizations and society. Some seemingly technical conditions, such as highway safety, do in fact rise to the level of public-health—and public-safety—crises. If some readers see the application of this tool as being social engineering and object accordingly, perhaps it is in some cases. Ultimately, society responds through lobbying, elections, established media, social media, and other forms of expression. It is up to society as a whole to determine its values and subsequent destiny. However, engineers should not shy away from being a part of such important discussions.

It was the epidemiological approach that led to the Haddon matrix. This matrix and its findings have served the motoring public well for decades. There were, in 1966, no federal regulations on automobile safety either in design or in performance. Indeed, it was in the early 1960s that a consumer advocate named Ralph Nader came to public prominence following the release of his eye-opening book *Unsafe at Any Speed* (Nader 1965) discussed in Chapter 2. Although the book is primarily remembered for its condemnation of the Chevrolet Corvair, the book and its author also lamented and documented the dismal state of general automobile-occupant safety at the time. One of the new NHTSA's tasks was to create national safety requirements for automobiles. The resulting early work focused on the sole factor of the product—or vehicle or agent. The other factors that could contribute to highway safety were not within NHTSA's control. For example, responsibilities for driver licensure and traffic enforcement exist with state and local authorities. Roadway design and construction is performed at primarily the state level with guidance from national groups, for example for highway design and test protocols ("AASHTO" n.d.).

NHTSA personnel considered the three phases of an accident and investigated how vehicle characteristics for each phase could improve highway-user safety. They considered which vehicle characteristics could help in which phases of an accident. A small sample of the type of findings made by NHTSA is shown in Table 4.11. This table shows

Phase		Factors			
		User	Product	Physical Environment	Socio-Economic Environment
1	**Pre-Accident** (FMVSS 1XX)		104: Windshield Wipers 105: Brakes 108: Lighting 111: Rearview Mirrors 126: Electronic Stability Control		
2	**During Accident** (FMVSS 2XX)		202a: Head Restraints 205: Glazing Materials 206: Door Locks 208: Occupant Crash Protection 214: Side-Impact Protection 216a: Roof-Crush Protection		
3	**Post-Accident** (FMVSS 3XX)		301: Fuel-System Integrity 302: Flammability of Interior Materials 305: Electric Vehicles: Electrolyte Spillage and Electric Shock		

TABLE 4.11 The Haddon Matrix with Sample Parts of the Federal Motor-Vehicle Safety Standards (FMVSS)

that requirements for such vehicle components such as windshield wipers, brakes, lighting, tires, and mirrors can help prevent accidents from taking place. When this is not possible, vehicle components and characteristics such as head restraints, glazing materials, door locks, occupant crash protection, and roof-crush resistance will help prevent or reduce the severity of occupant injuries. After an accident has taken place and there are survivors in the car, vehicle characteristics such as a robust fuel system to prevent an exterior fire, low-flammability materials to prevent the propagation of an interior fire, and the containment of spilled battery acid (electrolyte) with electric vehicles will provide further protection.

This breakdown of vehicle performance characteristics by accident phase formed the basis of NHTSA's Federal Motor Vehicle Safety Standards (FMVSS) (GPO 2018) which all light duty vehicles sold in the United States must meet with some exceptions. Those standards for the pre-accident phase are known as accident-avoidance standards. Those for the during-accident phase are known as crashworthiness standards. Finally, those for the post-accident phase are known as vehicle-integrity standards. The standards related to the pre-accident, during-accident, and post-accident phases are numbered as 1XX, 2XX, and 3XX, respectively. They are sometimes respectively referred to as the 100 Series, 200 Series, and 300 Series of the FMVSS as well.

Over the decades, individual vehicle-system requirements have been added to the original set of FMVSS. As technology became accessible, specifications for such attributes as anti-lock braking, airbags, and blind-spot monitoring have either been added or will be phased in over time. Although unknown to many, Haddon's epidemiological approach and application of that methodology through his matrix have no doubt save the lives of many people since they were put into practice.

The following example, Example 4.3, is offered as an example of the utility of the Haddon Matrix.

Example 4.3: On-Road Bicycling

Problem: Construct a Haddon matrix for the on-road operation of a bicycle during daylight hours. Nighttime operation need not be considered in this example. The hazard to be evaluated is that of an automobile-and-bicycle collision. Evaluate the three accident phases and the four factors typical of the Haddon matrix.

Solution: One possible resulting Haddon matrix is shown in Table 4.12. For ease of the discussion, the four factors human, vehicle, physical environment, and socioeconomic environment have also been labeled as (A), (B), (C), and (D), respectively. Using this labeling scheme, the physical-environment factors during the accident may be referred to as group *C2*.

Countermeasures that a human can undertake before an accident, group *A1*, include driving defensively and obeying traffic laws. Eye protection to preserve rider visibility and bright clothing to possibly enhance rider conspicuity may also prove helpful in avoiding an accident. Furthermore, both physical and mental fitness will help the rider in traffic. Many riding routes contain hills that automobile operators take for granted; bicyclists do not. Climbing hills can be a significant physical burden on a bicyclist. Extreme physical exertion can impair a rider both physically and mentally when the point of exhaustion approaches. In addition, the consumption of alcohol and the use of drugs—legal and otherwise—may also impair the judgment and abilities of a bicyclist.

Group *A2* is the human-crashworthiness aspect of bicycling. As with motorcycling, there is a limited amount of precaution that a bicyclist can take when confronted with

	Phase	Factors			
		Human (A)	Vehicle (B)	Physical Environment (C)	Socio-Economic Environment (D)
1	Pre-Accident	• Defensive riding • Obey traffic laws • Eye protection • Conspicuous clothing • Physical fitness • Mental fitness	• Conspicuity lighting ○ Front & Rear • Preventive Maintenance ○ Tires ○ Brakes ○ Steering	• Bike paths • Bike lanes • "Protected" intersections	• Bicycle Awareness • Laws on Automobile & Bicycle interaction • Law enforcement
2	During Accident	• Helmet • Gloves	• Rounded & Soft bicycle features	• Vehicle Design (non-aggressive automobile features)	• [None]
3	Post-Accident	• Motion-aware sensor or smart-phone application • ID & Alert bracelet • "ICE" contact on mobile phone • Rear-facing video camera	• First-Aid kit in bicycle pouch	• Ambulance & EMT • Hospital ER • Mobile-phone network	• Bystander assistance: ○ Call 9-1-1 ○ First aid ○ Redirect traffic

Table 4.12 The Haddon Matrix for On-Road Bicycling

moving automobiles. The use of PPE such as a helmet and gloves are frequently all that can be done. Unlike motorcyclists, less PPE is typically worn by the bicyclist because of the physical demands of pedaling. (Of course, off-road bicycling and MTB participants often wear more PPE due to the higher probability of a single-vehicle crash. This off-road PPE is worn to protect primarily against ground-rider impact and abrasion, not vehicle-to-vehicle impact. Much on-road motorcycle PPE is also to protect against ground-rider interactions.)

After an accident has taken place, the user can have taken steps at *A3* to reduce the severity of an injury suffered as a result. These steps include, for example, a smartphone application that perhaps receives a signal from a helmet-mounted sensor once an impact has taken place or determines that no movement through the smartphone's GPS system has been detected over a specified period. The phone will then dial a preprogrammed telephone number to report an accident even if no one else is nearby. Once assistance has arrived, emergency medical technicians (EMTs) can identify the injured, see if there are medical complications of which they need be aware, and notify the

"ICE" (In Case of Emergency) contact on the mobile phone. These items can reduce the time before treatment of injuries, prevent the administering of allergen to the injured, and increase comfort of the injured.

Regarding the vehicle, the bicycle, pre-accident measures to be taken (*B1*) include lighting at both the front and the rear—even during daylight operation—of the bicycle to improve vehicle conspicuity in traffic. PM to be undertaken by a human (the rider or a mechanic) will assure the function of tires, brakes, steering, shifting, and other necessary sub-systems on the bicycle.

There is little, unfortunately, that can be done regarding *B2*, crashworthiness on a bicycle. The best a bicycle designer, and user through customization, can do is to assure that the bicycle itself does no harm to the rider if hit by an automobile. As in the case of *B2*, group *B3* shows little is able to be done to reduce injury by the bicycle itself aside from the prior fitment of a first-aid kit to the bicycle.

There are some steps that can be taken in the physical environment before an accident takes place, *C1*. The community can create bike paths to separate bicyclists from traffic and the community can designate bike lanes. Bike lanes do not offer the protection of bike paths since bicyclists are still alongside traffic in bike lanes. In addition to these options, there exist some innovative roadway and intersection designs that could possibly be incorporated in certain places. One of these, the "Protected Intersection" (Gilpin et al. 2015), is designed to protect bicyclists at busy, urban intersections. This, and other schemes, should continue to be investigated as they develop and real-world experience with them is obtained.

Group *C2* is the *automobile* corollary to *B2* for the bicycle. Just as no bicycle features should injury the rider, no automobile features[20]—such as sharp corners or edges, hood ornaments, and non-foldaway mirrors—should further injure the rider.

Once the physical environment is called upon, perhaps by cellular-communications networks, to provide assistance to the injured rider (*C3*), the existence and proximity of ambulances, EMTs, and hospital emergency rooms can affect the extent of an injured rider's injury. In some cases, elapsed time between injury and initial treatment will be critical to patient outcome and recovery.

Socio-economic methods to help prevent bicycle accidents and injuries from even occurring (*D1*) include raising driver awareness of bicycles on the same roads. If this proves insufficient, traffic laws regarding such criteria as minimum lateral distance, for example 3 ft (~ 1 m), between passing cars and bicyclists had been passed in 32 states as of July 2018 ("Safely Passing Bicyclists Chart" 2019). Some may debate if such laws and their strict enforcements are effective in actually reducing bicycle-accident injuries. However, in the United States, the number of bicycle fatalities from motor-vehicle collisions fell from 852 in 2016 to 763 in 2017 ("NHTSA in Action: NHTSA Is Dedicated to Promoting Safe Behaviors on Our Nation's Roads," n.d.).

The absence of a socio-economic measure to be taken to protect a rider during the accident, *D2*, is either an admission that nothing can be done or a call to devise an effective countermeasure to meet that need. Given the age of bicycling and the number of intelligent people who have been either consciously or subconsciously involved with bicycle safety, it is unlikely that this particular gap can be filled in a simple manner. However, with a new

[20] In this analysis, the *other* vehicle, the automobile, becomes part of the physical environment.

activity, such a gap may indeed be fillable given thought and, perhaps, some research and cooperation between multiple parties.

Finally, *D3* is the intersection of socio-economic environment and the post-accident phase. At this element of the matrix, an injured rider may be assisted by passersby in the United States who may be able to call 9-1-1 (1-1-2 in some other countries), render first aid and reroute oncoming automobile traffic to prevent further rider injury.

Thus, using the Haddon matrix, society's system of addressing the hazard of motor-vehicle accidents with bicyclists is revealed. Although there are weaknesses in the overall system, many of the weaknesses are due to inherent characteristics of bicycle design and operation. The bicycle is small and lightweight. It is also human-powered, thereby limited to a continuous power output of well below one horsepower (0.7 kW) and typically speeds of less than 25 mi/h (40 km/h). These characteristics are simply incompatible with sharing space with an automobile of over 3000 pounds (1364 kg), 200 horsepower (149 kW), and at 60 mi/h (~100 km/h). It may well be that all that can reasonably be done has already been done, but engineers and society should stay alert to developments and technologies that can further advance on-road bicycle safety.

In the meantime, bicyclists, motorcyclists, and pedestrians remain "at-risk" roadway users due to their disadvantages when engaging with automobiles on the streets. Consequently, these users should be afforded special protections when wise and practical in policy and design matters. These protections can include segregation from or within traffic as illustrated by bike paths and bike lanes, respectively, in this example.

None of the groups of countermeasures shown in Table 4.12 should be considered as exhaustive. They merely represent the sample set of measures in each matrix element. There may even exist disagreement within a group completing a Haddon matrix on where within the matrix a particular countermeasure should go. Sometimes a countermeasure can correctly span more than one factor, for example. The most important aspect of this exercise is including reasonable or probable countermeasures somewhere within the matrix. Further discussions can help clarify proper placement if it is a group exercise or if results must be presented to a larger group.

4.14 Conclusion

This chapter's concepts will help a design or a product-safety engineer better understand product safety. Some material and observations presented are original. Some of these concepts have existed for decades, but are not well known or were widely dispersed in the literature. Regardless, these concepts have been assembled in one location to help design and product-safety engineers *design-in* desirable product characteristics from the outset—what is here called Design for Product Safety (DfPS).

The safety hierarchy is often quoted by people who have neither considered its true efficacy nor attempted its real-world application. Many such people may not even know *which* safety hierarchy they are discussing. The safety hierarchy is usually raised by those blaming a product for an injury. Yet, even the consensus safety hierarchy is not of much help once one realizes that the performance of a useful function by a product often requires the presence of a hazardous design element or characteristic. Most such hazards simply cannot be completely designed out of a functional product.

There are two methodologies for the evaluation of safeguard devices and safeguarding systems presented. The first of these evaluations can help determine the positives and the negatives for a given safeguard-device proposal. The second of these can help an engineer better understand the function of a *system* of interacting countermeasures used within an engineering design.

The concepts of dependency, uniformity, vigilance, and compatibility can help design engineers and product-safety engineers assess future potential problems with their products. Such problems can arise once consumers begin to use a product and become familiar with it. Issues can also arise from using a type of product when the operation of one particular product deviates from the operation of many similar products. Furthermore, people, in general, and product users, in particular, do not always behave in ways that engineers and regulators expect. Users may modify their actions in ways neither expected nor needed by engineers to achieve their product-safety goals. There are times, however, when such changes in user behavior can be predicted.

The author introduces an elementary two dimensional exploration of product safety risk to users versus the amount of user activity needed to operate the product. The purpose of the resulting plot is for the consideration of whether it is reasonable to expect the same level of product safety for highly user-active products that is emergent with low user-active products. In the high user-active product cases, the consumer can be a significant part of the observed product-safety level. Although design-and-manufacturing companies continue to provide product-usage guidance to help consumers, consumers ultimately decide for themselves whether or not to follow these recommendations.

In addition to user activity, different product types make different demands on the user. A simple method of evaluating user task load is provided. This information can be useful when trying to compare products—perhaps even to the design engineer who creating an innovative new product and is benchmarking an existing product for design guidance. The recognition of varying levels of user activity between products is also an admission of the need to consider the varying user task loads.

There is a brief discussion of two factors that may aggravate existing safety risks in products. The first of these is user qualification. Not all people are equally able to operate all products. In some cases, it will be necessary for engineers to specify the requisite physical and cognitive abilities for those using a product, for example, to help consumers decide if the product is appropriate for them or their dependent(s). The second of the factors is the determination of whether or not an at-risk population could be vulnerable when using the product. The consideration of these factors during design could help provide greater product safety to its future users by designing-in particular design attributes to facilitate its use by those who may be at-risk.

Finally, the Haddon matrix is introduced and discussed to help the engineering-design team consider a product, potential accident, and its sometimes-avoidable outcomes, as a whole. This matrix was developed for epidemiological use, but has application in engineering. Its consideration of accident phases and entity factors in the use of a product may help both the design engineer and the company identify where there may exist gaps in a comprehensive product-safety system that go beyond the product itself. There may be instances in which it will be in an entire industry's interests to work with regulators and legislators to create the elements found lacking in a Haddon-matrix analysis.

Many concepts were discussed and it may take some time for readers to reflect upon some of them and understand them. However, it is hoped that the awareness and consideration of the product-safety concepts presented here will help the engineers in charge of design better understand product-safety engineering and then proceed to create safer products for the consuming public.

References

"AASHTO." n.d. Accessed May 8, 2019. https://www.transportation.org/.

"ANSI/ISO 12100:2012 Safety of Machinery—General Principles for Design—Risk Assessment and Risk Reduction." 2012. Houston, TX: ANSI/ISO.

"ANSI/SVIA 1-2017: American National Standard for Four-Wheel All-Terrain Vehicles." 2017. Specialty Vehicle Institute of America (SVIA).

Bailey, Robert W. 1996. *Human Peformance Engineering: Designing High Quality, Professional User Interfaces for Computer Products, Applications, and Systems*. Third edition. Englewood Cliffs, NJ: Prentice Hall.

Barnett, Ralph L. 1994. "The Principle of Uniform Safety." *Triodyne Safety Brief* 10 (1). http://www.triodyne.com/SAFETY~1/SB_V10N1.PDF.

Barnett, Ralph L. 2020. "On the Safety Theorem." American Journal of Mechanical Engineering 8 (2): 50–53. https://doi.org/10.12691/ajme-8-2-1.

Barnett, Ralph L., and Peter Barroso. 1981a. "On Classification of Safeguard Devices (Part I)." *Triodyne Safety Brief* 1 (1). http://www.triodyne.com/SAFETY~1/SB_V1N1.PDF.

———. 1981b. "On Classification of Safeguard Devices (Part II)." *Triodyne Safety Brief* 1 (2). http://www.triodyne.com/SAFETY~1/SB_V1N2.PDF.

Barnett, Ralph L., and Dennis B. Brickman. 1985. "Safety Hierarchy." *Triodyne Safety Brief* 3 (2). http://www.triodyne.com/SAFETY~1/SB_V3N2.PDF.

Barnett, Ralph L., Gene D. Litwin, and Peter Barroso. 1983. "The Dependency Hypothesis (Part I)." *Triodyne Safety Brief* 2 (3). http://www.triodyne.com/SAFETY~1/SB_V2N3.PDF.

———. 1984. "The Dependency Hypothesis (Part II)—Expected Use." *Triodyne Safety Brief* 3 (1). http://www.triodyne.com/SAFETY~1/SB_V3N1.PDF.

Barnett, Ralph L., and William G. Switalski. 1988. "Principles of Human Safety." *Triodyne Safety Brief* 5 (1). http://www.triodyne.com/SAFETY~1/SB_V5N1.PDF.

Bridger, Robert S. 2018. *Introduction to Human Factors and Ergonomics*. Fourth edition. Boca Raton, FL: CRC Press. https://www.crcpress.com/Introduction-to-Human-Factors-and-Ergonomics/Bridger/p/book/9781498795944.

Constant, Aymery, and Emmanuel Lagarde. 2010. "Protecting Vulnerable Road Users from Injury." *PLOS Medicine* 7 (3): 1–4. https://www.ncbi.nlm.nih.gov/pmc/articles/PMC2846852/.

Embrey, David. 2005. "Understanding Human Behaviour and Error." Dalton, UK. http://www.humanreliability.com/downloads/Understanding-Human-Behaviour-and-Error.pdf.

Gilpin, Joe, Nick Falbo, Mike Repsch, and Alicia Zimmerman. 2015. "Lessons Learned: The Evolution of the Protected Intersection." Portland, OR. https://altaplanning.com/wp-content/uploads/Evolution-of-the-Protected-Intersection_ALTA-2015.pdf.

GPO. 2018. "FMVSS." 2018. https://www.govinfo.gov/app/collection/cfr/2018/title49/subtitleB/chapterV/part571.

Haddon, Jr., William. 1973. "Energy Damage and the 10 Countermeasure Strategies." *Journal of Trauma* 13: 321–31.

Hall, Stephen M., Stephen L. Young, J. Paul Frantz, Timothy P. Rhoades, Charles G. Burhans, and Paul S. Adams. 2010. "Clarifying the Hierarchical Approach to Hazard Control." In *Advances in Human Factors, Ergonomics, and Safety in Manufacturing and Service Industries*, edited by Waldemar (CRC Press) Karwowski and Gavriel (CRC Press) Salvendy, 1057–64. Boca Raton, FL: CRC Press.

Hart, Sandra G. 1986. "NASA Task Load Index (TLX); Paper and Pencil Package." Moffett Field, CA. https://ntrs.nasa.gov/search.jsp?R=20000021488.

________. 2006. "NASA-Task Load Index (NASA-TLX); 20 Years Later." *Proceedings of the Human Factors and Ergonomics Society Annual Meeting* 50 (9): 904–8. https://doi.org/10.1177/154193120605000909.

Hart, Sandra G., and Lowell E. Staveland. 1988. "Development of NASA-TLX (Task Load Index): Results of Empirical and Theoretical Research." *Advances in Psychology* 52: 139–83. https://doi.org/10.1016/S0166-4115(08)62386-9.

Kuczmarski, Robert J., Cynthia L. Ogden, Shumei S. Guo, Laurence M. Grummer-Strawn, Katherine M. Flegel, Zuguo Mei, Rong Wei, Lester R. Curtin, Alex F. Roche, and Clifford L. Johnson. 2002. "2000 CDC Growth Charts for the United States: Methods and Development." Washington, DC: Government Printing Office. https://www.cdc.gov/growthcharts/cdc_charts.htm.

McKnight, A. James, and H. Blair Heywood. 1974. *Motorcycle Task Analysis*. Linthicum, MD: Motorcycle Safety Foundation (MSF).

"Meet Bolo." n.d. Meetbolo.Com. Accessed February 7, 2020. http://meetbolo.com/.

Nader, Ralph. 1965. *Unsafe at Any Speed: The Designed-In Dangers of the American Automobile*. First. New York, NY: Grossman.

"NHTSA in Action: NHTSA Is Dedicated to Promoting Safe Behaviors on Our Nation's Roads." n.d. NHTSA. https://www.nhtsa.gov/road-safety/bicycle-safety.

NIOSH. 2013. "Prevention Through Design." https://www.cdc.gov/niosh/topics/ptd/.

________. 2015. "Hierarchy of Controls." https://www.cdc.gov/niosh/topics/hierarchy/default.html.

NSC. 1955. *Accident Prevention Manual for Industrial Operations*. Third edition. Chicago, IL: National Safety Council.

OSHA. 2010. *29 CFR 1926.1431 Hosting Personnel*. https://www.kokeinc.com/wp-content/uploads/2015/05/CraneandDerrickRequirements.pdf.

Rasmussen, Jens. 1983. "Skills, Rules, and Knowledge; Signals, Signs, and Symbols, and Other Distinctions in Human Performance Models." *IEEE Transactions on Systems, Man, and Cybernetics* 13 (3): 257–66.

Ross, Kenneth. 2015. "The Safety Hierarchy." *Strictly Speaking* 12 (2). http://productliabilityprevention.com/images/35_-_DRI_Strictly_Speaking_-_Safety_Hierarchy_May_2015.pdf.

"Safely Passing Bicyclists Chart." 2019. National Conference of State Legislatures. http://www.ncsl.org/research/transportation/safely-passing-bicyclists.aspx.

Sanders, Mark S., and Ernest J. McCormick. 1993. *Human Factors in Engineering and Design*. Seventh. New York, NY: McGraw-Hill.

Swain, A. D., and H. E. Guttman. 1983. "Handbook of Human Reliability Analysis with Emphasis on Nuclear Power Plant Applications, Final Report, (NUREG/CR-1278)." Albuquerque, NM.

Washington Post. 2002. "National Highway Transportation Safety Administration." http://www.washingtonpost.com/wp-srv/business/includes/nhtsa_primer.htm.

Wichter, Zach. 2019. "What You Need to Know after Deadly Boeing 737 Max Crashes." *New York Times*, April 30, 2019. https://www.nytimes.com/interactive/2019/business/boeing-737-crashes.html.

Woods, David. 2009. "Rasmussen's S-R-K 30 Years Later: Is Human Factors Best in 3's?" In *Proceedings of the Human Factors and Ergonomics Society*, 217–21. https://doi.org/https://doi.org/10.1177/154193120905300412.

CHAPTER 5

Hazards, Risks, Accidents, and Outcomes

Risk management is a more realistic term than safety. It implies that hazards are ever-present, that they must be identified, analyzed, evaluated, and controlled or rationally accepted.

—JEROME F. LEDERER[1]

5.1 Introduction

Thus far, readers have been indulgent and permitted the author to speak about safety, hazard, risk, and outcome without defining any of these words. Many of the words and terms used in the product-safety discussion so far have generally accepted definitions and meanings that permit society to speak about safety. However, it is important to define these words so that precise discussions for product-safety engineering can be undertaken.

5.2 Recent History of System Safety

There are several words that must be defined in order to advance the discussion of product-safety engineering, its challenges, and its ultimate goals. Although several of these words can effectively be used imprecisely, it serves the engineering community well to fully understand the similarities and differences between these words. The field of *system safety* has, fortunately, taken the lead in defining many of these words and developing their treatment methods over the years. Readers wishing to study the field of system safety in depth are encouraged to study one of the several fine system-safety textbooks available (Leveson 2011; Ericson 2016; Gullo and Dixon 2017).

Prior to dedicating the time and effort needed to memorize particular definitions for safety, its related words, and its concepts, it is helpful to study prior engineering

[1] Lavietes, Stuart. "J. F. Lederer, 101, Dies: Took Risk Management to the Sky." *New York Times*. February 9, 2004.

work leading to such definitions. Some safety practitioners are quick to use definitions that suit themselves and their immediate needs. This should be avoided whenever possible. Good sources of safety-related definitions include the U.S. Department of Defense and NASA. The basis for this reasoning is provided below.

The sources for definitions used here will start with the U.S. Department of Defense and its Military Standard on system safety (USA/DoD 2012). This document, MIL-STD-882E, is a milestone document and now in its sixth version.[2] It provides a strong starting point for discussing both system safety and product safety. The field of system safety received a tremendous boost, in both attention and resources, during the 1960s in the United States. In the years leading to this decade, there were many technologically aggressive aerospace programs which were also intrinsically hazardous. The crewed space missions of Mercury (sub-orbital and orbital), Gemini (orbital and rendezvous), and Apollo (orbital and lunar) were on-going at NASA through its aerospace contractors. Most of these same aerospace contractors were also defense contractors for the U.S. military and were tackling large-scale weapons-systems programs such as the Boeing Minuteman ICBM (inter-continental ballistic missile), the Mach 2 + Convair B-58 "Hustler" bomber, the variable-geometry ("swing-wing") General Dynamics F-111 TFX (Tactical Fighter Experimental) fighter-bomber, and many more incorporating cutting-edge propulsion, airframe, and electronics sub-programs. This integration was sometimes successful but only was rarely, if ever, on-budget or on-schedule.[3] This period of unprecedented demands upon research and development efforts, technologies, and practicing engineers and scientists often resulted sub-par delivered-system performance and sometimes compromised personnel and crew safety. This period was also the time when enormous cost overruns became commonplace with U.S. government defense contractors.

In the 1950s, the pioneering age of jet and rocket flight, scientists and engineers were struggling with the new technologies of gas-turbine and rocket engines, supersonic flight, and atomic/nuclear power. Politicians, including military generals and admirals, were also struggling with the "Cold War" at this time. The geo-political fears of the day led to innovative and aggressive combinations of these new technologies in the forms of advanced missiles, fighter planes, and bombers which would have been unthinkable only a decade before. The undertaking of such ground-breaking aircraft designs produced some successes and some failures. As a case in point of the latter, the U.S. Air Force never ran out of pilots, killed in aircraft accidents, after whom to name or rename Air Force bases.

In addition, on January 27, 1967, the lives of three NASA crew members were lost to fire during a pre-launch test. Astronauts Virgil "Gus" Grissom, Edward White, and Roger Chaffee died while they were seated in their *Apollo 1* (also called Apollo 204) command module atop their booster rocket by a fire that might not have lasted more than 30 seconds. The layout of the accident scene is shown in Figure 5.1 (Thompson 1967); the command module was approximately 300 ft (90 m) above the ground. This short-duration fire, however, consumed all of the available oxygen in their capsule

[2] MIL-STD-882 was followed by MIL-STD-882A (Ericson 2006).

[3] The B-58 bombing and navigation system was delayed for years and proved unreliable at first. The initial B-58 contract was awarded in 1952. The plane became operational in 1961. It was not until 1967, two years *after* the U.S. Air Force had started *retiring* the planes in 1965, that its bombing-and-navigation issues were finally sorted out (Converse 2012).

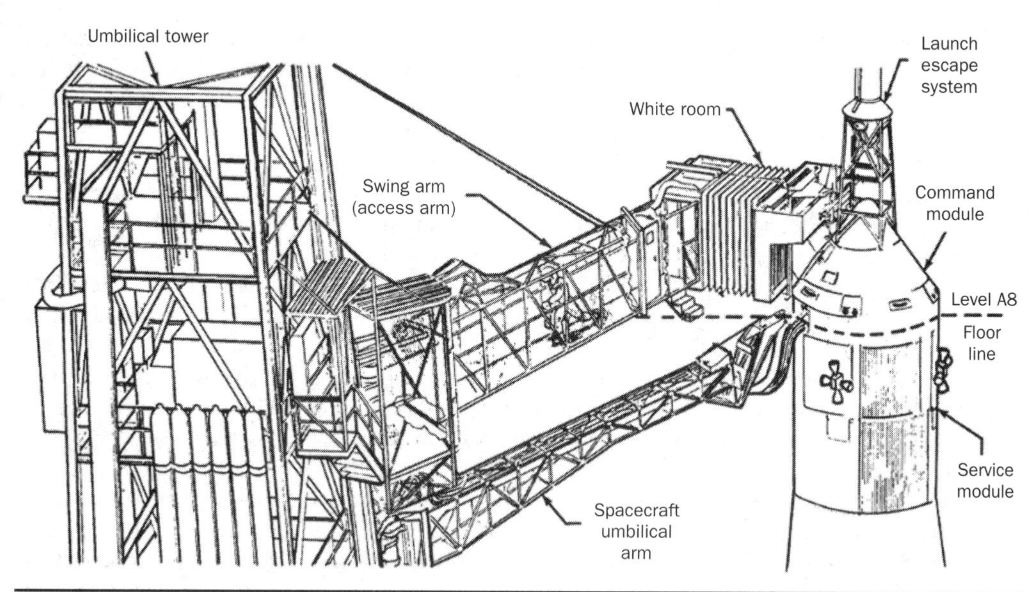

FIGURE 5.1 Apollo 1 accident scene. (Source: "Report of Apollo 204 Review Board to the Administrator, National Aeronautics and Space Administration," 1967.)

and also generated extreme temperatures and toxic gases. Among the findings of the Senate-investigation report of the accident (Anderson 1968) were:

- The testing underway was not considered to be hazardous by NASA since they had done this before without incident.
- The capsule had an atmosphere of pure oxygen (O_2) pressurized to 16.7 lbf/in^2 (115 kPa)—which is above atmospheric pressure.
- The entry/escape hatch was *inward* opening[4] and required approximately 90 seconds to unsecure for removal.
- There was no rapid way to depressurize the cabin of the command module so that the inward-opening escape hatch could be removed even once it was unsecured.
- The escape hatch could only be unsecured and opened from the *outside.*
- No ground-support crew members were in the immediate area.
- Ground-support crews were inadequately trained and equipped to respond to such emergencies.

[4]The design decision to use an inward-opening escape hatch on the Apollo command module is not as bizarre as it may first sound. Ironically, one of the Apollo-1 victims, Gus Grissom, barely avoided drowning on July 21, 1961 on his first space flight when his *Liberty Bell 7* Mercury space capsule sank soon after splash down in the Pacific Ocean. It was determined that its outward-opening escape hatch had opened prematurely ("Liberty Bell 7 MR-4 (19)" 2000). It was this mishap that led NASA engineers to ultimately reason that an inward-opening Apollo escape hatch would be safer than the outward-opening escape hatches used in the earlier Mercury and Gemini spaceflight programs. Following the U.S. Senate report, Apollo command modules had newly designed quick-releasing, outward-opening escape hatches.

- The initiation and propagation of fire and flame had not been given the attention to detail that it deserved through conscious materials selection and proper capsule housekeeping.
- There was not a fire-extinguishing system in the capsule.

Although no conclusive single source of ignition was identified, it is suspected that a small electrical arc ignited nearby flammable materials. Then, fed by the pure-oxygen environment, the fire rapidly spread and raised temperatures to over 1000°F (540°C). All three astronauts died from asphyxiation, not from burns. The hatch was not removed until 4 minutes 55 seconds after the fire was reported by the crew. The first doctor did not arrive at the scene for at least 11 minutes 55 seconds.

If this tragedy had any silver lining it is that many engineering-design and administrative-procedure changes were made following this accident. The entire aerospace industry, as well as the discipline of system safety, was transformed by changes made subsequent to this mishap. There is little doubt that such incidents and industry responses have produced such system-safety advances in the space and military-aerospace industries. These system-safety advances have led to significant improvements in the levels of safety that the public enjoys today from several industries including aviation, transportation, and energy. Without a doubt, these increases in safety engineering have saved many human lives in the years since the formalized beginnings of system-safety engineering.

The Apollo program resumed crewed space flights on October 11, 1968 when Apollo 7 launched for a successful 11-day Earth-orbit mission ("About Apollo 7, the First Crewed Apollo Space Mission" 2018). This space program, of course, went on to successfully land several astronauts on the lunar surface—and return them to Earth safely. Although an accident struck Apollo 13 on its way to the Moon, no lives were lost as a result of the event. NASA continued with a strong safety record until the Space Shuttle program experienced the loss of a ship, the *Challenger*, during its launch phase on January 28, 1986 costing the lives of the seven crew members. Again, on February 1, 2003, the space shuttle *Columbia* came apart killing its seven crew members upon re-entry to Earth's atmosphere. Each of these accidents was followed by official investigations whose boards issued reports, Rogers (1986) and Gehman (2003), respectively. Decades later, one observer of the disaster later wrote of his unheeded advice (McDonald and Hansen 2012) regarding the first of these two events.

5.3 Definitions

From the above, it is clear that the contributions of the aerospace industry have positioned it at the vanguard of system-safety technology. Therefore, system safety's definitions from MIL-STD-882E will be the starting point for this discussion. Words of particular interest in product-safety engineering are "safety," "danger," "hazard," and "risk."

Prior to defining particular words, MIL-STD-882E defines the discipline of *system safety* to be:

> The application of engineering and management principles, criteria, and techniques to achieve acceptable risk within the constraints of operational effectiveness and suitability, time, and cost throughout all phases of the system life-cycle.

From this, it may be concluded that the absolute safety of a system is not the target of system-safety engineering. There exist real-world constraints, such as technology and mission, which will prevent a system from ever being completely safe, hence this chapter's opening quote from a pioneer in system safety. Instead, what a system-safety engineer should strive for is "acceptable risk." The same is said for the design engineers of many consumer products.

Proceeding with definitions, the word *safety* is defined in MIL-STD-882E as "[f]reedom from conditions that can cause death, injury, occupational illness, damage to or loss of equipment or property, or damage to the environment." There are many products that are safe, or at least relatively safe. Take, for example, a common paper clip. There is little about a paper clip that could prove harmful, or unsafe, to a person who does not swallow it. It is *possible* that a bent paper clip could injure someone's eye, but that would be either a highly *improbable* event or a malicious act. Remember from the section on the classification of safeguard devices that considerations should generally be kept to events that are *probable* rather than simply *possible*. This book will also not cover malevolence in product use. It is also possible that materials used to either construct or coat the paperclip could be toxic if handled excessively or placed in the user's mouth while handling papers. This would be a case of a product having *health* consequences rather than *safety* issues, however.

Although the word *danger* is widely used in everyday speech, it is rarely used when discussing system-safety matters. Therefore, the word danger is not defined in the military system-safety standard. This word is used, however, in some industrial-safety literature (Hammer and Price 2001; Barnett and Switalski 1988). The latter pair of authors states that danger is the opposite of safety—or its antonym. For everyday communications, this is an acceptable working definition. If a product is not safe, it could be considered "dangerous." This word, danger, is used and addressed later in this book during the discussion of warning signs to product users, but this word will not be used outside of that context in this book.

The word *hazard* is defined to be a "real or potential condition that could lead to an unplanned event or series of events (i.e., mishap) resulting in death, injury, occupational illness, damage to or loss of equipment or property, or damage to the environment." So, a hazard is a condition—which if not avoided—could lead to injury or damage. Examples of hazards include kitchen knife blades, lawn-mower blades, snow-blower augers, and portable gasoline-powered electric generators fuels.

A kitchen knife, Figure 4.8, has a sharp blade that can cause severe injury if misused or if not respected. Fortunately, many people learn of this hazard early in life. A lawn mower, Figure 4.2, has a sharp blade which rotates at high speed during use that could inflict severe damage to a person's feet and hands if encountered. Fortunately, the lawn mower has a guard, usually the appropriately shaped chassis, to prevent human access to the blade from certain directions. The snow blower, such as that shown in Figure 5.2 has semi-exposed augers which also rotate during use and are exposed at the front due to its intended use which is moving deep snow in its path of motion. Snow blowers usually have a chassis which guards the user from its augers when the user is positioned at the rear or sides of the product. In addition, many snow blowers have separate controls of the dead-man[5] type to engage the auger-and-impeller drive and the wheel-drive systems.

[5] A dead-man type of control is one which turns off or idles a product once a user has left the user position. For example, many riding lawn mowers have a spring-loaded seat positioned over a switch to sense when a person is no longer sitting on the seat and will shut off the mower when this occurs. Similarly, snow blowers generally have spring-loaded levers to control the wheel and the auger drivetrains separately. It should be neither easy nor simple for a snow-blower user to reach the augers while still holding onto and activating the auger-drive lever.

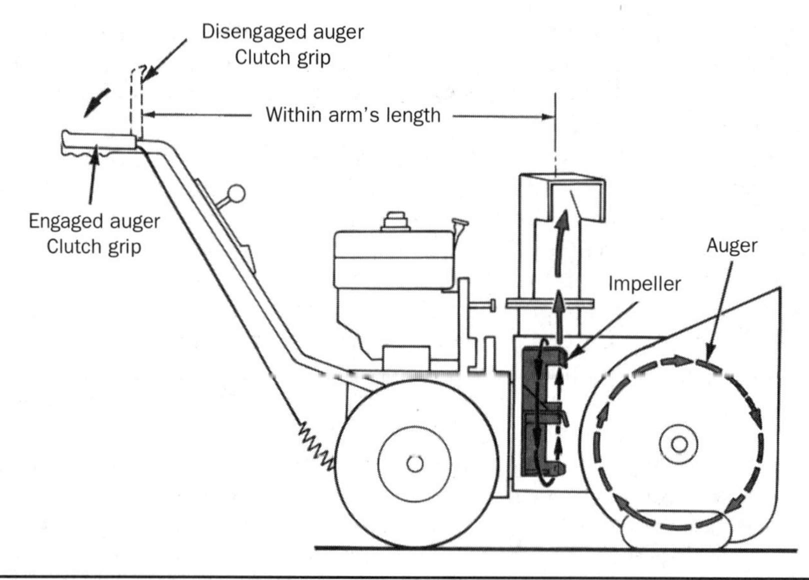

Figure 5.2 Walk-behind snow blower. (Used by permission of Triodyne Inc.)

These controls will disengage these power-train systems to eliminate, or vastly reduce, user exposure to the rotating-auger hazard once the user walks away from the snow blower and releases the control levers. Portable electric generators have the hazard of carbon monoxide (CO) from internal-combustion engine (ICE) exhaust. This is in addition to the fire and burn hazard of containing gasoline on the same small chassis as a hot ICE exhaust system.

Finally, the word *risk* is taken to mean a "combination of the *severity* of the mishap and the *probability* that the mishap will occur" by MIL-STD-882E. This definition raises the need for three more definitions: mishap, severity, and probability.

First, a *mishap* is an "event or series of events resulting in unintentional death, injury, occupational illness, damage to or loss of equipment or property, or damage to the environment…."[6] Second, *severity* is defined as "the magnitude of potential consequences of a mishap to include: death, injury, occupational illness, damage to or loss of equipment or property, damage to the environment, or monetary loss." Third and finally, *probability* is defined to be an "expression of the likelihood of occurrence of a mishap."

Each of these words may be examined by continuing the use of the portable gasoline-powered electric generator example from above. The hazard of the generator being covered is CO emissions from the gasoline-powered engine. The mishap, or accident and outcome, in this instance would be the CO poisoning leading to unconsciousness and, finally, death of the product users. As a design engineer for generators, in order to evaluate the risk of user death from CO poisoning, it will be necessary to take this hazard and consider the two aspects of risk associated with this hazard. These two aspects of risk are the severity and the probability of injury or death to CO

[6] Most people would simply call this an accident and this book will not differentiate between a mishap and an accident. An accident, however, will be differentiated from an outcome.

exposure. There is little that can be done to reduce the consequences of exposure to CO. Therefore, the potential severity of the injury from CO remains extreme. The other element of risk, probability, is the only other mechanism available to designers to reduce the risk of poisoning users with CO. Therefore, it is necessary to engineer ways to limit the probability of CO poisoning for generator users.

Over the last two decades, the U.S. Consumer Product Safety Commission (CPSC) must have paid particular attention to the rise of CO poisonings among survivors of natural disasters such as hurricanes which often cause massive and prolonged power outages. Such outages cause some people to purchase and use portable generators to restore electrical power for residential use. Some of these users are unfamiliar with the characteristics of power equipment and may use a generator in close quarters and under adverse conditions. The increase in CO poisonings and deaths following such natural disasters resulted in a Federal Regulation with labeling requirements for all portable generators sold in the United States (*Portable Generators; Final Rule; Labeling Requirements* 2007; *Portable Generators; Final Rule; Labeling Requirements (Correction)* 2007). This regulation required prescribed signs to be on both the product (Figure 5.3) and on the product packaging.

This was a measure to reduce the risk of CO poisoning through reducing the probability of fatal CO exposure by separating the hazard, CO, from the portable-generator users. Again, since the lethality of CO is beyond the control of designers, severity cannot be controlled. Only probability can be addressed. This was accomplished by reducing the CO-exposure probability through user admonitions on a required warning label attached to the generator and its packaging.

Although some portable generators up until the point of CPSC intervention had warning signs on them affixed by their manufacturers already, the Commission felt strongly enough that an effective, standardized sign was necessary that they intervened on the matter and have now have labeling requirements for the products. Therefore, although design engineers, and others, at manufacturers perhaps should have taken the lead—or maybe even a more-active lead—in risk reduction through warning signs, the regulators felt it necessary to step in, amplify, and standardize portable-generator warnings. The engineering-design process should *not* work this way. However, in the

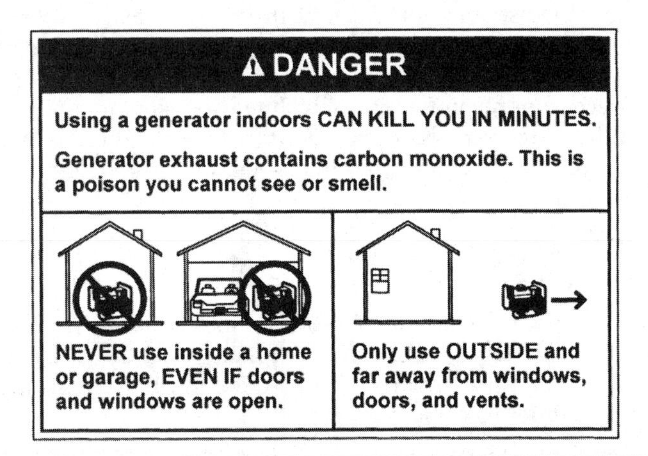

FIGURE 5.3 On-product warning sign required by U.S. CPSC. (Source: 16 CFR Part 1407)

end and so long as users follow warnings with which they have been provided, the risk of CO poisoning to portable-generator users has been lowered through a reduction in the probability of such a mishap to an acceptable level—at least for now. Recently, some portable-generator manufacturers have started making generators with CO-sensing abilities which will shut down the generator when CO levels are rise and are sensed around the generator (Hope 2019).

In addition to the word danger, the following words considered to be safety-related by many people are *not* defined in the military's system-safety standard: accident, incident, mistake, and error. Definitions for some of these words are found in Ericson's system-safety text. He defines an *accident* to be an "unexpected event that culminates in the death or injury of personnel, system loss, or damage to property, equipment, or the environment." There are no definitions provided in the reference for *incident* or *mistake*. An *error* is defined to be an "(1) Occurrence arising as a result of an incorrect action or decision by personnel operating or maintaining a system, and (2) a mistake in specification, design, or implementation."

From the above two paragraphs, it appears that mishap and accident are essentially synonymous. Both are unintended events possibly leading to injury, death, or damage. For purposes of this book, the word *incident* will be used to mean a "close-call" where injury or death does *not* take place. The word incident may also be used in the book to denote an event whose conditions are still under investigation. For example, if a manufacturer learns some, but not all, of the details involving what third parties are calling an "accident" with the manufacturer's product, the event may be categorized as an incident, properly albeit temporarily, until more information is collected or provided. This may rightfully be construed by some to be a subtlety—or a distinction without a difference. It may be. The word incident does not carry with it the connotations of an accident. However, in the mind of some, calling an event an accident or mishap immediately brings with it the mandatory allocation of blame to the various parties involved. This primal human urge is counterproductive, especially in the early stages of fact finding. As Leveson concludes, "[b]lame is the enemy of safety. Focus should be on understanding how the system behavior as a whole contributed to the loss and not on who or what was to blame" (Leveson 2011, 56). Whether an event is classified as an incident, an accident, or a mishap, what that event is eventually called should not affect the proper response to it. Organizational semantics must not be permitted to steer an investigation of an event in one particular direction merely due to the assignment of a noun. A "thermal event" sounds nicer than a "fire;" however, the responses to each should be similar. An *incident*, or any "close call," should also be pursued with the same determination as an *accident*. Perhaps it was only good luck that prevented the close call from being an accident resulting in injuries and deaths.

5.4 Hazards

In Section 5.3, readers saw that Ericson defines a hazard to be a "condition" that could result in a mishap, or accident, with injury or death as a result. The differences between hazards and risks will be illustrated in this chapter. It is important that readers understand the differences.

A hazard is an abstract term when used within system-safety or product-safety engineering. Hazards are generalizations of some condition, or design element, that could cause harm, either injury or death,[7] through a combination of other factors. These other factors, which more directly relate to potential injury, will be considered when discussing risk in following section.

For the moment, consider the following four conditions, or states:

1. A truck is parked on the flat and level top of a steep downhill slope, with its transmission in Neutral, with its parking brake OFF, and with a village at the bottom of the slope.
2. A passenger car is suspended by a hydraulic jack alone while a mechanic works beneath the car.
3. A gasoline can is placed next to a portable generator while it is in use in a garage with its door open.
4. A multi-blade, folding knife is used to peel an apple.

To begin our discussion of hazards, the above four conditions provide more information than is necessary for an introductory discussion of hazards. This extra information also confuses the concept of "hazard." Below are the updated four conditions with extraneous information pruned off:

1. A large mass is elevated
2. A large mass is suspended
3. A flammable liquid exists
4. A sharp edge exists

This distilled list helps to identify the precise *hazards*—or the essence of the conditions—without conflating them with any of their *risks*.

In the *first* case, the truck, the hazard presented is one of impact. Should the truck begin to roll downhill, then a person in the village may be impacted by the large moving vehicle. Anything beyond the situation of a mass being near a down slope leading to people goes beyond pure hazard; whether someone could or would be injured by the vehicle once put into motion is immaterial at this point.

In the *second* case, the elevated car, the hazard is similar. However, the hazard in this instance would be crushing by the suspended mass instead of impact by a moving mass.[8] In both of these first two cases, potential energy is converted into kinetic energy which could result in human injury. Many hazards are related to energy and their conversions and releases as observed by Haddon in Section 4.2.

In the *third* case, gasoline, the chemical energy of the flammable liquid can be converted into heat energy through ignition and is a fire hazard. This thermal hazard could cause injury, but other factors come into play before any injury would take place. For now, the gasoline merely has the potential to cause harm to people.

---

[7] Hazards can also lead to property and environmental damages, but these non-human injuries are not considered in the present discussion.

[8] *Crushing* would result from the *impact* of a falling piano being raised to another floor above a sidewalk, yet these three hazards—suspended load, crushing and impact—are frequently maintained as separate from one another. However, in a practical product-safety engineering sense, it makes no difference *how* a hazard is classified, so long as it is addressed. These could also be considered hazards arising from acceleration from the Earth's gravity.

In the *fourth* case, a knife, is a mechanical hazard—a sharp edge. Other factors such as the knife's folding feature or multiple blades are extraneous to the identification of the hazard itself. Knives have a known laceration, or cut, propensity associated with them—but this characteristic is also directly related to its utility. Knives exist to cut things; however, they also cut people. This is an example of a mechanical, non-energy hazard.

In each of these four instances of hazards, it is important to notice that *nothing* is happening and *nobody* is being injured by any of these hazards. A hazard is not an event and not a risk. These aspects will be discussed later. Each of these four conditions merely has the *potential* to cause injury provided that the right set of events take place.

The classification of hazards is sometimes an inexact practice. Depending upon education or preference, a hazard that could result in a burn to a user could be considered either a thermal hazard or a chemical-energy hazard. Again, as stated earlier, so long as all hazards are discerned, it makes little difference exactly how they are categorized. This leads to the next part of the study into hazards.

During the engineering design of any new product, it is important to identify the hazards that a product may present to users. This exercise is appropriately named *hazard identification*. In the case of a portable electrical generator, its hazards would include the fire hazard from the presence of gasoline, the exhaust emissions hazard from products of hydrocarbon (HC) combustion, and the electrical hazard from the generator output. Unless the generator is mounted in a precarious, elevated manner and has sharp edges, the generator is unlikely to present any acceleration or mechanical hazards.

Hazards arise from elements of a product's design, as has already been stated. However, this relationship is not necessarily one-to-one. A single product *design element*—which may be a component, a characteristic, or an emergent property of the product—may present multiple hazards. The rotating-blade hazard on a residential lawn mower is of mechanical nature. This single design element, in this case a component, produces hazards arising from a sharp edge—including laceration and amputation—and from a thrown object. Another lawn-mower design element is of the chemical type from the gasoline used for grass-cutting power. There are at least three hazards arising from the use of gasoline in the lawn mower. These hazards are burn/fire/explosion (BFE) from the heat of the engine, lack of oxygen (O_2) from CO emissions, and poisoning from the ingestion of the toxic fluid itself.

A model of this product design-element-to-hazard relationship is shown in Figure 5.4. A product design element is a characteristic or feature of the engineering design of a product that permits that product to perform its intended function. Examples of product design elements are the rotating blade of a lawn mower for cutting grass and the resulting static-stability characteristics of an off-road vehicle (ORV) from the ground clearance needed for operation on unprepared terrain.

Such product design elements may *not* be an individual component or a subsystem, but may instead be an *emergent property* resulting from the overall engineering design of the product. NASA states that *[s]afety is an emergent property of a [product] that arises when [product] components interact with each other, with the environment in which the [product] is used, and with the [product] operators*[9] ("Second Volume of NASA System Safety Handbook Released" 2015).

[9] The author substituted "product" for "system" in NASA's quotation.

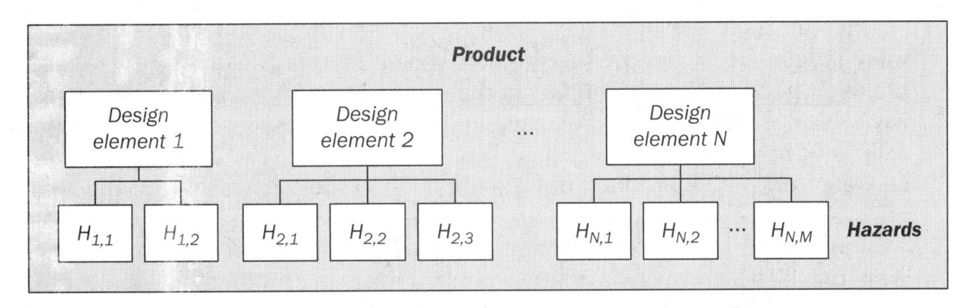

FIGURE 5.4 Product design-element and hazard model.

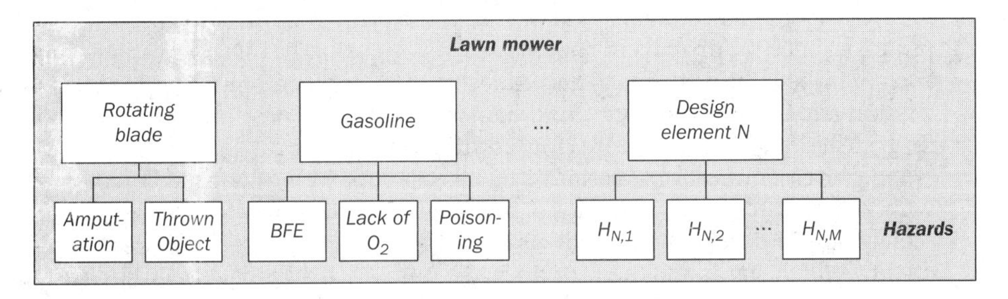

FIGURE 5.5 Product design-element and hazard model for lawn mower.

The Product Design-Element and Hazard model shown in Figure 5.4 shows the "branching" of a product's design elements to its hazards. Product design elements range from 1 to N. Hazards, $H_{n,m}$, for each of the N design element are numbered from 1 to M.

In one case, the Product, as represented by the shaded box, may be a residential gasoline-powered lawn mower from Figure 4.2. Using elements and hazards just mentioned for the lawn mower, Design Element 1 is the rotating blade and Design Element 2 is gasoline. For the rotating blade, the first hazard $H_{1,1}$ would be a sharp edge and the second hazard $H_{1,2}$ would be thrown objects, both mechanical hazards. For the second design element, gasoline, the first hazard $H_{2,1}$ is BFE, the second hazard $H_{2,2}$ is lack of oxygen (O_2), and the third hazard $H_{2,3}$ is poisoning; these hazards are all chemical in nature. There are numerous other hazards presented by the lawn mower, but these will not be covered here. The hazards described, and those not yet identified, are shown in a partially complete model of design elements and hazards presented in Figure 5.5.

For an analysis of a complex, real-world product, there may be many more product design elements than the two that were just briefly adduced. Since these design elements and hazards are contained within the "product" box, these are all properties of the product's engineering design. This propagation of both design elements and their hazards is evident by the placeholders on the right-hand side of Figure 5.5.

As stated at the start of this section, a hazard should be considered a qualitative entity. It is only when the risks presented by hazards are examined that the quantitative aspects of product-safety engineering are encountered.

5.5 Hazard Identification

Now that a basic understanding of hazards has been established, it is now both *possible* and *necessary* to identify hazards that can affect the safe use of products. This step, hazard identification, is also sometimes called hazard recognition.

Having an understanding of "hazard," it is now important for design and product-safety engineers to identify the product's hazards. It is not that no one already has a relatively good working definition of the word "hazard," but it will be easier to classify hazards once they have been clearly identified. It will also be easier for an engineering team to identify previously unidentified hazards—or to identify a set of hazards for a new and innovative product that has not yet been built—once the origins, or sources, of hazards have been studied.

In order to provide safe products to consumers, it is *necessary* to understand the ways in which a product may harm a person. For many products, these ways, or modes of harm, are already known. For example, with the two-slice toaster, burns and electrocutions rise to the top of the potential-accidents list. For the lawn mower, serious injuries include amputations and strikes from thrown objects. Even with established products, once the prominent hazards and risks—*risk* to be discussed soon—are known and controlled, the next logical and ethical step for engineers to take is to similarly identify and address these *less-prominent* hazards and risks. These may prove not so easy to find. Hence, another need arises for hazard identification.

Hazard identification is certainly much easier for well-established types of products with which an engineer, company, or industry has significant experience. Creating completely new and innovative products may require a "clean sheet of paper" approach to dealing with hazards and their risks. This requires that hazards, in fact, be clearly identified before they can be addressed intelligently during the product-design process (PDP) and before product delivery to consumers. This task will be simplified, to a degree, when these new product types are derivatives of product types already designed and manufactured by a company. However, the differences between the old and new products may give rise to new hazards. It is here that hazard identification is needed.

The hazard-identification process is summarized well by Ericson, "Hazard recognition is the cognitive process of visualizing a hazard from an assorted package of design information (Ericson 2016, 40)." These product hazards sometimes do not arise from a single mechanical component; they may, instead, be an emergent property of the product system. The term "design element" is used, as in Figure 5.4, et seq., to denote either a component or emergent property of a product.

As with most disciplines within engineering practice, the more experience a design or product-safety engineer has with a product or a type of product, the better equipped that engineer will be to identify that product's resultant hazards. This familiarity with a product will, of course, also better equip that engineer to quantify risks in the risk-management process covered later in the chapter. As Ericson also astutely observes "…that hazard recognition is somewhat an art more than a science… (Ericson 2016, 40)."

The easiest way to get started with hazard identification is to consult a listing of hazards that may arise from product, or system, design. There are numerous hazards checklists available to readers in various references including Goldberg et al. (1994), Ericson (2016), "Consumer Products Safety in Europe: Corrective Action Guide" (2011), and EC/DGE (2006). This book includes a simple product-safety hazards checklist as Appendix A.[10]

---

[10] This short appendix focuses only on consumer-product safety hazards and not the larger set of hazards often addressed by system-safety and industrial-safety engineers.

In addition to Appendix A, a simple listing of several types of design-element sources, hazards, and one their risks—or particular ways of becoming injured by the hazard—is included as Table 5.1. The hazards shown in the table are generalized and quite broad. These hazards could be broken down further into subcategories and, in some cases, should be. For example, design elements presenting mechanical hazards could be broken down into several categories including reciprocating and rotating, for example. The same certainly goes for the other categories of hazards as well. Electrical hazards could be grouped together and, perhaps, focused on by the electrical-engineering team. However, it is anticipated that several types of engineers will be working together to address hazards and risks throughout the PDP. Yet, for the purposes of this part of this chapter, the broad hazard categories remain illustrative to readers studying this subject matter for the first time.

Hazard identification, or recognition, is an important early part of product-safety engineering. It is straightforward and simple to start. However, unless product-design hazards are uncovered and made visible to design engineers, their resulting risks may go unnoticed and unaddressed.

This and the prior section discussed hazards and hazard identification. Table 5.1 serves as the segue onto the next section on risks. The shaded portion of the table includes a brief set of human design elements, sometimes beyond the control of engineering designers, that can affect ultimate product safety by leading to accidents and possible injuries.

5.6 Risks

The difference between a *hazard* and a *risk* can sometimes be subtle. It is, however, important to understand the differences between these two concepts. A hazard is a characteristic, or a state, of a product design element which is capable of injuring someone. A risk is a metric for a hazard, thereby making a hazard measurable. This book will only focus upon hazards and risks to human safety and, generally, not consider property-damage only (PDO) risks or accidents. Human-health risks will also not be covered at length.

Most adults know that a sharp kitchen knife can readily cut a finger, but being cut is the risk and not the hazard. The *hazard* in this case is the sharp-edge of a mechanical product component. The cut to the finger is the *risk* from that hazard. The hazard is more of a *theoretical* aspect to a product—that hazard may exist or it may not exist on the product. The risk, however, is more of an *applied* facet of product safety. If the product does present the hazard, in this case a sharp edge, to its user then the engineer must assess the risk of the hazard.

Risk is the next logical step forward from hazard and represents progression from the abstract to the concrete, as stated earlier. All *risks* arise from *hazards* and all product hazards arise from product design elements. However, while hazard is an abstract term that may be considered in a vacuum, risk must be evaluated within particular circumstances.

Risk is a real threat posed by an indefinite hazard. Because of this, risks may be difficult to estimate. In some instances, the values of different risks are compared to one another by design teams in order to identify those risks *most* in need of possible further reduction.

Design Elements	Hazards	Risk Examples
Mechanical	Impact	To suffer serious internal injuries from being hit by a truck
	Falling object	To be severely injured by a passenger car falling off of a jack
	Sharp edge	To receive a deep cut on finger while peeling an apple
	In-running nip point	To have a finger amputated when lubricating a chain-and-sprocket machinery drive
	Stability	To fall from working from a ladder placed on uneven ground and break an arm
	Fatigue	To fall from a ladder because of being physically exhausted and break a leg
	Fall	To be fatally injured by falling off of a roof while installing a satellite television antenna
	Suspended load	To die by being crushed by a falling beam at a construction site
	Rotating machinery	A person with long hair receives removal-of-scalp injury when hair is caught in an operating go-kart engine fan
	Thrown object	A person is hit in the face by a rock thrown from a lawn mower but requires only first-aid treatment
	Pinch/Pinch point	A machinery operator suffers broken ribs when a reciprocating work platform pins worker against wall
	Crush	A worker loses a hand while changing the dies of a punch press when the punch press "cycles"
	Choking	A toddler dies from suffocation after a small item lodges in and obstructs her throat
Electrical	Shock	A home owner suffers fatal electrocution while re-wiring a home
	Ignition	A family perishes in a house which burns to the ground because of electrical short circuit
Chemical	Burn/fire/explosion (BFE)	A gasoline can is placed next to a portable generator while it is in use in a garage catches on fire and burns the house down without loss of life
	Lack of oxygen (O_2)	A worker dies of asphyxiation after entering a confined space filled with nitrogen (N_2) gas
	Poisoning	To die from carbon-monoxide (CO) poisoning at home
	Corrosive	To get a mild irritation from splashing drain cleaner into the eyes
Other energy	Thermal	To receive a mild burn on leg from an exhaust pipe while riding a motorcycle
	Radiation (ionizing)	To contract Stage-I cancer because of extended exposure to radioactive materials
	Noise	To develop severe hearing loss from long-term exposure to loud workplace machinery

Table 5.1 Examples of Design Elements, Hazards, and Risks

Human	Maintenance error	To die in a house fire because smoke-detector batteries were not replaced
	Inadvertent activation	To amputate a mechanic's finger while working on an automobile because the mechanic's assistant thought it was permissible to start the engine when it was not
	Substance addiction	To fatally overdose from taking too much of a prescribed opioid
	Workplace violence	To be seriously wounded by an active shooter while at work

Table 5.1 Examples of Design Elements, Hazards, and Risks (*Continued*)

The engineer—either design or safety—of a product possessing a very-sharp edge must determine, or at the very least understand, the severity of a resulting cut and the probability of receiving such a cut from the product being designed and while used as intended or as reasonably expected. Only then can a suitable *system* of engineering countermeasures be employed to either reduce the severity of a cut and/or to reduce the probability of such an injury.

There are two fundamental components to risk. These components of risk are *severity* and *probability*.[11] Severity is the extent to which a hazard may injure a person. Probability is the likelihood of a hazard injuring a person.

Severity of injury runs the gamut from none to fatal. Similarly, probability ranges from nonexistent to certain. Not surprisingly, there has been little *explicit* quantification of risk within the field of product-safety engineering, complete with units of measure, as a function of severity and probability. Therefore, generally,

$$Risk = f\,(Severity,\ Probability) \tag{5.1}$$

Some authors have presented other, more-explicit relationships for risk as functions of severity and probability. For example, some authors (Ericson 2016; Hammer and Price 2001; Ertas 2018) use risk formulations of the general form:

$$Risk = Severity \times Probability \tag{5.2}$$

Not surprisingly, there are no specific units provided for either severity or probability.[12] There is, generally, also no guidance given for what level of risk is *acceptable*. This, too, is not surprising since acceptability of risk is based upon one's value system as described earlier in the book. There is unlikely to ever be a universally acceptable level of risk that still permits the design, manufacture, and function of a useful consumer product.

It is likely to prove unproductive to try to uncover a universal *precise* relationship between risk, severity, and probability. The exact relationship will likely vary between

[11] Some other authors and materials include a third component to risk. These additions include economic loss (Jensen 2020) and avoidability ("ANSI/ISO 12100:2012 Safety of Machinery—General Principles for Design–Risk Assessment and Risk Reduction" 2012). Such aspects will not be considered in this elementary discussion of risk.

[12] Some severity scales for system-safety engineering, such as in MIL-STD-882E, include the costs of some accidents in U.S. dollars. Some probability scales may be of the form on in one thousand or 1:1000; other scales include such guidance as "not likely," "likely," and "almost certain," for example.

both evaluators and situations under consideration. What is important to understand is that product risk can be affected by *either* or by *both* components of risk—severity and probability. It is also worth mentioning that risk is *directly*, rather than *inversely*, affected by increases in severity and/or probability. Thus, any increase in severity and/or probability will result in an increase in risk.

If an arbitrary hazard presents a risk of *A* as shown in Figure 5.6, then any increase in either severity or probability—or both—will result in increased risk. These increased risks are shown as dashed lines of increasing risk. If both probability and severity are increased, then risk increases as shown in line 1; if only severity or probability increase, then their consequent increases in risk are shown as lines 2 and 3, respectively.

Similarly, if a hazard presents at a risk level of *B* as shown in Figure 5.7, then risk is reduced when severity and/or probability are decreased. When both severity and probability are lowered, risk is reduced along line 1. When only severity or probability is decreased, then risk is also reduced along lines 2 and 3, respectively.

While it is important to understand how risk can be lowered or raised through changes in severity or probability, or both, it is equally important to recognize that two points of equal risk may occupy different places within Severity-Probability space. One such example is shown in Figure 5.8. In this example, two different combinations of severity and probability produce the same level of risk. In this case, point *C* represents a low-severity, but high-probability, combination. On the other hand, point *D* represents a high-severity, but low-probability, combination. Yet, both of these combinations of severity and risk produce the same risk level. A hypothetical dashed line showing a uniform level of risk is shown in the figure as well.

In order to further explore the relationship between hazards and risks, Table 5.1 is provided. This table consists of three columns, the first of which is the product design element. Subsequent columns are hazards and risk examples. The rows form

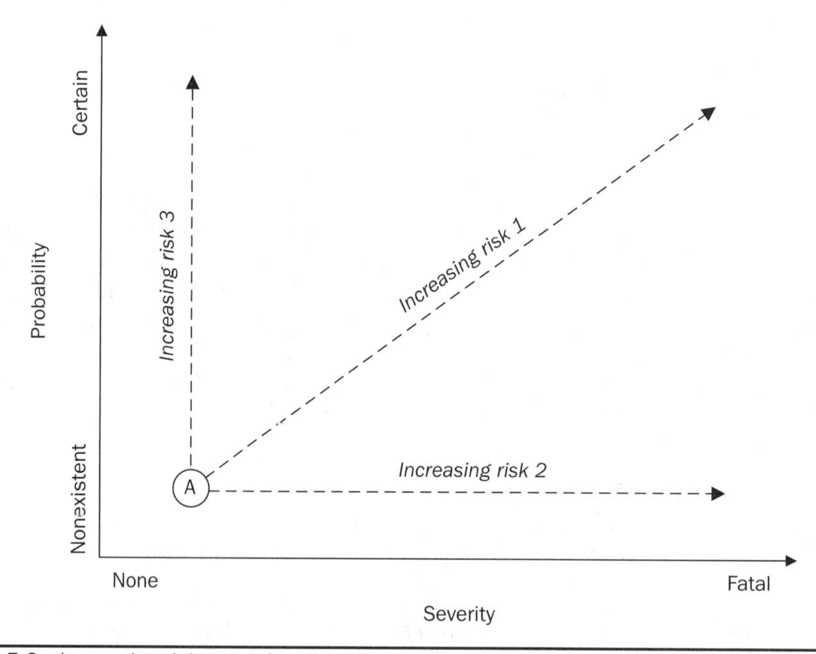

FIGURE 5.6 Increasing risks as a function of severity and probability.

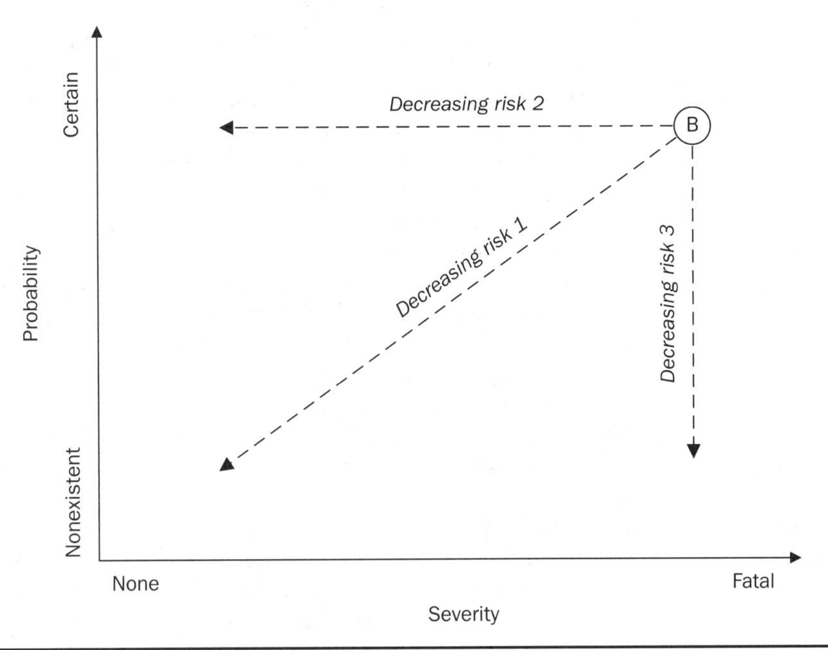

FIGURE 5.7 Decreasing risks as a function of severity and probability.

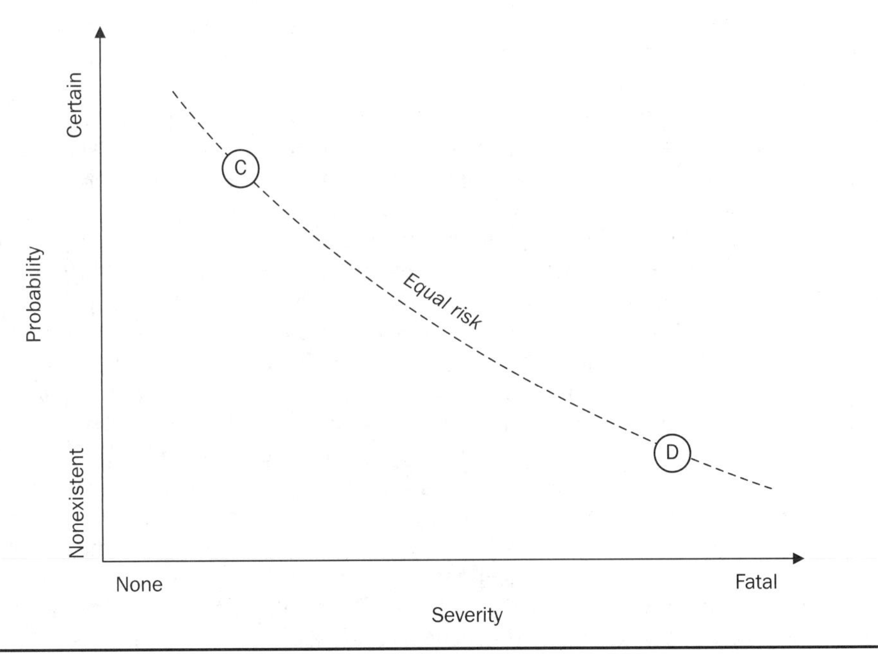

FIGURE 5.8 Equal risks as a function of severity and probability.

an abbreviated listing of hazards and risks which may present themselves to users of consumer products. Table 5.1's risk examples also include some outcome information in order to help demonstrate how hazards and risks differ from one another.

The first column, Product Design Element, is a generalized listing of categories of product characteristics or properties—hence, the term "elements"—contained in the product design. The titles shown represent placeholders for specific mechanical, electrical, chemical, and other-energy design elements used by the design engineer of the product to accomplish product certain tasks within its intended use. For example, a mechanical element to a product design could be an elevated load, a suspended load, or a sharp edge. A chemical element could be gasoline used to fuel a powerplant. From each product design element flow one or more hazards.

One example of a risk for each hazard is listed in Table 5.1. Hazards from mechanical product design elements include impacts, falling objects, sharp edges, in-running nip points, and others. The first three of these are hazards 1, 2, and 4 from the prior section. Hazard 3 from that section, the gasoline, is a hazard from a chemical product design element as shown in the table. There are, of course, other hazards from other types of design elements that may present themselves to users through products including those involving electricity and other energies. Only single examples of these are included in that table.

From the entries in Table 5.1, the reader can begin to consider further investigating the risk of a product hazard in a given situation. For immediate risk-estimation purposes, risk *severity* rankings will be limited to Negligible (N), Marginal (M), Critical (C), and Extreme (X);[13] risk *probability* rankings will be limited to Improbable (I), Remote (R), Occasional (O), Probable (P), and Frequent (F). These risk rankings for severity and probability are intuitive but are covered in detail later in this chapter. Negligible severity is much less than Extreme severity and Improbable probability is much less than Frequent probability. This above risk-evaluation rank imposition will ease the discussion of risk.

Earlier example risks put hazards into context. Consider the case of the parked truck pointed toward a village on a hill. It is relatively straightforward to deduce that a large vehicle in motion can easily kill or severely injure a person, or several people, so the potential harm is Extreme; this is the *severity* of the risk. Since the truck is pointed toward a village rather than an empty plain, the *probability* of injury would be considered Probable if the truck was in neutral and not braked but left there indefinitely. Given enough time, a strong wind, a severe earthquake, or another phenomenon could disturb the truck enough to get it rolling down the hill. However, if the vehicle's transmission was placed into a low gear or in "Park" along with the application of the park brake, the probability of injury from the runaway truck can be reduced to Remote.

In the case of the car elevated above the mechanic, the severity of a potential injury is also Extreme since a falling car can easily kill someone beneath it. Although working beneath an elevated car supported by a hydraulic jack, instead of jack stands, is never to be recommended, people do it regularly without consequence.[14] Therefore, the probability of injury could be considered Occasional, if not Remote, in magnitude.

As for the gasoline can near the operating portable generator, the severity of any injury suffered would be high if a house fire resulted, especially if no one was aware or awake to fight the fire and evacuate residents. Thus, this risk's severity is Extreme.

---

[13] The probability of "Eliminated" used in some risk evaluations will not be included here to simplify the analysis and since an eliminated risk might not be recognized during risk identification.

[14] This says much positive about the reliability and stability of many hydraulic jacks today. This is also an example of misuse *in kind* by using a temporary elevating mechanism as a semi-permanent one.

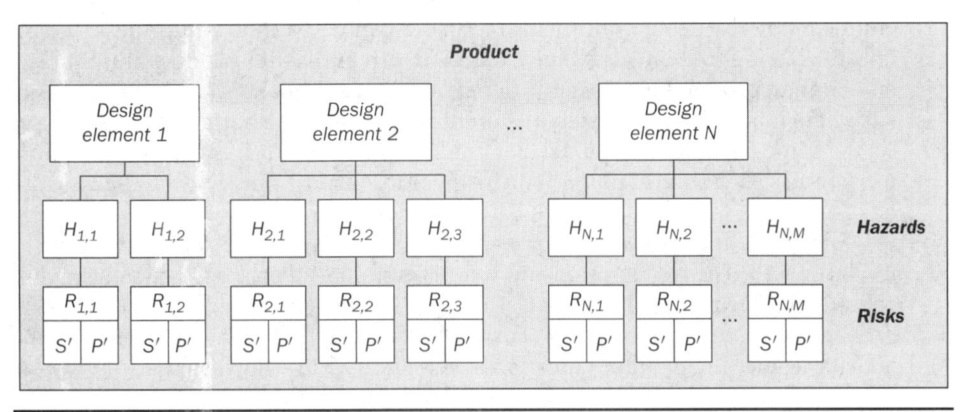

FIGURE 5.9 Product design-element, hazard, and risk model.

The probability of this event would likely be Remote due to the relatively low temperatures near modern generators in operation—unless the gasoline can is in indirect contact with the generator's exhaust system.[15] What might be more illustrative, in this example, is that the fire hazard and risk posed by the gasoline can may not be as large as the CO-poisoning risk presented by operating the generator in a semi-enclosed environment which could lead to a severity and probability combination of Extreme and Probable, respectively.

The final hazard example from the prior section involved a folding knife being used to peel an apple. At this point, the folding design of the knife is not important so long as the knife does not inadvertently fold on the user's fingers while peeling the apple. The severity of a cut finger, or fingers, during apple peeling should generally be considered Negligible since life-threatening injuries are unlikely from this activity. The probability of inflicting an injury upon oneself during apple peeling would depend upon one's experience and the attention paid to the task. Probability of getting a severe cut during apple peeling could also be considered Remote so long as an experienced adult devoting reasonable attention to the task was the knife's operator.

A generalized representation of risks from hazards is shown in Figure 5.9. In this figure, the content of Figure 5.4 is reproduced with the addition of a bottom row of risks from each hazard. The risks are labeled according to the hazard which presents that risk. The first hazard from the first design element is $H_{1,1}$ with its *largest* risk labeled $R_{1,1}$, and so forth, on through $H_{N,M}$ and $R_{N,M}$. Although one box is used for each titled risk, there are clear demarcations within each box for the severity and the probability of each risk. This helps to reinforce to the reader that risk is not a single quantity, but is instead a pair of two components that are to be estimated separately from one another. Each of the risks included in Figure 5.9 has a given, initial value for its severity, S', and for its probability, P'. For simplicity, the subscripts for S' and P' are not used and are presumed to be those of the risk $R_{i,j}$ from which they originate.

These values for S' and P' are those for a particular product and are the results of engineering-design decisions made by product-safety, design, and manufacturing

[15] It is also assumed for this discussion that the openings of the gasoline can are properly sealed with suitable capping devices.

engineers by the design-and-manufacturing company and its suppliers. It is important to consider the effects of the PUE model in Figure 1.3 when evaluating risk levels. Similar products made by different design-and-manufacturing companies and suppliers may have completely different S'-P' pairs. Furthermore, the same product by the same designer/manufacturer could have different S'-P' pairs if a manufacturing process is highly variable with respect to output. This will be discussed in the next section.

Taking the lawn-mower hazard example from Figure 5.5 one step farther results in Figure 5.10. The two hazards and five risks from two product design elements have been reproduced there. The new figure shows the risks labeled and with severity and probability values derived below. For the first risk, amputation, the severity of that risk is Critical since finger amputation is a severe injury, but—however bad—will not likely result in total disability. The probability of an injury from that risk is Remote due to decades of engineering-design experience and a readily discernable hazard—a sharp, spinning blade—to many people. The severity and probability of injury from a thrown object, the second risk, are Critical and Remote, respectively, as well. This is because, if the lawn-mower operator exercises even modest care, then few objects will be encountered during mowing and, even if hit by a thrown object, the object will probably not impact an area of a bystander's body, such as the eyes, to result in total disability.

The BFE risk severity is, of course, Extreme. Injuries from burning gasoline can be quite severe if not fatal. The probability, although not impossible, remains unlikely as indicated by a Remote rating due to consumer awareness of gasoline storage and spillage issues. The risk from a lack of oxygen (O_2) is Extreme in severity due to the terminal effects of extreme carbon-monoxide (CO) poisoning. The probability assigned in this example is Remote because the engine burning the gasoline is mounted on a mobile platform and is meant to be used outdoors. In addition, many people using lawn mowers are well aware of the unpleasant, if not fatal, effects of using a lawn mower in a small space. The poisoning risk due to gasoline is considered to be Marginal, neither Extreme nor Critical, since any ingestion would likely result temporary illness rather than death. In addition, the probability of ingesting a significant quantity of gasoline is limited to Improbable by gasoline's foul taste and odor. Placeholders are retained in the figure to represent as-of-now unidentified design elements, hazards, and risks from lawn mower use.

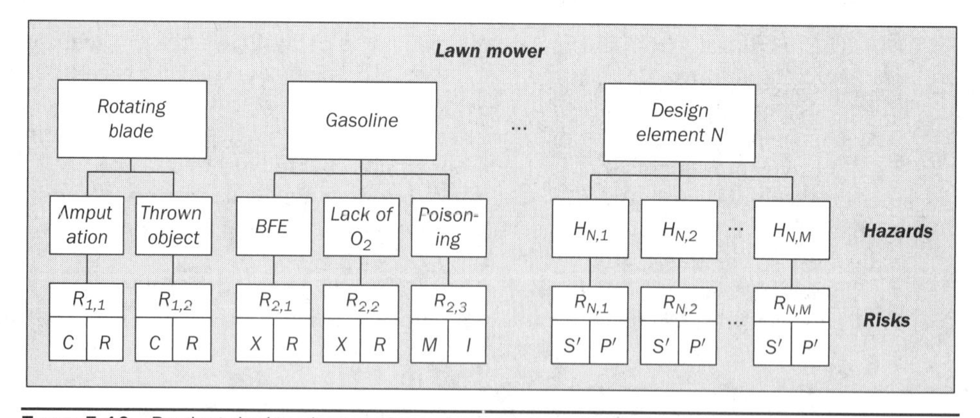

FIGURE 5.10 Product design-element, hazard, and risk model for lawn mower.

Each of these hazard risk severity and probability estimates are shown in Figure 5.10. It is likely that some readers will disagree with severity and/or probability estimate values concluded by the author. Such disagreement is unavoidable in the field of product-safety engineering. In addition, the author no doubt has a particular manifestation of, or design for, a residential lawn mower in mind and also makes unstated presumptions that, perhaps, differ from those of some readers. It is for this reason that analyses contained in this book are completely generic in nature and do not necessarily apply to any particular actual product. The author is not doing the "deep dive" that would be required to evaluate a detailed engineering design of a real-world product that will be delivered to actual consumers.

For real-world products, it is necessary to keep in mind the multiple-risk possibility of any hazard during the design, development, analysis, and testing portions of engineering design. For purposes of this book, and for simplicity's sake, only the *most-grave* risk will be considered for each hazard. Doing otherwise would overly complicate discussions and figures and detract from the important concepts. However, real-world applications for product design may well warrant parallel multiple-risk evaluations for single hazards.

The risks of the hazards, in the forms of *couplets* of severities and probabilities, from the previous section have been *estimated*[16] by the author above for the lawn mower. A diverse group of people will undoubtedly generate a varied set of severities and probabilities for risks. Therefore, although an effort is made to estimate risk—or to quantify it—the assignment of precise levels, or values, to risk components does not, *by itself*, give credibility to any risk-estimation endeavor. As with any aspect of product-safety engineering, the results should be reasoned conclusions rather than quick and convenient opinions. There will likely be disagreement, by someone, with the results of any risk-estimation exercise. This is another area of product-design engineering where hard work by practicing product-safety and design engineers is necessary and should be followed by the objective, engineering-ethical consideration of all pertinent factors.

5.7 Accidents and Outcomes

Some readers may expect a treatment of risk management or risk assessment at this point. The discussion on the nature of risk and consequences, however, is not yet complete. Although a product's design elements present hazards which have risks, the extent to which these risks manifest themselves is influenced by other factors—both *before, during*, and *after* an accident takes place. What is generally of interest to people is not only really an accident, but the final *outcome* of an accident. The accident is an event and its outcomes are a set of final "destinations" possible from the accident.

Chapter 1 presented the PUE model in Figure 1.3. This model indicates that all three components of the PUE model—the product, the user, and the environment—affect product safety. The Haddon matrix, Table 4.9, further expands on these components by dividing Environment into the Physical environment and the Socio-Economic environment. The dimension of "phase," with respect to an accident, is also introduced through

[16] No detailed analysis of either risk level or acceptability was conducted because there is no specific lawn-mower design presented having the much-needed specifics to do so. Again, risk estimation will soon be covered in this chapter.

the Haddon matrix. This section will explore how these may all be used to further examine design elements, hazards, risks, accidents, and accident outcomes.

Consider the product design-element and hazard model of Figure 5.4. This model shows the relationship between a product design element, or characteristic, and the hazards that flow from those design elements. No accident has yet happened and none of the accident's final results have yet been determined.

Some authors have constructed models for the events leading up to an accident, or mishap. Ericson (2016), for instance, identifies three necessary stages which take a hazard on to a mishap. In between the two endpoints, there is a hazard-actuation step. There is also a transition phase where a hazard state—a condition—transforms into a mishap state—an event.[17] This model, and others, have served the system-safety community well.[18] This book's author has, however, found that such a model does not address all aspects the product-safety accident and outcome issues that he faced when practicing engineering design. Without discounting existing and useful accident models, he was realized that there are sometimes substantial differences between system safety and product safety when it comes to critical aspects such as product complexity, physical environments, socio-economic environments, user qualification, pass/fail criteria, resource requirements/efficiency, and more. The large-scale aerospace systems designed by system-safety engineers are generally far more complex than traditional consumer products—even consumer products at the complex end of that spectrum (see Figure 1.2). Aerospace systems are also generally well specified for their physical-use environments. The organizational operators of such large-scale systems know and accept that the operation and maintenance of such systems will be quite expensive. They, therefore, dedicate the effort needed to properly specifying and operating the equipment or systems.

Consumer products are, on the other hand, routinely used in environments for which they were not designed. The operation of a large weapons system is unlikely to be affected by an unruly, inebriated crowd of people. This is not always the case with consumer products. In addition, those operating the large, complex systems *will be* educated and trained to cope with the complex needs of using that system. They will also, probably, be working regular shifts under suitable working conditions. They will be subject to reprimand—if not dismissal—for not following proper operating protocols, including the use of PPE. This is not so with some consumer products that may be used at night, after a long day's work, for instance. There is usually no administrative oversight in recreational activities to ensure that prudent operating rules are followed and that PPE is worn. There may be no penalties—aside from injury or death—for the misuse or abuse of a product.

The successful operation of large-scale systems usually culminates in an event-free experience. Experiencing no problems is a good thing with large, expensive equipment. Consumer goods sometimes provide the same boring experience, for example a toaster; however, some consumer products are purchased by users to provide excitement after a boring day at work. Furthermore, many designers and manufacturers of large-scale systems dedicate voluminous resources to assure, to the extent practical, the safety of

---

[17] Ericson's model is more detailed than this and deserving of a more-thorough explanation for those interested in studying hazard and risk further.

[18] Product-safety engineers should always use the tools and methods most appropriate for the task at hand.

these systems. This large and important engineering effort is built into the procurement price and operating costs of these systems for their consumers who either expect or demand such safety efforts. The consumers who buy many products do so on the basis of price or performance alone. This is no *excuse* for any design-and-manufacturing company to cut short its safety efforts. However, it could be a *reason* for some companies to try and use the money saved by not staffing product-safety personnel to further lower cost or increase product performance.

No ethical engineer should ever be convinced that a customer's blasé attitude toward her or his own safety is a legitimate excuse for poor product-safety characteristics. But the fact remains that many consumer-product development programs simply cannot support the expense of the extensive safety-engineering efforts going into large systems. This is true both in terms of money and in terms of schedule. The monetary expense is self-explanatory. The scheduling problem would arise from the time needed to successfully perform all of the traditional system-safety analyses used in large-scale systems. Many consumer products by different companies compete with one another by being the first product to market through rapid engineering design and development cycles. Some of the large-scale system analyses may also simply be unnecessary for many consumer products due to their relative simplicities. The system-safety methods are used in large-scale systems because their complexities can quickly overwhelm even the most capable of minds. In other words, the added cost of extensive system-safety engineering efforts on simple consumer products may ultimately add no value to the product as measured through ultimate user safety.

In summary, the large-scale system often operates in controlled environments with skilled workers under vigilant supervision and with proper financial support from the operating organization. Some consumer products are sometimes used in uncontrolled environments by operators, who are sometimes unskilled or even impaired, and without supervision. It is for these reasons that the author set out to construct an accident model that better represents the obstacles that he faced as a manager of product-safety engineering at a large design-and-manufacturing firm. Generally, the products delivered by his employer could be purchased by anyone with the funds to secure them. Operator training and PPE use was encouraged, but never mandatory nor could it be given the legal and regulatory framework within which these products were located.

The resulting Product-accident outcome model is included as Figure 5.11. Some of this model looks similar to Figure 5.9, the Product Design Element, Hazard, and Risk model. As with the earlier figure, the shaded box at the top of Figure 5.11 still represents the product being evaluated. Also, as before, included in this box are the product's design elements, hazards, and risks. Toward the bottom of the same shaded box is a horizontal dashed line that partitions the product's engineering design into two regions:

1. As designed and when used as intended
2. As delivered

The *first* of these areas is the region representing the product's design where the design engineer and the design-and-manufacturing company have produced detailed engineering drawings, analyzed the prototypes, tested the prototypes, and iterated through successive design improvements so that the product could be manufactured and distributed. The company also took measures to be "sure" that the product was

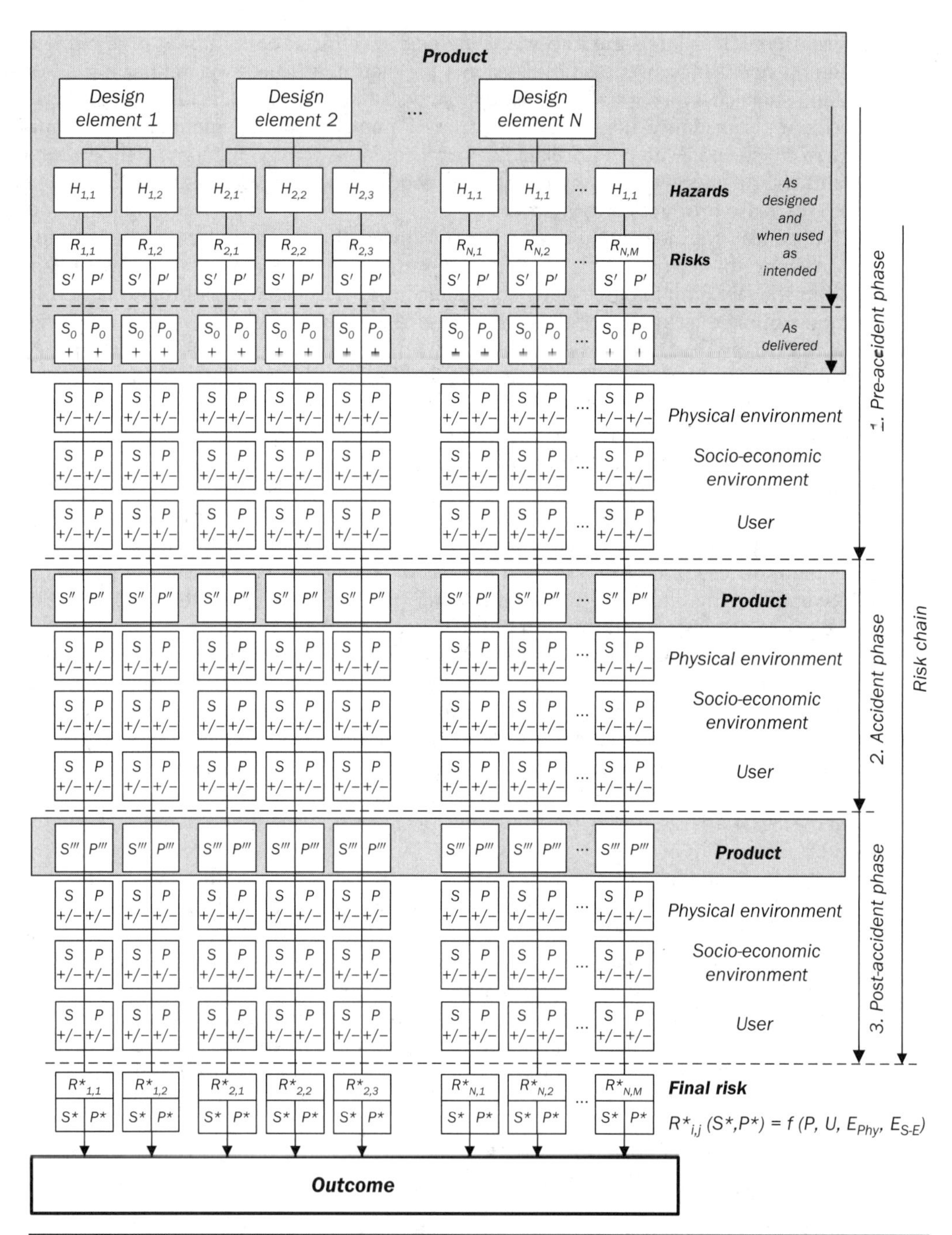

FIGURE 5.11 Product-accident outcome model.

manufactured as designed. This area is the "design intent" of the product. Presumed within the design intent is that the product would be used as intended by the company.[19]

In an ideal world, this product hazard and risk discussion of the top, shaded box would go no farther. Real-world factors, however, such as manufacturing flaws and unexpected product conditions, may make the product that was *designed* for the consumer and the product that was *delivered* to the consumer two different products. This is the reason for the *second* region in the top, product box below the dashed line. Design and manufacturing flaws, which may later prove to be safety defects, could affect one or more risks from using the product. Therefore, an additional risk-modification level is added to the bottom of the product box. The severities and probabilities, for these risks, are called S_0 and P_0, respectively.[20] Since it is *unlikely* that the design and manufacturing deficiencies leading to the creations of S_0 and P_0 will *lower* product risk, the boxes for these parameters only include a "+" sign to indicate potential increases[21] in injury severity and/or probability. This top, shaded box prescribes the *product* and its design including design elements, hazards, risks, and severity-probability pairs for the pre-accident phase of an accident event as described by Haddon (Haddon, Jr. 1983) as shown by the first downward arrow at the middle right of the figure.

Also included in the pre-accident phase, phase 1, of Haddon's methodology are the influences of the other PUE-model components. The product has just been covered. The other three factors, user and environments, are considered next. These, however, will be reversed in order so that the environments—both physical and socio-economic—are discussed before the user. The reason for this is to facilitate the evaluation of Figure 5.11 since these environments can have an effect on the actions ultimately taken by a product user.[22]

Immediately beneath the product box are rows for physical environment, socio-economic environment, and user. With some products, each of these factors can affect the risk of using a product—either by decreasing or increasing severity and/or probability of injury. This is why each risk box for these factors has both a "+" sign and a "−" sign. For some products, especially those with low levels of user activity as determined through Figure 4.12, the effects of these factors may be negligible. Yet, for other highly user-active products, these effects on overall risk to a user can be significant. For example, with an electrical product, its use in a damp or wet physical environment can greatly increase risk. In the case of the portable electric generator, using it within an enclosed space greatly increases its risk to home residents. In the case of an off-road vehicle (ORV), the accumulation of ice on a trail could contribute to an

[19] A product which is quite versatile, or whose company has a marketing department and a sales force either showing or encouraging virtually unlimited product usage, may encounter problems since there is no readily identifiable "intended use." This makes the company allegation that product misuse or abuse by a consumer led to an accident and injury a difficult one for a company to support if they ever need to.

[20] These risks are also devoid of subscripts in the model for greater clarity.

[21] While it is *possible* that design and manufacturing flaws could possibly decrease risk, it is presumed here that it is *highly improbable* for this to happen. In addition, there may be cases in which such a design or manufacturing flaw can make no difference in overall product risk. However, since unknown and uncontrolled engineering design and manufacturing variances are to be avoided for product-safety reasons, it is presumed that such deviations *only* negatively impact product risk and consumer safety.

[22] The user, physical environment, and socio-economic environment do not necessarily need to appear in a particular *order* so long as their effects are properly included and addressed within the model.

accident by increasing the probability of running the ORV off the road and hitting a tree or falling into a crevasse. As for socio-economic environmental factors, if a recreational product is being used by or near a large, rowdy, risk-seeking crowd, social facilitation may encourage the user to operate the product recklessly. This would increase the risk to users—and to bystanders, perhaps. The user may instead, despite the social pressure to provide thrills to the onlookers, decide to operate the product in a quite-careful manner. This would lower the risk of the product to people including the user, any passengers, and bystanders. All of the factors included in the Product-Accident-Outcome model—the product design, design-and-manufacturing variations, environmental factors, and user behavior—may each have an impact, positive, negative, or negligible, on the overall risk posed by a product to a user in its pre-accident phase of use.

There is *no* attempt made with this simple Product-Accident-Outcome model to specify the precise relationships among severities, probabilities, factors, phases, and potential final outcomes. Many of these may vary significantly for different products and types of products. It is up to the engineers, and other personnel at the company, working on the products to understand, to consider, and to address these influences and their effects on ultimate product safety as far as practical.

Phase 2 of this model, as with the Haddon matrix, is the accident event involving the product. The vertical accident-phase arrow starts where the pre-accident phase arrow ends in Figure 5.11. The automotive industry has engineered crashworthiness properties into its vehicles so that when an accident does take place, there are measures through design elements within vehicle that have the ability to protect users or to reduce injury when accidents do take place. Of course, it is always better to avoid accidents through good engineering-design and manufacturing practices. It is not, however, always possible to prevent accidents.

Just as on-highway vehicles have seat belts and airbags to protect vehicle occupants in a crash, a product may possess a designed-in feature to help reduce the severity of an accident. An example of one such effort could be the component identified by the arrow in Figure 5.12(A). The long, tubular element on the printed-circuit board (PCB) is installed in a fuse holder. Such a PCB could be part of a television in a home. This element is an elongated glass tube which is filled with a fire-extinguishing chemical agent. The sealed glass tube, or "bulb," is also coated with an electrically conductive material so that the bulb functions as a conductor, hence its placement in a fuse holder. Should an element on the PCB overheat and cause excessive heat or a fire, the bulb is designed and manufactured to shatter at a precise temperature as shown in Figure 5.12(B). When the bulb shatters, the fire extinguishant contained within the bulb is released to put out a small fire.[23] This bulb is just one way that engineering design can be used to reduce the *severity* of an accident event at the time of the accident. Such a device can do nothing to reduce the *probability* of the event since those factors are well beyond the control of the bulb and its engineering designers likely working for the supplier. However, such an element can stop a fire, thereby reducing the severity of an accident and, perhaps, preventing injury altogether. This is an example of engineering design reducing product risk during the accident phase of use.

[23] The proper sizing and placement of such a fire-extinguishing bulb on a PCB is important to the function and efficacy of the bulb.

Figure 5.12 Fusible fire-extinguishing bulb. (A) Extinguishing bulb installed on PCB. (B) Extinguishing bulb shattering from heat and releasing extinguishing agent. (Used by permission of the JOB group. JOB is an internationally registered trademark of the JOB Group.)

Since these during-the-accident risk severities and probabilities are designed into the product during engineering design, there are no "+" and "−" signs to S″ and P″. These properties have already been established by through engineering design and will not change during the accident event.

In the case of a television, there is probably little that the physical environment, socio-economic environment, and user can do during the accident phase to influence overall product risk. As mentioned earlier when discussing the pre-accident phase, other products and product types may differ significantly in potential effects during this and the post-accident phases of product use.

Presume for the moment that the small fire which was just mentioned had erupted on the television's PCB and that it had, in fact, been extinguished by the shattered bulb. This may not be the end of risk to the television owner. Even *after* an accident event has taken place, the final outcome may not yet have been determined. In the case of an automobile, even if a single-vehicle accident is survived by the vehicle's occupants, if that car bursts into flames because of a large gasoline spill, followed by ignition of the pooled gasoline, then the occupants still face significant injury *after the accident*. Automobiles are tested for post-crash integrity as a result of such accident scenarios. Similarly, with the PCB above, again imagine that a small fire erupts and is successfully extinguished by the bulb. Perhaps damage to the PCB circuitry had already taken place. A short-circuit electrical path could have been created through heat damage to the PCB from the small, short-duration fire. Although the fire which caused the bulb to shatter and release the extinguishing chemical has been put out, that circuit might otherwise remain energized, or "live," due to the short circuit since the television remains plugged into the residential electrical outlet. This condition could lead to a post-accident fire. However, due to the design and construction of this particular extinguishing bulb, it acts as a fuse and—when shattered by heat—*permanently*[24] disconnects the electrical power to the PCB. This bulb's action acts as a post-accident safety mechanism that is *designed into* the television, at some added cost, by the engineering team. Unlike the bulb's fire-extinguishing agent that lowers the *severity* of an accident-phase injury, the bulb's fuse function reduces the *probability* of a post-accident phase fire in the product through design-engineering foresight.

Just as the during-accident risk severity and probability values do not change during the accident event, the post-accident phase risk severity and probability do not change. Therefore, again, there are no "+" and "−" signs to S″ and P″ for the risks during this phase.

Moving on to the other post-accident factors affecting risk, had the fire not been extinguished by the bulb, the existence of a residential smoke or fire detector in the home could reduce the probability[25] of an injury through is audible alarm. This would be one effect of the physical environment in the post-accident phase as would the physical distance to the local fire department and the resulting on-scene response time.

As elementary as it seems, the socio-economic environment could also help reduce the propagation of any fire, and any resulting injury, through social pressures to keep a home clean. Good housekeeping practices, such as not having flammable materials

[24] Some thermal cutoff (TCO) devices may reestablish current flow once the PCB has cooled while other TCOs could melt and fuse together to permit electrical current to again flow to the PCB.

[25] It could be argued that the smoke detector can reduce the severity of an injury since it could reduce a fatal-burn injury to a smoke-inhalation injury.

scattered throughout the house, especially by the television, can help reduce the damage and injury, or severity, from a fire. In addition, next-door neighbors could see the fire and help evacuate people from the burning residence. However, such socio-economic environmental influences are not ones upon which an engineering-design team should count. Their design-and-manufacturing duty is still to prevent a fire from ever erupting on the PCB in the television. The user can help in product-risk reduction through assuring that the smoke or fire detector in the home is functioning properly, perhaps by using the "test" button found on many units and by regularly changing its batteries.

The above hazards, risks, severities, and probabilities show influences of the factors and the phases used in the Haddon matrix upon the final risk and accident outcome of product use. Engineers affect the product characteristics during the product-design phase (PDP), users can make prudent product-use decisions, and the environments can affect either the product or the user—or both. These influences are present, in varying degrees, during the three phases of product operation: pre-accident, accident, and post-accident.

The final risks of the product are shown near the bottom of Figure 5.11 above the "Outcome" box and are the result of the "risk chain" vertical line at the far right of the figure. This final risk is a theoretical construct intended to demonstrate how the ultimate outcome from an accident event involving a product may be affected by factors beyond the control of the product-design engineers—although those engineers absolutely play a vital role in both preventing an accident and affecting accident outcome. There can only be theoretical considerations and no practical absolute levels assigned to the final risk's severities and probabilities affected by the entire risk chain. However, to represent these entities, the final risk from the *i*th design element's *j*th hazard is denoted by $R^*_{i,j}$. Adding the two necessary risk components results in the notation for final risk of $R^*_{i,j}(S^*, P^*)$ where S^* and P^* represent final-risk severity and probability, respectively. This final risk and potential accident outcomes are influenced by the product, the user, and the physical and socio-economic environments.

As mentioned earlier in this section, the actual item of interest to product-safety engineers and professionals is generally not the accident itself; of typical interest is the final outcome from the accident. The two are not identical since it has been shown that the post-accident phase of product use can be affected by the PUE elements. An accident event, although bad, can be made even worse before it is all over. The severities and probabilities of some post-accident risks can be increased through poor engineering design, user actions, and environmental factors. But it has also been shown that intelligent design decisions made by engineering staff can prevent an accident event from producing an injury outcome.

Certain consumer products will, of course, have will have a wider range of outcomes than others. The extent of user activity, from Figure 4.12, may also significantly factor into the accidents and outcomes as well.

5.8 Risk Estimation

Risk, and its components, have been discussed without having presented a methodology for estimating it. Although this risk-estimation process is sometimes called risk assessment (USA/DoD 2012), the risk-management model shown in the next section uses the latter term differently. The section after the present one will help to differentiate these various "risk" terms. Risk estimation will be the focus of the discussion here.

Risk estimation is the stage at which risk severity and risk probability are combined to form a *single* value for risk *level*. Risk estimation is the process for determining the risk levels for pairs of severity and probability. For example, if two risks share the same severity of injury should the risks manifest themselves, but they have different probabilities for occurrence, then these to risks may have different levels of risk, depending upon the risk-estimation methodology used.

The risk-severity levels provided in the prior section will now be given criteria to help with risk estimation. These prior four levels of severity—Negligible, Marginal, Critical, and Extreme—are listed in Table 5.2 alongside their numeric severity levels (1–4) and potential consequences or resulting injury extents. A Negligible severity level, 1, would be for a minor-injury potential requiring nothing beyond a bandage, perhaps. A risk-severity level of 2 would be for a Marginal injury potential that could, potentially, require an emergency-room (ER) visit—without in-patient hospitalization, but which would be followed by a complete and rapid recovery by the injured person. A Critical injury, 3, would be serious and require the hospitalization of the injured party. In addition, permanent partial disability would be the result of the injury. An Extreme risk severity, 4, would result in fatal injuries or permanent total disability, including brain death. Depending upon the product, the risk-severity levels may only apply to the product users themselves. For other types of products and usages, the risk severities may extend to bystanders, residents, and passengers. It remains up to the engineering-design team to make the appropriate, ethical determination on the potentially affected population and the effects thereon.

In a manner similar to risk severity, the risk-probability levels are listed and described in Table 5.3. Within this table, the five levels of risk probability—Improbable, Remote, Occasional, Probable, and Frequent—are assigned the numeric levels 1, 2, 3, 4, and 5, respectively. An Improbable risk-probability level, 1, denotes that the risk will rarely, if ever, actually occur. The next most-probable risk level is Remote, or 2. At this

Description	Severity Level	Consequences of Injury
Extreme (X)	4	Fatal, permanent total or significant disability
Critical (C)	3	Serious, permanent partial disability, requires hospitalization or extended recovery period
Marginal (M)	2	Moderate, could require ER visit but not hospitalization, full and rapid recovery
Negligible (N)	1	Minor, requiring first aid only

Table 5.2 Risk-Severity Levels

Description	Probability Level	Probability During One Product-Unit Lifetime
Frequent (F)	5	Will occur often
Probable (P)	4	Will occur several times
Occasional (O)	3	Might occur once
Remote (R)	2	Unlikely to occur
Improbable (I)	1	Unlikely to occur—almost impossible

Table 5.3 Risk-Probability Levels

		Risk Severity			
		Negligible	Marginal	Critical	Extreme
Risk Probability		1	2	3	4
Frequent	5	Medium	High	Severe	Severe
Probable	4	Medium	High	Severe	Severe
Occasional	3	Low	Medium	High	Severe
Remote	2	Low	Medium	Medium	High
Improbable	1	Low	Medium	Medium	Medium

Table 5.4 Risk-Estimation Matrix

level, it is still unlikely that the given event will occur, but it would not be totally unexpected, however rare. The Occasional risk level, or 3, would indicate that the event might, perhaps, occur once in each product's lifetime. If such an event *could* be expected to happen multiple times with each product unit, then the risk probability would be Probable and be assigned a numeric value of 4. The Frequent risk-probability level of 5 is assigned to those risks that *should* be expected to happen with regularity.

Now that both a risk's severity and probability have been assigned values using Tables 5.2 and 5.3, it is possible to combine the two measures into a single metric through the use of a risk-estimation matrix. For this matrix, the risk-severity range, from 1 to 4, is placed horizontally while the risk-probability range, from 1 to 5, is placed vertically. This representation is shown in Table 5.4 as the Risk-Estimation Matrix. This combination of risk severity and probability produces twenty (20) risk-estimation matrix elements. Each of these elements is assigned a risk estimate of Low, Medium, High, or Severe.

Revisiting the lawn-mower hazards and risks discussed earlier and shown in Figure 5.10, the risk severity and probability of amputation were determined to be Critical and Remote, respectively. Using the earlier hazard notation, this lawn-mower example hazard could be represented by the following which denotes, by subscript, the first design elements' first hazard, the blade:

$$H_{1,1} = H_{Rotating\ Blade,\ Amputation}$$

Using the same notation now for the risk from this hazard produces the following:

$$R_{1,1} = R_{Rotating\ Blade,\ Amputation}$$

Remember that the risk severity and probability for the blade-amputation risk were Critical and Occasional, respectively. Supplying these two arguments to $R_{1,1}$ and looking up the corresponding risk estimate from Table 5.4 produces:

$$R_{Rotating\ Blade,\ Amputation}(Critical,\ Remote) = Medium$$

Completing the second hazard from the first lawn-mower design element produces the following:

$$H_{1,2} = H_{Rotating\ Blade,\ Thrown\ Object}$$
$$R_{1,2}(Critical,\ Remote) = Medium$$

Looking now at the second design element, the gasoline, the corresponding risks, severities, probabilities, and risk estimates are:

$$H_{2,1} = H_{Gasoline,\ BFE}$$

$$R_{2,1}(Extreme,\ Remote) = High$$

$$H_{2,2} = H_{Gasoline,\ Lack\ of\ O2}$$

$$R_{2,2}(Extreme,\ Remote) = High$$

$$H_{2,3} = H_{Gasoline,\ Poisoning}$$

$$R_{2,3}(Marginal,\ Improbable) = Medium$$

Thus, the risk estimates for the lawn-mower hazards and their risks were determined by using the criteria and methodology contained in Tables 5.2, 5.3, and 5.4. The risk criteria used in these tables are based upon a combination of two existing risk-estimation methodologies. These methodologies are the military system-safety standard (USA/DoD 2012) and the European Commission's guide for corrective actions on consumer products, including product-safety recalls ("Consumer Products Safety in Europe: Corrective Action Guide" 2011).[26]

The risk-estimation methodology—contained in Tables 5.2 through 5.4—is intended to be *descriptive*—not *authoritative*—in its nature. Its use permits a simple and straightforward discussion of risk estimation in a classroom. Product-safety and design engineers should use the classification regiment most appropriate—or even *required*—for their products. Such a methodology could come from an industry or be an internal company process. The author remains certain that other qualified and ethical product- and system-safety engineers and safety professionals will disagree with the scheme presented here. Disagreements may be with the severity levels, the probability levels, their "mappings" to risk-estimation levels, or all three. Such a discussion is healthy and is to be expected with such an important topic with varying perspectives by the numerous parties involved.

There are sometimes criticisms of a methodology used to reach a risk-estimate result. However, the real criticism sometimes is not of the *method*. The true criticism may be that the party objecting did not get the *result* that was expected, wanted, or needed to further their respective interests. The risk-estimation methodology presented here has not been approved of by any group. It is, instead, the mixture of two established methodologies. Mixing two things together can sometimes create a mess that is less than the sum of its two parts; however, neither of the two available methodologies was ideally suited for a product-safety engineering course. One important problem with some established risk-criterion ranking schemes is that, for classroom purposes, there are no data available to quantify the probability of a risk from a product—especially if the product used in that classroom example is fictitious. This resulting risk-estimation methodology was created to fill that void. Due to this need and with the admitted lack of peer review, there is substantial room for criticism if this methodology is used in engineering practice. However, this method might just as easily be perfectly

[26] From personal experience, the author found neither of these methods ideally suited to classroom use in a product-safety engineering course.

appropriate for some situations. Consequently, criticisms of this methodology should be of its criteria, not its results.

As a corollary, some parties may be all too happy with any agreeable result regardless of the methodology used to reach it. This is irrespective of whether that party is a designer/manufacturer, regulator, attorney, or consumer. This is why engineering ethics is such an important part of engineering design and *Design for Product Safety (DfPS)*. Simply changing which risk-probability level is *assigned* to a particular risk does not change its actual potential harm despite possibly *hoping* that it will. Changing the estimate of a risk, though manipulating its severity and/or probability, actually changes neither and doing so can hide a true hazard from further scrutiny. Doing so is also tantamount to eliminating crime by legalizing everything. Any re-classified hazard risks will still happen with the same severities and frequencies once users begin operating the product. This is why honest and ethical discussion and reasoning is so important to true product-safety engineering. The risk estimation that has been discussed is one important element of a larger process called risk management and is covered next.

5.9 Risk Management

The broad and potentially baffling topic of risk management is now discussed. This topic can be confusing due to the proliferation of *risk* terms. These terms include risk management, risk analysis, risk assessment, risk evaluation, and risk reduction. There are more such terms. These will be defined. If they cannot all be recalled precisely by readers, remember that it is much more important that these risk measures be *performed* than that they be *memorized*.[27]

The engineering-design process, coupled with engineering ethics, requires that a product having design elements leading to product hazards must have the risks from those hazards evaluated and addressed, if necessary, in some way. This overall task is called the risk-management (RM) process. It is composed of several sub-processes. It will be presumed that those performing risk management will fully consider the multiple factors and phases of the product-accident outcome model in Figure 5.11 while performing that task. Despite methodological rigor in the risk-management process, there are many points within the process at which valuations—sometimes in the form of judgment calls—must be made. These valuations will, the author hopefully says, be made by engineers using solid engineering ethics once numbers and physics are no longer helpful. It is at such points in the process that assessments made by applying the value systems for different people and parties may well diverge and cause disagreement with final risk-management results. However, good design and product-safety engineers should always strive to make ethical decisions throughout the risk-management process.

The risk-management model shown in Figure 5.13 is similar in some ways to the one used in one international machinery-risk standard ("ANSI/ISO 12100:2012 Safety of Machinery—General Principles for Design—Risk Assessment and Risk Reduction" 2012). Yet, the risk-management model shown in this figure differs from the ISO-12100 model.

Within the industrial-machinery industry for which ISO 12100 exists, machinery is often sold and delivered to a commercial manufacturing firm. That company is

[27] The author, himself, confesses to sometimes needing a diagram to keep straight some of these terms.

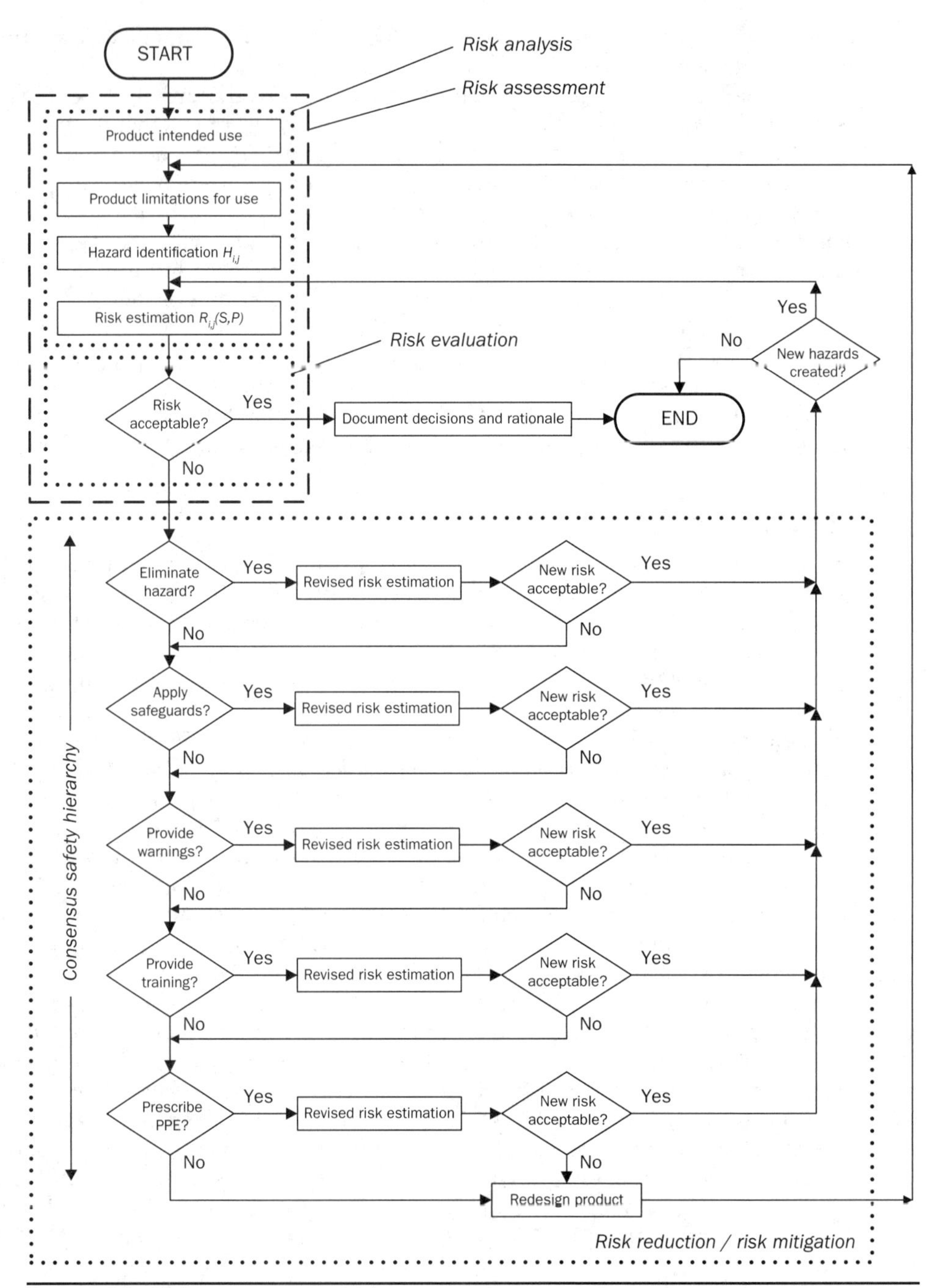

Figure 5.13 Risk-management model.

considered the "user," not the operator at the machine's controls. The ISO standard and its model are intended for a design-and-manufacturing company making such machinery for industrial use. In the ISO model, that *user* is responsible for implementing the additional layer product-safety measures for which the consumer-product design engineer is responsible. These measures to be implemented by a commercial user include administrative controls, engineering controls, education and training, and prescription of PPE as shown in the hierarchy of controls in Figure 4.3.

For the risk-management model in Figure 5.13, the engineering-design task includes the potential prescription of PPE to be worn while consumers use a product. This is because the "user" is the purchaser and operator of consumer products. There is no party between the design-and-manufacturing company and the end user of the product that affects product use. [28]

The engineering designers of consumer products should consider steps, such as those shown in Figure 4.1, prior to delivering the product to their consumer—the end user. Therefore, the prescription and use of PPE must necessarily be considered within this book's risk-management model. In addition, the use of the five-step consensus safety hierarchy replaces the use of the three-step *inherently safe design* measures from section 4.2 and the ISO standard.[29]

The risk-management process begins with *risk analysis*. In this sub-process, design or product-safety engineers consider the intended use of the product along with any limitations for use of the product. This establishes a spectrum of product applications and conditions for those applications. Some products will have a few, very-specific uses and also be used under limited conditions; for other products, both the intended use and limitations for use may be significantly broader and harder to pinpoint during the risk-management process. Risk analysis continues with the identification of the product hazards presented by the design elements of the product and is followed by risk estimation which was covered in the prior section. During risk estimation, the level of risk from $R_{i,j}(S,P)$.[30] Through studying the design of the product, its intended uses, its limitations, its users, and its usage environments, the engineering team will be able to assess the level of risk posed by a hazard to a user. Risk analysis consists of the combination of product intended use, its limitations for use, hazard identification, and risk estimation. This sub-process is shown by the dotted box surrounding these components at the top left of Figure 5.13.

The next step after risk analysis is *risk evaluation*. At this stage, the risk level determined in risk estimation is evaluated to see if it is low enough to be considered *acceptable*. The determination of what level of risk is acceptable is, of course, key to risk management. As the reader will see, the risk-management process iterates until all risks from all hazards are reduced to acceptable. There is no level of risk—aside from zero, perhaps—with which all parties involved in product safety—engineers, companies, regulators, advocates, attorneys, insurers, and consumers—will be happy. A design-and-manufacturing company, its engineering personnel, and its attorneys will generally be tolerant of a higher level of risk than an injured person and her/his attorney.

[28] There are exceptions to this, for example, a company that rents or otherwise temporarily provides a product to end users.

[29] Inherently safe design is achieved "by avoiding hazards or reducing risks by a suitable choice of design features for the machine itself and/or interaction between the exposed persons and the machine" ("ANSI/ISO 12100:2012 Safety of Machinery—General Principles for Design—Risk Assessment and Risk Reduction" 2012, 20).

[30] This notation represents the risk severity and risk probability of the *j*th risk of the *i*th design element as used in Figure 5.9.

Advocates and regulators are obligated to apply constant pressures on designer/manufacturers to lower risk, as well. Again, engineering personnel must always act in a manner consistent with their ethics, despite any disagreement from others—within or outside of the company.

In the field of system safety, there is generally a person or group responsible for approving the final risks presented by large, complex systems. It may be a military or government official, or it could be a group or an individual subordinate to one of these two people. Either way, the government contractor is fortunate enough to have a single party by whom risk-management efforts can be approved. It would be unrealistic to presume that this final approval will be easy to get; however, there is no ambiguity in this relationship and the final Pass/Fail decision.[31] Yet, it is the lone product-safety engineer who must make risk-evaluation judgment routinely without access to a supreme entity who can unilaterally approve of an acceptable level of risk. In the end, the exercise of critical thought when objectively assessing a problem or a potential problem, consultation with knowledgeable people, and the application of engineering ethics must give guidance on risk acceptability.

Design and product-safety engineers do not work in a vacuum, however. With many products, there are levels of performance and safety that have been established—and may even be expected—over years of production. Some levels of safety may even be explicitly required. There are sometimes useful criteria regarding performance and safety characteristics, too, in some industry standards and regulations. Another chapter in this book discusses why there is little concrete guidance provided to engineers working on new, innovative products and why some products and companies are more affected by criterion scarcity than are others.

The risk-evaluation process is also shown within a small dotted box on the left of Figure 5.13. The sequence of risk analysis plus risk evaluation is known as *risk assessment*. This is shown by the dashed box at the top left of Figure 5.13.

If the level of risk truly is acceptable, then the risk-management exercise is over—at least for that particular hazard. All that must be done is to properly document the decision about the design and any rationale and data used that may be helpful if asked about it at a future date. It may be difficult to remember precisely why a decision—especially one considered to be insignificant at the time—was made many years after the fact. It is, therefore, wise to document decisions—especially if there were particular reasons why a conclusion was reached that may not be obvious to casual observers at a later date. The simple example just mentioned only considered a single risk arising from a single hazard. Real and complex products require that each hazard risk be so assessed.

In some cases, the level of risk is not acceptable initially. In these cases, it is necessary to go through *risk reduction* where, obviously, risk is reduced. Risk reduction is also known as *risk mitigation*. This risk-reduction process is shown within the large dotted line in the middle and to the bottom of Figure 5.13. In this stage, the consensus safety hierarchy from Chapter 4 is applied, in principle, to help reduce product risks to consumers. Readers will see that the first step is, logically, the consideration of using a less-hazardous design feature. As with the earlier discussion of the safety hierarchy, there is often no alternative available that is less hazardous but still meets functional requirements. However, if hazard elimination or hazard replacement is possible and is

[31] The author concedes that no substantial system-safety approval is a simple decision.

undertaken, then another risk estimation and risk evaluation are performed afterward to confirm risk reduction. If risk is now acceptable, then the process—for *this* risk—concludes unless another risk has been created, in which case, the risk-management process returns to risk estimation for the new hazard.

If the first step, hazard elimination, does not sufficiently reduce risk, then the second step of the safety hierarchy is applied. That step is the application of safeguards. If safeguards are applied, then another risk estimation and risk evaluation are conducted. As with the hazard-elimination step, if safeguard application sufficiently reduces risk, then the process is done unless more hazards have been created.

If applying safeguards does not reduce risk to acceptable, then the third step, providing warnings to product users, is considered. If warning product users of hazards and risks is *truly* able to sufficiently reduce risk, then—unless new hazards have been created—the process is over. Otherwise, the next step in the safety hierarchy, the provision of user education or training, is applied.

Risk estimation and risk evaluation are conducted if education and training is used to reduce hazard risk. If risk can be reduced to an acceptable level through these risk countermeasures, then the risk-reduction phase is complete unless a new hazard is created through the effort. The risk reduction moves onto the final priority in the consensus safety hierarchy if risk remains elevated.

Lastly, the recommendation for use of PPE is to be considered for additional risk reduction. There are two important aspects of this consideration. With the first consideration, the engineering-design team must be realistic about the true level of protection that both *can* and *will* be provided by PPE. Each of these is itself a separate sub-consideration:

A. The level of protection that can be provided by PPE when it is worn and
B. The consideration that PPE may not be worn by some users—especially if the PPE is cumbersome or inconvenient to use

Each of these sub-considerations should be evaluated by the engineering team when recommending the use of PPE and estimating its efficacy.

With the second consideration for PPE use, the engineering-design team should ask if its recommendation to use PPE is simply "off-loading" a product-safety engineering responsibility onto consumers that could be better addressed through additional engineering design. However, there are indeed products and instances where PPE use may remain recommended, if not necessary, regardless of additional engineering-design efforts, for example, the use of helmets when operating certain vehicles.

The use of PPE to establish a given level of risk could be considered an *active* safety measure since actions by a consumer are required to realize this level of safety. The product safety provided through engineered product design elements can be considered *passive* safety measures since this level of safety will exist regardless of user actions. It is, of course, preferable to use passive safety measures over active safety measures. This is, appropriately, reflected in this application of the consensus safety hierarchy.[32]

It should be hoped that, after the recommendation for PPE use, the level of product safety will have been reduced to one which is acceptable to most people. If this is

[32] It is also implicit in the hierarchy of controls.

so, then—barring the creation of additional hazards—the risk-reduction process concludes. If PPE use proves ineffective or cannot be used to sufficiently reduce risk level, then product redesign is necessary to further reduce product risk. The risk-reduction process proceeds back to the top and into the risk-analysis box where limitations for use may change as a result of the product redesign. Once there, the risk-management process continues as before. Iteration continues for each risk until each risk is successfully reduced to an acceptable level.

It goes without saying that a product should not be released for production and distribution until and unless the design and product-safety engineers are satisfied with the resulting risk levels posed by all hazards. Again, there may not be consensus about the risk assessment between designer/manufacturer and regulators, but the engineering team should be satisfied with the product's safety characteristics both technically and ethically. This decision should not be influenced by business pressures from management or executives.

There obviously have been and will continue to be products that have not or will not follow such a strict risk-management process. At least, not formally anyway. Products designed decades ago and continue to be made today may not have proceeded through such a sequence of specific analyses. However, it is possible that intelligent and capable engineers have, over the years, carefully considered each of these risk-management tasks and iterated through engineering-design improvements—even through *ad hoc* analyses—to lower the risk levels to acceptable for products delivered to consumers.[33] Such engineering efforts may just not have been organized and documented in a compelling manner. At this point, please recall the notice from Chapter 0—*nothing in this book is either a necessary or a sufficient condition for producing safe products*. Such steps and methods may be helpful, but no method presented assures either success or failure in producing safe products. However, since the focus of this book is to guide design engineers for design-and-manufacturing organizations producing innovative products, a formal risk-management must be covered and is encouraged.

5.10 Conclusion

The discussion of hazard, risk, accident, and outcome presented in this chapter is different than those seen in some books. Several of those books involved system-safety engineering on large-scale, quite-expensive systems. Some of these systems are aerospace or weapons systems or both. These large systems are operated by responsible organizations with the resources to set up, operate, maintain, and educate/train personnel. Despite the fact that many of these large-scale systems operate under extreme conditions, such as in space, underwater, in the desert, and in the arctic, these systems were designed and tested under these conditions. Therefore, these systems were intended for such uses and are, at least to some degree, controlled—although still extreme.

This chapter, and this book, is focused on the design, development, and testing of consumer products which may be used in completely uncontrolled environments by unskilled operators. Some users of these products may not understand how to use the product, may have a false understanding of how to use the product, or be unaware of the product's hazards and risks. Such a user may be completely unqualified to use a

[33] This may *not* be the case if the particular product had been plagued with safety issues including injuries and deaths, safety recalls, adverse publicity, and many lawsuits.

particular product and could also be impaired through alcohol or drug consumption. Environmental influences sometimes contribute to accidents and injury outcomes as well. This book is intended for the design and product-safety engineers involved in creating new and innovative product designs. As such, this chapter expands upon material of interest to engineers wishing to *design-in* positive safety characteristics to their products through the concept of *Design for Product Safety (DfPS)*.

The original hazard, risk, accident, and outcome models presented here are of much less value to a pure safety analyst evaluating a completed system design than to an engineer actively developing and refining an engineering design to increase its safety for its consumers. Consequently, the Haddon-matrix factors and phases are used to illustrate their effects as well as point out how a design engineer can potentially increase the safety of the final product before the product design has been "frozen" for manufacturing.

It is for these reasons that this book's treatment of product, design element, hazard, risk, severity, probability, accident, and outcome are unique. They are intended to illustrate the possible influences and effects that may confront a product and its user in the future. Perhaps, through the recognition of these factors and their influences, a design engineer will be able to anticipate a Product, User, Environment (PUE) issue and address it to prevent future injuries from a product.

Another vital function of this treatment of these topics is to present one part of the complex, nearly insurmountable challenge that is the successful and safe engineering design of innovative products. Some external parties may be perfectly content with second guessing the work of design engineers and design-and-manufacturing companies. This type of work is quite difficult—especially when conducted by ethical engineers. The true *values* of the employing company can either accelerate or retard the efforts of these engineers.

Many critical factors for the safe performance of consumer products are beyond the direct control of the engineering-design team. All engineers must always do their very best work—both technically and ethically. The author has never met an engineer who wished to harm users—a group which included their friends and families. In addition, their employers must permit and *encourage* their engineers to do their very best work—both *technically* and *ethically*.

All parties should be free and honest enough to admit than an injury resulting from an activity combining recreational products, criminal activity, alcohol, and firearms can be classified as "stupid." On the other hand, *all* parties should also be free and honest enough to admit to accidents which were avoidable through additional engineering-design effort. All parties are "in this" together. It is the author's hope that these disparate parties can find ways to constructively work together.

Even at the end of this chapter, there is still no answer to the pivotal question of "How safe is safe enough?" The author is hopeful that, at least now, the reader sees just how difficult that question is to answer. A timely, but unfortunate, health example is provided through current events.

At the present time, individuals, groups, governments, and societies around the world are struggling to come to grips with the true threat of the COVID-19 *pandemic* ("Coronavirus Disease 2019 (COVID-19) Situation Report—29" 2020). Various organizations are grappling to estimate the true severity and probability presented by the Coronavirus. This is evidenced by the wide assortment of people seen in public and by their countermeasures to either prevent becoming infected or to prevent their spread of the virus. Some don masks and gloves; others simply practice "social distancing." It is hoped, that soon, these same groups will be able to make the important risk evaluation of: "Is it safe to resume normal, pre-COVID-19, life?" Although the virus presents a health risk rather than a safety risk, as with any risk evaluation of import, there will *not* be a uniformly agreed upon decision at any point in time on how to respond to the virus.

Returning to consumer products, the conflicting answers produced by various value systems from parties both outside of and within a company make the absolute resolution of the "safe-enough" question impossible. At some point, someone or some party will need to make a decision on behalf of the design-and-manufacturing company as to whether or not the product is acceptably safe to sell. By the time companies make their decisions, it is necessary for their product-safety engineers to have made known their reasoned *conclusions* on product safety. Once sold, regulators, consumer advocates, attorneys, and consumers will make their views—either conclusions or opinions—known.

It will never be possible to please everyone. It is, however, possible to satisfy oneself with having done the best technical and ethical engineering work possible. If so, there is simply nothing more that can be asked of the engineer.

This chapter provided the reader with some insight into hazards, risks, accidents, and outcomes as well as an introduction to the risk-management process. The remainder of the book will help the reader understand other important aspects of the product-safety engineer role. Chapter 10 will provide the product-safety engineer, in particular—and all engineers, in general—with some tools to apply and to help accomplish the DfPS goal, namely, safer consumer products.

References

"About Apollo 7, the First Crewed Apollo Space Mission." 2018. NASA Website. 2018. https://www.nasa.gov/mission_pages/apollo/missions/apollo7.html.

Anderson, Clinton P. (US Senate). 1968. "Apollo 204 Accident." Washington, DC.

"ANSI/ISO 12100:2012 Safety of Machinery—General Principles for Design—Risk Assessment and Risk Reduction." 2012. Houston, TX: ANSI/ISO.

Barnett, Ralph L., and William G. Switalski. 1988. "Principles of Human Safety." *Triodyne Safety Brief* 5 (1). http://www.triodyne.com/SAFETY~1/SB_V5N1.PDF.

"Consumer Products Safety in Europe: Corrective Action Guide." 2011. Brussels, BE.

Converse, Elliott V. III. 2012. "History of Acquisition in the Department of Defense. Volume I. Rearming for the Cold War 1945–1960." Washington, DC.

"Coronavirus Disease 2019 (COVID-19) Situation Report—29." 2020. Geneva, CH: World Health Organization (WHO). www.who.int › docs › situation-reports › 20200219-sitrep-30-covid-19%0A.

EC/DGE. 2006. "Directive 2006/42/EC of the European Parliament..." Brussels, BE.

Ericson, Clifton A. 2006. "A Short History of System Safety." *Journal of System Safety*. 2006. https://system-safety.org/ejss/past/novdec2006ejss/clifs.php.

———. 2016. *Hazard Analysis Techniques for System Safety*. Second edition. Hoboken, NJ: John Wiley & Sons.

Ertas, Atila. 2018. *Transdisciplinary Engineering Design Process*. Hoboken, NJ: John Wiley & Sons. https://doi.org/10.1002/9781119474654.

Gehman, Harold W. Jr. 2003. "Columbia Accident Investigation Board Report, Volume I." Washington, DC.

Goldberg, B. E., K. Everhart, R. Stevens, N. Babbitt III, and L. Stout. 1994. "System Engineering 'Toolbox' for Design-Oriented Engineers." Marshall Space Flight Center, AL: NASA.

Gullo, Louis J., and Jack Dixon. 2017. *Design for Safety*. First edition. Hoboken, NJ: John Wiley & Sons. https://doi.org/10.1002/9781118974339.

Haddon, Jr., William. 1983. "Approaches to Prevention of Injuries." In *AMA Conference on Prevention of Disabling Injuries*, 31. Miami, FL.

Hammer, Willie, and Dennis Price. 2001. *Occupational Safety Management and Engineering*. Fifth edition. Saddle River, NJ: Prentice Hall.

Hope, Paul. 2019. "New Safety Feature on Portable Generators Could Save Lives, Consumer Reports' Tests Show." *ConsumerReports.Org*, November 2019. https://www.consumerreports.org/portable-generators/new-safety-feature-on-portable-generators-could-save-lives-consumer-reports-tests-show/.

Jensen, Roger C. 2020. *Risk-Reduction Methods for Occupational Safety and Health*. Second edition. Hoboken, NJ: John Wiley & Sons. https://www.vitalsource.com/products/risk-reduction-methods-for-occupational-safety-and-roger-c-jensen-v9781119493976?term=roger+jensen.

"Liberty Bell 7 MR-4 (19)." 2000. Kennedy Space Center. 2000. https://mediaarchive.ksc.nasa.gov/media/history/mercury/mr-4/mr-4.htm.

Leveson, Nancy G. 2011. Engineering a Safer World: Systems Thinking Applied to Safety. Cambridge, MA: The MIT Press. https://mitpress.mit.edu/books/engineering-safer-world.

McDonald, Allan J., and James R. Hansen. 2012. *Truth, Lies, and O-Rings: Inside the Space Shuttle Challenger Disaster*. Gainesville, FL: University Press of Florida.

Portable Generators; Final Rule; Labeling Requirements. 2007. USA: Federal Register. https://www.cpsc.gov/Regulations-Laws—Standards/Rulemaking/Final-and-Proposed-Rules/Portable-Generator-Labels.

Portable Generators; Final Rule; Labeling Requirements (Correction). 2007. USA: Federal Register. https://www.cpsc.gov/Regulations-Laws—Standards/Rulemaking/Final-and-Proposed-Rules/Portable-Generator-Labels.

Rogers, William P. 1986. "Report to the President by the Presidential Commission on the Space Shuttle Challenger Accident." Washington, DC.

"Second Volume of *NASA System Safety Handbook* Released." 2015. Sma.Nasa.Gov. 2015. https://sma.nasa.gov/news/articles/newsitem/2015/10/16/second-volume-of-nasa-system-safety-handbook-released.

Thompson, Floyd L. 1967. "Report of Apollo 204 Review Board (N82-72199)." Washington, DC. https://history.nasa.gov/Apollo204/content.html.

USA/DoD. 2012. "MIL-STD-882E—System Safety." Washington, DC.

CHAPTER 6

A Product-Design Process

Climate is what we expect.
Weather is what we get.

—ANDREW JOHN HERBERTSON[1]

6.1 Introduction

It is through the disciplined application of a systematic and comprehensive product-design process (PDP) that a design-and-manufacturing company will be best able to deliver the product that was originally conceived along with all of its intentions. Such company intentions will include product function, cost, performance, weight, reliability, durability, and safety.

This chapter will explain what is meant by a PDP and the Engineering Design, Development, and Testing Phase (EDDTP) within a PDP. The details of each will be explained. The use of such a process and its phases should be a portion of a significant corporate effort to design and manufacture new products, both efficiently and effectively, that are fit for intended consumer uses.

When coupled with a conscious product-safety engineering effort, as covered in the Chapter 7, the PDP and EDDTP should decrease the number of hazards and the levels of risks for products delivered to consumers. Through an integrated PDP and a thorough EDDTP, the numbers of both engineering and product-safety problems should be reduced.

In many cases, the products being engineered by this type of an integrated design process are mass produced in a factory, or several factories around the globe—not handcrafted by artisans in a workshop or a studio. This mass production makes possible economies of scale which increase the accessibility and reduce the cost of these products to consumers. This same mass production also has the effect of quickly multiplying the number of product units that may have a design or manufacturing flaw when the EDDTP does *not* work as intended. Engineering design-and-manufacturing corporations should, therefore, dedicate the resources necessary to their PDP so that products consumers *get* are, in fact, the products that consumers *expect*. In this chapter, the features of one particular PDP and one particular EDDTP are discussed.

The PDP followed by a given design-and-manufacturing company need not be identical to the example used in this chapter—please, remember Chapter 0. This is why

[1] https://quoteinvestigator.com/2012/06/24/climate-vs-weather/

the title of this chapter begins with an "A" instead of a "The." Each organization's particular PDP should be adapted to suit the product, the corporation, the industry, and other variables, parameters, and requirements that may be unique to the product(s) being designed and manufactured. However, many effective PDPs are likely to have numerous common characteristics.

Although this is a book about product-safety engineering, it is fundamentally an engineering-design book with a focus on product safety and its dependence upon applied engineering ethics at many points within the PDP. For maximum effectiveness, engineering design should take place within a structured—and yet still flexible—EDDTP. Such a systematic approach to design should enable engineers to design, analyze, test, revise, and finally release the finest, and safest, products. The ultimate goal of this exercise is to "design-in" safety to a product. This is what is meant by *Design for Product Safety*[2] or *DfPS*.[3] One such product-design sub-model, the EDDTP, will be presented in this chapter. The next chapter will discuss how the product-safety engineering *function* is integrated into the overall PDP to achieve the DfPS objectives.

The goal of DfPS and of having a disciplined PDP with an integrated, or systematic, product-safety function is to design safety *into* a product. This concept is not new, but at a recent product-safety conference,[4] it was being mentioned frequently by both speakers and attendees—especially by U.S. CPSC commissioners—as a unique concept. This idea makes much sense to both regulator and designer/manufacturer alike. A disciplined PDP has a greater probability of delivering safe products to the public than an undisciplined one. An integrated EDDTP has the ability to identify potential product-safety issues sooner than a non-integrated process. Such a disciplined PDP will also provide higher-quality, more-reliable products than an unstructured engineering-design process. Consumers, of course, also win in terms of both price and product safety.

6.2 A Generic Product-Design Process

The PDP model presented here is not unique, nor is it the only PDP model that is capable of producing good results when followed. This model is presented here in order to show an example of one workable PDP model. It will provide "access points" for the product-safety engineering (PSEg) *function* to be discussed more fully in the next chapter. The discussion here is not intended to help *create* a comprehensive product design and development process; it is merely to illustrate how such a process might function.

In order to look at the EDDTP, it is first necessary to see where this aspect of engineering design fits into the overall PDP. Figure 6.1 shows the various phases of a PDP for a product. It is not too different fundamentally from design-engineering treatments

[2] It was not until the finalization of this book that the author became aware of a paper using the term "design for product safety" (Biancardi 1970). However, that paper is focused on product liability from an insurance company's perspective. Although there is discussion of the engineering-design process in the paper, its coverage is rudimentary and incomplete. Also, consistent with insurance-company parlance, this paper uses the euphemism "loss" for damage, injury, or death.

[3] Design for Product Safety (DfPS) should not be confused with Design for Safety (DfS) which is an industrial-engineering approach to (re-)design manufacturing processes and workplaces to improve occupational-worker safety and health as well as to "save money in the long run" as described by OSHA ("Design for Safety" n.d.).

[4] ICPHSO Annual Meeting, Washington, DC, February 2019.

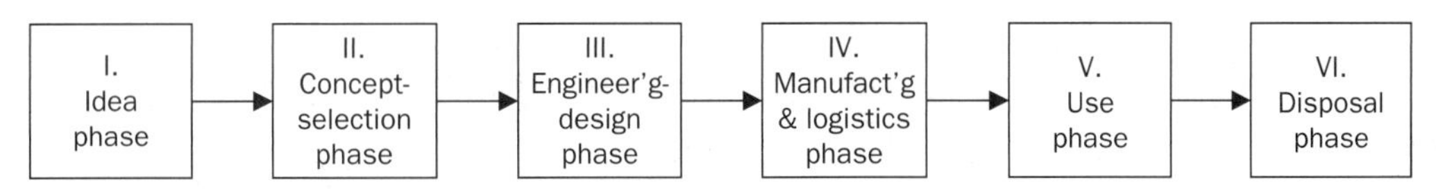

FIGURE 6.1 Phases of a generic product-design process.

in numerous other books (Ertas 2018; Hales and Gooch 2004; Pugh 1996), but also not identical to them either. A PDP often breaks down into the following six phases:

I. ***Idea (or ideation) phase***: This is the phase where an idea is born either by market demand or by corporate inspiration. Sometimes a customer comes to the designer/manufacturer with their needs for a new product. There are some instances when a technology must be invented or further developed in order to make possible a product under consideration.

II. ***Concept-selection phase***: It is at this point that the idea from Phase I takes on either a definite shape or another more-concrete form. It is no longer merely an abstract idea. In some cases, market research has provided input to design engineers and project managers about the size of market demand for the product, the price that the product can command in the market, an estimate of the manufacturer's cost to design and manufacture the product, and customer preferences such as "must-have" and "nice-to-have" characteristics. At this point in the design process, the decisions will start being made to determine properties such as general size, weight, configuration (e.g., the number and the arrangement of seats or wheels), feature set, and other aspects of the product. This may sometimes be called the "conceptual design," but here that design *embodiment*, within Phase II, is called the *preliminary design*. Budgeting and scheduling of the project, along with allocating the staffing possessing the requisite skills, are planned out at this phase of product design.

III. ***Engineering-design phase (EDP)***: If a product passes Phase II, the engineers can begin work on designing, analyzing, testing, and then redesigning prototype products. The preliminary design from Phase II is refined, or developed, into a more-detailed *intermediate* design. This intermediate design is then further developed into the *final* design which is ultimately released to Phase IV for production. The current design will be called the *detailed* design whether it is a preliminary, intermediate, or final design. The progression in time of the engineering designs for a product under development in Phase III is shown in Figure 6.2. Product analysis and testing will take place and the results of these activities will be fed to the engineering-design team for necessary changes, or redesign. Parts and assemblies will be designed and, ultimately, released by the Purchasing Department to Manufacturing Engineering or to a supplier to make. All of the released engineering drawings will result in the parts and processes included in the Bill of Materials (BOM) for the product. The engineering-design phase envisioned here is something that would be used for a durable consumer good rather than for computer-software products or process-manufactured goods such as chemical solutions. The product being considered is one which will need to be mass manufactured in a factory setting.

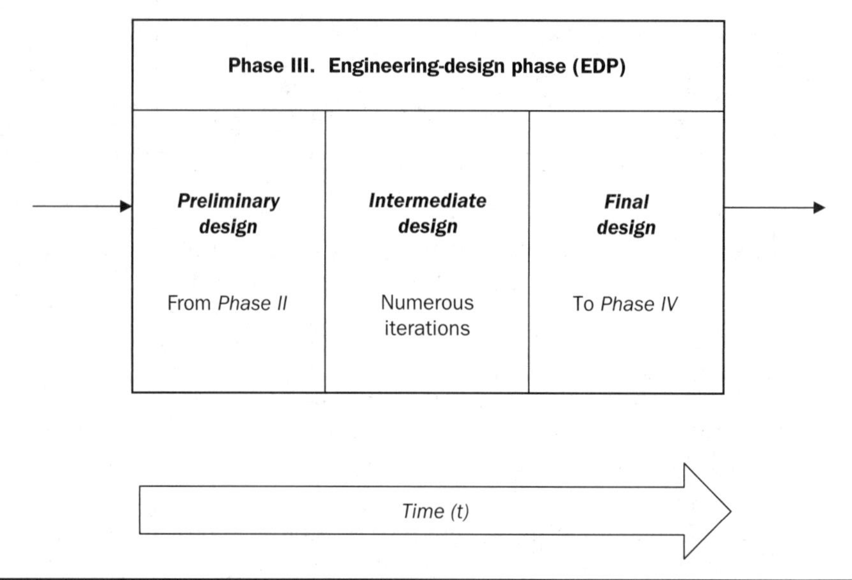

FIGURE 6.2 Progression of engineering detailed designs in phase III of a generic PDP.

IV. ***Manufacturing and logistics phase***: Long ago, many design engineers had not been too concerned about the actual manufacture of the products that they were designing. This has changed in recent decades for much the better. Now, manufacturing engineers work alongside design engineers to be sure that designs are reliably manufacturable at reasonable costs. In addition, the tooling needed either to form or to injection mold parts is expensive and often requires significant lead times for scheduling in order to meet Manufacturing's time windows for production. Some larger products require additional attention to logistical hurdles as the products leave the factory and are transported to distributors, dealers, or retailers. At times, products must be designed and assembled so that they will fit into a specific container or onto a limited-width semi-trailer or shipping container. Manufacturers of durable consumer goods know that damage during shipment can be a large cost and may lead to other operational and logistical problems as well.

V. ***Use phase***: This is the phase after the product leaves the control of the manufacturer and retailer. It is during this phase that a product provides the function for which the consumer bought the item. That function could be the toasting of bread, the cutting of lumber, the clearing of snow, or the provision of recreational entertainment. After leaving the retailer's control, the product may be subject to a wide range of uses, users, and operating environments—despite whatever might have been the product's stated intended use. Many of these products will not be used in an occupational or otherwise-supervised setting where designer/manufacturer's recommendations and legal employment rules and regulations are enforced. In addition, not all of these uses, users, and environments may be recommended by the designers of the product—in fact, some may be explicitly warned against. During this Phase V, it is important for the designer and manufacturer to actively collect information on the performance of the product in the field. It is important to actively search for indications of product-safety conditions or problems that could lead to user injury. This phase and aspect of the engineering-design spectrum is covered in detail by Chapter 11.

VI. ***Disposal phase***: Once the product has reached the end of its useful life, it must be retired. This can range from an extremely easy task for simple products to an extremely complex process for elaborate systems. A product's retirement intricacies depend upon the product itself. For many products, simply disposing of it properly is all that is called for. Some products may merely be put in the trash; others call for simple recycling. In certain cases, toxic substances within the product call for specialized disposal and recycling, for example, products with batteries or with electrical circuitry in them. For mammoth systems such as nuclear reactors, decommissioning is a major factor and is considered from the moment of design inception.

As a design for a product concept and its design matures, it progresses linearly from one phase to the next. An engineering design moves from the Idea (Phase I), through the Concept-Selection Phase (Phase II) onto the Engineering-Design Phase (Phase III) onto the Manufacturing & Logistics Phase (Phase IV) onto the Use Phase (Phase V) and, finally, onto the Disposal Phase (Phase VI). This is one possible product life-cycle progression.

When looked at as generic engineering, all of the above phases are generally straightforward and may be found in numerous engineering-design textbooks and other resources in some form or another, including those listed earlier and these (Dominick et al. 2001; Pahl and Beitz 1996; Ulrich and Eppinger 2000). While there are numerous good books on engineering-design methodologies, none has included significant content on product safety.[5] When looked at from an occupational safety and health (OSH) perspective (Hammer and Price 2001), Phase IV, the manufacturing phase, is of greatest interest because workers can become injured in the short term from either industrial accidents or can be harmed by long-term exposure to environments or repetitive tasks. From an environmental-health point of view, Phases V and VI are of the greatest interest. If a product use produces emissions, such as CO (carbon monoxide), HC (hydrocarbons), and NOx (nitrogen oxide) in the case of a product with a gasoline-burning engine, then those emissions can affect human and environmental health in the long term. Similarly, if a product is difficult or impossible to recycle, humans and the environment may suffer damage. This recyclability issue would be one of interest to those practicing product stewardship ("A Guide to Maximizing the Use of Existing Product Stewardship Programs," n.d.; "Core Competencies for the Product Stewardship Professional" 2014). Once a product reaches Phase VI, Disposal, that product should be designed to pose little safety risk to people.

If looking at the post-sale safety aspects of products—such as product-safety recalls, it is Phase V that is of most interest. It is during this Use phase that a product can injure a person through short-term exposure to a product hazard or through an accident involving the product. This is also discussed in Chapter 11.

---

[5] Nor have any of these books focused on the topic of product safety either.

6.3 A Specific Product-Design Process

The prior section laid out a generic PDP without details for each phase contained within the PDP. However, since this is an engineering-design book at heart, the focus will be upon Phase III—the engineering, or product, design phase. It is at this phase where most of the engineering-design decisions affecting product function and product safety will be made. Therefore, a more-detailed Phase III is necessary and is now offered.

Rather than simply calling this version of Phase III the Engineering-Design Phase as was done before in the generic PDP, this phase will now be given a much-more descriptive title: the *Engineering Design, Development, and Testing Phase (EDDTP)*. This updated Phase III reflects the realities of real-world engineering-design efforts. Also, since Phase VI, the disposal phase, generally does not factor into the product-safety engineering efforts studied in this book, that phase will not be considered further in this book's discussions.[6]

A revised version of Figure 6.1 with a detailed Phase III, and without a Phase VI, is included as Figure 6.3. The Engineering Design, Development, and Testing Phase (EDDTP) in the new figure reflects the true nature of engineering design. There is an initial design based on Phase-II input arrived at through design synthesis.[7] One or more prototype products are constructed for laboratory-testing and field-testing purposes. At the same time, the initial design will be subjected to engineering analysis. The results of the testing and analysis will be provided to the engineering designers to make the necessary changes to the product design. The changes to the initial design become the current detailed design for the product. This is the act of *product development*. This design-iteration loop continues until the designed, analyzed, and tested product's characteristics convergence with those of the desired product. These product characteristics include performance, dimensions, mass, durability, reliability, manufacturing cost, and, certainly, safety.

In addition to the addition of the EDDTP, the updated figure shows the potential need to develop a new technology for the new product. Technology development is always risky to the success of an ambitious product-project plan. It is unwise to depend upon an unproven technology, such as a new electric-battery type, to be ready at a given point for *insertion* into EDDTP on a particular date. There are numerous examples of partial or complete failures using this approach to product-development engineering. One such failure is the Convair B-58 Hustler bomber example which is included as an example in this section.

The above description briefly mentioned the components of this EDDTP. Each component of the phase will be expanded upon below. As stated earlier, the detailed design is the current design of the product within the current design iteration of the EDDTP.

The detailed design is constructed as prototypes which are then tested—*both* in a laboratory and in the field. A prototype, often called a "mule" if the product is a vehicle, will represent the current design in meaningful physical ways. It may look nothing like the

---

[6] If a particular product's Disposal phase does contain product-safety implications, then Phase VI should be re-introduced to the overall specific PDP and EDDTP discussion.

[7] *Synthesis* is putting together; *analysis* is taking apart. Design synthesis is the combining of elements and the selection of element parameters such as dimensions, materials, and quantities. Design *analysis* is the application of mathematics and physics to evaluate the sufficiency of decisions made during design *synthesis*. Analytic abilities are needed for competent synthesis.

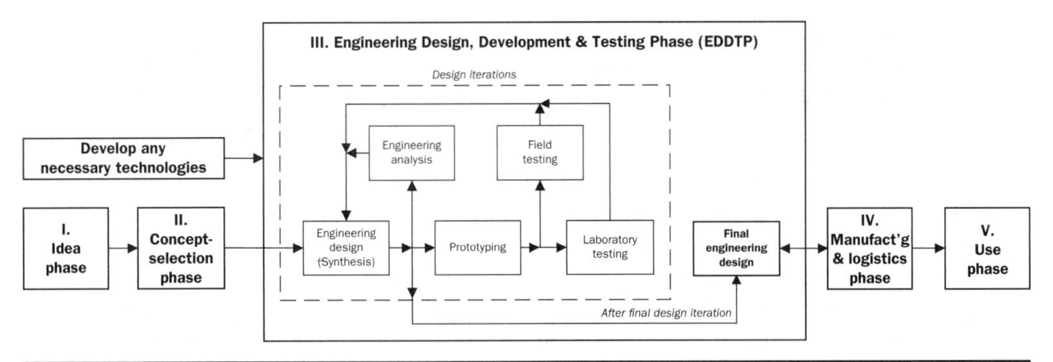

FIGURE 6.3 Phases of a specific product-design process (PDP).

intended final product. Many prototypes are rather ugly in appearance but yet still serve to demonstrate the critical characteristics of the product design at that time. Prototypes are often missing components that themselves are being designed to be tested at a later date. Some prototype parts are "cannibalized," or stolen, off of other products just to have something in place until the proper component prototypes arrive for testing. These prototypes are usually not made on the production line, but are instead produced by prototype-shop technicians or by contractors specializing in small-volume production. It is important that the prototype be functionally representative of product characteristics such as power source, power delivery, structure, durability, reliability, and safety. In the case of a land-vehicle design, the prototype should accurately represent the current state of the chassis, suspension system, steering system, service-brake system, fuel-storage system, fuel-delivery system, engine/powerplant, power-transmission system, power-plant-cooling system, electrical system, occupant restraints, and vehicle-stability.

The laboratory-testing element of the EDDTP covers many necessary tests to either determine, verify, or validate product performance. Much laboratory testing is at the component level in the case of large vehicles, although some equipment may permit the testing of the entire product. However, some laboratory testing may be of the "accelerated" type. For example, if a particular frequency range is known to be harmful to a vehicle chassis, that frequency range alone can be fed into the vehicle's suspension system, through appropriate test equipment, to accelerate the "aging" of the vehicle chassis and permit the evaluation of the durability of the chassis and its componentry in a shorter time period. Similarly, there are accelerated-aging tests for product response to ultra-violet (UV) light exposure for material stability and to exposure to salt spray for corrosion resistance. There are sometimes *ad-hoc* test procedures for products and product components either prescribed by a recommended practice, required by a standard, or used by an industry or designer/manufacturer.

Field testing is an integral part of product development since, for new and innovative products, there are no established industry tests or knowledge bases of product use. All of this must be created by the innovative design and manufacturing firm. By using the new product as intended—and perhaps beyond its intended limits—the company is able to determine the suitability of the product and its components. Frequently, products will be instrumented as they are tested in order to measure forces, moments, accelerations, and strains on product elements. Instrumentation examples include strain gauges and accelerometers on chassis components, suspension components, and temperature probes at various locations on the vehicle. After a round of field testing, a prototype is often disassembled, or "torn down," by technicians to inspect components for wear and other damage. Test engineers will also record their observations of product use during field testing. These observations may include items such as vibrations, noises, vehicle-handling issues, discomfort, component failures, component failures, and suggestions for improvements. Some of the information from field testing will be used by those doing the engineering analysis of the product as inputs to computer and other analyses. In a well-integrated, functioning EDDTP, all elements of the product-development effort will be communicating with one another and sharing the results of their work to further successful product development.

That current detailed design will also be further analyzed through calculations and software modeling by such methods as multi-body dynamics (MBD), finite-element analysis (FEA), and computational fluid dynamics (CFD). For a vehicle, the suspension system consisting of control arms, springs, shock absorbers (dampers), axles, steering arms, wheels, and pneumatic tires are modeled subject to varying vehicle speeds, maneuvers, and terrain profiles. The reaction forces and moments produced by the MBD software will often be used as inputs to an FEA model of the vehicle chassis to locate "hot spots," or locations where stresses are too high and must be reduced. These high-stress areas are typically denoted by the colors yellow and red in a computer-generated output image of the component stresses. Hence, their moniker. "Cool" colors such as green and blue usually designate low-stress areas in FEA output images. Also, in the case of a vehicle, CFD may be used to model the aerodynamics of the vehicle passing through air for such characteristics as drag, ventilation, and accumulations of heat and toxic gases. Modern on-highway vehicle design is extremely sensitive to aerodynamic drag as increasingly strict mileage requirements mandate aerodynamic-drag reduction at almost any cost. The efficiency of an internal-combustion engine (ICE) cooling system is important for both performance and safety considerations. Insufficient cooling may affect the utility of the vehicle when in consumer hands and the build-up of high temperatures—that could lead to fires—is a product-safety concern. The build-up of poisonous gases, such as carbon monoxide (CO), in pockets around the vehicle is a safety concern for vehicles, especially those having an "open cockpit" design such as convertibles and motorcycles. The results from laboratory testing, field testing, and engineering analysis should be made available to the design engineers for additional design refinement.

Once the product is where the design engineers want it, that detailed design becomes the final engineering design, or simply the final design. Then, that final design is "released" to manufacturing engineering for production in Phase IV.

The connector between Phases III and IV is a *double*-headed arrow because the manufacturing personnel should be heavily involved with the detailed design of the product. They are the ones, after all, that will be tasked with efficiently producing the final product. Manufacturing engineers and personnel can provide design engineers with valuable information about the costs, complexity, scheduling, variability, and viability of the processes necessary to produce the product.[8] A few minutes spent listening to a manufacturing engineer can save design engineers from a mountain of issues which could, perhaps, lead to product-safety problems.

---

[8] The author was involved with a vehicle having one plastic component whose tooling could only be accommodated by one injection-molding press in the world due to the part's volume requirement.

Example 6.1: The Development of a Necessary Technology for an Advanced Product

The end of World War II (WWII) in 1945 set the stage for the start of the "Cold War." This rapid "chill" in foreign relations between WWII allies led to the rapid escalation of fear and distrust between world governments. This fear and distrust were aggravated in the new atomic age initiated by the dropping of atomic bombs on Japan by the United States at the end of WWII. This Cold War also saw the start of the "Arms Race" where countries were developing new defensive and offensive weaponry, whose capabilities had never before even been dreamt of, in order to either protect themselves at home from their adversaries or to project their own power abroad.

This atomic age was accompanied by two other burgeoning technologies: jet propulsion and solid-state electronics. Weapons could now be more powerful than ever before imagined—and those weapons could now be delivered more quickly and with greater precision than ever before. Students of history, especially of military history, already know of many of the multitude of bizarre—and even frightening—weapons conceived of, developed, and sometimes even deployed during this surreal time in world history. Fortunately, none of them was ever used.

There was often an extreme mismatch between the demands on new weapons systems and the capabilities of the technologies of the day to meet these demands. For example, there was a constant struggle to fly at supersonic speeds[9] to quickly penetrate enemy air defenses and deliver destruction upon the foe. Supersonic flight consumes fuel rapidly; consequently, aircraft range is seriously shortened by demanding such high-speed performance. Although aerial refueling can supplement aircraft range, this must be expected to be impossible to accomplish in enemy-contested airspace during wartime.

One particular Cold War example is the United States Air Force (USAF) Convair B-58 Hustler (Converse 2012, 457, et seq.). This aircraft is shown in Figure 6.4. It was developed in the 1950s as a high-speed bomber to deliver nuclear weapons upon America's foes. It was designed to be as fast as the fastest of fighter planes of the time—Mach 2+—over 1300 mi/h (2090 km/h).[10] This requirement led to numerous technical difficulties. Two of the greatest hurdles were aircraft range and aircraft guidance and navigation.

The *first* of these two obstacles, un-refueled aircraft range, was never solved and persists even to this day with modern aircraft. A partial solution to the B-58's problem was the design and development of a large centerline pod (not shown in Figure 6.4) which was a combination fuel tank and nuclear bomb carried beneath the aircraft fuselage. It added to aircraft range, but the range of B-58 was never sufficient for its intended mission—bombing targets in the Soviet Union from bases in, or controlled by, the United States. The B-58 was constructed and deployed nevertheless. Perhaps there was just no other option to using the aircraft given the perceived enemy threat at the time and the amount already spent on the bomber program.

---

[9] Greater than the speed of sound, Mach 1, or 767 mi/h (1235 km/h).

[10] To this day, the USAF has only operated one other supersonic pure-bomber aircraft and that is the North American Rockwell B-1B, in current operation, with a top speed of Mach 1.2 (https://www.boeing.com/defense/b-1b-bomber/), well below the maximum speed of the B-58.

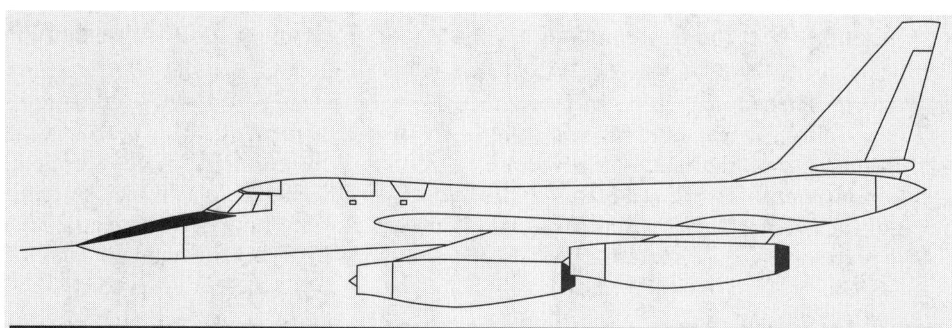

FIGURE 6.4 Convair B-58 Hustler.

The *second* obstacle, aircraft guidance and navigation, was eventually solved—at least to the satisfaction of the USAF. However, the ultimate guidance and navigation system only became available in 1967, *two years after* the U.S. Secretary of Defense had directed the retirement of the B-58 fleet (Knaack 1988, 390). This aircraft was simply too fast for the radar and guidance systems of its day given the numerous design constraints imposed such as weight, spatial volume, power consumption, and reliability. Electronics technology had not yet reached the level of maturity and dependability necessary for a weapons-system program as ambitious as the B-58.[11]

Today, there may exist greater alignment between weapons system demands and technological capabilities than in past decades. Electronics, and communications, have advanced to a state where they are perhaps well beyond other aspects of weapons systems in technical complexity. Aircraft, both military and civilian, are now flying *slower* than their counterparts of 50 years ago. This is no doubt due to numerous factors, including stealth technologies and fuel economies. There is less incentive to sacrifice other aircraft capabilities, such as range, in order to fly as fast as possible so long as the aircraft remains difficult to detect by enemy early-warning radar systems. The turbojet-powered commercial airliner from the 1960s completed air routes in *shorter* times than current turbofan-powered airliners due to different operating characteristics of the two types of gas-turbine engines. The current airliner, however, is much quieter and uses less fuel than its counterpart from decades before. Electronics are, however, much more advanced than they have ever been with no end to further advancements in sight.

In the case of the B-58, it was simply not possible to develop the necessary systems for full mission completion due to the limitations of that period's technologies. Today, the same as then, it is a risky business plan that depends upon new technologies to be developed in time for use in new and innovative products. If the failure to develop needed technologies negatively impacts the level of product safety provided to users, it is wise to *not* release a product that could be compromised with respect to performance, reliability, and safety.

[11] This failure in the weapons-systems acquisition process was among the earliest in what has become commonplace for U.S. Department of Defense systems. Much of this is probably unavoidable given the extreme performance requirements of modern weapons systems. Yet, it appears that huge delays and extreme cost overruns are now considered acceptable and to be expected in weapons-system contract administration.

Although depending upon emerging new technologies may be risky in new-product development, some of these technologies, once mature, may help product-safety engineers reduce risks, which are currently almost impossible to control, for future consumers.

6.4 The Role of Suppliers within a PDP

Many parts to today's products are manufactured, not by the designer and manufacturer of the entire product, but by suppliers of the design-and-manufacturing company. This design-and-manufacturing company is often referred to as the original-equipment manufacturer or OEM.[12] In order to reduce capital investments and management overhead, OEMs often resort to using suppliers, or vendors, to provide systems, subsystems, component parts, and materials for the OEM's final product.

Many OEMs and suppliers forge long-term strategic relationships that are mutually beneficial. However, OEMs who outsource to suppliers too many of their vital competencies risk losing the expertise and manufacturing ability for capabilities essential to the OEM's existence and future growth. In addition, many suppliers have their own suppliers, or sub-suppliers to the OEM. These arrangements can go on for multiple levels. The multiple levels of this supply chain bring with them their own headaches, but they will not be deeply explored here.

These parties, OEMs, suppliers, and sub-suppliers are shown in Figure 6.5 below the EDDTP, Phase III, in the diagram. The Purchasing and Quality (Quality Control (QC) or Quality Assurance (QA)) Departments generally act as liaisons between the EDDTP and the suppliers and their suppliers. Not only do supplier-provided systems, subsystems, and components have to meet specifications to be verified by QA/QC personnel, they also must meet the cost and scheduling requirements of the purchasing agent in order to meet the OEM's overall goals. Purchasing agents will work with suppliers regarding delivery requirements for Phase IV, Manufacturing and Logistics, to meet Operations' requirements. These interactions are illustrated in the figure.

In the case of a vehicle power-distribution module for a vehicle, the OEM might contract the design, development, testing, and delivery of the module to the *first* Tier-1 supplier. That first Tier-1 supplier may contract the design, testing, and delivery of integrated-circuit (IC) boards to one or more Tier-2 suppliers. Those Tier-2 suppliers may contract the delivery of resistors, capacitors, and transistors to one or more Tier-3 suppliers. Similarly, the OEM may contract the design, development, testing, and delivery of the fuel-injection system to a *second* Tier-1 supplier who will develop a supply chain of Tier-2 suppliers who have their own Tier-3 suppliers. Supplier tiers may exceed three levels depending upon the OEM's product, for example, a Tier-4 supplier could provide raw materials to a Tier-3 supplier from which to manufacture its components.

As shown in Figure 6.5, there can be a direct line of action between Tier-1 suppliers and Phase IV, the manufacturing phase, of the PDP. The supplier will often directly ship systems, sub-systems, and components to the OEM's manufacturing facility. These production-ready elements may or may not be subject to the same stringent QC/QA evaluations that the supplier products were when they underwent an initial process

[12] Although the term "OEM" is mainly used in automotive and heavy industries, the use of that term here facilitates the discussion by using this abbreviation here instead of "design and manufacturing company."

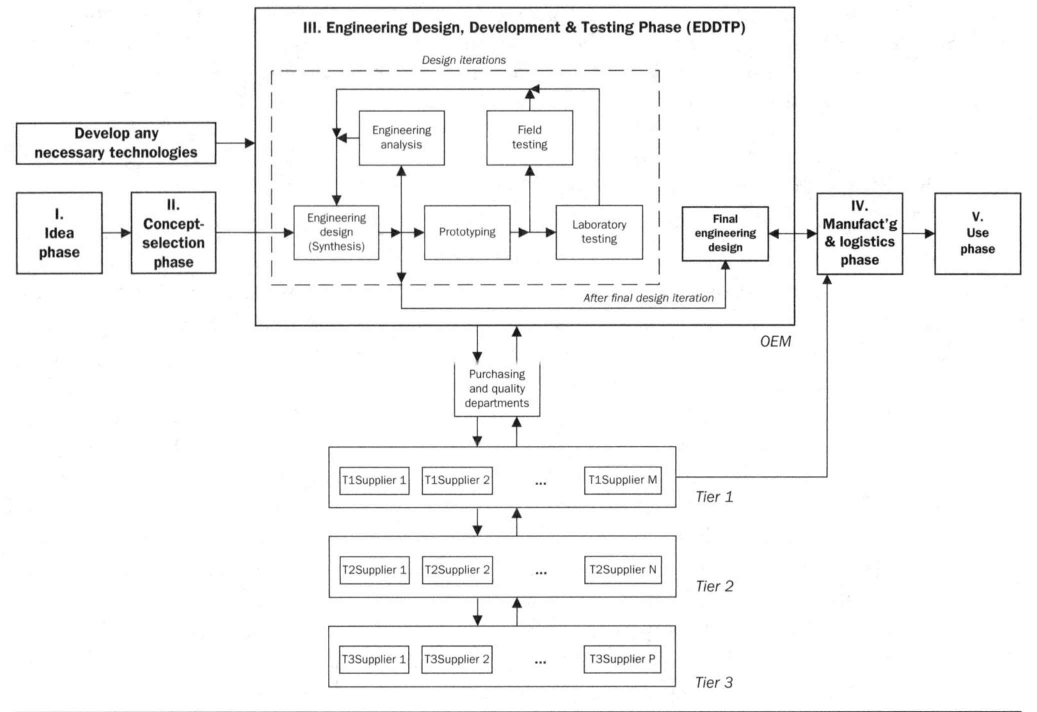

FIGURE 6.5 Phases of a specific product-design process (PDP) showing the role of suppliers.

similar to what is called a production-part approval process (PPAP). In a PPAP, supplier parts are inspected, evaluated, and tested by the OEM's QA/QC personnel prior to approving them for future production use. It is important that the OEM stay on top of these part shipments and their fitness for use. The author is aware of instances where the PPAP part delivered by the Tier-1 supplier was *not* the same as the production part due to the *supplier* having changed the Tier-2 supplier in the interim. If an OEM uses, for an extended time, the same part provided by its Tier-1 supplier, that OEM can *expect* that part to change over time as stated earlier in the book. The Tier-1 supplier can be expected to constantly update its supply chain just as the OEM does.

6.5 Gate Reviews

The PDP is shown generically in Figure 6.1 and, in a more-specific manner, in Figure 6.3. As a product design moves from one phase to another—or even to a more-advanced point within a phase—a gate review of the engineering design should be conducted. Purposes for a gate review are many. The gate review provides a chance to check the current product design against specifications such as speed, power, weight, strength, durability, reliability, feature set, mileage, comfort, operator visibility, product safety, and more.[13] A drawing showing where such gate reviews would be conducted is shown as Figure 6.6.

[13] Since this chapter is a generalized treatment of engineered-product development, the product-safety aspects of a gate review are delayed until the next chapter.

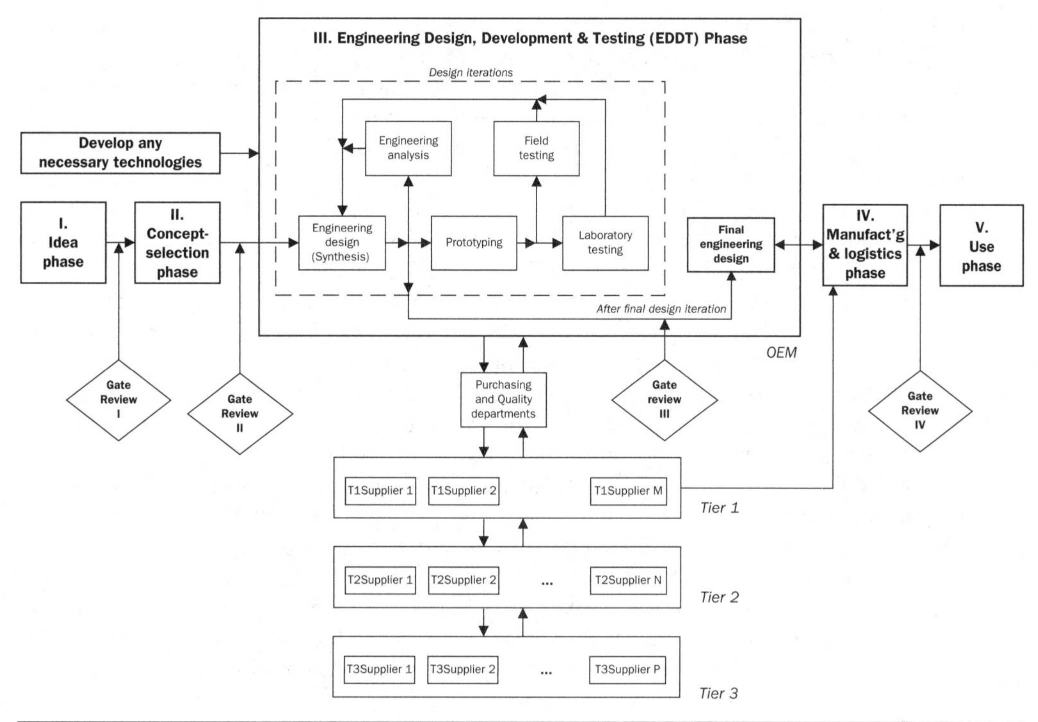

FIGURE 6.6 Phases of a specific product-design process (PDP) showing gate reviews.

As the general idea for a product advances from Phase I to Phase II, Gate Review I is conducted. The product design lacks significant detail at this early PDP phase, but it is worthwhile, nonetheless, to conduct such a gate review to be sure that all parties involved are aligned with the product direction and ultimate goals, as well as a plan to achieve them. These involved parties include engineering, project management, manufacturing, accounting, sales, marketing, legal, technical support, and other corporate functions.

The above PDP creates a phase-and-gate product-development process which is similar to a stage-and-gate[14] process. This type of stage-and-gate process is used widely within the field of project management (Cooper 1994).

Some authors use the terms "preliminary design review" and "critical design review" for places in a PDP or an EDDTP where the current design is evaluated (Ericson 2016). Another author numbers the design "audits" rather than name them (Pugh 1996). This book will follow Pugh's lead in order to simplify the discussion of the PDP.

Once the product has a firm conceptual design, the second review, Gate Review II, is conducted to be sure that all of the company's parties are in agreement including engineering, sales, accounting, compliance, product safety, and others. It is at this point that funds are released and "metal is cut," or parts start being designed and fabricated by design engineers.

Once the product program reaches Phase III, the EDDTP, there will be revisions to the detailed design as product development continues. It is recommended that, at

[14] https://dictionary.cambridge.org/dictionary/english/stage-gate.

numerous points within the EDDTP, design reviews are conducted to track the program's progress against its schedule—which is likely a tight one. These design reviews may simply be called Design Review 1, Design Review 2, and on. During these EDDTP design reviews, the critical performance metrics—such as size, weight, speed, and cost—are compared to those prescribing the program from Phase II. If the program is lacking in any way, a course of action should be determined and agreed upon to get the program back on track. Product safety will be a characteristic that the product-safety or design engineer will monitor and, if found to be lacking, work to remedy. Such safety reviews, which take place during design reviews, will be discussed in the next chapter.

One final and crucial design review will take place prior to releasing the product for manufacturing. This review is Gate Review III. It is the final review of the product. If a product passes gate III, then that product is good for consumer use once it has been manufactured. All reservations from the company about the product must be resolved by this point. These include product performance, features, and safety. The Operations Department and manufacturing engineers take over from here. Product design is done.[15]

Due to the complexities of manufacturing across broad products and industries, little will be said about Phase IV. If the design engineers and the manufacturing engineers have been collaborative, the transition from design to production should go smoothly. It rarely does, but this is often due to the large number of moving parts during production.[16] Once manufacturing is complete, or when a given amount of product has been produced, there is sometimes a Gate Review IV decision that must be made. It may be wise to "embargo," or not ship, the units produced until the manufacturing engineers have satisfied themselves that the assembly lines have been working smoothly and that the product is manufactured as intended. If there has been a production problem that may compromise product performance—or product safety, then now is the time to correct it. It is not uncommon for products to need "re-work" prior to shipment. If the product passes Gate Review IV, then the product is "released" and shipped to distributors or retailers for consumers to purchase and use.

The final phase of the PDP is the use phase or Phase V. This final phase is covered in Chapter 11 of this book in great detail.

6.6 Conclusion

What has been presented in this chapter is but one example of an integrated PDP. Certain presumptions about the product being designed, as well as the industry in which the design-and-manufacturing company exists, have been made by the author. Consequently, this is *a* product-design process, not *the* product-design process. Engineers, companies, and industries should always use the tools and methods most suitable for their tasks. However, a rigorous and systematic PDP, regardless of form, should consistently deliver the finest results for the company and the safest products for the consumer.

Also shown in this chapter is a detailed and expanded Phase III in a PDP. It is the Engineering Design, Development, and Testing Phase (EDDTP) that will be further

[15] Design engineers should remain available to assist Manufacturing as assembly, and other, issues inevitably arise during Phase IV.

[16] The author believes that manufacturing engineers, and other Operations personnel, are extremely underappreciated for the vital work that they perform. Unless manufactured, no design—however brilliant—will either reach a consumer or earn a company one cent.

expanded upon in the next chapter. The EDDTP contains the vital design, analysis, laboratory-testing, and field-testing functions needed to develop a product's design. It is iterative in nature and employs a wide range of skills from many people. Out of the EDDTP, comes the final product design—good or bad—that will be used by many people. All engineering design is complete after this point.

This chapter is not intended to be a stand-alone treatment of the vast field of engineering design. Several authors have undertaken work on the engineering-design process that is both greater in detail and better than this author's treatment of the topic. Those readers interested are directed to the references provided in this chapter. This chapter merely lays the framework for looking further into product-safety engineering. The next chapter will illustrate how the *function* of product-safety engineering can be integrated into a PDP and the EDDTP.

References

"A Guide to Maximizing the Use of Existing Product Stewardship Programs." n.d. Boston, MA: Product Stewardship Institute. https://www.productstewardship.us/.

Biancardi, M. F. 1970. "Design for Product Safety." In *Combined National Farm, Construction & Industrial Machinery and Powerplant Meetings*, 8. Milwaukee, WI: SAE International. https://saemobilus.sae.org/content/700679.

Converse, Elliott V. III. 2012. "History of Acquisition in the Department of Defense. Volume I. Rearming for the Cold War 1945–1960." Washington, DC.

Cooper, Robert G. 1994. "Third-Generation New Product Processes." *Journal of Product Innovation Management* 11 (1): 3–14. https://doi.org/https://doi.org/10.1111/1540-5885.1110003.

"Core Competencies for the Product Stewardship Professional." 2014. Product Stewardship Society. https://www.productstewards.org/.

"Design for Safety." n.d. Osha.Gov. Accessed August 24, 2019. https://www.osha.gov/dcsp/products/topics/businesscase/designsafety.html.

Dominick, P. G., J. T. Demel, W. M. Lawbaugh, R. J. Freuler, G. L. Kinzel, and E. Fromm. 2001. *Tools and Tactics of Design*. New York, NY: John Wiley & Sons.

Ericson, Clifton A. 2016. *Hazard Analysis Techniques for System Safety*. Second edition. Hoboken, NJ: John Wiley & Sons.

Ertas, Atila. 2018. *Transdisciplinary Engineering Design Process*. Hoboken, NJ: John Wiley & Sons. https://doi.org/10.1002/9781119474654.

Hales, Crispin, and Shayne Gooch. 2004. *Managing Engineering Design*. Second edition. London, UK: Springer-Verlag.

Hammer, Willie, and Dennis Price. 2001. *Occupational Safety Management and Engineering*. Fifth edition. Saddle River, NJ: Prentice Hall.

Knaack, Marcelle Size. 1988. "U.S. Air Force Aircraft and Missile Systems, Vol. II, Post-World War II Bombers." Washington, DC: Office of Air Force History.

Pahl, B., and W. Beitz. 1996. *Engineering Design: A Systematic Approach*. London, UK: Springer.

Pugh, Stuart. 1996. *Creating Innovative Products Using Total Design: The Living Legacy of Stuart Pugh*. Edited by Ron Clausing, Don P., Andrade. Reading, MA: Addison-Wesley.

Ulrich, Karl T., and Steven D. Eppinger. 2000. *Product Design and Development*. Second edition. Boston, MA: Irwin McGraw-Hill.

PART 2

Application

CHAPTER 7

Product-Safety Engineering

When your values are clear to you,
making decisions becomes easier.
—ROY DISNEY[1]

7.1 Introduction

When some people think of design engineering, they imagine the physics and mathematics governing the designs produced by engineers such as aircraft, automobiles, bridges, dams, manufacturing processes, mineshafts, and electronic equipment. As earlier portions of this book have demonstrated, the role of product-safety engineering (PSEg) goes beyond the simple engineering involved in the design process—even though such "simple engineering" may involve mathematics and physics that are be quite complex. The physical world makes few personal demands of many engineers because the mathematics and the physics both state that some characteristic or state either exists or does not exist within an engineering design. A column structural element will either support a specific load, or it will not support that load, according to an accepted and proven formula. Such objective truths are really not open to debate. What is open to debate, however, is just how much stronger than minimally necessary that column must be to safely provide its function to product users over a reasonable lifetime.

It is the social world—and the engineer's role in it—that prove taxing to the design engineer when the truth moves from being either black or white to becoming a shade of gray as with the column example. The design engineer must decide what shade of gray is acceptable for a design's characteristics such as performance, weight, durability, and safety. In order to help the design engineer assure that product safety is being *designed into* a product early in the design process, a product-safety engineer (PSEr) should be integrated into the company's product-design process (PDP) and the Engineering Design, Development, and Testing Phase (EDDTP) within that PDP. This chapter will discuss this integration. Chapter 10 will discuss in detail some of the tools that may be used during the EDDTP by the product-safety engineer.

---

[1] Colan, Lee. "A Lesson from Roy A. Disney on Making Values-based Decisions." *Inc.*, July 24, 2019.

7.2 Perceptions of Product-Safety Engineering

The author is reticent to reply when he is asked what he does by a new acquaintance. Whenever the author says "product safety," the other party invariably thinks of OSHA, hard hats, and "Safety First!" Although, as mentioned previously in the book, industrial safety is a truly important engineering and administrative field for protecting people in the workplace, it is nothing similar to product safety with respect to much subject matter.[2] It is not the new acquaintance's fault for misunderstanding product safety since relatively few people, even within engineering, understand this field and recognize it to be distinct from industrial safety or even system safety. It is hoped that this chapter and book will help fill this knowledge void.

As readers will have already realized, the role of product-safety engineer in this book is not one of simple compliance engineering—this will be further demonstrated in Chapter 10. The active PSEr will not be satisfied with using a checklist taken from the requirements of standards, regulations, and industry agreements. Chapter 8 will demonstrate that such standards and regulations often do not contain the crucial guidance needed for effective and safe product design—especially when the product is new or "cutting edge" in nature. The PSEr must be looking for known, existing hazards with unacceptable risks *plus* new hazards arising from innovative designs and emerging technologies. Such hazards and risks are not always obvious and some of the ultimate risks may depend, to a degree, upon user behavior and the use environment in addition to the engineered product itself. It is the engineering team's task to identify, evaluate, and reduce these risks, as necessary, through thorough analysis and testing programs in the EDDTP.

Some product-design engineers may feel that there is no need for intrusion by a PSEr into the EDDTP of the PDP if the product that the design engineers are developing is primarily based on existing product designs and that those prior designs have always been "safe enough." Even when such beliefs by design engineers are essentially true, there should always be the pursuit of continuous improvement in levels of product safety as new knowledge and capabilities are gained from development, analysis, testing, new technologies, and real-world experience. The levels of safety for products should increase over time and not stagnate at a given point because it was "safe enough" a few years ago.

The PSEr must also never *earn*[3] the position of being the "no police" within the PDP. It is easy for the PSEr to simply say "no" to design engineers who wish to push boundaries with exciting new products. What is harder—but far more productive—is to work with the product-design teams by participating in the PDP and EDDTP to help *solve* any product-safety problems. This requires that the PSEr be a fully competent engineer in her/his own right.

Again, actively seeking product-safety issues prior to their manifestations will make a PSEr even more effective and valuable to the design-and-manufacturing

---

[2] One recent book (Goodman 2020), however, indicates what many in industrial safety have observed for years: that this field, too, is misunderstood, that it is subject to arbitrary and sometimes ineffective "solutions," and that many people believe that it takes no skill to practice it. As a result of this, Goodman says that the industrial-safety person may be "someone's freaking cousin (p. 50)." The present book's author expects more from product-safety engineers and those who hire them.

[3] Product-safety engineers may have such a title *thrust* upon them by those having a different threshold for acceptable risk than the PSEr, but it is hoped that this unofficial title is never truly *earned* by the PSEr.

company by reducing design costs, scheduling problems, design-engineer irritation by requiring re-design(s), and product-safety problems once the product has been sold. Through the effective integration of the PSEr, or simply even the PSEg *function*,[4] within the PDP, some instances of product redesign can be avoided. Of course, this will also help prevent post-sale issues with products from arising and, perhaps, leading to product-safety recalls and their associated costs—both financial and human.

7.3 Product-Safety Management

In recent years, there has been renewed attention to the field of product-safety management (PSMt). This has likely been aided by increased attention to related fields including product stewardship and sustainable engineering. Although this renewed attention is undoubtedly good for society, it will be shown that product-safety management is not the same activity that is product-safety engineering.

Due to the present state of manufacturing in the United States, many large companies no longer design and manufacture their own products. Or at least, many companies are not intimately involved with the design and manufacture of the many products which they retail. Some companies may not be involved with product design whatsoever. They may merely contract with an overseas company to deliver *X* thousand "widgets" at a given price by a given date.

It is at this time that the focus of this book is reiterated. This book in intended to help the engineers at *design-and-manufacturing* companies provide innovative and safe consumer products in a mass-production setting. Furthermore, the book is not intended to help "bottom-feeder" companies who wish to do the "bare minimum" by importing a product that is simply—or barely—*legal* to sell. Such companies are likely to have neither the resources nor the fortitude to establish a competent and disciplined engineering department which values product safety.

For some companies heavily reliant upon off-shore engineering, PSMt is more of an *accounting* exercise than an *engineering* one. They may have no control over the design process other than providing a "thumbs up" or a "thumbs down" on a final design—perhaps based on appearance alone. Some companies may only be able to barely search the available regulatory mandates and industry standards—if they exist—for mandatory conditions and voluntary requirements in their best-faith, but myopic, efforts to produce safe products.

For example, the result of a PSMt effort could be a large spreadsheet containing boxes that must all be checked off in order to sell a product. There will likely be numerous boxes showing that certain chemicals have either been limited in concentration, not exposed to users, or eliminated completely. Routine callouts include lead (Pb) and asbestos controls and the containment of small batteries. There can be many *health* factors contained in the PSMt documents, but these documents may also have requirements on *safety*-related product characteristics such as hard surfaces, hot

---

[4] The product-safety function is the set of tasks that would be performed by a product-safety engineer. In situations with limited resources, a design engineer may be tasked with performing this function along with others. Companies should carefully consider the best ways to protect the safety of their product's consumers.

surfaces, sharp edges, and static stability. Some of these characteristics may, again, be regulatory or industry criteria, but some may also be from internal corporate-policy documents that are followed for a corporation's products.

Although these compliance measures are good, they simply are insufficient for some products. Even the most-inexperienced of engineers—if reasonably ethical—will be sure that no sharp edges appear on a product. The addition of a check box showing the successful elimination of sharp edges on a PSMt spreadsheet really accomplishes nothing beyond documentation of a process. People thinking otherwise are often fooling only themselves. For example, the All-Terrain Vehicle (ATV) national standard has always included restrictions on sharp edges to ATV features in a certain area of the vehicle ("ANSI/SVIA 1-2017: American National Standard for Four-Wheel All-Terrain Vehicles" 2017), but sharp edges were never a true concern for those people highly critical of the safety characteristics of ATVs. Many critical engineering-design decisions will never be detectable—let alone fixable—with a crude and retrospective compliance exercise. It may be that certain product characteristics, or product design elements, cannot be measured easily and may not have widely regarded acceptable performance levels. Therefore, those involved with PSMt are encouraged to critically examine the levels of product-safety that can be either detected or assured through simple compliance exercises. Some PSMt efforts may simply be too far removed from the engineering-design work or too late in the design process to actually influence a design, or *design in* safety, to a product.

There have been some books written about "product safety" (Pine 2012; Seiden 1984). However, these books have dealt with the topic in the more superficial, business-oriented perspective of PSMt rather than through engineering. The landscape of product-safety engineering literature is rather barren.

Several decades ago, a trio of books was written by a singular author, Willie Hammer (Hammer 1972; Hammer 1980; Hammer 1993). These are good books but they appear to have not been followed by similar works from other authors. Hammer's work is similar to some of the work in this book. He covers hazards, risks, errors, standards, manufacturing, and maintenance that either have or will be covered by chapters in this book. Hammer goes further to include product-liability concerns and other regulatory and legal concerns such as subcontractors and vendors which are intentionally left out of this book. His books conclude with the presentation of hazard-analysis methods of the type found in system safety books such as those of Ericson (2016) and Gullo and Dixon (2017). Some of these methods are included in Chapter 10 of this book.

Other works from the era include one by a publisher's editorial staff (*Managing for Products Liability Avoidance* 1996) which covers much products-liability[5] material, but also covers some other important aspects of necessary engineering steps—albeit without detail. One edited book (Christensen and Manuele 1999) was written by a collection of authors who had a good idea, "safety through design,"[6] but perhaps spread the topic over too many industries to be particularly useful to practicing design engineers.

Yet another book on product-safety management (NSC 1997) walks through elements of a PSMt program. The pages dedicated to product design touch upon some

[5] The term "products liability" is generally used in legal references, whereas the term "product liability" is usually used by engineers and the general public.

[6] An alternate use of that term.

useful considerations, but—after mentioning two three-tiered safety hierarchies—revert to reliability, compliance, certification, and industry standards followed by a military system-safety standard, now MIL-STD-882E (USA/DoD 2012).

The author is hopeful that this book extends the knowledge base of product-safety engineering by including materials such as those in Chapter 4. These are concepts which, when coupled with concepts and methods in Chapters 5 and 10, will help the design engineer create safer products to protect future consumers.

7.4 The Product-Safety Engineering Function

Some large design-and-manufacturing companies may have a dedicated product-safety engineering department with product-safety engineers dedicated to product groups or even to individual development programs. Smaller companies may not have the resources of larger companies and may implement a product-safety engineering program in a different fashion. What is important is not *how* a product-safety engineering program is structured, but simply *that* a product-safety engineering effort is included during the particular PDP and that it leads to safer products for consumers. In addition, that PSEg effort must be recognized by company executives as sufficiently important to properly staff with competent engineers having access to the required resources to fulfill their duties.[7]

The product-safety engineering function may be performed by a product-safety engineer (PSEr), possibly a product-safety manager (PSMr), or by any engineer involved in the PDP, including design engineers. Those non-engineering personnel performing product-safety *management* could include a purchasing agent, a compliance specialist, and a manager. The PSEr and the PSMr will be close to, and involved with, the design engineering in a way that those performing traditional PSMt will not. Because of this intimate relationship with the engineering-design team, the PSEr and PSMr will be able to *identify* design and product-performance problems early in the EDDTP. The PSEr and PSMr will also be able to help *solve* these problems, often without sacrificing product performance, before they become larger issues later in the life of the product.

Just as it is unfortunate that some educators, executives, and engineering managers expect knowledge of engineering ethics to magically manifest itself within the essence of a student or young engineer, it is equally incorrect to expect that a deep understanding of product-safety engineering will suddenly blossom in that same person—or in an otherwise occupied design engineer. It is sad that so little material on product-safety engineering is available. Consequently, an engineer interested in knowing more about product-safety engineering may need to study alone to better understanding the topic through considering the true nature of product safety. Although it is hoped that this book will prove helpful, there will be a need for the engineer to study material and to dissect product-safety problems to best understand how product safety can be *designed into* a product and, finally, be delivered to a consumer. This will take time and effort. Unless an employer is willing to demonstrate to a budding product-safety engineer that product safety is a corporate *value*, it may well be that a young engineer is committing career suicide—although this pursuit is ethical—by spending time in an attempt to further

[7] Ethically, product-safety engineering is an engineer's duty to society and consumers rather than a business task.

understand product-safety engineering. It is an employer's responsibility to provide an environment to engineers where ethical conduct is supported, expected, and rewarded.[8] Although the PSEr may function as the *conscience* of a design-and-manufacturing company; it is up to the executives and values of a company to decide what product-safety path its engineering team ultimately takes.

There are many decisions in which the PSEr will participate. A few of these decisions will be relatively straightforward. Many of them, however, will be challenging because there will be no clear-cut answer. Again, there will be people who will disagree with some of these more-difficult decisions. It is here where a PSEr must use all of the technical skills available to her/him, and may need to ask for additional help in some cases. But it is also here where the employer of that PSEr will demonstrate the true values of that company. The following is an example (see Example 7.1) of a relatively simple design-philosophy decision that might confront a PSEr in the real world.

Example 7.1: Prevention of Misuse versus Fitness for Intended Use

Prior to the recent latest ANSI standard on All-Terrain Vehicles (ATVs) ("ANSI/SVIA 1-2017: American National Standard for Four-Wheel All-Terrain Vehicles" 2017), there was no United States requirement for these vehicles to have reflectors mounted on their sides. Despite this, several ATV manufacturers included side reflectors on all of their ATVs, while other manufacturers did not. Canada has required side reflectors on ATVs for some time. Those manufacturers not supplying American ATVs with side reflectors did provide them on their Canadian ATVs.

To put this matter into perspective, the ATV manufacturers entered into a Consent Decree with the U.S. Consumer Product Safety Commission, through the Federal Court, in 1988 ("Final Consent Decree" 1988). This was after many people were injured riding three-wheeled ATVs. These three wheelers are sometimes called All-Terrain Cycles (ATCs) to differentiate them from the four-wheeled vehicles. It was from this decree that all ATV manufacturers selling ATCs in America agreed to no longer sell the three wheelers.[9] This three-wheeler ban, however, was only a single part of the overall requirements and restrictions placed on ATV makers.

ATVs are off-road vehicles (ORVs) and, consequently, they were not designed for use on paved surfaces such as road, streets, and highways. Very few ATVs were equipped with differentials and other features that enhanced on-road use. In order to stay within the terms of the Consent Decree, some ATV makers were extremely careful to prevent their vehicles from being used in any on-road setting. This included the exclusion of side reflectors on ATVs since side reflectors are required of on-road vehicles by many state laws. This was perhaps seen, by some ATV makers, as one step to prevent ATVs from being used where they were not intended. An on-road hazard could be a collision with an automobile. ATV use on public roadways was not the intended use of the vehicle. This group of ATV manufacturers worked to prevent the *misuse* of the vehicle.

Although the above may be true, other ATV manufacturers did install side reflectors on their ATVs. Although this other group of ATV makers included reflectors on

[9] But those ATC manufacturers agreed to stop selling them without admitting any deficiencies of the vehicles.

[8] At times, it may be necessary to reprimand or punish unethical behavior contrary to organizational *values*.

their vehicles that would have facilitated making an ATV "street legal," their reasoning might have been that such reflectors increase the safety of ATVs and their operators by making the vehicle more conspicuous to the lights on other vehicles in darkness even in an off-road setting since many other ORVs have headlamps.[10] This second group of ATV manufacturers worked to make their vehicles better suited for their *intended* uses.

Two questions are pertinent after reading of these two different approaches to protecting consumers of one particular product through designed-in product safety. They are:

1. *Specifically*, although there initially existed a choice of whether or not a design-and-manufacturing company provided its ATVs with side reflectors, which group of ATV manufacturers was correct during the reflector-optional period?
2. *Generally*, is it more important to try to reduce a risk level (lower the probability of ATV impact with a car or truck on public roads) from the misuse of a product, or to reduce risk level (lower the probability of ATV being hit by another ORV at night) from the proper use of the product?

Learning to address, if not answer, such questions as these is a part of the role of the product-safety engineer.

[10] There was also a significant and pragmatic reason that an ATV designer/manufacturer would decide to put side reflectors onto all of its North American ATVs. (ATV sales in Mexico were relatively small.) Since Canada required these reflectors on ATVs sold there and, because an ATV company should want to sell its products in both Canada and the United States, it simply made operational sense to manufacture a single model of ATV that could be sold in both countries. Doing otherwise would have led to two separate models, or model numbers—each with its own Bill of Materials (BOM). Both production planning and distribution become more difficult when certain product variants can only be sold in certain countries. Therefore, an ATV company could benefit from only producing and distributing a single-BOM product for both countries, since the material cost of side reflectors themselves is negligible.

Much of a PSEr's time may be spent working with engineering-design teams on new product projects. However, it is important to not lose focus of the overall range of company products. The above ATV example was an issue that affected all models of ATVs produced by a particular designer and manufacturer. It is an issue larger than would likely be one ATV design project for one model year. Such a decision affects all ATVs for the company. If all of a PSEr's time is spent on projects, then there is no time available to consider the larger-scope product-safety issues.

There needs to be the opportunity to "catch one's breath" and look at the larger picture of the products made by a company as well as all of the product-safety issues which they face. It is here where a properly supported and staffed product-safety effort—or even department—would have a product-safety manager or director to help the product-safety engineers carefully ponder current product-safety strategies and the need to possibly revise them or even design new approaches for the future. An understaffed and under-resourced product-safety effort at a design-and-manufacturing company may well lead to disaster for the company in today's increasingly litigious and regulated business climate.

7.5 Safety Reviews

A PDP was presented in Chapter 6. The overall PDP was described and contained inputs from engineers, technicians, executives, product managers, sales and marketing people, manufacturing personnel, logistics personnel, quality personnel, suppliers, sub-suppliers, and consumers. Initially, a simple box was used to show a generic phase for engineering design in Figure 6.1. However, in Phase III, the EDDTP showed a detailed model of design engineering within a specific structured and disciplined product-development system in Figure 6.3. The importance of engineering-design reviews, or audits, was also covered earlier as were the points within the PDP and EDDTP at which they could occur. This design-review process showed the major engineering design reviews which are shown as "gates" in Figure 6.6. These are called gates because they signify the passage of the product design from one phase to another.

There will also be "minor" design reviews that may be signified by nothing more than a number, such as Design Review 1, Design Review 2, and Design Review 3, for example. These design reviews may also be named for particular significant events, such as "Verification Review." These same design-review—and gate-reviews—events are also excellent places and times to review the product-safety characteristics of these products. Such reviews of the product-safety characteristics may be called *safety reviews*.

It is at these safety reviews that the safety-engineering function be conducted in the open although the bulk of the PSEg work will already have been done. At other points within the EDDTP, the PSEr or PSMr can work with the engineering team on issues and then converge on a solution to a product-safety problem. However, at the design reviews, the product-safety engineer should be expected to assess the product-safety characteristics of the product under development and to deliver that assessment to attendees of the review. The resulting safety-review documents are explicit design-review documentation covering product safety and should become part of the official product-design project documentation and retained as required by company policy, by legal counsel, or by law. It is at a safety review where a PSEr can state if there is a problem that has not yet been solved or given an action plan to resolve. This is especially true for the Gate-III design and safety reviews. This is the last place that product-safety can be designed into a product. Any product-safety concern, unless it just arose, should already have been tackled well before the day of the Gate-III review. Project and program managers do not want to have any black marks or blemishes on their project or their employment record. They will generally be quite helpful to a PSEr in resolving a problem. It is during the time approaching a formal design review that the PSEr has the greatest ability to fix any problems with a product. These safety reviews are shown in Figure 7.1 which also shows points at which PSEg functions can be integrated into the EDDTP.

The considerations for a safety review are covered in other chapters of this book. For example, the concepts of hazards, risks, risk levels, and risk management from Chapter 5 are essential for a safety review. In addition, Chapters 8, 9, and 10 contain information and methods that may be employed in the design-review effort. One particularly useful tool during the PDP and EDDTP may be the Product-Safety Matrix (PSMx) demonstrated in Chapter 10. The product-safety engineer, or design engineer, conducting the safety review must understand both the abilities and the limitations of each tool used in the safety review. As with any engineer, the conductor of the safety review should always be searching for new methods and new ways of using existing methods in her/his work.

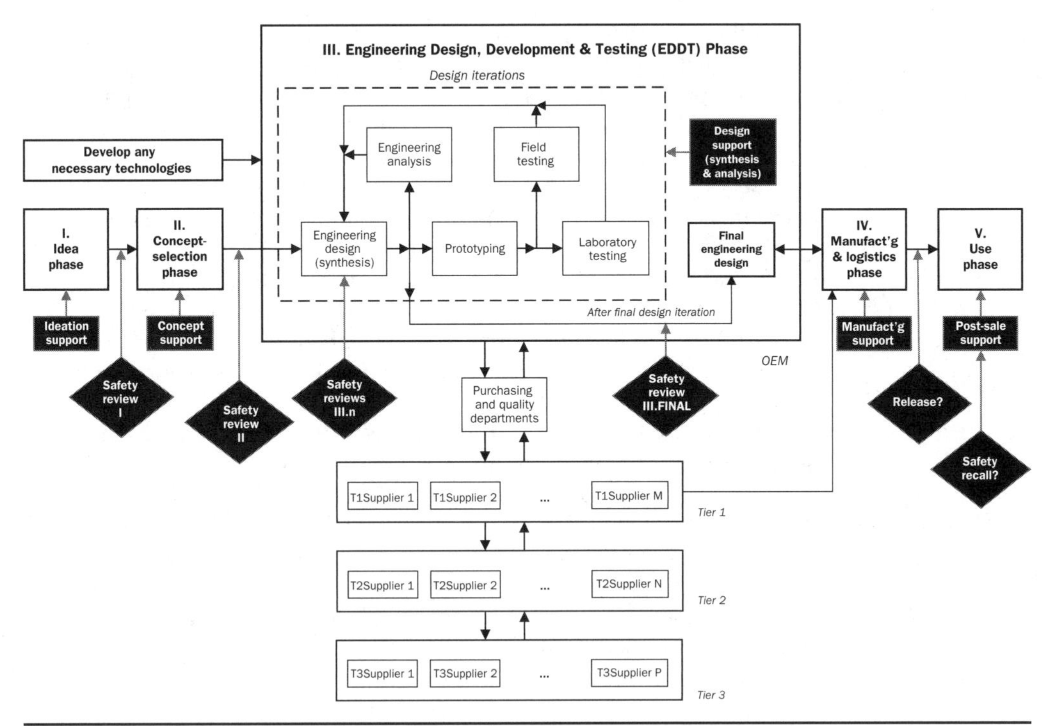

Figure 7.1 Phases of the product-development process (PDP). (Product-safety engineering functions shown in black diamonds and rectangles.)

In the author's experience, there has been occasion when an obscure project's design engineer has come to the PSMr with the need for a rapid, and perhaps even superficial, safety review before manufacturing the product in question. In other words, this person was told by a superior that they needed to "check off a box" in a company-PDP document to say that the PSEr or PSMr said that the product was "O.K. to sell." Such a design engineer often already violated the corporate policy of always following the established PDP since that design engineer had not included the product-safety engineer in the meetings leading up to production. Also, and just by the way, the start of production for the obscure product is *next week*. These people were not truly interested in product safety; they were simply hoping to be able to avoid any unnecessary scrutiny of their product—including safety reviews. They were only interested in getting their product out the door through a *pro forma*, rubber-stamp safety-review sign off, rather than systematic and thorough safety-review procedure, and then moving on to a new project or getting a promotion for making money for the company. Depending upon the design engineer's management, the design engineer may not be solely culpable for this type of behavior.

In essence, these people wanted the PSEr to sprinkle "holy water" on the project and to make the proclamation that the product was indeed at least "safe enough." To this, one PSMr might respond with the utterance: "I do not have a magic wand. All I have is a bandage." Fortunately, one particular product-safety manager does not recall an instance when a truly unsafe product was headed for production, but cavalier attitudes exhibited by some engineering personnel in such cases, could have led to serious consequences given the right set of circumstances. Such attitudes demonstrate a disregard for corporate policies, behavioral norms, and engineering ethics. Despite this, sometimes a corporate-reward system of promotions, pay raises, and annual bonuses work precisely as intended to produce the exact results that they entice—greater company revenue. The values of the organization are revealed through such action-reward cycles. Ethical engineers, as well as other organization personnel, deserve to have an environment where improper behavior is not tolerated and where product safety is a *value* and not a *fear*.

In the course of conducting safety reviews, a PSEr may need to consult with a variety of engineering and non-engineering personnel alike. This is illustrated in Figure 7.1 where the PSEr can support product development in each phase of the EDDTP. Potential or known challenges to product safety can be raised in the ideation and concept phases of the EDDTP. There will be much activity by the PSEr in the EDDTP of Phase III. Manufacturing may need PSEr support if problems arise during manufacturing, especially if those problems affect the final design which passed through Gate III. Chapter 11 will cover the post-sale support of products.

It is wise for the PSEr to consult with the engineering-program manager before gate reviews which are major events. Throughout the assorted design reviews within the EDDTP, it is quite helpful to speak one-on-one with engineers performing critical tasks on the new product. It is from such conversations that valuable information about the state of the project may be revealed. Due to cost, schedule, and other pressures from outside—and even within—a project, a solitary engineer may be reluctant to open speak of product-safety concerns that she or he has. This type of information is critical for the PSEr to know since, without it and if people are afraid to speak the truth, the safety review may be nothing more than a well-documented façade on a potentially unsafe product.

Among the most important people with whom PSEr must speak are the test engineers.

7.6 Test Engineering

In Chapter 6, the important role of analysis and test engineers in developing products during the PDP was shown to be an integral part of the EDDTP. This is clearly shown in Figure 6.3. Within this figure appear two boxes that are critical to innovative-product development. These are laboratory testing and field testing. They will be cumulatively called "Test Engineering" to simplify the discussion which follows.

Also shown in the figure is "Engineering Analysis," or simply "Analysis." Analysis is the application of much of engineering's "book learning," the formulae, theories, and software applying those formulae and theories. Such software includes finite-element analysis (FEA), multi-body dynamics (MBD), and computational fluid methods (CFM). Make no mistake, analysis is important work and, without it, engineering could not have contributed to people and societies in the significant ways that it has over the centuries. However, not all phenomena of interest have well-developed theories and formulae. Also, it simply may not be useful to develop detailed computer models for every single engineering-design issue. For some industries, such model development is worthwhile and is undertaken—although some of those models may not be useful beyond a specific application or set of conditions. The complexity of products, coupled with a compressed product-delivery schedule, makes it virtually impossible to fully assess a product without a systematic test-engineering program. In fact, even when theory exists, the wise design-and-manufacturing company will validate the theory through the testing of the actual product. Many engineering-analysis engineers freely admit to this need as well, regardless of how good of an analysis they probably performed. Parameters going into models may be incorrect and not every important product-performance factor makes its way into an engineering-analysis model. Therefore, test engineering remains a *vital* link in the chain that leads to delivering safe products to consumers.

Analysis is the evaluation of a product through mathematic equations based on physical relationships for various properties of products and their components, for example, the stress analysis of a mechanical member used in a product. Knowing the size and material of the part, along with *anticipated* loads for that part, enables the design engineer or stress analyst to calculate whether the part will not fail mechanically and work as intended. These calculations can sometimes be done by hand for simple parts, but much of this analysis is performed by computer through FEA due to the complex shapes and materials often needed to reduce both weight and manufacturing cost.

Engineering analysis is typically the skill taught to undergraduate students during their educations to become engineers. To a large degree, analysis is what makes engineering "engineering." Analysis takes background knowledge in science and mathematics to learn and to apply. Analysis is one of the reasons why many students become engineers. Yet, much great theoretical work has been backed up by experimental, or test, work, and its results. Therefore, test engineering has always had an important role in advancing engineering analysis.

Many, if not most, good engineering programs at colleges and universities, at least within the field of mechanical engineering, supplement the engineering-analysis portion of their curriculum with a senior-year "Capstone" *design* course. In the Capstone-design course, students typically work in teams on various design efforts where needs are considered, design options explored, conceptual designs proposed, and a final product prototype built. Many of these projects rightfully include analysis and testing phases. This is, of course, a good thing. However, this brief experience may not sufficiently demonstrate to students the importance of test engineering. Teams of students in the Capstone design project often test their prototype after the construction of the final design. As seen in the EDDTP diagram, Figure 6.3, the testing—both laboratory and field—is conducted during the engineering design's development. Therefore, testing results can help shape the final project. Of course, if time permits, a Capstone-design project can incorporate knowledge gained from their testing program back into product redesign.

There are generally two types of objectives to test engineering. Many tests are performed to be certain that the product meets a design requirement or regulation; the other, to be certain that the product's present design is suitable for its intended use. These types of tests and testing are known as *Verification* and *Validation*, respectively. Consequently, this subfield of test engineering is sometimes called *Verification, Validation, and Testing* (VVT). This is the part of test engineering of interest. There are numerous references for additional VVT information (Engel 2010; Lake 2014; Balci 1998; Balci et al. 2000).

The field of test engineering has come a long way from its early days and has evolved into a full-fledged engineering discipline. This is as it should be. One early test-engineering book is more of a popular read than a how-to guide for engineers (Hilton 1974). It does, however, show some interesting testing of products that the general public probably took for granted at the time. The book demonstrates just how important product testing is for establishing and checking the real-life safety characteristics of everyday products.

Test engineers for mechanical, electrical, chemical, materials, and other types of products are to be thanked for their vital work and contributions to product safety. Testing many mechanical products often involves gasoline and diesel fuels and exhaust emissions as well as hydraulic, pneumatic, and electrical components. It is a dirty, noisy, and hazardous job. Test engineers, especially when working off-site, often must deal with high and low temperature extremes, dry or wet weather, windy conditions, remote locations, bad work hours, uncomfortable modes of travel, poor lodging, bad food, and the overall general foulness of such work. It is not for the faint of heart. Yet, test engineering is one of the most important tasks for engineering safe products although it is only sometimes acknowledged for its contributions.

Despite its lack of glamor, the importance of test engineering cannot be overlooked because it, sometimes alone, can both prove the performance—including durability, ultimate strength, failure modes, and more—for a given component of a product and—perhaps more importantly—make visible problems with a product that no engineer has ever considered. A component that has passed verification testing may prove faulty under validation testing. The need for field testing is especially critical for innovative products that break new ground; these products may have no use history associate with them from which to draw knowledge. Components considered well designed by the analysis-engineering team, may prove to be severely inadequate when exposed to real-world conditions with all use factors in play.

Regardless of capabilities of design engineers, the amount of computational analysis performed, or the years of experience with a type of product, product testing—both in the laboratory and in the field—is vital to both assess and assure the safety of products used by consumers through testing. One example of neglected test engineering will be discussed in Chapter 11. Supplier lapses in their own PDPs are also sometimes discovered through the OEM's product testing.

Most people generally understand the concept of testing. Testing is the evaluation of a *constructed* product or product component through laboratory equipment or through actual use. There are generally fewer design assumptions for test engineering than for engineering-design analysis. For example, design analysts must assume a given load for a part. Test engineers often test without regard to many such engineering-design assumptions. They use the product as intended—and sometimes beyond—to determine if and how the product or product component fails. Sometimes test engineering shows that the loading *anticipated* by the design engineer during analysis is too low. Without an adequate margin of safety built into the product or part, that product or part *will* fail when the part's load exceeded the part's strength. However, this failure, within the test-engineering portion of the EDDTP, is a good thing. Test engineering has performed its job. It has discovered a flaw in the engineering design *before* the product was manufactured in quantity, delivered to retailers, and sold to consumers and potentially injured someone. The design engineer may literally "go back to the drawing board" and redesign the product with this new test-engineering information in hand. After this, the product will be re-analyzed and re-tested to investigate if significant problems still remain in the new design.

Test engineering also serves by assisting analysis personnel in the development of product mathematical models and computer simulations, especially those for complex subsystems and components. The real-world results provided by test engineers permit the analysis engineers to either validate[11] their models and simulations or to recognize that one or more aspects of the product itself are incorrectly represented, or perhaps not represented at all, in the model used for computer simulation. This is an iterative engineering process of continuous improvement and should be a byproduct of any rigorous and disciplined PDP.

Discussion: One Woman and Nine Months—or Nine Women and One Month?

The author will raise a rhetorical example known to many: since it usually takes approximately nine months for a woman to deliver a healthy baby, can the process be shorted to one month by employing nine women? The answer is, of course, no and the question ridiculous. This line of reasoning is, however, possibly valid for product-development efforts, especially within test engineering. Consider the following example.

Assume that a company designs and manufactures vehicles of some kind. Of course, that company must test those vehicles during and after designing it. The company knows that the vehicle must be tested extensively, but real-world scheduling requirements permit only so much time for the test-engineering validation effort. If a company wants to evaluate the condition of a car after 100,000 miles (160,000 km), does the company test one vehicle to 100,000 miles (162,000 km) or test 10 vehicles each to 10,000 miles (16,200 km)? Which answer, if either, is the correct one?

Imagine that this simple vehicle has one hundred (100) components. Assume that each component has only one critical property such as dimension or another important characteristic. Also assume that each of these critical characteristics only has three states: nominal—or as designed and intended, –X percent, and +X percent.[12] Both of the assumptions just made are gross oversimplifications. Each component may have several critical characteristics. A simple bolt has both dimensional and structural characteristics, each of which may have serious safety consequences if there is deviation from the specification.[13] The statistical distribution of characteristics, with respect to nominal, is continuous and neither discrete nor uniform as offered. Furthermore, one hundred components is actually far fewer than any real "parts count" from the BOM of any modern vehicle. Even with these simplifications, there are 3^{100}—three raised to the one-hundred power—possible combinations of components and variations that may be appear in a vehicle being designed and manufactured. The calculated number is 5.1×10^{47} and is large enough to practically approach infinity for a design engineer. However, the test-engineering question remains.

Should a test-engineering program test one vehicle over many miles—or test several vehicles over fewer miles? If only one vehicle is tested, then only one combination, or sample, out of the huge number of combinations is tested. That is far from optimal.

[12] The precise value of X is unimportant and would likely vary for each component.

[13] It is sometimes incorrect to presume that just because a part was "not to specification" that a defective condition exists. Even if product safety is potentially reduced through deviations from the engineering drawing and its specifications, the product may still be safe—albeit less safe than intended.

[11] The term "validate" is used here to indicate that a mathematical model or computer simulation has been compared to real-world testing results and that the computer outputs agree with the actual product outputs to an acceptable degree over a range of conditions. Thus, the software combination is considered to be suitable to use under its validated conditions. Such a model cannot be considered "valid" under conditions differing from those under which test results were provided to the analyst.

However, even with 10 samples, there is a universe of component-characteristic combinations still going untested.

What is the answer? It turns out that there is no universally correct answer to this. If the vehicle design turns out to function flawlessly when in use, then a correct test-engineering procedure, coupled with good engineering design, was employed. Unfortunately, the converse is also correct. However, for many products, it is risky to test only one sample of the product for the reasons mentioned above. Testing more samples of an engineering design is usually superior to testing only one unit. It is sometimes the case that, upon a test product's "tear-down" or disassembly after testing, engineers will discover certain components or areas of a product that are of greater concern than others.

In some of these cases, a specially designed test may be designed and conducted or accelerated-life testing could be used to examine the long-term effects of certain aspects of vehicle usage. Such accelerated testing could be the laboratory testing of a vehicle-suspension system where the suspension components are only subjected to known-harmful loading or loading frequencies. Such test-engineering tools help maximize the testing time window by artificially speeding up the aging process of a suspension system. This, of course, requires expensive laboratory-testing equipment and highly skilled test engineers and technicians, but may be necessary to fully test products within a given timeframe. Product and industry experience are of great assistance with the test engineering of products.

Regardless of the design and analysis competence of a design-and-manufacturing company, test engineering is able to reveal much about a product design such as design assumptions that were false and product failure modes which were never before even considered. Due to the time-proven importance of test engineering to the product-safety engineering effort, it is important for the product-safety engineer to work closely with test engineers throughout the EDDTP. The PSEr may be able to learn about, or even just better understand, a phenomenon being seen during testing that may not be well conveyed through other channels of engineering communication—especially if Test Engineering reports to Design Engineering and tests have uncovered a problem that could compromise the schedule or cost of a product program.

It should be clear to readers just how vital test engineers are to the success of a PDP. These engineers are able to identify, and then help solve, product problems before the products reach consumers. They, along with manufacturing engineers, are often severely underappreciated for their contributions. They are also an invaluable resource to a PSEr at safety review time.

7.7 Forensic Engineering

Conscientious engineers and companies are ever vigilant for product hazards and their risks. However, in some cases, real-world use experiences by consumers reveal hazards, risks, accidents, and outcomes that were previously unknown to a company's, or even an industry's, engineers. Perhaps an accident took place in which a unique combination of product, user, and environmental conditions merged. Unfortunately, people might have been injured or even killed in such events.

Attorneys are necessarily involved once a lawsuit has been filed following some serious accidents. These are usually product-liability actions. As a result, there are often claims made by plaintiffs and their legal counsel about the safety characteristics of the products involved. Not all of these allegations of unsafe conditions are necessarily true. Yet, product-safety engineers must remain alert to both design and manufacturing problems with their companies' products. Therefore, the PSEr must take seriously allegations of unsafe products. What is equally important at this time is that product-safety and design engineers not start chasing the "ghosts" of unsubstantiated product-safety allegations.

In order to actually improve product safety, it is vital to separate fact from allegation by working to find and understand facts of harmful accident events and their outcomes. Without doing so, attempting to remedy a product with an alleged—yet unproven—problem may fail to accomplish anything of value. In addition, such blind "engineering" may result in unintended consequences that may, in fact, compromise product safety overall.[14] Engineering-design changes and their effects form a closed-loop system, such as that discussed in Chapter 4. Without knowing and understanding the cause-effect relationship of a proposed, purely reactive design change, there is the possibility of creating even more or larger risks to product users. Until a problem is sufficiently understood, little can truly be done to correct it. Before understanding a problem, additional engineering effort may be little more than guesswork.

It is during such times that forensic engineering—often as the result of a lawsuit—can be a useful tool for a product-safety effort within a company.[15] Although this topic could have been included in Chapter 10 which addresses the typical tools used by product-safety engineers, its unconventional nature and utility make it a prime candidate for inclusion in this chapter. There are widely ranging skills and motivations of those offering forensic-engineering services to the legal community. Because of this, examining the abilities and credibilities of the consulting engineers delivering the results of testing, experimentation, and analysis are vital for any of their recipients. This is true for the work of the engineers retained *both* by plaintiffs and by defendants. Some "engineers" are actual engineers, some are technicians, some are mechanics, some are present or former law-enforcement officers, while others have varied backgrounds. In addition, some accident situations require skills beyond traditional engineering, mechanical engineering in many cases, to fully comprehend and address. For example, electrical engineering, biomechanics, and other fields may be needed to understand an accident and its injury mechanism(s). This discussion, however, will focus on mechanical engineering.

Ever since automobiles have been equipped with airbags,[16] there have been allegations of deficient performance in vehicle-airbag systems. These allegations from plaintiffs have included the failure to equip with airbags vehicles from earlier model years, failure for airbags to deploy, airbags deploying too aggressively, and more.

One possible scenario is that of an automobile driver whose vehicle was hit from the side by another vehicle. This driver is unrestrained, or not wearing her/his seat belt. The airbag does deploy during the accident. Yet, the driver hits the vehicle windshield and is seriously injured despite airbag deployment. Disregarding the driver's failure

---

[14] In some situations, it might be prudent for a manufacturer to stop the sale and advise against the continued use of its products until more is known about a problem for the good of the consumers.

[15] It is unlikely that a PSEr will have direct contact with this activity conducted through attorneys. The PSEr may only get information well after a legal matter has been resolved.

[16] Airbags are sometimes called Supplemental-Restraint Systems (SRS) to indicate that they are designed and tested to be used in conjunction with seat belts.

to heed the sound advice of always wearing seat belts when driving, an after-the-fact observer of the accident aftermath might be troubled about the driver striking the windshield despite airbag activation. Could this be an improperly designed or malfunctioning airbag or airbag-control unit? Should something be done immediately to prevent future recurrences of such injury? These questions, and others, are valid and need to be asked. Such questions should also come to the minds of ethical engineers upon seeing this accident outcome. The reader and engineer alike have not yet, however, have seen a full and factual analysis of this particular accident scenario.

This accident scenario was one of several types of accident events studied by a group of authors in a technical paper on motor-vehicle *occupant kinematics*[17] (Bready et al. 2002). A trio of diagrams, shown as Figure 7.2, illustrate the details of the accident just described. Shown in Figure 7-2(1) is the moment of contact when the first vehicle of interest, a sport utility vehicle (SUV) traveling horizontally to the right, is struck on its right-hand side by the second vehicle. Two arrows are provided on this vehicle in this subfigure; one to show the velocity vector of the first vehicle—the lower arrow—and one to show the velocity vector of the first vehicle's unbelted driver's head—the upper arrow. The first vehicle is hit by another vehicle traveling upward and to the left to the viewer. The subfigure's arrow attached to the second vehicle shows its velocity vector. The vehicles are each in motion and traveling at an angle to one another when the second vehicle hits the right-rear portion of the first vehicle.

Shown in Figure 7.2(2) is the motion and positions of the first vehicle and its driver as seen from a *stationary observer above the accident* at four instances in time superimposed upon one another. These times are: the moment of impact (0 seconds), after 0.1 seconds, after 0.2 seconds, and after 0.3 seconds. The motion of this vehicle, as a result of the impact, is clockwise rotational with continued translation to the right of the figure. The dynamics of the *vehicle* involved can roughly be determined through calculations using elementary scientific laws such as the conservation of momentum which can be deduced from Newton's Laws of Motion (Newton 1687). Greater precision of the path, velocities, and accelerations of this vehicle can be determined through the use of computer-modeling software. However, this detailed analysis is unnecessary for the unrestrained *driver* for much of the accident. The driver is not attached to the vehicle and, consequently, is unaffected by the vehicle's motion. In essence, this driver is a projectile and will continue in her/his current direction and velocity until hitting something that interferes with that motion. This is from Newton's First Law of Motion. Serious or fatal injury can result from the driver's impact with the vehicle's interior surfaces and affected by their characteristics such as hardness and shape. Unfortunately, the vehicle's windshield is the first object that *reconnects* the driver and her/his head with the vehicle. This reconnection is, sadly, quite abrupt—a head strike to the windshield.

This described motion of the driver *relative to the vehicle* is shown in Figure 7.2(3). This is as viewed from an observer above but attached to the first vehicle rather than remaining stationary as in subfigure (2). Therefore, the position of that vehicle never changes because the viewer is moving with that vehicle. However, this reference frame clearly illustrates the motion of the driver's head with respect to the vehicle and the airbag which is shown in its deployed, or inflated, state. The airbag could well have inflated

[17] Occupant kinematics is the study of the motion of motor-vehicle occupants—drivers and passengers—within that vehicle during an accident and how those occupants interact with vehicle-interior surfaces.

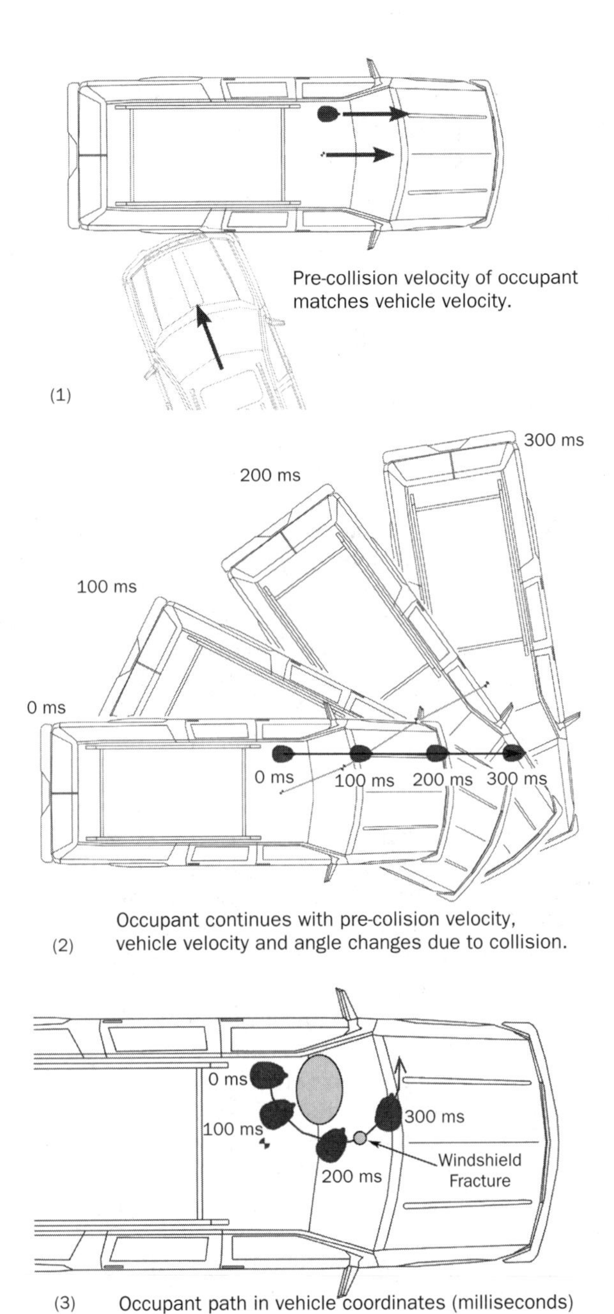

FIGURE 7.2 Kinematics of automobile and unrestrained driver during side-impact collision. (Used by permission of Collision Safety Engineering, L.C.)

as intended and as designed, yet the driver, through the projectile motion involved from not wearing a seat belt, would have gone completely around the airbag. Had the driver been wearing a seat belt, the driver's resulting constrained-passenger kinematics would likely have led to driver-airbag contact. This is not to say that the driver *would* have left the accident scene without injury but, if the driver was seat belted, the airbag would at least have had the opportunity to positively affect the accident's outcome.

This simple example of sound forensic engineering is just one manner of how specialized engineering work from outside sources can help a company's PSEr better understand what truly happened in what is likely a serious accident. Know that there was no indicated[18] failure in the airbag system's control, deployment, and protective characteristics. Therefore, from these results, there is no problem that must be solved. Good companies will continue to improve their products' levels of safety regardless of any glaring deficiencies. This is a simple and elementary, yet vital, part of PSEg.

Some liability-averse people within a design-and-manufacturing company may fear that safety improvements made to this year's products will show that last year's products were "unsafe." No one can ever promise that only nice things will be said in a court of law about any product and its predecessors. There may indeed be such negative remarks made by the counsel for an injured party about a company's products. However, the ethical pursuit of engineering makes the incremental improvement of products year-over-year a *necessity* and not just an option. This is a good thing and is something to be fully embraced and not avoided. Remember that this book is about *improving* product safety, not about *avoiding* product liability. Engineering, as a whole, should proceed along the same lines. There may be times when corporate in-house or outside counsel may argue with the above guidance, but always remember that the engineer's set of ethics differs from that of the attorney.[19] Would engineers feel better saying that over a period of time, say 10 years for example, the products on which they worked continually became safer, or that no safety improvements were made during their 10 years of working on the product? *Enlightened* corporate and trial counsel will recognize the same thing, fortunately.

Normal FMVSS (GPO 2018) crashworthiness testing does not cover the unique accident situation in this case. Other simpler accident scenarios are examined in the federally mandated testing program to evaluate the performance of vehicles and vehicle systems. Although the utility of forensic engineering is not generally taught in school, those engineers involved in the automotive industry are well aware of the *potential* significance of the results from forensic engineering—when properly done. Unfortunately, not all forensic engineering is done using scientifically dependable methods. Necessary conditions for good forensic engineering include using proven science, sound analysis, and justifiable conclusions. In some cases, representative testing, thorough documentation through photography, high-speed videography, and data collection/analysis are also needed in order to reach valid conclusions. Unproven methodologies and baseless opinions by engineering consultants alone do not count as valid forensic engineering. This caveat notwithstanding, the PSEr should always be open to receiving helpful information about possible problems, as well as guidance on potential solutions to these problems, should they prove to be real. It is for this simple reason that forensic engineering is included within this book.

---

[18] This is not to say that there could no longer be a problem with this airbag system. This merely indicates that, from this portion of one single accident-scenario investigation, there was no apparent airbag-system shortcoming from the engineering work conducted.

[19] This is a strict interpretation of the attorney's minimal ethical requirements. Many fine attorneys will go well beyond these minima.

7.8 Corporate Product-Safety Policy

The PSMr, PSEr, and any associated staff will be helped by a corporate policy stating the *values* of the corporation regarding product safety. Such a public and widely circulated policy should, of course, confirm the organization's commitment to designing, manufacturing, and delivering safe products to its consumers. Furthermore, the policy will permit, expect, and even demand that its employees conduct themselves in manners which will result in such safe products.

Yet, however heartwarming such a corporate product-safety policy might be, such a policy alone is insufficient for employees to feel safe—let alone empowered—to act according to the policy's edict to deliver of safe consumer products. If managers and executives of corporations, after having issued a sweeping corporate product-safety policy, continue to manage programs and reward participants according to a set of rules contrary to the policy, then all is for naught. Behavior such as this vividly demonstrates that product safety is not a value. In some cases, assuring product safety may not even be a *priority*. Instead, perhaps the *illusion* of product safety is the priority. It is ultimately the responsibility of upper management to show—to both their employees and their consumers—that they are willing to do more than simply make promises about safety and their products.

Being certain that their employees are encouraged to design and manufacture safe products is an effort going well beyond simply posting nice words on the corporate website. It will take more than sending a company-wide email with pleasant-sounding platitudes carefully crafted by the corporate Legal Department. A conscious and continued commitment by a high-ranking person[20] who is willing to stick her/his neck on the line is often needed. This person will need to be tough with thick skin to continue pursuing *values* and withstand the pressures put upon that person by those others pursuing *priorities* such as Profit and Loss.

Lacking any such high-level "white knight," those involved with securing the safety of the company's products and customers must still continue to act ethically. Being a PSEr or PSMr is not for everyone. It is an outwardly unrewarding task. Unless a company has gotten itself into *deep* trouble with regulators or the public, a PSEr will never become an engineering "superstar"—probably at least not permanently. Once the sense of urgency has passed, it may be that the permanent corporate value of profitability will replace the fleeting, temporary priority of product safety. At some points in their careers, product-safety engineers will need to ask themselves precisely what *their* values are. Are their values represented by the goals of promotion, financial security, or not making waves? If so, these individuals should change to another engineering field. Not only are such people not doing themselves any good by remaining in product-safety engineering, they could be endangering the public by doing so.

The corporation's true product-safety policy is not what is posted on its website, however prominently, nor it is the words spoken by an executive to industry friendly writers. Instead, a company's true product-safety policy is *embodied* in the products that result from the actions taken—or permitted—by the people within that organization.

---

[20] It has been the author's experience that a committee, however well intentioned, cannot match the effectiveness of an identifiable and accessible figure when it comes to representing a single, urgent corporate goal. Furthermore, committees usually cannot provide the immediate feedback needed by engineers working on time-sensitive matters.

If design engineering is the primary source of both safe and unsafe products, then having misguided or silenced engineers designing and manufacturing products emerges as that company's value and becomes its actual product-safety policy. Although the company may continue to make money, their customers may pay a high price as a result. A company's *true* product-safety policy is written on—and in—the product that goes out the factory door. A design-and-manufacturing company should not tell its customers what its company's policy is; the company should demonstrate that policy to them through products.

7.9 What Is Not Helpful

This book's chapters have included numerous aspects useful to the practice of product-safety engineering. It is in this section that some actions or information considered by many to be useful in product-safety engineering are shown to be useless or even counterproductive.

One particular activity that has not proven to be especially helpful to a PSEr is the placement of a monetary value upon a human life. There is absolutely no attempt made within this book to put a valuation upon a life. Because of the fact that almost one hundred people are killed on a given day in motor-vehicle traffic accidents in the United States, there most definitely is a *finite* value to each human life. The American economy and society continue to function despite this high daily human toll.[21] Thus, the loss of a generic human will not cause the world to stop. Yet, to each individual, her/his own life is generally of infinite value. Consequently, there will be no financial examination into the worth of a human life. The PSEr should treat each life—and its wellbeing—as extremely valuable and irreplaceable.

The author has concluded that this perspective is the correct one for a PSEr, admittedly through use of his own value system. That position has served him well both personally and professionally. A design-and-manufacturing company may have its own valuation of human life—insurance companies certainly do, but they prefer to label a "death" as a "loss" instead. The author also has no knowledge of when a company putting a dollar amount on human life—and having that decision become public knowledge following a product-safety debacle—has worked out well for that company. In fact, history is replete with counterexamples of conducting this exercise and acting upon it. Two counterexamples are offered below.

In the *first* counterexample, one author writes (Hammer and Price 2001, 1) that in 1893: "...[A] railroad executive said it would cost less to bury a man killed in an accident than to put air brakes on a car. The railroad executive probably was not an evil or malicious man. ...[B]ut he considered the safety of other people only in monetary terms." This example was from long ago when many people, including engineers, considered accidents to be unavoidable. The engineering profession has properly progressed to a place where many accidents are considered predictable and, therefore, preventable. Safety engineering is a result of this new perspective within engineering and other associated fields.

[21] The American social value system has effectively stated that this level of human loss is acceptable, in general. There continue to be transportation-safety advocates pushing for fewer traffic-accident deaths, but the body of law produced by duly-elected officials has not stepped in with changes to laws and regulations.

The *second* counterexample offered is from more-recent history, although well before the time of many readers. The product under review is the 1971–1980 Ford Pinto. There are multiple sources for details of this single example (Dowie 1977; Lee and Ermann 1999). The narrative supplied by one author (McGinn 2018, 149–61) will serve as the basis for discussing the engineering-design details of this vehicle. Among the problems with this car was the vehicle's fuel tank and its filler-neck connection. Dowie alleges that Ford Motor Company brought to market a vehicle with a known propensity to leak fuel due to rear-impact collisions. Of course, a fire could, and sometimes did, arise from the leaked gasoline and could lead to severe injury and death. Dowie continues and estimates that, in 1977—while the vehicle was still in production—between 500 and 900 people may have been burned to death as a result of the Pinto's engineering design characteristics. McGinn says that Ford considered several changes to the Pinto to improve its post-crash integrity, never exceeded $8.00 per car, but that "...Ford apparently believed that shutting down production and retooling to implement changes would be too expensive." He goes on to say that vehicle-safety improvements suggested in 1971 were delayed until the model year 1973 Pinto. McGinn examines a litany of other avoidable engineering-design failures of the Ford Pinto's development program that were brought to the attention of Ford management who then decided against a rapid response to any of them for financial reasons.

Other problems with the development process of the vehicle include that this vehicle's test engineers had not started conducting production-vehicle crash testing until one month *after* the vehicle was offered for sale to the public. This would be indicative of a huge failure in the PDP used by the company. The Ford Pinto story also involved whistleblowing by a Ford engineer—an engineer who felt compelled to follow her/his sense of engineering ethics. Interested readers are encouraged to read about this failure as well as another more-recent large, high-publicity problem involving motor vehicles, but this time, concerning its tires (Greenwald 2001).

Ford did not come out and say that a human life is work *X* dollars. However, their upper-management engaged in a cost-benefit calculus in which the benefit of public safety lost out to the cost of a vehicle-safety improvement. It is likely that any reader so disposed will be able to uncover numerous examples of similar behavior, decision making, and resulting fallout regarding a variety of products.

7.10 Conclusions

Outwardly, the job of the PSEr is not a high-profile career. Product-safety engineers will never be recognized as the "superstars" of a company or even an engineering department. Consequently, the rewards of these engineers must come from having done the right and ethical thing as a significant part of the conscience[22] of the organization.

As physicist Freeman Dyson wrote, "There are no prima donnas in engineering" (Dyson 1979, 114). In that vein of thought, there is no room for large egos in the field of engineering, especially when it comes to product-safety engineering. Mistakes will continue to happen within the PDP and EDDTP as long as many, many decisions are being made in a short amount of time—such as in the engineering design of a truly innovative

[22] It is hoped that many other company personnel also make and support ethical decisions.

product for a competitive marketplace. All of these decisions will never all be correct. Fortunately, many of the incorrect decisions should not significantly affect the product-safety characteristics of the new product. However, when a mistake is made which does adversely affect product safety, it is best to recognize it, admit it, and then move to correct it *quickly*. Large egos, potential promotions, and thoughts of other rewards may get in the way of, thereby obstructing, objective judgment at this crucial time. Having already made a mistake, an engineer, product manager, and a company are best served by fixing the error, learning from it, and avoiding its future repetition. A good PSEr can, if permitted by the engineering organization and the product-development process, help prevent and identify mistakes by a design-and-manufacturing company. If these poor decisions have already been made, then the PSEr can help the company quickly resolve problems as will be explained in Chapter 11.

Becoming a successful PSEr calls for three conditions. The *first* condition is that the product-safety engineer *must* be a competent *engineer*. Unless this person is a truly a skilled engineer, then both her/his acceptance by other engineers and her/his ability to actually contribute to the engineering solution to the problem will suffer. A product-safety "professional" will likely not have the technical background needed to significantly contribute to an engineering problem's solution.[23]

The *second* condition is that this engineer must have the confidence—driven by the person's sense of engineering ethics—to work within the gray area that is "product safety" engineering. Bertrand Russell once said, "The trouble with the world is that the stupid are cocksure and the intelligent are full of doubt." Young, burgeoning product-safety engineers should learn to reach well-reasoned conclusions that can withstand logical debate from those in disagreement. Reasoned conclusions, however, will never be sufficient for some detractors who have a different value system or, even, ulterior motives. Illogical argument will always exist, but it should not drive important deliberations. The PSEr must, in the end, come to grips with the simple fact that no decision she or he will ever make will find agreement from everyone.

The *third* and arguably most important condition necessary for a product-safety engineer's corporate success is having an environment which truly *values* product safety for its customers year after year. This requires both visible support and conscious commitment from design-and-manufacturing company executives. There should also be a viable career path available to maturing product-safety engineers if the company truly values product safety. Although in the following quote, Miller speaks of "system safety" personnel, his observations are equally valid for product-safety engineers (Miller 2020, 15):

> If a career path is provided, but not within the system safety organization, then the high-potential safety professionals will leave system safety to pursue a career outside of safety. Safety then becomes background knowledge that the candidate has acquired and may use in their next job. Organizationally, this drives a continuing gap in the talent required to perform, audit, and assess the tasks that ensure no unreasonable risk to the general public. This is not a sustainable environment for the safety organization or the enterprise.

---

[23] This sentence is not intended to say that a product-safety professional can play no role in the safety of consumer-product users. Such a person certainly can make contributions to problem solutions.

References

"ANSI/SVIA 1-2017: American National Standard for Four-Wheel All-Terrain Vehicles." 2017. Specialty Vehicle Institute of America (SVIA).

Balci, Osman. 1998. "Verification, Validation, and Accreditation." In *IEEE Winter Simulation Conference*, edited by D. J. Medeiros, E. F. Watson, J. S. Carson, and M. S. Manivannan, 41–48. Piscataway, NJ: IEEE. https://doi.org/10.1109/WSC.1998.744890.

Balci, Osman, William F. Ormsby, John T. III Carr, and Said D. Saadi. 2000. "Planning for Verification, Validation, and Accreditation of Modeling and Simulation Applications." In *IEEE Winter Simulation Conference*, 829–39. Piscataway, NJ: IEEE. https://doi.org/10.1109/WSC.2000.899688.

Bready, Jon E., R. Nordhagen, T. Perl, and M. James. 2002. "Methods of Occupant Kinematics Analysis in Automobile Crashes (2002-01-0536)." In *SAE World Congress & Exhibition*, 6. Warrendale, PA: SAE International. https://doi.org/https://doi.org/10.4271/2002-01-0536.

Christensen, Wayne C., and Fred A. Manuele, eds. 1999. *Safety through Design*. Itasca, IL: National Safety Council. https://www.nsc.org/faculty-portal/safety-through-design.

Dowie, Mark. 1977. "Pinto Madness." *Mother Jones*, 1977. https://www.motherjones.com/politics/1977/09/pinto-madness/.

Dyson, Freeman J. 1979. *Disturbing the Universe*. New York, NY: Basic Books.

Engel, Avner. 2010. *Verification, Validation, and Testing of Engineered Systems*. Edited by Andrew P. Sage. Hoboken, NJ: John Wiley & Sons. https://www.wiley.com/en-us/Verification%2C+Validation%2C+and+Testing+of+Engineered+Systems-p-9781118029312.

Ericson, Clifton A. 2016. *Hazard Analysis Techniques for System Safety*. Second edition. Hoboken, NJ: John Wiley & Sons.

"Final Consent Decree." 1988. Washington, DC: U.S. District Court (DC).

Goodman, Sam. 2020. *Safety Sucks: The Bull$h!* in the Safety Profession They Don't Tell You About*. Middletown, DE: Hominum, LLC.

GPO. 2018. "FMVSS." 2018. https://www.govinfo.gov/app/collection/cfr/2018/title49/subtitleB/chapterV/part571.

Greenwald, John. 2001. "Inside the Ford/Firestone Fight." *Time*, May 2001. http://content.time.com/time/business/article/0,8599,128198,00.html.

Gullo, Louis J., and Jack Dixon. 2017. *Design for Safety*. First edition. Hoboken, NJ: John Wiley & Sons. https://doi.org/10.1002/9781118974339.

Hammer, Willie. 1972. *Handbook of System and Product Safety*. First edition. Englewood Cliffs, NJ: Prentice Hall.

———. 1980. *Product Safety Management and Engineering*. First edition. Englewood Cliffs, NJ: Prentice Hall.

———. 1993. *Product Safety Management and Engineering*. Second edition. Northridge, IL: ASSE.

Hammer, Willie, and Dennis Price. 2001. *Occupational Safety Management and Engineering*. Fifth edition. Saddle River, NJ: Prentice Hall.

Hilton, Suzanne. 1974. *Beat It, Burn It, and Drown It*. Philadelphia, PA: The Westminster Press.

Lake, Jerome G. 2014. "V & V in Plain English." In *INCOSE*, 1134–40. Brighton, UK. https://doi.org/https://doi.org/10.1002/j.2334-5837.1999.tb00284.x.

Lee, Matthew T., and M. David Ermann. 1999. "Pinto 'Madness' as a Flawed Landmark Narrative: An Organizational and Network Analysis." *Social Problems* 46 (1): 30–47. https://www.jstor.org/stable/3097160.

Managing for Products Liability Avoidance. 1996. Second edition. Chicago, IL: CCH Incorporated. www.cch.com.

McGinn, Robert. 2018. *The Ethical Engineer: Contemporary Concepts & Cases*. Princeton, NJ: Princeton University Press.

Miller, Joseph D. 2020. *Automotive System Safety: Critical Considerations for Engineering and Effective Management*. Hoboken, NJ: John Wiley & Sons. https://www.wiley.com/en-us/Automotive+System+Safety%3A+Critical+Considerations+for+Engineering+and+Effective+Management-p-9781119579700.

Newton, Isaac. 1687. *The Principia (Philosophiæ Naturalis Principia Mathematica)*.

NSC. 1997. *Product Safety Management Guidelines*. Second edition. Itasca, IL: National Safety Council.

Pine, Timothy A. 2012. *Product Safety Excellence: The Seven Elements Essential for Product Liability Prevention*. Milwaukee, WI: Quality Press.

Seiden, R. Matthew. 1984. *Product Safety Engineering for Managers*. Englewood Cliffs, NJ: Prentice Hall.

USA/DoD. 2012. "MIL-STD-882E—System Safety." Washington, DC.

CHAPTER 8

Engineering-Design Guidance

What is a committee? A group of the unwitting, picked from the unfit, to do the unnecessary.

—RICHARD HARKNESS[1]

8.1 Introduction

The goal of Design for Product Safety (DfPS) is to *design-in* as many positive product-safety characteristics to new products as possible during in the product-design process (PDP). This entails coordinated work within the product-design team including engineers involved in design, analysis, and testing of the new product. Positive design characteristics should be built into a product while design, analysis, and test are still on-going rather than after large volumes of some parts are ordered from suppliers, before tooling is "cut" for other parts, and certainly before the product is sold to consumers. This requires the effective use of the time available for engineering-design work during the Engineering Design, Development, and Testing Phase (EDDTP). This consideration involves both productivity and thoroughness.

Engineering students and engineers are generally quite conscientious and want to design and manufacture products with high integrity, whether that is function, quality, durability, reliability, or safety. No one wishes to be professionally associated with lousy products. Neither do engineers want to overlook anything during the design process. When inexperienced engineering students are faced with a design project, such as the senior-year Capstone-design program, even they will generally look at comparable products and perform other research in order to do their "due diligence" for the upcoming design effort. This is known as competitive benchmarking and is performed in many companies to assess the competition and learn from them. It is always good to see what other companies have done regardless if those attempts were necessarily successful. Knowing what does not work in PSEg is also useful.

[1] https://www.forbes.com/quotes/7941/

If a new design engineer at a company is assigned to a product-design program, it is likely that she or he will do the same as the students above. In addition, that young engineer will also likely speak with the more-experienced design engineers at the company for their guidance from past experiences. This outreach to established engineers usually proves useful within an established design-and-manufacturing company; however, if the subject company is newly established, a "start-up" for instance, then the type of information sought by the novice design engineer may not be available due to the immaturity and inexperience of that organization.

Another place where many students and engineers expect to find design guidance to help establish a prescribed, expected, or even required level of safety in a new product design is in documents such as standards. This book will use the term *engineering-design guidance* to mean information, recommendations, advice, and requirements for engineers synthesizing new products. The design engineer may also consider industry and company design practices. This exercise is sometimes helpful, but it is certainly not always so. Again, the more innovative a product is, the less likely an engineer is to find suitable guidance from existing documents and knowledge bases.

Consequently, this lack of useful product-design guidance will be especially true for the types of products and companies on which this book is focused—the innovative design-and-manufacturing company of mass-produced consumer goods within a fast-paced, competitive, and regulated industry. Being first to market with new ideas is a profitable road to success for such companies. Therefore, it is likely that some product-design ideas on which the company's engineering team is working are without precedent. Quite logically, engineering-design guidance is hard to find for products that have never before existed. An existing industry standard will probably not apply with certainty. The newness of an idea should mean that, although standards for related products might provide some level of design guidance, these standards may fail to provide guidance in critical areas of the innovative design. Some of these areas, no doubt, affect the safety of product users. The design engineers will necessarily be designing without explicit guidance from either the industry or the regulators. Even when industry standards or regulatory requirements exist, they often provide *necessary* conditions rather than *sufficient* ones. Many of these sufficient safety conditions are ridiculously easy to achieve and would routinely be done by any competent and ethical engineer. The simple example citing the banning of sharp edges on products was discussed in the last chapter. Thus, especially in the case of innovative products, the design engineer's true question of "How safe does my product need to be?" remains unanswered even through established standards and regulations. It is this type of design quandary that this book seeks to help engineering students, engineers, and even non-engineers understand and address.

8.2 Sources of Design Guidance

There are numerous sources from which engineers can seek design guidance. Of course, there are standards. There are also regulations. There are also other sources. Some of these will be explored in this section, others may be legal issues that are better addressed by qualified legal counsel, not by the author.

Before diving too deeply into standards and the like, a couple of simple definitions will aid the discussion. In their book on engineering standards, two authors

(Greulich and Jawad 2018) do a fine job of explaining the various types of engineering standards, regulations, and their roles. They use three definitions:

1. *Rule*: A single specific requirement that must be met...
2. *Procedure*: A set of rules regarding how a task or function is performed...
3. *Standard*: A set of rules and/or procedures recognized as authoritative in a particular area of interest.

These are workable definitions for this chapter's treatment of standards here. This book takes issue with the presumption of "authoritative" for standards, since not all standards may enhance product safety if followed.[2] Greulich and Jawad add several types of engineering-design guidance documents. They include Voluntary Consensus Standards, Codes, Governmental and Industry Standards, and Regulations.[3]

Voluntary Consensus Standards (VCSs) are those standards written by a competent standards-development group. In the United States, such a group may be recognized by the American National Standards Institute (ANSI) to develop an "American National Standard." The establishment of an ANSI standard requires the consideration of views from a diverse set of "directly and materially affected interests" ("ANSI Essential Requirements: Due Process Requirements for American National Standards" 2010, 24). Such interested parties could include: manufacturers, regulators, commercial users, recreational users, law enforcement, labor groups, and transportation providers. There need not be unanimity among parties—only *consensus*.[4] Unless other agreements exist, voluntary standards are simply that—voluntary. A VCS is generally a good set of practices to follow, but they do not have direct legal penalties if not followed.[5]

A *code* is by definition "a system of principles or rules" ("Code" n.d.). Generally, codes are a large set of rules or procedures aggregated into one document or several documents. Greulich and Jawad state that codes are "often a body of standards groups together for ease of reference" (Greulich and Jawad 2018, 38). Codes protect and enable the daily lives of many people through various of building and electrical codes and through electronic-device and electronic-communications codes.

There are numerous industries with specific standards and codes with great impact upon public safety and health, examples include ("Boiler and Pressure Vessel Code 2019 Complete Set" n.d.) and ("NFPA 70: National Electric Code(R)" 2020). Since its inception in 1914, the ASME Boiler and Pressure Vessel Code (BPVC) has grown to eleven sections covering such topics as rules of construction, materials, non-destructive evaluation (NDE), welding, and nuclear components. Such codes may be either national or international. Many have heard of Underwriters Laboratories (UL) standards as well. UL[6] is an accredited ANSI standards developer.

---

[2] Engineers are cautioned against blindly following standards merely because they are standards which have been approved by *a* standards-development committee. The ethical engineer must consider if public benefit results from the requirements of a standard or any other form of engineering-design guidance.

[3] Please remember that the author is not an attorney and readers are advised to seek competent legal counsel on any legal or regulatory matters.

[4] The author has been on standard-development committees which have had long discussions about the precise meaning of "consensus." The standards-development organizations provided committees with no guidance on the meaning of the word.

[5] It is possible that a company could be *indirectly* penalized by a punitive courtroom verdict for failing to follow a voluntary consensus standard.

[6] https://www.ul.com/

Governmental and Industry Standards (GISs) are those that do not follow such an open ANSI-like process for gathering input from affected parties. This permits a standards developer to maintain tight control on the requirements of the standards. Some such standards are produced by government agencies to benefit the government or by companies within an industry to possibly protect the interests of that industry. Some of these GISs also help the industries contracting with the government or industry.

One example of GISs is what was once called the Society of Automotive Engineers (SAE). They now go by SAE International,[7] but their standards are seen on such automotive-industry products as motor oil and motor-vehicle lighting assemblies. As another example, the Department of Defense has a governmental system-safety standard (USA/DoD 2012). In addition, NASA has a handbook for Fault-Tree Analysis (FTA) (NASA 2002) and the Naval Surface Warfare Center produces a handbook on reliability-prediction procedures ("Handbook of Reliability Prediction Procedures for Mechanical Equipment" 2011). Although the second of these government documents is not called a standard, it is often used as one.

Finally, *regulations* are rules or laws from a government or a regulatory agency. One well known example is the U.S. Code of Federal Regulations (CFR)[8] and cover such topics as Transportation and Labor. In some jurisdictions, governments may issue orders in the form of *directives*. Certain products sold in specific parts of the world are required to meet or exceed particular directives, for example, machines in the European Union and some other countries are required to meet directives which include the one for machinery (EC/DGE 2006); other directives cover topics such as electrical potential (voltage) and restricted hazardous substances.

There are sources of design guidance in addition to those just discussed. Among them are agreements and internal organizational rules and guidelines. Agreements may be made with other parties involved with similar products or within an industry regarding product-safety levels or characteristics. Agreements can also be made with regulators. Such agreements are non-binding in nature[9] and are only as strong as the honesty and integrity of each party involved. For example, the designer/manufacturer of a regulated product may promise regulators that certain design specifications will be followed by a company. This agreement will necessarily be short-lived and, for permanence, may need to be followed by a regulation—but such a stop-gap measure can prove useful to company and regulator alike, albeit for a limited period, since it can quickly address a problem. Such agreements may either be public or private. They may either be verbal or commemorated in writing through an instrument such as a memorandum of understanding.

Agreements may also be within an industry. Companies may see that having a particular agreement among companies is in the best long-term interests of that industry. For example, an industry may agree to limit the top speeds of vehicles in an effort to dissuade regulators for doing that for them. Not all members of an industry always agree to go along with their industries, however. Some dissenting companies may service a particular niche within an industry and, consequently, be unduly burdened by such an industry agreement.

[7] https://www.sae.org/
[8] https://gov.ecfr.io
[9] The violation of such agreements will often lead to ill-will between parties and potential future consequences.

Some companies also have their own internal processes, with design specifications, on product performance. Therefore, even though not required, some companies will demand that their product meet internal design requirements. However, as stated earlier in the chapter, such design guidance may be of little good for a ground-breaking new product. The design engineer will still need to find a basis for making critical detailed-design decisions affecting product safety.

The above is a generalized and brief discussion of some forms of engineering-design guidance. This field can become quite complex and depend heavily upon the product or the industry at hand. Readers involved with such products or industries are strongly advised to research their areas of interest and consult with legal counsel, if necessary.

8.3 Approaches to Design Guidance

No one should presume that the creation of a new industry standard will suddenly make a significant difference in how a given product is being designed. It is likely that the new standard will be created by a committee at least partially composed of the current designers and manufacturers of that product. It is also likely that the current designers and manufacturers believe that their products are completely safe. Given these two conditions, the planned new industry engineering-design standard may largely reflect the current state-of-the-art which consists of the products currently being designed, manufactured, distributed, and sold. Thus, there may be little visible impact from the new standard—at least as far as current, established companies already in the market are concerned. This is because the established manufacturers are already doing what they think should be done regarding the engineering design of the product. They, through establishing a standard, are merely requiring that other new companies do what they themselves are already doing in order to fairly compete with them. Many manufacturers will be disinclined to advocate for a standard with requirements that their existing products cannot meet. They may fear that such a new standard will show their extant products in a bad light and, thereby, provide ammunition to litigants in the future.[10]

There, of course, can be exceptions to the above potential effects from the development of new standards on established manufacturers. However, in some cases, there would need to be consensus that the current products within the scope of the new standard are lacking in some significant way. Lacking such a consensus, there would need to be coercion—perhaps coming from regulators, attorneys, or the media. The designs for the products from new-manufacturer entrants to the market could well be affected, however, since they may not have an *explicit* baseline from which they can measure their product designs. However, many new entrants may already be using competitive products from established manufacturers as *implicit* competitive benchmarks for their new designs. So, even in this case of new market entrants, the impact of a new standard may not be significant.

[10] This, of course, is not a valid reason to avoid making continual product-safety improvements as mentioned elsewhere in the book.

One area in which standards could do *harm* is for products that do not exist. A *specific* standard for a non-existent product is unlikely to be created and, probably, should not be created. During the creation of such a standard, the effects of the inductive reasoning necessary for the cause-and-effect relationships *imagined* during standards development might often prove wrong. This is one reason why test engineering is so important to the development of safe products. Those who have been involved in product development and, especially test engineering, understand that many preconceived ideas of what works and what does not work may be shown to be incorrect once a working design prototype is produced and used.[11] Therefore, such speculation on what is necessary or what is to be avoided is a perilous activity without proper testing in the real world.

There may be instances in which creating *general* standards for types of products is prudent. This is when specifications for *types* of products are developed. An example of such a standard is one that is currently under development for "connected products" ("Subcommittee F15.75 on Connected Products" 2019). These are consumer products which are connected or can be controlled by wireless means, such as a smartphone. Such connected products include baby monitors, physical-fitness trackers, security cameras, household thermostats, and entry-door locks. A trendy term for the proliferation of such products falls is the "Internet of Things" (IoT). It is likely that, upon completion, a standard will outline product-design specifications that must be met or exceeded[12] by this class of products. It is important to avoid creating design-restrictive standards in any field. Therefore, due to the rapid developments in technology regarding the IoT, it is important to focus more upon the end requirements rather than upon the means of achieving them. Therefore, rather than telling a designer/manufacturer *how* to design a safeguard for a connected product, for example with technology X, it is more desirable to specify the necessary level of *performance* for the product and let the designer/manufacturer achieve that level as it thinks best. It will be important for any such standard for products, that are as-of-yet under signed, have allowances for exceptions when a given performance goal cannot be achieved without negatively affecting product safety in a demonstrable way. Standards that specify how something must be designed are known as *design standards*, while those that only specify the performance or protection of the design are known as *performance standards*. With few exceptions, performance standards protect better protect the welfare of the public by allowing design engineers to employ new technologies to meet requirements. This permits potentially greater levels of protection to product users as well as, perhaps, lower prices.

The author has heard some people mistakenly utter the words: "Any standard is better than no standard." The next discussion will show this reasoning to be false.

---

[11] The author has first-hand failure experience with speculating on what would and would not work on new products.

[12] The term *minimum standard* is an oxymoron and should not be used. A standard is generally a set of requirements. These requirements may be *exceeded*—but they must at least be met.

Discussion: Standards Development—Goodness, Utility, Competence, and Self Interest

Before going any further, it is necessary to dispel any reader presumptions that all standards are good, useful, and written by competent people having pure motives.[13] It is good that many standards meet the above criteria, but not all standards do. The question of the usefulness—or utility—of a standard will be addressed later in this chapter. However, problems with the other three criteria for standards just mentioned: goodness, competence, and proper motivation, are demonstrated by counterexamples below.

Not all standards are good. The term "good" here will mean that the standard is complete. That is to say that the standard accomplishes its intended purpose. In a case of failed standard-development effort, the U.S. Department of Transportation's National Highway Traffic Safety Administration (NHTSA) produced one particular Federal Motor Vehicle Safety Standard (FMVSS)[14] that is tremendously lacking in important safety-related content. A low-speed vehicle (LSV) is an on-highway vehicle that travels no faster than 25 mi/h (40 kmh). The LSV is regulated, at a national level, through FMVSS 500 ("Standard No. 500; Low-Speed Vehicles" 2018). In it, FMVSS 500 requires that all LSVs be equipped with: headlamps, front and rear turn-signal lamps, taillamps, stop lamps, reflex reflectors, an external mirror, a parking brake, a windshield, a conforming VIN (Vehicle Identification Number), seat belts, a design providing sufficient rearward visibility, and a pedestrian-alert sound generator. The only test procedure in the standard is for measuring the top speed of the LSV. Those people familiar with vehicle standards will realize that there is no test procedure for service-brake performance—a common section within vehicle standards. One possible reason for FMVSS 500 not having a service-brake performance test is that the standard does not require an LSV to even possess a service brake. However, there is a requirement for LSVs to have parking brakes, but no performance criteria are provided for parking brakes such as operation or ability to hold the vehicle at on a prescribed slope for a prescribed period of time. As a result, a large rock placed downhill of a parked LSV's tire would suffice as a parking brake—so long as it was provided with the LSV upon purchase. Fortunately, another standards organization has an LSV ("SAE J2358_201611: Low Speed Vehicles" 2016) standard that is more comprehensive in nature, although this standard's speed criteria are slightly lower, at not more than 20 mi/h (32 kmh), than FMVSS 500.

Society expects, and deserves, that those developing standards be competent. Standards are commonly developed by committees. In the case of an American National Standard, ANSI must approve of an organization prior to it developing an ANSI standard. The ANSI-accredited organization becomes a Standards Development Organization (SDO) ("Domestic Programs (American National Standards) Overview" 2019). The SDO then forms a committee to draft a standard. This committee, of course, consists of members with varying levels of competence. Some SDOs are formed by industry trade groups such as those companies making a particular type of product. In many instances, the companies within a trade group do indeed possess the necessary skills to develop a useful and complete standard. However, this is not always the case. Take for, example, the ladder industry. This industry is represented by the American Ladder Institute (ALI).

[13] A standard may also potentially be unnecessary.

[14] From the perspective of this chapter, it is unimportant whether this is a "standard" since it is a Federal Motor Vehicle Safety Standard or a "code" or "regulation" since FMVSS appears within the U.S. Code of Federal Regulations.

The ALI, which was formed in 1923 ("About ALI Standards" n.d.), and writes and maintains the nine American standards for ladders. Given the expertise possessed by the industry and their personnel, one would expect that ladders and their uses would be well developed through equations of the fundamental principles at work during ladder use. That person would be mistaken. Looking back at the near century that has elapsed since the formation of the ALI, it was not until 1996 that someone—not directly within the ladder industry—did a static analysis of a simple ladder (Barnett 1996). In it, Barnett evaluated ladder set-up and loading conditions and their effects on ladder "slide out," a serious ladder-use problem that often leads to accident events with serious injuries. Ladder slide out is the slipping of the ladder's feet on the ground away from the wall or structure upon which the upper end of the ladder is leaning. Although quite insightful, Barnett's ladder analysis is nothing that could not be done by a second-year engineering student who had taken an engineering-statics course. Yet it took over seventy years for anyone to formulate *and publish* these straightforward, yet useful, relationships with serious public-safety implications.

It is also in the public interest for SDOs to operate for the benefit of society. At least one government act ("Clayton Antitrust Act of 1914" 1914) precludes such industry practices as price fixing and limiting competition. Most SDOs are diligent and avoid such illegal activities. Remember that the NSPE's Code of Ethics (NSPE 2018) requires that engineers work with "honesty, impartiality, fairness, and equity" to "[h]old paramount the safety, health, and welfare of the public." Working to squelch competition is not in the interest of the *financial* welfare of the public. Any engineers involved in such activities may also see these as being dishonest and unfair. The U.S. Department of Justice has investigated the IEEE for anti-competitive activities during standards development. "The US Department of Justice is looking into whether a group of companies at a key standards organization attempted to exclude technologies from certain firms in creating a new Wi-Fi connectivity standard" (Sisco and Nylen 2018). This investigation is not over and all parties should be presumed innocent of any wrongdoing; however, this investigation shows that it may be possible for corporate or industry interests to be confused with—or even displace—those of the public.

The above discussion is pertinent for several reasons. One, many readers either are or will be involved with engineering-design programs. Two, some readers either are or will become involved in developing standards. The ethical responsibilities of an engineer are not limited to the design process. The standards-development process is also equally subject to engineering ethics. Unfortunately, for some engineers developing standards, at some point, it might become necessary to "step away from the table" and graciously bow out of the standards-development process if committee members are not working for the good of the public. If things go quite wrong, it may be necessary to inform the Department of Justice about anti-competitive activities taking place under the guise of standards development.

Once engineers become involved in standards development, some may justifiably ask themselves: "Whose interests do I represent when I attend a standards-development meeting?" This is an important engineering-ethics question. The explicit answer may depend upon the standards-development organization. Some ANSI standards-development committees are sponsored by industry organizations, or *companies*. The members of these industry groups are other organizations including companies. In the case of SAE standards, the standards-committee members are the *individuals*.

In a perfect world, it would make no difference whether a company or an individual was the member of a committee developing a standard. This is not an ideal world, however, and this difference could become a salient matter. Imagine that a member of a standards committee, an engineer, is told by corporate management to support a particular part of a standard that the engineer believes to be contrary to the welfare of the public. To whom does this engineer owe allegiance—the employer or the public? At this point the engineer must exercise applied engineering ethics and decide on the right thing to do. It can be a hard choice—with penalties for *either* choice, perhaps. Careers can be harmed, but conscience can also be damaged.

In the case of SAE standards-committee meetings, the chair of the meeting always reads an opening statement saying that each committee member was selected to serve on the committee on the basis of *personal* qualification. They are not to act in the interests of any organizations. This is truly a great ethics statement which should certainly get the attention of the ethical committee members, at least. The author in no way believes that corporate powers do not influence some decisions made by some committee members during SAE standards-development efforts; however, this standards-committee model—and ethics statement—are worthy of mention and, perhaps, imitation.[15]

Discussion: Types of International Technical Standards

The International Organization for Standardization (ISO)[16] has established a variety of international standards applicable to the design of machinery. Without going into great detail, the classification of their standards into separate groups will be instructive to readers. Not all of ISO's standards apply to the same *aspect* of machinery—although each of these standards does, indeed, apply to machinery.

Some ISO standards apply only to general aspects of machinery and are called "basic safety standards" (Kelechava 2017). These standards are called Type-A standards. Type-A standards provide "basic concepts, principles for design and general aspects that can be applied to machinery" ("ANSI/ISO 12100:2012 Safety of Machinery—General Principles for Design—Risk Assessment and Risk Reduction" 2012, vii). Such standards will provide design guidance at the conceptual level that may be too general for some detailed-design applications.

Another type of ISO standard applies to safeguarding machinery. These ISO standards are more specific than Type-A standards. This type of standard is called a Type-B standard. ISO 12100:2012 states that Type-B standards address "one safety aspect or one type of safeguard that can be used across a wide range of machinery." Type-B ISO standards are further broken down into two categories: Type B1 and B2 standards. Type B1 ISO standards address particular machinery-safety characteristics, while Type B2 standards address machinery guarding and safeguarding.

The final type of ISO-machinery standard is Type-C which addresses a specific type of machine. The criteria contained within a Type-C will be limited to that group of machines.

It is also worth mentioning another type of standard which is common in Europe. This is the "harmonized standard" ("Harmonized Standards" n.d.). Some harmonized standards become ISO standards, but not all are. In the case of machinery, a harmonized standard will have been developed by CEN (European Committee for Standardization,

[16] https://www.iso.org/home.html

[15] Despite numerous attempts, the author was unable to secure permission to reproduce SAE International's short committee-opening statement here. He believes that this is unfortunate both for the reader and for SAE International.

or in French *le Comité Européen de Normalisation*). The term "harmonization" means that the standard has been aligned, or harmonized, with the requirements of an EC directive. In the case of machinery, the Machinery Directive (EC/DGE 2006) applies, but does not cover all aspects of compliance necessary to CE-mark a machine.[17] Consequently, in order to meet the Essential Health and Safety Requirements (EHSR) in the Machinery Directive, a harmonized standard will have accepted tests, procedures, and criteria to show compliance with particular EHSR sections. Since both the directives and the harmonized standards have been developed by the official entities of the European Union, then a directive's safety-compliance requirements met using a harmonized standard are essentially unassailable by other parties. Despite the numerous downsides to the European CE-mark system as seen by those in the United States, the use of harmonized standards, when possible, is an upside to that system.

8.4 Application of Design Guidance

Although extremely liberating to design engineers, a "clean sheet" design program can be a problem once Concept Selection, Phase II of the PDP in Figure 6.3, has been completed. Now, critical decisions are made on design details for the new product—such as speed, weight, length, strength, temperatures, and more. Since there will be no design restrictions, or "carry over," from prior designs, engineering designers have a broad spectrum of design-parameter values from which to choose. It is likely that one or more of these detail-design parameters will have an effect on the ultimate product safety delivered to users of the new and innovative product.

A majority of design engineers want to provide a safe product to users. Yet, there are few human-safety metrics that can be directly applied during the PDP. For example, when designing a hand-powered winch[18] which could be used on a trailer for a small boat, there will be hazards and risks associated with its use that can result in serious user and bystander injury. If designing an All-Terrain Vehicle (ATV), there is no rationale to saying that an ATV of 600 pounds (272 kg) is safe, but an ATV weighing 601 pounds (273 kg) is not safe. Therefore, design engineers must look to other sources for design guidance during the PDP and in the EDDTP. Fortunately, there are existing industry standards for each of these products, ("SAE J1853-2014: Hand Winches—Boat Trailer Type" 2014) and ("ANSI/SVIA 1-2017: American National Standard for Four-Wheel All-Terrain Vehicles" 2017)[19], respectively.

Although engineering-design guidance is provided in each of these standards, key design decisions will need to be made that are not prescribed within these standards. The case of the boat winch will be discussed in greater detail in Chapter 10. In the case of the ATV, engineering-design details such as the rider ergonomics and human factors are not spelled out within the standard, although there are footwell-area design specifications in the form of geometric design requirements. There is no maximum

---

[17] Other CE-marking directives also apply to machinery if they are to be imported to countries requiring the CE mark. CE marking is an extensive engineering and administrative effort that is too lengthy to address in this chapter. Those interested in studying CE-marking requirements are directed to https://ec.europa.eu/growth/single-market/ce-marking_en.

[18] Such as that shown in Appendix B.

[19] The original 1990 version of this ATV standard limited an unladen, dry ATV's weight to 600 pounds (272 kg).

speed limitation unless the ATV is intended for use by youths. In addition, there are no dynamic-stability requirements, although there is a minimum static pitch-stability requirement.[20] The ATV is an example of a product with a history of protracted regulatory conflict between the U.S. Consumer Product Safety Commission (CPSC) and ATV design-and-manufacturing companies. These actions began in the 1980s when three-wheeled All-Terrain Cycles (ATCs) were popular. ATC riders were experiencing accidents and injuries. Without admitting that ATCs were defective, ATV[21] manufacturers at the time agreed to stop the sale of ATCs and to enter into a 10-year consent decree ("Final Consent Decree" 1988). This consent decree outlined several requirements for ATV manufacturers to follow and included such areas as warning signs, advertising, rider training, and the establishment of the first ATV standard ("ANSI/SVIA 1-1990: American National Standard for Four Wheel All-Terrain Vehicles—Equipment, Configuration, and Performance Requirements" 1990).

As a result of this standard and the consent decree, there are requirements that ATV manufacturers must meet to sell their products in the United States. There have also been state laws that ATV makers have optionally met for pragmatic, business reasons that include limitations of wheel and tire sizes for operation on state-maintained trails. There are several ATV-safety issues under the jurisdiction of states and these include helmet, underage-rider, and impaired-operator laws and enforcement of laws. It is noteworthy that the ANSI/SVIA standard *never* refers to itself as a "safety" standard; it is an equipment standard.

Engineering-design guidance may come from multiple sources—and it frequently does in the real world. For ATVs, *partial* design guidance has come from an industry standard initiated by a temporary legal agreement. Many industry standards are voluntary in nature. Compliance with the above ANSI/SVIA 1 standard was mandatory, not because ANSI had developed it, but because the 1988 Consent Decree mandated that ATV makers develop an industry standard and agree to follow it. In the intervening years, the ANSI/SVIA 1 standard was made mandatory and permanent through national legislation ("The Consumer Product Safety Improvement Act (CPSIA)" 2008) as the original Consent Decree and its subsequent renewals became unwieldy for the CPSC as new ATV importers began entering the U.S. ATV market. The situation was only complicated for ATV makers as different states passed different laws and, if the ATV maker wanted to also sell the product in Canada, further considerations were necessary both at the national and at the provincial levels. Some ATV makers made a single North American model of ATV while some other makers had separate U.S. and Canada models as discussed earlier in the book. The Canadian ATV industry developed their own standard ("COHV 1-2018: Canadian Off-Highway Vehicle Distributors Council Standard for Four-Wheel All-Terrain Vehicles" 2018), but that standard incorporates much of the American ATV standard. In some instances, it was necessary for ATV makers to produce a kit to modify a Canadian ATV for national or provincial language requirements regarding product-safety facilitators such as warning signs and manuals in the French language.

Despite the proliferation of regulations and the permanence of the Consent-Decree requirements, there are many critical design decisions, some with product-safety

---

[20] Static lateral-stability criteria for ATVs have been handled by individual agreements between the U.S. CPSC and ATV manufacturers. The author does not necessarily advocate establishing dynamic-stability criteria for ATVs, but uses this gap to again show that not all standards are necessarily complete in their coverages.

[21] ATVs are, by definition in all versions of ANSI/SVIA 1, equipped with only *four* wheels—not three or six.

implications, to be made for an ATV that are simply not covered by any of the documents. The engineering-design challenge for many consumer products is simply too complex and multi-dimensional to be completely addressed through even several standards and regulations. Again, if a product is innovative in nature and thereby incorporates features and characteristics never before imagined, then these standards and regulations will be even less helpful for design-engineering. Even after shortcomings in standards and regulations become evident, it may take five or more years before changes to address these gaps can be approved by a committee. By that time, the designs of affected products may have moved well beyond the changes made by standards developers and regulators. It must also be remembered that, even though a design-and-manufacturing company might belong to an industry group which develops standards, it is in that company's business interests to keep silent about their new product developments so that competitors remain unaware of their upcoming products. Thus, it seems unlikely that even an industry standard will ever keep pace with a company producing innovative products with an aggressive business plan.

8.5 Effects of Design Guidance

Before going further on the utility and need for *external*[22] design guidance, which includes standards, regulations, or any other agreements, it is useful to examine the environment in which such guidance exists—or does not exist. A simple set of matrices will be constructed to help identify differing design-guidance needs by an engineering designer and product segments as well as different potential risks to consumers from these interacting factors—designers and products.

Prior sections of this chapter have already touched upon divisions within the design-and-manufacturing spectrum and the product-market spectrum. In order to keep the conversation simple, each of these two spectra, or *factors*, is divided into two populations. The design-and-manufacturing, or simply *Designer*, spectrum consists of *Established* designers and *New* designers. The product or market, or simply *Product*, spectrum is similarly divided into *Established* products and *New* products.

These two segments of each of these two factors, product and designer, can be arranged in the form of a 2×2 matrix. Such a matrix, for some unspecified product property, is shown in Table 8.1. This product-designer matrix shows the combinations of *Established* and *New* products with *Established* and *New* designers.

As shown at the bottom right of the table, the resulting 2×2 matrix has columns for established and new products and rows for established and new designers. Such a matrix may be constructed for any product property of interest.

The first property examined will be that of the availability of design guidance, it is understood that, for established products, guidance for design engineers exists—to some degree. A product-designer matrix is completed to show this relationship for design guidance and appears as Table 8.2. The matrix elements beneath Established products are white and read "Guidance." Those elements under New products are shaded and read "No guidance." This shading indicates a higher level of risk to consumers from a new product—where neither design guidance nor prior product experience exist—than

[22] The term *external* differentiates this design guidance from *internal* guidance available only to someone within a given organization from its experienced engineers.

[Property]		Product	
		Established	New
Designer	Established		
	New		

Table 8.1 The Product-Designer Matrix for a Property

from an established product. This is because there is no prior product-safety knowledge available to guide the engineering-design team. There, generally, are neither standards nor useful regulations for such new products or markets.

This same matrix-based approach is now applied to the engineering ability that exists within established and new design-and-manufacturing companies. Such established companies possess institutional engineering ability from years of staff engineering experience. New companies often do not possess such technical skills and experience. Consequently, the matrix shown as Table 8.3 is produced. The row of matrix elements for Established designers are white and read "Proven Ability." The row elements for New designers are shaded and read "Unproven ability." The shading indicates a higher level of consumer risk to users of products made by new designers and manufacturers than by established ones. This is due to the lack of knowledge from which new-entrant companies sometimes suffer.

Table 8.2 shows that there is a lower risk to consumers of an established product than to consumers of a new product. Similarly, Table 8.3 highlights that there is a lower risk to consumers of products designed by established companies than by new companies.

The superimposition of Tables 8.2 and 8.3 is shown as Table 8.4 which shows the combination of the two properties: design guidance and engineering ability.

Design Guidance		Product	
		Established	New
Designer	Established	*Guidance*	*No guidance*
	New	*Guidance*	*No guidance*

Table 8.2 The Product-Designer Matrix for Design Guidance

Engineering Ability		Product	
		Established	New
Designer	Established	*Proven ability*	*Proven ability*
	New	*Unproven ability*	*Unproven ability*

Table 8.3 The Product-Designer Matrix for Engineering Ability

Design Guidance + Engineering Ability		Product	
		Established	New
Designer	Established	*Guidance + Proven ability*	*No guidance + Proven ability*
	New	*Guidance + Unproven ability*	*No Guidance + Unproven ability*

Table 8.4 The Product-Designer Matrix for the Combination of Design Guidance Engineering Ability

The summation of the two risk levels from the prior tables shows that, in addition to there being consumer risk from new products and new designers, there is an elevated level of risk posed to consumers when those two factor segments combine. The lowest level of risk exists for established products by established designers. Elevated, yet moderate, risk exists for new products by established designers and for established products by new designers. As would be expected, Table 8.4 shows that the highest level of risk arises when new products are produced by new designers.

There are two key takeaways from the construction of these matrices. *First* is that engineering-design guidance is more important for new-product designers than for established product designers. *Second* is that engineering ability may not be available at a new design-and-manufacturing company. This lack of engineering expertise may be especially important within those companies, either new or established, which only import and distribute products while neither designing nor manufacturing them. Additionally, some established and experienced designers and manufacturers may already be doing all that will be required by any likely new industry standard or regulation. While it is true that standards-development organizations and regulators may be increasing consumer safety by creating standards to force compliance by technically inept importers, it may be wrong to presume that product safety will necessarily be increased for *all* users of such products. The safety levels of products from established design-and-manufacturing companies may already be sufficient and not see a resulting increase from further standards or regulation.

Admittedly, there are limitations of the product-designer matrix which is used as a demonstrative aid. These include the assumptions necessarily made within it. For instance, it is presumed that established design-and-manufacturing companies follow any and all design guidance given and that they also utilize the full technical abilities that they possess. For example, established companies may willfully *not* follow standards and regulations—or other accepted design practices. In addition, they may also choose to assign inexperienced design engineers to new products, not provide sufficient product-development scheduling, and also ignore previous lessons that should have been learned with prior products. Also, in some situations, it may be possible to extract some useful design guidance from somewhat-similar products—so there may not exist a complete vacuum in design guidance for all new products. It has also been presupposed that new manufacturers lack the engineering ability of established manufacturers. This is not always the case. Sometimes, expertise can be *purchased* by new manufacturers through the hiring of experienced personnel and consultants who might have come from the established designers and manufacturers. In some other cases, the only true expertise for a new product may come from a small set of researchers who created a "start-up" company. Such information may not be available to any established design-and-manufacturing company.

Another presumption made by the public is that the standards developer—or a regulator—is an organization that is competent to make such a standard. Following this presumption is another which is that any standards produced are, in fact, good standards that truly advance the safety of product users. This is not always the case.

There is no universal engineering-ethics truth that all standards or regulations *must* be followed if product safety is the goal. If simple compliance is the goal of the "product-safety" design exercise, then, yes, all standards and regulations must be blindly obeyed. Some standards and regulations are so old that certain portions of them are no longer understood and those who created them may no longer be alive. Some provisions of some standards or regulations might have been the result of some political compromise that is no longer pertinent or even remembered. This is one danger of seeking political solutions to technical product-safety problems.

An example of the first situation, where there was no logic for a particular requirement, is the author's experience in developing industry standards for off-road utility vehicles. Some existing standards for lower-speed vehicles contained in them a cargo-bed loading curve to show the load's necessary center-of-gravity height to be used for static-stability measurements. When the author asked the other standards technical-committee members about the justification and source of this table, he was met with silence. No one knew. Yet, this table had continued to be used in various successive standards for years.[23]

One example of the failure of regulations known personally by the author is that of headlamps for on-highway motorcycles. By the United States Federal Motor Vehicle Safety Standards (FMVSS) 49 CFR 571.108 ("Standard No. 108; Lamps, Reflective Devices, and Associated Equipment." 2018, S10.17.1.2.2), twin motorcycle headlights cannot be separated horizontally by more than 200 mm (7.87 inches). Yet, the definition of "motorcycle" within the FMVSS ("Definitions" 2018) permitted motor motorcycles to be as wide as passenger cars, so long as the "motorcycle" only had three wheels. This headlighting requirement was realized during the design development an on-highway "trike" for sale as a "motorcycle." The vehicle looks and *drives*—rather than *rides*—like a convertible passenger car, with side-by-side bucket seating, but having a single rear wheel instead of two. This design was unlike some competitor models that look more like a motorcycle with two front wheels. Therefore, in the pursuit of greater vehicle conspicuity and driver nighttime visibility, the headlights were intended to be at the front corners of the vehicle as is done on passenger cars. The headlight-separation restriction in FMVSS 108 for motorcycles was soon discovered, however. Due to this legal requirement, the design team had to mount "headlights" near the centerline of the vehicle and then mount "auxiliary" lights at the vehicle's front corners thereby resulting in four forward-directed white lights. This was only true of the vehicles being sold in the United States. Canada had no similar headlight restrictions in its motorcycle regulations. Thus, the Canadian vehicle's headlights were mounted on the front corners of the vehicle, as with a passenger car, resulting in only two forward-directed white lights. The source of the 800-mm maximum-separation requirement in the U.S. FMVSS could never be discovered. However, its mere inexplicable existence may compromise highway safety in certain instances. Even after working through the headlight issue on the trike motorcycle, the regulatory landscape still needed to be "smoothed out" through lobbying efforts regarding vehicle registration, operator licensure, and required helmet use at the state and provincial levels for this vehicle.

One further example is one where manufacturers saw a problem and those manufacturers and their regulators were able to work together to solve the problem for the benefit of product users. As a result of the All-Terrain Vehicle (ATV) Consent Decree ("Final Consent Decree" 1988), the ATV manufacturers developed the first industry standard for ATVs ("ANSI/SVIA 1-1990: American National Standard for Four Wheel All-Terrain Vehicles—Equipment, Configuration, and Performance Requirements" 1990). This first version was published by ANSI in 1990. In that standard, ATVs were divided

---

[23] The author made sure that this table never made it into standards with which he was involved.

into categories depending upon their uses. Categories were *G* for general use, *S* for sport use, *U* for utility use, and *Y* for youth riders who were under 16 years old. Although headlighting and tail lighting were required of all models in adult-ATV categories (G, S, and U), headlights and tail lights were explicitly prohibited on Category-Y ATVs[24] to prevent children from riding ATVs at night, when there was perhaps greater risk for them to operate ATVs than during the daylight. Category-Y ATVs were also to be equipped with speed-limiting devices. At the time, the late 1980s, these were probably reasonable demands by regulators and reasonable concessions by the ATV industry given the amount of negative media attention directed at that industry. However, two decades later, the consensus within the ATV industry was that providing head, or frontal, lighting would be beneficial for operators of category-Y ATVs. Just as daytime running lights (DRLs) on on-highway vehicles help other drivers see those vehicles, it was believed that DRL-type lighting on youth ATVs would help other vehicles—both on-highway and off-road—better see the youth operators under low-light conditions. This is especially true when operating the Youth ATVs on trails in dense forest where daylight is limited. The added lighting can assist other trail users in seeing approaching Youth ATVs.

It was this motivation that led the ATV industry to approach industry regulators at the U.S. CPSC to discuss the idea of permitting frontal and rear lighting on youth ATVs. The industry explained that these lights were not intended to encourage the nighttime operation of the Category-Y ATVs. However, if adults and children happened to be riding appropriate ATVs together in a group when the sun set before they could return back from riding, then the addition of lights on the Youth ATVs would both let other riders see the youths and let the young riders see better in the darkness than without lights.

The ATV industry reached out to the CPSC and discussions between these two usually adversarial parties took place. Ultimately, the two sides agreed that lights could be added to Category-Y ATVs—although nighttime use was still to be discouraged by manufacturers and the speed-limiter requirement remained in effect. The permissibility of head and tail lights was incorporated into the next version of the SVIA 1 ATV standard ("American National Standard for Four Wheel All-Terrain Vehicles ANSI/SVIA 1-2010" 2010). Both sides should be commended for their actions and their spirit of cooperation which resulted in a better product for a group of vulnerable consumers.

It is highly unlikely that *all* parties involved in ATV safety were pleased by the end result of permitting head and tail lights on Youth ATVs. Perhaps not all ATV manufacturers agreed; perhaps not all CPSC commissioners and staff concurred; perhaps not all consumer advocates were pleased, but this lack of unanimity is common when developing standards. That is why those efforts are called "consensus" standards.

As with all conceptual models, the product-designer matrix presented in this section is a simplification of reality. What is important are the concepts of elevated risk levels associated with new products and designers when compared with established products and designers. There will be exceptions depending upon product, designer and manufacturer, technologies employed, and, as always, the integrities of the people involved.

The continued development of new standards is likely to have an effect on established products primarily. New products will be little impacted by published standards because they will likely be beyond these scopes of the standards. Just as there

---

[24] Section 4.16.2, Requirements for Category Y Vehicles.

are multiple types of standards, as seen in the international-standards discussion, the effects of different standards may also vary.

Companies who only produce, but do not sell, export, distribute, or otherwise market their products directly, may not feel much if any responsibility for the design and performance of their products once sold to importers across an ocean. These importers are sometimes businesses, or entrepreneurs, who simply want to quickly sell a product at a profit. Such importers should be considered to be designers with little-to-no engineering expertise in the product-designer matrix. Some companies just find a low-cost manufacturer of a product that can be sold at a quick profit. The true designer and manufacturer of that product, and their design staff, may be quite isolated from its sale and use.

In addition, some start-up entrepreneurial companies are unaware of the true resources required to develop robust and safe products. The pointing out of this "hurdle," in the form of reality, may be seen by some readers as an attempt to damp the romance and allure of an individualistic visionary with an idea and a dream. However, this book may be able to actually help these entrepreneurs realize and address their obligations to the public that the manufacture and sale of products bring. These obligations include the responsibility of tracking and, perhaps, remediating product problems after the sale of the product as will be shown in Chapter 11.

Unfortunately, even with established designer/manufacturers, some companies and engineering staffs simply may have neither the time nor the skills—and have never been forced—to formalize their methodologies regarding the design and manufacture of their products. This situation, fortunately, is becoming increasingly rare due to competitive pressures and discerning consumers. For many consumer products in the United States, there exist no standardized or required product-design processes. The consequence is that some comparable consumer products available to users may be the result of different PDPs, some of which are more meticulous than others.

To reiterate, there will be a particular perspective used throughout the book. The products being designed, manufactured, and distributed internationally will often be new or innovative durable mass-produced consumer-type products in a fast-paced, competitive, and regulated industry. Such a successful organization will be a "lean" operation by necessity. Since the company is a lean organization, there will not be excess(ive) labor resources, financial resources, and time.

Product-safety issues are qualitatively and fundamentally different than P&L (Profit and Loss) or marketing issues. Although disagreement on either type of issue can result in friction between parties, since people can be injured or killed by product-safety issues, marketing issues (e.g., red vs. blue products) are relatively inconsequential.

When engineering-design guidance is unavailable or insufficient, other sources of guidance must be used. Without explicit criteria on product design or performance requirements, engineering ethics becomes a major source of guidance for engineers during PDP decision making.

8.6 Conclusion

Hammer and Price (Hammer and Price 2001, 74) wrote:

> A famous author once wrote: The American has always reacted to the setting of standards the way Count Dracula responds to a clove of garlic or a crucifix.

This author has been unsuccessful in identifying *that* particular famous author. This lack of attribution notwithstanding, some people are torn when thinking of standards. On the one hand, the setting of a standard, although voluntary, strips design engineers of some freedom of choice in design engineering. On the other hand, some of these same engineers may be fearful of not having structure in the form of specific design criteria for their designs.

It is hoped that this chapter helped to dispel fears of the former condition. Many standards are not overly oppressive and may only formalize what is already being done with a properly designed product or within a responsible industry. Regarding the latter condition, even established engineering standards may be of little use when making critical detail-design decisions—especially with new, innovative products. The design engineer will need to rely upon technical skills as far as possible and, then, rely upon her/his engineering-ethics principles to make many product-safety decisions.

It is also hoped that this chapter illustrates that the any-standard-is-better-than-no-standard mindset is untrue. Standards made by incompetent or misguided committees can do harm either directly, through design restriction, or indirectly, through the stagnation of design innovation which could lead to better levels of product-safety for consumers. However, some parties in the product-safety debate will always prefer an *activist* approach, rather than merely an *active* one.

Some good is being done with innovative products at an industry level such as the IoT-standard effort mentioned earlier although it is still in its infancy. However, an innovative company will likely want to get a particular product to market well before an industry standard is eventually published.[25] This notwithstanding, intelligent standards-development efforts are to be applauded when observed.

This chapter has also introduced a simple product-designer matrix which may help demonstrate that, even when standards do exist, their effects are not uniform across all products and design-and-manufacturing companies. Those design companies that are the most likely to do a good job with product design are also those who least need any help that even a good design standard may provide. The corollary is also true, but such help from industry standards may not exist for all products—especially the innovative ones. Qualitative levels of potential risk to consumers from various product properties can be assigned to products and design-and-manufacturing companies using this matrix's four elements.

This chapter discussed, in an elementary fashion, sources of design guidance for the development of new products. There are various sources available to design engineers and these sources included standards, codes, regulations, and accepted engineering-design practices.

There are numerous sources of design guidance available to the engineer. These include standards. However, it is hoped that readers understand that not all standards are created equal. Not all standards have equal effect either. Some standards might be successful if the only desire is to make improvements to established products by inexperienced designers and manufacturers. As products become more innovative, the ability to *positively* influence their product-safety characteristics diminishes rapidly. Such limitations of standardization efforts should be understood by all parties involved

[25]Some other companies remain hesitant to release a product for which there is no standard.

so that unrealizable expectations for product-safety increases are not created. Therefore, due to the lack of much specific design guidance, the need for an engineer to apply her or his own engineering ethics remains a vital part of answering the product being "safe enough" question.

References

"American National Standard for Four Wheel All-Terrain Vehicles ANSI/SVIA 1-2010." 2010. Irvine, CA: Specialty Vehicle Institute of America (SVIA). www.svia.org.

"ANSI/ISO 12100:2012 Safety of Machinery—General Principles for Design—Risk Assessment and Risk Reduction." 2012. Houston, TX: ANSI/ISO.

"ANSI/SVIA 1-1990: American National Standard for Four Wheel All-Terrain Vehicles—Equipment, Configuration, and Performance Requirements." 1990. Irvine, CA: Specialty Vehicle Institute of America (SVIA).

"ANSI/SVIA 1-2017: American National Standard for Four-Wheel All-Terrain Vehicles." 2017. Specialty Vehicle Institute of America (SVIA).

"ANSI Essential Requirements: Due Process Requirements for American National Standards." 2010. New York, NY: ANSI.

"Boiler and Pressure Vessel Code 2019 Complete Set." n.d. Asme.Org. Accessed November 29, 2019. https://www.asme.org/codes-standards/find-codes-standards/bpvc-complete-code-boiler-pressure-vessel-code-complete-set.

"Clayton Antitrust Act of 1914." 1914. Washington, DC: U.S. House of Representatives. https://history.house.gov/HistoricalHighlight/Detail/15032424979.

"Code." n.d. Merrian-Webster.Com. Accessed November 29, 2019. https://www.merriam-webster.com/dictionary/code?src=search-dict-box.

"COHV 1-2018: Canadian Off-Highway Vehicle Distributors Council Standard for Four-Wheel All-Terrain Vehicles." 2018. Markham, ON: COHV. https://www.cohv.ca/.

"Definitions." 2018. Washington, DC: US DOT/NHTSA. https://www.govinfo.gov/app/collection/cfr/2018/title49/subtitleB/chapterV/part571/subpartA.

EC/DGE. 2006. "Directive 2006/42/EC of the European Parliament..." Brussels, BE.

"Final Consent Decree." 1988. Washington, DC: U.S. District Court (DC).

Greulich, Owen R., and Maan H. Jawad. 2018. *Primer on Engineering Standards*. Expanded t. Hoboken, NJ: John Wiley & Sons.

Hammer, Willie, and Dennis Price. 2001. *Occupational Safety Management and Engineering*. Fifth edition. Saddle River, NJ: Prentice Hall.

"Handbook of Reliability Prediction Procedures for Mechanical Equipment." 2011. West Bethesda, MD: Naval Surface Warfare Center (NSWC).

"Harmonized Standards." n.d. Ec.Europa.Eu. Accessed April 16, 2020. https://ec.europa.eu/growth/single-market/european-standards/harmonised-standards_en.

Kelechava, Brad. 2017. "ISO Type A-B-C Structure for Machinery Standards." Ansi.Org. 2017. https://blog.ansi.org/2017/10/iso-type-abc-structure-machinery-standards-ansi-b11/.

NASA. 2002. "Fault Tree Handbook with Aerospace Applications." Washington, DC.

"NFPA 70: National Electric Code(R)." 2020. Quincy, MA: National Fire Protection Association (NFPA). https://www.nfpa.org/codes-and-standards/all-codes-and-standards/list-of-codes-and-standards/detail?code=70.

"SAE J1853-2014: Hand Winches—Boat Trailer Type." 2014. Warrendale, PA: SAE International.

"SAE J2358_201611: Low Speed Vehicles." 2016. Sae.Org. 2016. https://www.sae.org/standards/content/j2358_201611/.

Sisco, Joshua, and Leah Nylen. 2018. "DOJ Probes Role of Special Interest Group in New WiFi Standard." Mlexmarketinsight.Com. 2018. https://mlexmarketinsight.com/insights-center/editors-picks/antitrust/north-america/doj-probes-role-of-special-interest-group-in-new-wifi-standard.

"Standard No. 108; Lamps, Reflective Devices, and Associated Equipment." 2018. Washington, DC: US DOT/NHTSA. https://www.govinfo.gov/app/collection/cfr/2018/title49/subtitleB/chapterV/part571/subpartBwash.

"Subcommittee F15.75 on Connected Products." 2019. Astm.Org. 2019. https://www.astm.org/COMMIT/SUBCOMMIT/F1575.htm.

"The Consumer Product Safety Improvement Act (CPSIA)." 2008. Cpsc.Gov. 2008. https://www.cpsc.gov/Regulations-Laws—Standards/Statutes/The-Consumer-Product-Safety-Improvement-Act.

USA/DoD. 2012. "MIL-STD-882E—System Safety." Washington, DC.

CHAPTER 9

Product-Safety Facilitators

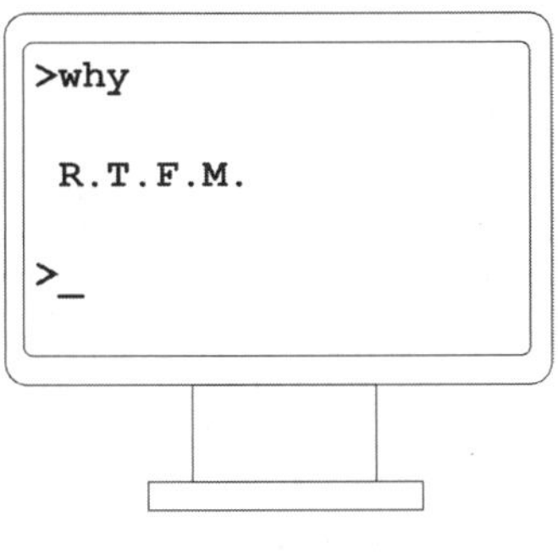

—LINPACK[1]

9.1 Introduction

In a departure from what might be considered logical, this book places this chapter on *facilitators before* the chapter on product-safety engineering methods, Chapter 10. Product-safety facilitators are items such as a product's manuals, warning signs, and other safety-related materials. It might be more expected to order the chapters in a book on product-safety engineering in the reverse order. However, there is a reason that these chapters are ordered as they are. That reason is that design engineers should fully understand the capabilities and limitations of facilitators *before* they stop reducing the risks of their products. Engineers should not expect facilitators to magically fill "gaps" that additional engineering-design work could better "fill." Such gaps may continue to exist in some product designs and their facilitations, but engineers should do their ethical best to reduce risk subject to technological practicality and economic feasibility. Having done their best, engineers can simply do nothing more—but they must have done their best. Then, and only then, might it be safe to advise users to "Read The Manual."

9.2 Facilitators

In a perfect world with perfect products, all product hazards and their resultant risks could be controlled or reduced to a level such that the physical characteristics of the

[1] An early interactive computer-software package for performing linear-algebra calculations (Dongarra et al. 1979).

product alone would be sufficiently safe for all consumers to use it without further regard. This is rarely the case, however, in reality. A common situation is that intelligent and ethical engineering design has certainly lowered product risks to consumers, but to only a certain level. Beyond that, hazards and risks must be further controlled through other materials from the product's designer and manufacturer. These materials are known as *facilitators* since they function through "facilitating human performance" to "help ensure an acceptable level of human performance" (Bailey 1982, 232). Product-safety facilitators have been written about by others (Bailey 1996; Peacock and Laux 2005). Facilitators include educational materials, warning signs (or labels[2]), and manuals whether these manuals are for "owners," "users," or "operators." A more-complete list of facilitators is shown as Table 9.1. The facilitators listed in Table 9.1 can be delivered in multiple fashions. Traditionally, many facilitators can be delivered with—or on—the product itself. However, technology has enabled facilitators to be delivered to product users in a variety of ways through a variety of media. Table 9.2 contains a list of possible facilitator-delivery mechanisms. Such delivery options permit users to get product-facilitation information before, during, and after product use—even while in remote locations.

The above tables are but a handful of facilitator types and delivery mechanisms. This chapter will discuss several of these types of facilitators as well as different approaches to effectively reach consumers with product-safety information. As will be shown, some methods are conventional, even boring, but some routes taken by some organizations, either with unique products or unique consumers, have been original and decidedly "outside of the box." One purpose of this book is to help design and product-safety engineers address product-safety issues that have not already been solved and are not considered to be routine. Therefore, this chapter will not be limited to the *pro forma*, or "status-quo" based, approach toward product-safety instructional materials used on many products that have been produced by multiple companies for decades. Instead, innovative products with challenging new safety challenges are the purpose of this book and this chapter.

9.3 Facilitators and Risk Reduction

As already stated in this chapter, an ideal product would be completely safe just by itself through its physical, mechanical properties produced through a good engineering-design process. This will be called the product's "physical design."[3] There would be no need for guards, facilitators, or any other risk-reduction countermeasures. Sadly, this is rarely, if ever, the case in the real world. Consequently, there is often a need to further reduce product risk beyond the product's physical characteristics through the design of facilitators during the product-design process (PDP), shown in Figure 6.3,

[2] The term "warning label" will be regarded to be the same as "warning sign." Preference will be given to "warning sign" since the American national standard Z535.4 uses that term. Through connotation, however, the term "warning sign" might be imagined to be a metal placard bolted to a fence post while a warning label might be a small self-adhesive decal. They both share the same requirements through that standard, however.

[3] The term "physical design" is used to describe the characteristics of a product without any facilitators. Facilitators should indeed be a part of the PDP and designed alongside the product, but facilitators are considered separately in this instance.

Educational materials
Procedures
Checklists
Tutorials
Warning signs
Manuals
Hang tags and other point-of-sale (POS) materials
Marketing materials including advertisements
Photographs
Drawings
Videos
Augmented Reality (AR)
Packaging materials
User-Product Interface (UPI)

Table 9.1 List of Facilitators

Printed material provided with the product
Printed material applied onto the product
Multi-function display (MFD) in the product
DVD
Website
E-mail
Social media
Smart-phone apps (applications)
On-site training and trainers
Demonstration devices

Table 9.2 Delivery Methods for Facilitators

when a new product is being designed. After phase II of the PDP, a conceptual design is forwarded to Phase III of the PDP, the Engineering Design, Development, and Testing Phase (EDDTP). Once within the EDDTP, design engineers create an initial design for the new product.

Imagine an engineering design called "Product 1." Its risk level versus time decreases as Product 1 passes through the various engineering-design tasks over time. This is shown in Figure 9.1. The dashed horizontal line across the figure indicates the maximum risk level that a product may exhibit and still be considered acceptable. This risk level is marked as R_{Accept}. Product 1, manufactured by Manufacturer 1, starts off at time $t = 0$ with a risk of $R_{Initial}$. This is shown as Point A. Since the Product 1 has not yet been refined through logical engineering design, the initial risk is greater than the risk level that is considered to be acceptable.

$$R_{Initial} > R_{Accept}$$

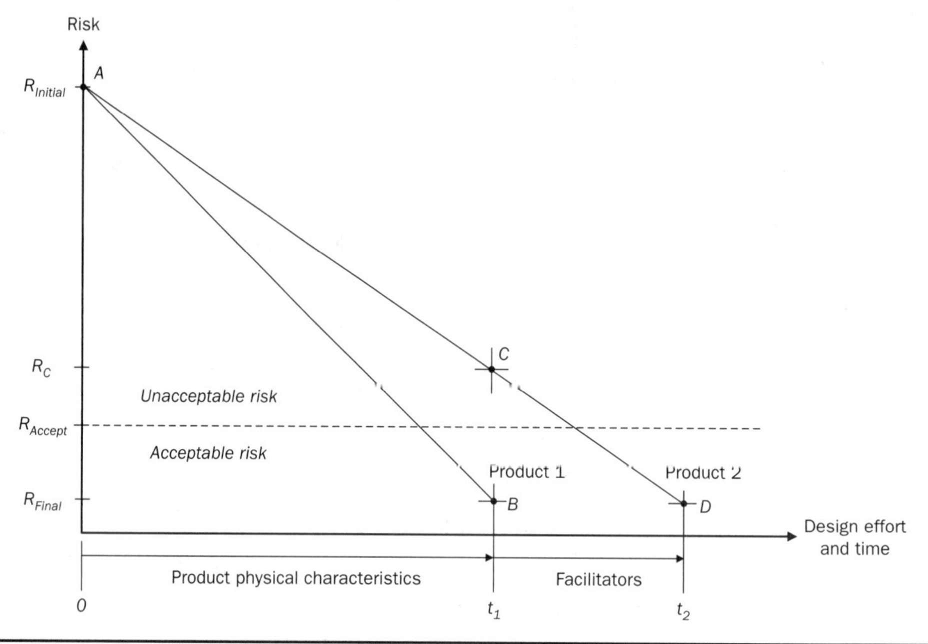

FIGURE 9.1 Reducing risk to acceptable.

Therefore, risk must be reduced by the designer, Manufacturer 1, and is reduced through engineering-design work during the EDDTP shown in Figure 6.3. From Figure 9.1, it can be seen that, after the engineering design of the physical characteristics of Product 1 has been completed at time $t = t_1$ shown by Point B, the risk has been reduced to R_{Final}. It is evident from this figure that R_{Final} is less than R_{Accept}.

$$R_{Final} < R_{Accept}$$

Therefore, no further risk reduction is *necessary* through the application of facilitators by Manufacturer 1. That is not to say that facilitators should not be applied to further refine Product 1. The use of facilitators may indeed even further reduce the risks of using Product 1 but are not considered for Product 1 in this discussion.[4]

Consider the engineering design of another product, Product 2, this time designed by Manufacturer 2. It, too, starts off at time $t = 0$ at Point A with an initial risk of $R_{Initial}$. Again and, of course, $R_{Initial}$ for Product 2 is larger than R_{Accept}. Therefore, engineering-design methods are applied by Manufacturer 2 to reduce the risk of Product 2 to acceptable. The problem is that, at time $t = t_1$, the risk level, R_C, shown by Point C, is still greater than the acceptable risk.

$$R_C > R_{Accept}$$

[4] In real-life applications, facilitators should be considered and used when appropriate for products.

Consequently, Manufacturer 2 must rely upon the use of facilitators to reduce the risk level of Product 2 downward to acceptable. Facilitators are designed and applied after t_1 resulting in the final product design and its risk level at time $t = t_2$ which is R_{Final}. Since R_{Final} is less than R_{Accept}, no further risk reduction is necessary by Manufacturer 2 for Product 2.

Quantitatively, there may be no difference between the risks from Products 1 and 2 since the risk levels of both products have ultimately been reduced to the same acceptable level by their respective manufacturers. Qualitatively, however, the two products do differ from one another in a significant manner. Manufacturer 1 was able to reach risk acceptability for its product through its engineered physical attributes alone. Manufacturer 2, on the other hand, necessarily had to rely upon facilitators to reach acceptable risk for its product.

For some products, especially those with high levels of user activity (Figure 4.12), it may be necessary to design facilitators carefully in order to deliver a safe product to consumers. Although this may be the case, it is still important that design engineers not rely upon a facilitator as a "crutch" when further engineering design could further reduce risk both practically and feasibly. Engineered characteristics may be more effective than facilitators for addressing many product risks. Yet, facilitators may remain necessary to help consumers keep themselves safe when using some product types.

9.4 The Design of Facilitators

As just discussed, not all products may be able to meet acceptable-risk levels through physical engineering-design efforts alone. Some products, by their very natures, must rely upon facilitators to reduce user risk down to acceptable. Peacock and Laux conclude that:

> Ideally, systems should be designed for effective, efficient, and safe use without the use of facilitators. This ideal is largely true for simple systems and experienced users...But these ideals are usually unrealistic in practice, so facilitators will always be needed.
>
> [I]t is unlikely that we will ever accommodate all the users all the time. But we can try by formally integrating the development of facilitators into the system design and evaluation process.

The importance of facilitators to safe product use is recognized and addressed in this chapter. Facilitators should not be created after the fact. They should be designed by the product-safety engineer, or other design engineers, along with the product, whenever possible,[5] in order to maximize their effectiveness for users of products. The design of such facilitators is covered in this chapter.

Since the reason for providing consumers with extra product-safety information is to lower the injury risk of product use, the goal of these facilitators is to effectively convey important product-safety content to consumers in an efficient and effective fashion. Innovative products may call for innovative methods of communication with

[5] There will remain valid instances when the *final* design of facilitators may take place after the engineering design of the product's physical characteristics due to considerations involving feasibility and practicality.

consumers. The reader should approach this chapter, and this book, with this perspective in mind.

The list of facilitators shown in Table 9.1 includes many of the facilitators used in common products purchased by consumers today. The last facilitator listed is the user-product interface (UPI). The UPI is not covered in depth in this book due to its breadth and depth, but interested readers are directed to the following references on the topic: Bridger 2018; Sanders and McCormick 1993; Bailey 1996; Tillman et al. 2016; Tilley 2002; "MIL-STD-1472G—Human Engineering" 2012; Barnett 1994.

This chapter will continue with on-product warning signs, proceed onto the ubiquitous owner's manual, and then go beyond that to cover other facilitators sometimes used to further enhance product safety. It is hoped that this chapter will be seen as a refreshing treatment of this topic and those materials that are sometimes lost, sometimes tossed (thrown away), sometimes ignored, and sometimes poorly written. There exist standards in the United States for both warning signs and manuals, the ANSI Z535[6] set of standards. These standards should be considered by a design and manufacturing firm as a mere starting point and not as an end point with sufficiency presumed for having followed an industry standard. One size may not fit all and specific products may require uniquely tailored solutions to best inform product consumers of their purchases. It is good to keep an open mind when *designing* this portion of the product. An integrated engineering-design process will incorporate the design of warning signs, manuals, and other *soft*[7] countermeasures once product-safety properties have reduced the product's hazard risks as far as practical. During the EDDTP, there may be design iterations on the facilitators themselves. However, these product-safety materials may be every bit as important to a consumer as the product's physical characteristics—yet warning signs, manuals, and other materials should never substitute for incomplete or poor engineering-design efforts or abilities.

9.5 Warning Signs

Consumers and, indeed, humankind have had a love-hate relationship with warnings, in general, and warning signs, in particular. People often do want to know what the rules are so that they can protect themselves; however, some of the same people may not want to follow those rules or recommendations even when following them is in their own self-interests.

Barnett poses what may be history's first recorded instance of a stern warning to people (Barnett 2001). Adam was placed into the Garden of Eden by God[8]. Adam was instructed to maintain it and eat whatever he wanted therein—with one exception. He was not permitted to eat from the tree of the knowledge of good and evil because God told Adam that "the day that thou eatest thereof, thou shall surely die." The serpent later came to Eve and convinced her otherwise by telling her that Adam and she

[6] The Z535 Committee on Safety Signs and Colors was formed by ANSI in 1979 by combining the Z53 Committee on Safety Colors and the Z35 Committee on Safety Signs.

[7] Administrative safeguards are considered soft countermeasures since humans will be protected only if humans follow the procedures. Hard countermeasures include physical safeguards—such as guards, seat belts, and electrical grounding—that should not depend upon humans following rules.

[8] Genesis 2, 3.

would "be as gods, knowing good and evil." Eve ate that forbidden fruit, as did Adam. Ultimately, both Adam and Eve were cast out of Eden as a result of their failure to heed that initial warning to humans. Since the fall[9] of woman and man, humanity continued to have problems with warnings. However, as shall be seen, it is not always the fault of the consumer.

In summary, Adam and Eve were instructed by *the absolute authority*, God, what they could do and should do. They were also told, in no uncertain terms, about what *would*, not *could*, happen—death—if they violated the rules. God even used the word "shall" instead of "may." They were also instructed on what the hazard was as well as how to avoid the consequences, namely, avoid eating the forbidden fruit of that particular tree by eating anything else. There appears to have been no shortage of other food in Eden. Barnett provides a warning sign for the tree in his work for illustrative and humorous purposes (Barnett 2001, 1). Continuing in this spirit, a warning sign conforming to the present warning-sign standard has been created here for demonstrative purposes. This warning sign is shown in Figure 9.2. Even if such a sign had been posted nearby, it may well have made no difference since both Adam and Eve were warned face-to-face by God Almighty. Some people would probably like to think that they would have obeyed such an in-person exhortation from God. It will never be known just how many well-intentioned people truly would have behaved thusly.

Although Barnett's example is facetious, his discussion does illustrate nicely the need for warnings in some situations and the resulting warning sign, embodies the essential elements of a good warning sign.[10] The author is indebted to Barnett for setting up this example so nicely.

9.5.1 The Need for Warnings

Some people are of the opinion that there is no need for warnings since warnings do not work. This is definitely an extreme view, yet it seemingly persists among some of the general, and even professional, population. For example, the kitchen knife from Figure 4.8 has a well-known hazard, an exposed sharp edge, and well-understood risk which is laceration injury when the edge contacts flesh. When not in use, kitchen knives should put away in a drawer, into a butcher's block, or into a sheath. Adults are cognizant of the hazard and risks of knife use. Therefore, children are to be kept away from such products until such time as they gain the understanding and the facility of adults and can use knives responsibly.

The Book of Genesis provides no indication that the specific tree and its forbidden fruit looked any differently than any other tree or fruit in Eden. Therefore, it was *necessary* for God to—in fairness to humans—convey to Adam that this fruit was unmistakably different from the other fruit in the garden despite there being no known difference in outward appearance to other fruit in Eden. The fruit of that tree had a now-known *hazard* and *risk* associated with its consumption. God told Adam of the consequences, but God also told Adam how to avoid those consequences. Namely, do not eat it!

[9] Perhaps Adam and Eve did not fall so much as jump.

[10] The terms "warning sign" and "warning label" will be used rather interchangeably in this book. Although the thought of a sign could be one of a metal placard bolted to a fence post and a label might be thought of as a small self-adhesive decal, they both share the same formatting and content requirements.

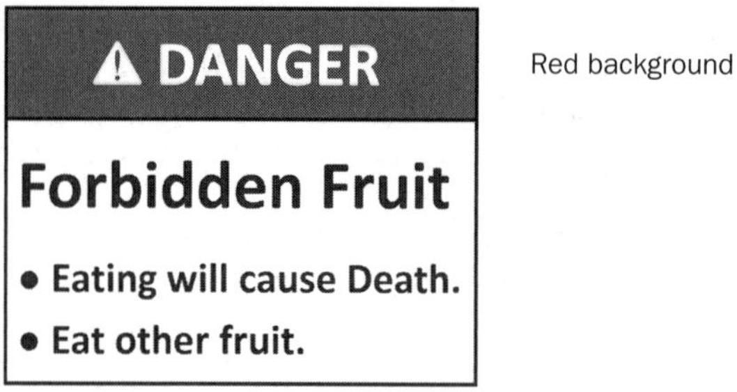

▲ DANGER

Red background

Forbidden Fruit

- Eating will cause Death.
- Eat other fruit.

Figure 9.2 Facetious warning sign on the tree of the knowledge of good and evil.

Similarly, if a consumer product was to emit a colorless, odorless—yet lethal—vapor such as carbon monoxide (CO) that is both toxic and displaces oxygen (O_2), then a warning regarding poisoning to users would be necessary, if not by standards or regulations, by engineering ethics. This resulted in the warning sign shown in Figure 5.3 which must be on all such new portable electric generators sold in the United States [Portable Generators; Final Rule; Labeling Requirements (Correction) 2007]. Many consumers already knew of the hazards of internal-combustion engine (ICE) exhaust gases. However, not everyone who ended up using a portable generator evidently either knew about the nature of exhaust gases or fully understood how to avoid the accumulation of such ICE exhaust-gas emissions. From the death tolls following disasters such as hurricanes and floods, many otherwise-knowledgeable people mistakenly presumed that the general public knew more of the dangers of CO than it actually did.

Products that have hidden hazards with risks that include severe injury or death should also provide information for consumers for the safe use that product. It could be wise to also include property damage-only (PDO) risks and those requiring this information if for no other reason than customer satisfaction. Such consumer information may appear in a warning sign, in a manual, on an MFD, or in other product-safety facilitators.

9.5.2 Elements of a Warning Sign

Most readers have probably seen a product bearing a warning sign that can unequivocally be categorized as *stupid*—providing no useful information to a sentient being or perhaps containing a poor instruction with unknown intention. There are even books dedicated to this phenomenon (Green, Dierckins, and Nyberg 1998; Hunt 2002; even Barnett 2001, 2) has a listing of some examples of poor warnings. Some bad warning signs are perhaps poor translations of content originally in a different language. Fortunately, some of these suspect warnings result in no serious injury if they are not understood or cannot actually be followed. In the United States, there is a standardized way to construct warning signs that works acceptably well for many products and their consumers. Yet, designers and manufacturers should be aware that a one-size-fits-all approach may be sub-optimal for communicating product-safety information to consumers in some cases.

Since 1991, the United States has had a modern standard for use in warning signs which is ANSI/Z535.4. Prior to this time, an older style of warning signage was used

Figure 9.3 Obsolete style of warning sign.

such as that shown in Figure 9.3 and is now obsolete. This style of warning sign can still be seen in many places today. There is generally no requirement that the old-style of warning signs be updated on products and at facilities. New products should follow the new standard, however. Consequently, this obsolete style of signage will not be covered here, but is briefly discussed and displayed merely to help readers understand that this old style of sign was common in the past but is no longer used.

The National Electrical Manufacturers Association (NEMA) developed the American standard for on-product warning signs was most recently published by ANSI as Z535.4 (NEMA 2017a). This standard specifies many aspects of a compliant warning sign. Included in the latest version of the standard are requirements on:

- Color
- Signal words and the safety-alert symbol
- Different "panels" in the warning sign for different kinds of information
- Borders
- Letter styles
- Letter sizes
- Sign placement
- Sign protection
- Expected life
- Maintenance and replacement
- Safety symbols
- Various lay-out formats, e.g., horizontal, vertical, two-panel, three-panel, and more

The text of the warning sign, or the particular text to be used, is not explicitly included in the standard. However, there are requirements about the need to state the type of hazard, the consequences of that hazard if not avoided, and ways to avoid that hazard.

The standard provides flexibility to designers to draft suitable and effective wording for their situations or products. Guidance is given regarding the order of the messages, which will be covered below.

Guidelines are provided for the formatting of the text in a warning sign as well as the recommendation to use "headline style" messages to consumers appear in the standard and in such references as those given for human factors and ergonomics. This writing style includes the following elements:

- It drops unnecessary articles ("the," "a," and "and") and pronouns ("the," "this," and "they") in the written messages to enhance rapid readability by consumers. Also dropped are forms of the verb "to be" ("are," "is," and "were").
- It uses text which is left-justified, not right-justified or centered.
- It frequently employs bullets, "•," to help the user identify different messages and read the warning sign quickly.
- It uses an *active* voice is used to be both concise and forceful. For example, rather than using the *passive* voice and saying "Be sure to turn the power off of the device when you are done using it," a warning sign might state "Turn power off when task is complete." The latter wording is shorter and considered more effective in quickly communicating information to users than the former wording.

Bridger notes that "[p]eople are inclined to carry out tasks in the order in which they receive them" (Bridger 2018, 527). Therefore, if a pre-operational inspection of a product is suggested by its manufacturer, then a warning sign should state first: "Inspect product before use." This wording is preferable to: "Before product use, inspect product" and is good practice.

Both upper- and lower-case letters may be used, but mixed-case lettering (using both upper- and lower-case letters) is preferable. Some people find "all caps" text harder to read than mixed-case text (Poulton 1967). Therefore, a mixture of upper-case, or capital, letters along with lower-case, or small, letters is generally preferable on warning-sign messages. A single, particular word may be set in UPPER-CASE letters—or **emboldened**—in order to emphasize it, however. The standard prescribes the use of a sans-serif typeface,[11] such as Helvetica or Calibri, for text in warning signs.

Generally, warning-sign text, outside of the signal-word panel, may be either black-on-white (black letters on a white background) or white-on-black. There is a limitation to either black or white letters and backgrounds in the standard. For many people, the black print on a white background is preferable to white letters on a black background (Bailey 1996, 425). The size of the text within warning signs should permit the user to read the sign from a safe viewing distance. In other words, the letters should not be so small that a consumer must draw close to the warning sign which should be the vicinity of the hazard being warned against. The standard provides a table of letter-size recommendations versus safe viewing distance under different of conditions. There is a column for minimum letter height (point size[12]) for favorable viewing conditions, a recommended letter height for *favorable* viewing conditions, and a recommended letter

[11] A typeface differs from a font. A font includes both a typeface and a size. Therefore, Times Roman 12 is a font. It includes a typeface (Times Roman) and a size (12-point).

[12] 1 point ≈ 1/72 inch (0.35 mm). The point is not an exact unit since fonts having the same typeface size may differ in height.

size for *unfavorable* viewing conditions. Unfavorable viewing conditions include viewing the warning sign at a bad angle or under poor lighting conditions.

Those designing warning signs should secure a copy of this standard, and other related standards, to get all of the details contained in the documents. There are simply too many specifics to cover here in this section. The basics of warning signs which can be covered here, however, are addressed below.

There are four required elements to an effective warning sign according to ANSI Z535.4, some of which were briefly mentioned above. These elements are:

1. Signal Word and the Safety-Alert Symbol
2. Identification of the Hazard
3. Identification of the Consequences
4. How to Avoid the Hazard

Sometimes it is possible to add a fifth element to a warning sign, but it is dependent upon the product and hazard risks involved. This is not part of Z535.4, however. For the sake of completeness, a fifth element of a warning sign can be:

5. How to Mitigate the Injury (once an accident event has already taken place)

Signal Word and the Safety-Alert Symbol

In order to comply with the ANSI/Z535.4 standard for warning signs, the first requirement is the selection of an appropriate *signal word* for the sign. This sole signal word serves as an indicator to the consumer of the severity of a hazard if it is not avoided. Although, to the layperson, there may be little difference between **DANGER**, **WARNING**, and **CAUTION**, ANSI Z535.4 provides distinct classification criteria for these three levels of hazard. Also included in the standard are criteria for **NOTICE** and **SAFETY INSTRUCTIONS**, but few people would likely mistake these last two signal words for the first three when it comes to severity of a potential injury.

The criteria for each of these five signal words are listed below.

- **DANGER** is used to denote a hazardous situation which *will* result in serious injury or death if not avoided
- **WARNING** is used to denote a hazardous situation which *may* result in may result in serious injury or death if not avoided[13]
- **CAUTION** is used to denote a hazardous situation which *may* result in minor-to-moderate injury if not avoided
- **NOTICE** is used for non-physical injury matters
- **SAFETY INSTRUCTION** is used to provide safety-related information to consumers such as operational instructions or procedures

The first three signal words above, **DANGER**, **WARNING**, and **CAUTION**, differ from the last two signal words because the first three indicate a personal injury could

[13] If serious injury or death is less than "almost certain," then WARNING is to be used.

Figure 9.4 The safety-alert symbol.

result from not avoiding the hazard. Consequently, when these three signal words are used on a safety sign, they *must be* accompanied by the safety-alert symbol (SAS). The SAS is reproduced in Figure 9.4 and has perhaps been seen at some time by most readers. Its purpose is to inform those viewing the warning sign that a serious hazard could be encountered if the warning sign's instructions are not followed. Research by Jensen and McCammack support that many people are able to comprehend that even a "CAUTION" signal word, without the SAS, represents a threat greater than mere property damage (Jensen and McCammack 2003).

The extent of personal injury is established by using the following scale:

- **Death**—Loss of life
- **Serious Injury**—Serious burns, permanent loss of limb or function (amputation, blindness, deafness, disfigurement, ...), and other injuries needed extended medical treatment or resulting in long-term pain
- **Moderate or Minor Injury**—Those injuries less that the two above categories including temporary pain, mild lacerations, and irritations

In the event that more than one injury may result from not avoiding the hazard, the *more- or most-serious* potential harm is used for signal-word determination.

In addition to the SAS, ANSI Z535.4 incorporates another visual aspect which is the particular color schemes used with each signal word. One example of an acceptable presentation of each signal word is shown in Figure 9.5. Each of the five boxes in this figure shows a proper use of the signal-word panel for a warning sign. Although background colors are not discernable in black-and-white print, they are listed alongside each panel. Lettering shown in Figure 9.5 panels is either black or white. As can be seen, the SAS appears with each of first three signal words. The SAS is tight to the left of the signal word with its bottom even with the bottom of the signal-word text. Just one of the several possible combinations of colors, the SAS, and signal words are shown in Figure 9.5. Details are elaborated upon within that standard. In many instances, this signal word panel will appear at the top of the entire warning sign, but there are permissible variations in positioning this panel on horizontally formatted warning signs. Again, refer to this standard for specifics and options for compliance.

"Panel" is the term used by ANSI Z535 to denote a separate area of a warning sign. For example, there are panels for signal words, for text, and sometimes for pictograms, or symbols. Each panel will have a black border surrounding it.

Looking back at the warning sign for the forbidden fruit in Figure 9.2, the signal-word panel contains DANGER with the SAS consistent with the probability and severity of the risk—certainty and death, respectively—for eating the forbidden fruit. The

Figure 9.5 ANSI Z535.4 signal-word panels.

white letters and the SAS are on a red background with a black border as required for ANSI Z535.4 compliance.

Identification of the Hazard

Once the severity of not avoiding a hazard has been established through the proper selection of a signal word, the next element of a warning sign is to let the consumer simply know what the hazard is. For example, some hazards will be obvious to many people, such as a rotating lawn-mower blade; other hazards, such as CO exhaust gas from the portable electric generator, may be unknown and undetectable to users. Therefore, it is imperative to let consumers know many of the hazards they might encounter during the use of the product. Chapter 5 included a discussion on both hazards and their associated risks. A partial list of hazards against which a user might need to be warned is included as Appendix A.

This hazard-identification information will be contained in a warning-sign panel separate from the signal-word panel. Depending upon the format of the warning sign, other information beyond hazard identification may appear in this second, or other, panel. The hazard to be avoided may be identified either through a signal word, a safety symbol, or a combination of both. Such an optional safety symbol is shown as Figure 9.6. This figure shows the well-known symbol for an electrical hazard. ANSI Z535.4 does permit symbols-only warning signs as long as the symbol "has been demonstrated to be satisfactorily comprehended…or there is a means…to inform people of the symbol's meaning" (NEMA 2017a, S11.2). Ross wisely recommends that "[i]f the product has symbol-only labels, the manual should describe the meaning of all symbols" (Ross 2011). Having said this, the rest of the chapter will focus on warning signs that include verbiage.

Several factors may affect the final design of a warning sign. These factors include:

- The sufficiency of words to concisely convey a hazard to users
- The availability of a recognizable symbol to convey the hazard

Figure 9.6 Safety symbol for electrical hazard. (Source: Henning, Torsten. 2006. Commons. Wikimodia.Org. 2006.)

- The amount of space that the warning sign will ultimately take up on the product
- Whether the warning sign will need to have a companion translation alongside the native-language text[14]
- And other parameters

Unlike international warning signs, ANSI Z535.4 does not *require* the use of a safety symbol to identify a hazard to users. *International* warning signs fall under a different set of standards. There are four standards, ISO 3864-1, -2, -3, and -4, on international warning signs published by the International Organization for Standardization (ISO).[15] These standards will not be covered in detail. Those designing warning signs for international users should consult these references. Due to the variety of languages that product consumers speak around the world, international warnings rely less upon words than do North American warnings. Instead of words, international warnings depend upon internationally recognized safety symbols to convey product hazard and risk information to product users. There has been *harmonization*, or coordination, between the ANSI and ISO formats in recent years, but this will also not be covered here. Instead, the focus of this book will be on product-safety warning signs on American and Canadian[16] consumer products.

Although it is true that Z535.4 does *permit* the use of safety symbols alone in warning signs to identify hazards, a verbal description of the hazard is generally expected in the United States and Canada. In fact, written messages in both English and French may be required in Canada. In instances where space for a warning sign is quite limited, it may be necessary to provide a symbol-only warning sign out of purely practical spatial constraints. This should generally be done when no other option exists.

[14] It has been the author's experience that adding French translations to English text in warning signs approximately doubles the amount of space (area) needed by English-only warning signs. This is largely due to the French language's need to retain articles such as "le," "la," "les," and "des." Be aware of similar issues when designing multi-lingual warning signs.

[15] www.iso.org

[16] Many U.S. standards are adopted, either officially or unofficially, by Canadian parties. Sometimes changes are made to American standards for Canadian adoption, for instance, the inclusion of a French translation on many materials. Canada often has specific requirements for bilingual (English and French) materials.

Revisiting the warning sign for the forbidden fruit, Figure 9.2, the hazard "Forbidden Fruit" is explicitly stated. There is only one internationally known symbol for not eating forbidden fruit, but it cannot be reproduced here without violating the intellectual-property rights of a famous technology company. There is also a symbol that is used in some warning signs to prohibit eating and drinking in some public places, but Adam and Eve were permitted to eat—just not from *that* tree. Therefore, in this case, no symbol is used in the warning sign to help identify the hazard to the consumers.[17] When a symbol is used, it generally appears in its own panel to delineate it from other portions of the warning sign.

Identification of the Consequences

The proper selection of the signal word for the warning sign requires that the *severity* of a potential injury be determined at that first step. Next, the hazard itself must be identified to consumers. However, it is at the present step that details of a potential injury be conveyed to the product user. Rather than simply telling a user that they could be killed, the user must be told how death will come. It is helpful to think of the identification of the hazard as the *cause* and the identification of the consequences as the *effect*.

In the case of Figure 9.6, which is for an electrical hazard, a product user must not only know of the type of hazard, but also of how that hazard could manifest itself and injure that user. In the case of a household circuit-breaker box, disassembling the electrical panel and touching the mains electricity could result in death from electrocution. In other cases of electrical hazards, touching a "live" electrical circuit may not result in death, but perhaps could electrically shock and, thereby, startle an unsuspecting user resulting in a fall from height, in the case of a ladder user, or could cause a fall down a stairway if access was positioned in a stairwell.

Such requirements of a warning sign will be covered in Example 9.1. Again, Chapter 5 discussed both hazards and their risks and Appendix A contains a partial list of potential product hazards.

How to Avoid the Hazard

Having told the product user how an injury or death might be caused, it is then necessary for a designer and manufacturer to inform the user of how to avoid such consequences. This is done through a warning sign's imperative statements below the identification of the consequences, but frequently in the same warning-sign panel.

Revisiting the case of the two-slice toaster, the way to avoid the electrical hazard is to only use the toaster in a dry environment and to never reach into the toaster (with anything conductive) when it is plugged into an outlet. The thermal hazard can be avoided by keeping the area around the toaster clear from flammable objects and by avoiding contact with any heated parts of the toaster until they cool to near room temperature. As far as handling hot toast, which is the way many people like their toast, tongs may be used to remove the hot toast immediately after it has popped up. Yet, the use of the wrong tongs—for example, steel tongs that conduct electricity—to reduce the probability of burn injuries only increases the chances of electrocution, the first hazard. The consideration and prevention of unintended consequences is

[17] The author must presume that Adam and Eve were literate.

a large part of product-safety engineering. Engineers should remember Newton's Third Law of Motion (Newton 1687, 17) which can be summarized as *Action = Reaction*. Quickly implementing a safeguard for product-safety reasons may inadvertently lead to a safety problem in another area of the product or later in the usage of the product. The application of any safeguard device, including facilitators, should always be done with care.

Example 9.1: Toaster Warning Sign

In the case of the two-slice toaster, if a warning sign were to be put on that product, it could warn of the hazard of electricity (cause) and also state the possibility of electrocution (effect). Such a warning sign is shown in Figure 9.7. In this figure, the consequences of not avoiding the hazard, electric shock, are indicated. Those consequences are death or serious injury. Prior to the listing of consequences, the signal word, WARNING, is shown along with the SAS since death of the user is a possibility if the user is electrically shocked. A pictogram is used in a separate panel to quickly convey an electrical hazard even upon a casual glance from the average consumer. The warning sign concludes with a short list of ways to avoid the electrical-shock hazard. These include never reaching into the toaster unless it is unplugged, not using the toaster in a wet environment or with wet hands, never using a toaster with a damaged electrical cord, and always using a GFCI[18]-equipped electrical outlet when using the toaster.

Such a warning might be positioned on the toaster's electrical cord. Although the warning sign could be positioned anywhere along the electrical cord, it would perhaps be best positioned near the location of the most-likely risk to users: at the toaster-plug and electrical-outlet interface. It is also at this location that the warning sign could be most visible and conspicuous. Although this location might be superior to others with respect to product safety through its conspicuity, this location would likely also be the most obtrusive to those concerned with kitchen aesthetics. In addition, even this location does not preserve the ability of the warning sign to be seen by toaster users. Such a located toaster warning sign could easily be obscured by a coffee maker, for example, placed in front of that warning sign. Of course, homeowners can always remove a warning sign considered to be visually obnoxious.

It might be just as meaningful to those inexperienced with toaster operation to place a warning sign for the heat hazard (cause) and the potential for burn injuries (effect). Few toasters bear such outwardly visible warning signs although some have such warnings on the bottom of the toaster, but that is where few people will ever see them. If a heat-hazard warning sign was to be on the electrical cord, that location is not located near the area of high temperature and would not be an ideal spot for such a warning message. Fortunately, many toaster manufacturers have reduced the amount of high-temperature surface area through thermally insulated toaster designs. This has lowered the probability of contact with a hot surface although hot surfaces still exist on toasters.

[18] GFCI: ground-fault circuit interrupter.

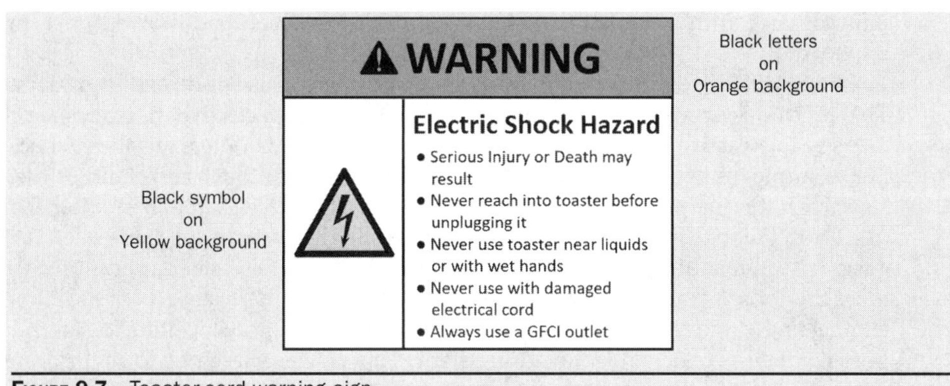

Figure 9.7 Toaster-cord warning sign.

ANSI Warning-Sign Summary

In the case of the warning sign presented earlier in this chapter for the forbidden fruit, Figure 9.2, the four elements to an ANSI Z535.4 warning sign are present.

1. The proper signal word, **DANGER**, with the safety-alert symbol ⚠, both in white and both on a red background for *certain* death as a result of violating the warning sign. This all appears within the signal-word panel.
2. The hazard has been clearly identified—forbidden fruit—but no symbol is used in this instance.
3. The consequences are clearly spelled out for consumers: "Eating will cause Death." The leading character in the word "death" was optionally capitalized for added emphasis.
4. The consequence-avoidance measure is provided plainly in an active voice: "Eat other fruit."

In addition,

- Optional bullets ("•") have been added to the text to help the consequence and the alternative action stand out within the warning sign.
- Items 2, 3, and 4 are contained within a single panel.
- All elements are all contained within one overall warning-sign border.
- Had a safety symbol been used for hazard identification, it would have appeared in its own panel within the warning sign.

This figure is indeed an elementary example of a warning sign. The consequence is simple, straightforward, and certain. The hazard identification and means to avoid are quite clear and concise. Perhaps there has never been a need for a more-terse warning sign than in this situation. So, although this example is both simple and facetious, it still serves to show how to construct a warning sign compliant with ANSI Z535.4. Remember that if it is possible to limit the severity of an accident and its resulting injury, then a fifth potential warning-sign message, which is described next, may be included.[19]

ANSI Z535.4 requires that a word message be "concise and readily understood." This precludes overly wordy messages and generally requires that messages to product users be in an active voice and dispense with certain words unless absolutely necessary. For example, rather than saying "The user should never touch the chain of the chain saw when it is in operation," a warning sign may state "Never touch moving chain."

The decision to include languages other than English in warning signs is left up to the designer and manufacturer of the product in ANSI Z535.4. The standard rightfully recognizes this topic as "an extremely complex issue." As with many aspects of product-safety engineering, design engineers and their companies are in a good position to determine the needs of their product and its warning signs. This view is shared by Ross (*In Compliance* 2016). Therefore, unless specified by law, the use multiple languages in warning signs is optional and the responsibility of the engineering-design team.

The ANSI Z535.4 standard contains many other criteria and suggestions for warning signs including those regarding typeface and character height given a particular reading distance, formatting of warning-sign panels, the durability of the warning-sign material and writing, and more. Again, those people involved in creating warning signs are advised to obtain a copy of the standard for full details.

How to Mitigate the Accident Outcome

There is a *fifth* potential element to a warning sign. While not contained in ANSI Z535.4, there are instances when this fifth, additional warning-sign element is appropriate. This is so in two cases:

1. The accident has already taken place and is complete. The injury has already been sustained. No further accident prevention is possible but the *long-term* severity of the injury can still be affected by user actions.
2. If the accident and its resulting injury take place slowly and are still ongoing. In some cases, lessening the injury outcome through user actions may still be possible.

In the first instance, it may be possible to *mitigate* the injury's severity after the accident through appropriate subsequent actions by the injured or other parties. In the case of ingesting a toxic or corrosive substance, it may be advisable to not induce vomiting from the victim on the product's label or warning sign. In the case of an injury involving the high-pressure injection of hydraulic fluid beneath the skin of a heavy-equipment maintenance person, it is imperative that the person receive immediate medical care even if the injury only appears to be superficial in nature to the recipient. These two accidents have already taken place so no further accident prevention is possible. However, a warning sign leading people to "not induce vomiting" or "always seek immediate medical attention" following such accident events could mitigate an injury already sustained.

In the second instance, if an operation leading to an injury happens at a relatively slow pace, it may be possible to prevent further injury during the accident event itself. For example, imagine that a garage-door opener/closer is closing the garage door and

[19] In the judgment of the author.

a person stumbles and falls onto the garage floor such that this person's neck is in the path of the closing door. Also imagine that the obstruction sensors for the garage door opener/closer do not function properly by stopping and reversing the closure of the garage door. If this fallen person cannot move out of the way of the closing garage door, then that person will have her/his neck trapped through some force applied between the floor and garage door being powered by the opener/closer device. If that person remains under the door for long enough, or if the door continues to close farther during this period of time, that person may suffer greater injury due to a larger degree of crush from the door or die from asphyxiation. A warning sign for the garage-door opener/closer could tell bystanders to press the Open/Close button again in order to reverse the direction of the garage door force. Such action could prevent a more-serious injury and even keep someone from dying. This example presumes a couple of things. For one, it is presumed that pressing the Open/Close button again will actually reverse the direction of the garage door. If not designed in this way, the garage-door opener/closer controller could continue issuing the same command to the actuating mechanism to close which would be contrary to the bystander's intent. If pressing the button will *not* accomplish the direction reversal of the door, the warning sign for the opener/closer could say to use the emergency door-release handle to uncouple the garage door from its drive mechanism instead. In this example,[20] it might be possible to limit further personal injury or avoid it altogether through a warning sign.

There may be few examples of this kind in typical engineering practice, but the future of innovative products, technologies, and substances has yet to be written. It is prudent to keep this fifth potential warning-sign element in mind as new products and their product-safety facilitators are being designed.

Discussion: Visibility versus Conspicuity

When driving through a construction zone on a public roadway, a frequent sight is the construction worker or flag person who will often be wearing an orange, yellow, and/or reflective vest. These are called high-visibility, or "hi-viz," vests. This is a misnomer. *Visibility* is, strictly, the ability to see. For example, the pilot of a fighter jet must be able to see in all directions to detect the approach of hostile threats. This is not always possible, but high visibility remains a goal for many aircraft designers. The same goes for motor vehicles, especially commercial trucks and school buses. Drivers have a need to see as much as possible around their vehicles in order to avoid running over a pedestrian, worker, or schoolchild. Therefore, engine-hood lines, windows, mirrors, and on-board cameras are carefully placed and adjusted to maximize driver visibility.

On the other hand, some people actually mean "conspicuity" when using the term "visibility."[21] *Conspicuity* is the ability to be seen or noticed—the state of being conspicuous.

[21] Although, some dictionaries include the ability to be seen as "visibility," this book will use that term to indicate the ability to see. The term "conspicuity" will be used for the ability to be seen or noticed. This is more than just a distinction without a difference. The qualitative differences between the two is sometimes important to product-safety engineering since these two characteristics of a product are addressed in different ways.

[20] This example was developed by a student team in the author's product-safety engineering course as part of a semester project. Because the author had not received permission from all team members and because of FERPA [Family Educational Rights and Privacy Act (20 U.S.C. § 1232g; 34 CFR Part 99)] restrictions, these students cannot be credited for their work by name.

Examples include the above "hi-viz" vest, automobile daytime-running lights (DRLs), always-on headlamps on motorcycles, and orange off-road vehicle (ORV) whip flags required in some riding areas. Conspicuity devices are any ones that help a person, or hazard, to stand out from its environment. Environment in this case includes such factors as weather, time of day, motion, and other on-going events and distractions.

It is vital to remember that conspicuity is a *relative* property. For example, when driving down a highway under construction at night without street lamps lighting the area, the reflective stripes on the orange barrels can be quite conspicuous when illuminated by a vehicle's headlights alone. This is an instance where the orange barrels and reflective stripes are quite effective. On the other hand, during the daytime with many barrels and barricades visible, numerous construction workers, and several pieces of heavy-construction equipment all in motion, a single construction worker wearing a hi-viz "safety" vest may not be conspicuous at all. Instead, a ninja dressed completely in black might be much more conspicuous than a yellow-and-orange-clad construction worker within this setting. Not only does the black stand out—or *contrast*—in a sea of orange, but a ninja is not something that you would expect to see in a construction zone. The construction worker is an expectation of drivers navigating the cones, barricades, and traffic-control signage despite the worker's donning of orange-and-yellow clothing. Thus, construction workers are not always noticed by drivers inside of a construction zone. They are visible, yet they may not be conspicuous.

When driving an automobile toward a setting sun, an approaching vehicle with its headlamps OFF may be *more* conspicuous than one with its headlamps ON. Just as with the ninja example, a dark object may more conspicuous in a well-lighted environment than a bright object. Therefore, there are multiple aspects to contrast and conspicuity.

It is simply not enough to wear a bright color or to print a warning in red, orange, or yellow. Getting someone's attention may well be a function of the situation, the user, and the use environment. A given user's task load, from Chapter 4, and the resulting haste may also hinder communicating a warning that a hazard may deserve. Consequently, the busier the driver is negotiating the obstructions and traffic within the construction zone, the harder it will be to attract that driver's attention and effectively communicate a hazard. Therefore, merely because a warning sign is placed upon a product, do not assume that the product's user will see and then understand and follow those directions. The warning sign may be poorly positioned, poorly sized, and even poorly written. The signal-word panels in ANSI Z535.4 compliant warning signs tend to attract the attention of users through the use of bright colors—red, orange, and yellow. These colors do enhance warning-sign conspicuity, in general, but alone cannot warn of product risks under all circumstances.

9.6 Manuals

Products often come with instructional books or papers. These instructional materials will be called "manuals" in general for the purposes of this book. This word will be used to refer to owner's manuals, owners' manuals, owner manuals, product manuals, operating manuals, reference manuals, instruction manuals, user manuals, and so forth. However, service manuals and the like are excluded from the use of "manual" here since these traditionally contain technical information for the maintenance and repair of products beyond the undertaking of many consumers and often do not include the instructions and warnings regarding the safe operation of the product. In this section of the chapter, *only* the product safety-related content in manuals will be discussed. Other portions, such as preventive maintenance, warranty registration, operation of non-safety related features, and specifications will not be covered.

When lecturing to students or other professionals, the author will sometimes ask the group to name any general fact.[22] A fact is something that is *always* true. Naming such a general fact is not a difficult exercise. There are many facts from which to choose: the sun comes up in the East and sets in the West; 2+2=4; and such. The audience has little difficulty with this task. However, when the same audience is asked to name any one safety *fact*, there is generally silence. Occasionally, an audience member will cite some non-actionable slogan such as "Safety First" and "Safety is everyone's responsibility." Sometimes an audience member answers with "read the manual." This answer would be correct if all manuals were, in fact, well written and accurate. However, they are not. Thus, this cannot be a fact.

This question-and-answer exchange illustrates two important points in product-safety engineering:

1. There are few—*if any*—safety facts.
2. People presume that a product's manual is good and correct.

The first point is probably correct.[23] The second point, however, is only correct some of the time. Many readers have probably seen a product manual that was horrible—hard to read, hard to understand, incomplete, incorrect, or otherwise flawed or even unusable. Maybe the figures could not be understood. Therefore, it is the job of product-safety engineers and professionals to provide good, correct, and useful content to product users so that telling users to "read the manual" is a worthwhile exercise for users.

A standard was developed by ANSI as the American standard for product-safety information including manuals, instructions, and other materials accompanying the product. Not covered are sales, marketing, and audio/visual materials. This standard is published by ANSI as Z535.6 (NEMA 2017b). Among the purposes of this standard are:

- "[Establishing] a uniform and consistent visual layout for safety information in collateral materials for a wide variety of products"
- "[Assisting] manufacturers in providing safety information in collateral materials" and
- "[Promoting] the efficient development of safety messages in collateral materials."

There are four primary types of safety messages to be found in ANSI Z535.6-compliant manuals. These are:

1. Supplemental Directives
2. Grouped Safety Messages
3. Section Safety Messages
4. Embedded Safety Messages

[22] The same opening technique borrowed from Barnett and discussed in Chapter 2.
[23] A lengthy philosophical debate could perhaps ensue at this point.

For each of these message types, ANSI Z535.6 provides guidance regarding purpose, content, location, and format. Property-damage sections of the standard will not be covered here.

First, Supplemental Directives include directing users to unique product-safety information, directing users to product-safety information contained in on-product warning signs, informing users of the importance of the manual or section of the manual, and warning users of the potential consequences of not reading the manual. The content of a supplemental directive is general in nature and could include warnings such as "Failure to read manual may result in serious injury or death" and "Read all warning signs before use." Again, generic product-safety information is provided by this type of message and can refer readers to other portions of the manual. Such information is sometimes in the manual at the bottom of the front cover, inside the front cover, or on the outside of the back cover. A supplemental directive may appear with the Safety-Alert Symbol without a panel border. Such a directive is shown in Figure 9.8.

Second, Grouped Safety Messages can be thought of as the traditional "safety section" of a manual. This permits users to see much, or all, of the safety-related material in one place within the manual. This information may be grouped according to hazard type, phase of product use, location on the product, or by risk severity or probability. This grouped information can also be referenced by on-product warning signs, or by supplemental directives, to quickly provide product-safety information to users. As with warning signs, the grouped safety messages should provide information to users about hazards, consequences, and avoidance. Furthermore, these messages may be combined into a separate manual or be aggregated into a specific section of the manual. If the latter route is chosen, this information should appear before any procedures to which the messages apply. For example, if a shock hazard exists from a product's battery, then the user should be given safety information on hazard avoidance, such as "Disconnect the battery," before service information on the electrical system is provided. Grouped safety messages should be listed in the manual's table of contents for ease of reference. Formatting requirements for this safety information specify that the manual or manual section be clearly labeled with "Safety Manual," "Safety Information," or similar heading so that the user can easily identify the information's purpose. There is significant layout flexibility afforded to grouped safety messages.

Third, Section Safety Messages provide safety-related information pertaining to a given procedure or topic in a section that is not related to a single step of a procedure. For reasons of brevity, it is permissible to omit hazard, consequence, and avoidance information if this information is obvious or can be inferred. But if this is not the case, that safety information should be included or may be referenced elsewhere in the manual. As with grouped safety messages, safety information in section safety messages should be provided prior to the procedures to which they apply. There are three permissible layout options for section safety messages. Although there is no guidance given by the standard regarding preference among these four methods, Jensen and Jenrich have done some research preferences of formatting options (Jensen and Jenrich 2008). One of these format options appears as Figure 9.9. In this case, what is being conveyed is an electrical hazard. The SAS is used since the signal word "WARNING" is employed.

Fourth, Embedded Safety Messages provide safety information within procedures or within other instructions. Content should include hazard, consequence, and avoidance information unless the user can easily infer such information. Such information

Always Read, Understand, and Follow the manual and warning signs.

Failure to do so can lead to Serious Injury or Death.

Figure 9.8 Supplemental directive from ANSI Z535.6.

ELECTRICAL HAZARD

Always read the safety precautions in this manual before attempting maintenance and service on the product.

Always disconnect the battery before servicing the product's electrical system.

Failure to do so can result in Serious Injury or Death.

Figure 9.9 Section safety message from ANSI Z5335.6.

can also be left out if it is available elsewhere in the manual or if including it could cause the message to become too repetitive or lengthy. Due to its specificity, an embedded safety message should be included at the point at which it is needed. Embedded safety messages should be written so that they become a part of the procedure and, therefore, less likely to be skipped or not seen. Embedded safety messages may be included without special formatting if it is clear from the verbiage that the content is safety related. Alternatively, a signal word either with or without the SAS may be used.

The same ANSI Z535.4 criteria for the signal words "DANGER," "WARNING," and "CAUTION"—as well as the need for the safety-alert symbol, SAS—apply to ANSI Z535.6 materials. Therefore, if "WARNING" is used in a message, then the SAS must appear.

The sections of ANSI Z535.6 are not altogether new. Ross observes that "[t]hese different kinds of messages have been in use for decades (a military standard from many years ago required a safety section in instruction manuals for products sold to the military)…" (Ross 2005). ANSI Z535.6 simply harmonizes and centralizes this material.

Although the ANSI Z535.6 standard is intended primarily for text-based messages, the use of appropriate symbols, drawings, and photographs is permissible. It is, generally, good practice to reproduce *all* warning signs found on product within the manual. Therefore, if safety symbols are in the warning signs, then those symbols should appear in the manual as well. The duplication of warning signs within the manual exposes the user to on-product warning signs even when away from the product. It is also a good practice to provide—free of charge—electronic access and downloads of product manuals to consumers. Downloadable manuals permit users to look through the manual in many locations even if the printed manual was left at home, became damaged, or was lost.

Making these electronic manuals searchable for words and phrases is helpful to users wanting to know more about a particular safety topic. Furthermore, it may be wise, if the product permits, to provide a storage location for the manual in or on the product itself. This further facilitates the delivery of the product-safety messages to the user. If the product is to be used outdoors, a weather/water-proof location or storage device for the manual will likely help preserve the manual for future use.

As with warning signs, those involved with creating manuals should obtain a copy of the current ANSI Z535.6 standard. There is simply too much within it to adequately cover ANSI-compliant manuals and other collateral materials without the standard.

The objective of product-safety engineering and the design of product manuals should *not* be compliance with ANSI Z535.6 or any other standard—although this and other standards are good places to start and should, by themselves, rarely hinder the design of effective manuals to enhance the safety of product users. The true objective of product manuals should be the effective communication of important product-safety information to product users.

Discussion: When (and Where) to Warn

Although much guidance is provided by ANSI Z535.4 on the construction of warning signs, there is no guidance given on when to warn and, conversely, when not to warn. This situation may lead to the phenomenon of "overwarning," but it is not expanded upon in this book due to space consideration. Those readers with interest in overwarning are pointed toward the following references on the topic: Cowan 2001, 2016; Barnett and Brickman 1985; Barnett and Switalski 1988; Ross 2019; Viscusi 1988, 1996; Robinson, Viscusi, and Zechhauser 2016; Craver 2019; Mohan 2020. However, many ethical engineers will likely be most fearful of not providing sufficient warning to product users.

Similarly, ANSI Z535.6 instructs design-and-manufacturing personnel how to construct a manual, but does provide criteria for what safety information is acceptable to put into the product manual versus what must appear on the product itself through a warning sign. Numerous warning messages may appear in the manual, but these messages may repeat on-product warning signs or be altogether new safety messages not seen by the user elsewhere. Thus, this pair of standards, ANSI Z535.4 and Z535.6, does not answer some critical questions facing earnest design engineers.

For many products, it is smarter and ultimately safer to design warning signs for the new or the inexperienced users of a product. Many skilled users of that or similar products may already know how to use it. Some skilled users of the product may have already formed their own habits—either good or bad—on using the product. The behavior of these people may be difficult to affect with a simple warning sign. However, conscientious unskilled people may well be seeking important information on safe product use. It is incumbent on engineers to provide these users with such necessary information. Some of the information may be on a warning sign; other information, in the manual. At times, it is sufficient, or even necessary, to refer a user to the manual. This is especially true if the information is lengthy *or* is needed only the first time that the product is used (Ross 2019). His words to product-safety personnel are to:

> [D]etermine whether the safety message should be seen by a user each time he or she uses the product, in which case it goes on the product, or can just be read in the manual before first using the product, and then the user can refer to it when necessary.

From this, the design-engineering team now has some preliminary guidance on the locations to place product-safety messages. The above guidance may not be practical for all products. The product-design team will need to evaluate each product on its own merits and make ethically driven decisions on what safety messages are necessary and on where they should be placed.

Prior to making these decisions, however, those people involved should consider the power and the pitfalls of the *inductive* inferences that some product users are quick to make. One definition of inference is "the act of passing from one proposition, statement, or judgment considered as true to another whose truth is believed to follow from that of the former."[24] Another, simpler, definition is "the act of reasoning from factual knowledge or evidence."[25] Inferences are made through either deduction or induction.

Deduction is reasoning based on truths. As a result, deductions are always true. For example, if one's pet is a dog and all dogs are mammals, then one's pet is a mammal. Thus, since the reasoning was based on fact, the resulting deduction is true.

Induction is reasoning based on observations that are generally true. Therefore, inductions are *not* always true. For example, if all of the old men one has seen had gray hair, then one might conclude that all old men have gray hair. Without arguing the criterion for old—perhaps over 60 or 70 years of age—this result of induction is false. Some old men have dark hair while others may have no hair whatsoever. Since this inductive reasoning was based only on evidence observed to date, there exist many counterexamples. The logical process was not based on fact and is proven false as a result.

The importance of this logical fallacy, inductive inference, is that many consumers employ such reasoning when they use some products. It is easy, and perhaps even reasonable, for product users to presume that the design-engineering team has taken a uniform approach to warning them about product risks. Barnett has titled this uniform approach to product safety as *The Principle of Uniform Safety* (Barnett 1994). He summarizes this as "Similarly perceived dangers should be uniformly treated." This is a reasonable presumption for users of highly engineered products. Through a product's warning signs, users may well assume that if any one risk is warned against, then every risk that is equal to or greater than that particular risk is also warned against. Take for example a product that bears an on-product warning sign that indicates that failure to follow the warning sign may result in a severe cut to a user's finger. The user of that product may then, through inductive reasoning, assume that all other risks at least as severe are also warned against on the product. Therefore, the user reasons that she or he has been warned about all risks that could result in cuts to toes, amputation of limbs, or death have been warned against as well.

As a result of this faulty potential reasoning by some users, Barnett states that the "[u]se of a few warnings to protect [users] from the most severe dangers is better than using so many that all are lost as leaves in a forest." If too many lower-level risks are warned against, then consistency with uniform safety will result in the proliferation of warning signs. The number of resulting warnings may be too large to produce an effective system of warning signs for product users. This is contrary to the goal of product-safety engineering. When this is the case, educating and training people may become necessary to sufficiently reduce product-user risk.

---

[24] https://www.merriam-webster.com/dictionary/inference
[25] https://www.thefreedictionary.com/inference

9.7 Manuals: Case Studies

What follows in this section are three case studies of unique approaches to product-safety facilitation, through manuals, for hazardous products and activities. Although not necessarily incorrect, these manuals are unconventional and depart from the standards and traditional practices found in typical manuals today. Each manual is unique in its own way. The first manual is quite explicit in its hazards, risks, and outcomes using text tailored to its risk-taking audience. The second manual uses a non-traditional medium to convey its message to a specific audience. The third manual uses prescriptive, advisory, and consultative approaches in speaking with its intended audience who operate in an uncontrolled, dynamic environment but still wish to reduce accidents and injuries.

The above-mentioned deviations from typical product-manual content enhance the *potential*[26] effectiveness of the manuals in promoting user safety. Creators of product manuals—and any other facilitators—should remain open to using unconventional methods and media in achieving their goal which is improved product safety for users. Such new methods do not necessarily have to *replace* existing approaches, they may simply *supplement* those already in place. Not all *effective* product-safety manuals must necessarily look like ANSI Z535.6-compliant examples.

Case Study 9.1: Explicit Hazard-and-Risk Communication in a Bicycle Manual

One problem with the current state of warning signs and owner's manuals is that the exhortations, or guidance, contained within them have often become quite vague, lacking details of the hazards and the true accompanying risks. They can all start to look alike to engineers—let alone users. It might be argued that additional specificity about hazards, risks, accidents, and outcomes would be useful for a product user to know and to consider during product use especially with highly user-active products when plotted in Figure 4.12.

It is not uncommon to see a warning sign stating that *failure to follow* the admonitions just mentioned *may result in serious injury or death.* Although, presumably true, the warning manual or sign does not always state the particular risks and subsequent potential injuries *explicitly*. It is understandable why many warning signs cannot go into great or greater detail on hazards and risks—there is limited space and some content is better suited to the product's manual. Yet, in the much-larger manual for a product, there may exist the same scarcity of detail regarding product risk.

For example, the designer/manufacturer of a product that has one particular accident type or failure mode, may simply say that "serious injury or death may result" in their manual without elaborating on the type, or types, of accidents and injuries that may result. Even if a product user already knows of the potential for one particular accident, the designer/manufacturer could choose to go further and identify other types of potential accidents to the consumer and also tell of several serious potential outcomes, or consequences, of that accident situation. In some instances, such a company could even

[26] Of course, any safety-related material, however well written, can be intentionally and completely ignored by product users.

go so far as to demonstrate the accident event to users through figures and video content included with the product or available at a corporate, industry, or social-media website.

Such demonstrations could include staged accidents of that type showing the violence, release of energy, or injury potential from such accident events. Since a company is already telling users about something, why not show them? Such demonstrative facilitation efforts, when done with thought, should do no harm[27] in keeping users from acting against their own interests—and the interests of others who may be adversely affected, such as passengers and bystanders.

The author is unaware of *many* designer/manufacturers going to such extremes with the product-safety materials for their products. However, the author is aware of *one* company, in particular, that does an exceptional job of going beyond the "status quo" of typical owner's manuals. In their manual ("Bicycle Owner's Manual" 2018), the particular bicycle company explicitly states that bicycle riders attempting stunts—at which they are not skilled—run the risk of injury, paralysis, and death. When speaking about "Freestyle & Downhill" bicycling, their manual further states:

> The stakes are high if you screw up. Realize too late that you aren't up to the challenge, and you run the risk of major injury or even—say it aloud—death, paralysis. In short, extreme riding carries a high degree of fundamental risk, and you bear the ultimate responsibility for how you ride and what you attempt to pull off. Do you want to avoid these significant risks? Then do not ride this way.

To many members of the general public, the possibility of injury or death are enough to deter risky behavior. However, this bicycle company may believe that many of their customers who may be tempted to attempt jumps and other stunts with a bicycle might be risk-tolerant young adults or adolescents. Perhaps this population has a cavalier attitude toward injury or death. Perhaps some of them do not fear breaking an arm or even dying given the bravado which typically accompanies youth. Yet, if contemplated, the ignored thought of spending 50 years in a wheelchair may be sufficient to get some users to reconsider their actions before trying a stunt seen on an internet video. A fate such as tetraplegia may be worse than death to these individuals. The manual mentions paralysis several times throughout its pages. This owner's manual also states that the use of particular PPE "might save your life or keep you out of a wheelchair."

Even casual readers of this manual should quickly realize that this is unlike the numerous bland manuals that have been seen—not read—before. Perhaps this differentiation will be enough to get that reader's attention and implant, even if only for a short time, those warnings about risk, accidents, and outcomes.

Would a designer and manufacturer of luxury automobiles marketed to the sophisticated social set use such wording in their product manuals? This is highly unlikely. However, this bicycle manufacturer may recognize, understand, and speak to the mindset of their particular consumers effectively through this tailored approach to facilitators. This manufacturer's courageous effort is to be applauded. Other designers and

[27] Such demonstrations could, in some cases, help improve product safety through showing various aspects of an accident scenario to users. This same measure, however, would also heighten the product-liability concerns of corporate legal staff and their trial attorneys. Again, this is a book about product safety, not product liability, and corporate values would no doubt factor into a company's decision to undertake such a facilitator effort.

manufacturers might consider if such an unconventional approach would benefit the users of their products.

This example also demonstrates that a one-size-fits-all approach to warning-sign and manual verbiage may not be in the best interests of users for products across the spectrum of products.

Case Study 9.2: The Comic Book as an Operation and Maintenance Manual

The United States Government started creating comic books, now known as "graphic novels," during World War II. This was an attempt to help the war effort through stimulating patriotism in its citizens and, perhaps, by distributing propaganda contained in the pages of the comics. This period was also a time when corner "drug stores" had book racks and magazine stands often displaying comic books. Therefore, it is not farfetched to see how the government saw the comic book as an attractive medium through which to reach people with its messages.

The Department of the Army continued the use of the comic-book format to educate its troops on the operation and maintenance of its new M16A1 automatic rifle in 1969. The M16 replaced the M14 rifle during the Vietnam-war era of the 1960s. The M16 entered service in Vietnam in the mid-1960s and was immediately plagued with fouling problems from gunpowder residue which could render the rifle unusable. Obviously, an unusable rifle put a soldier in grave danger during a firefight. All of this made the thorough and regular cleaning of the M16 rifle of critical importance. It is difficult to imagine more-important subject matter to effectively convey to product users than this—which could truly be a matter of life and death.

The manual was produced by the U.S. Army for its soldiers using the M16 automatic rifle as a 5 inch × 7 inch (12.5 cm × 18 cm) comic book (U.S. Army Materiel Command 1969). The cover and two pages from this manual are reproduced as Figures 9.10 and 9.11, respectively.

The sample pages shown were carefully chosen so as not to offend some readers. Many of the manual's pages feature a curvaceous young woman, sometimes wearing a beret and a low-cut shirt, suggestively offering operational and maintenance tips to U.S. Army soldiers reading the manual. Those were different days, no doubt, than today.[28] Despite these anachronisms and possible offenses, this manual serves as an example of a product-safety facilitator that appeared to work well with its intended audience at the time—and under the trying conditions—it was used.[29] The U.S. Army appeared to know their audience and tailored a mechanism to effectively transfer the necessary knowledge to them. This is information which would increase those troops' chances of survival during wartime. This example is not one of a safety manual, per se. However, there are few "higher bars" than this situation where conveying critical information to users could truly be the difference between life and death. Therefore, it is an example that is worthy of remembrance and potential use in the future for unique products.

---

[28] The manual also contains racial stereotypes of then-enemy soldiers.

[29] The author attempted to find studies on the efficacy of this manual. No academic work was found, although several reviewers/commenters on websites selling reprints of the manual indicated that—*from personal experiences*—this manual truly was an effective means of quickly educating, and subsequently reminding, troops of important aspects of M16 rifle operation, maintenance, and safety.

FIGURE 9.10 Cover of U.S. Army manual for the M16A1. (Source: U.S. Army Materiel Command, 1969.)

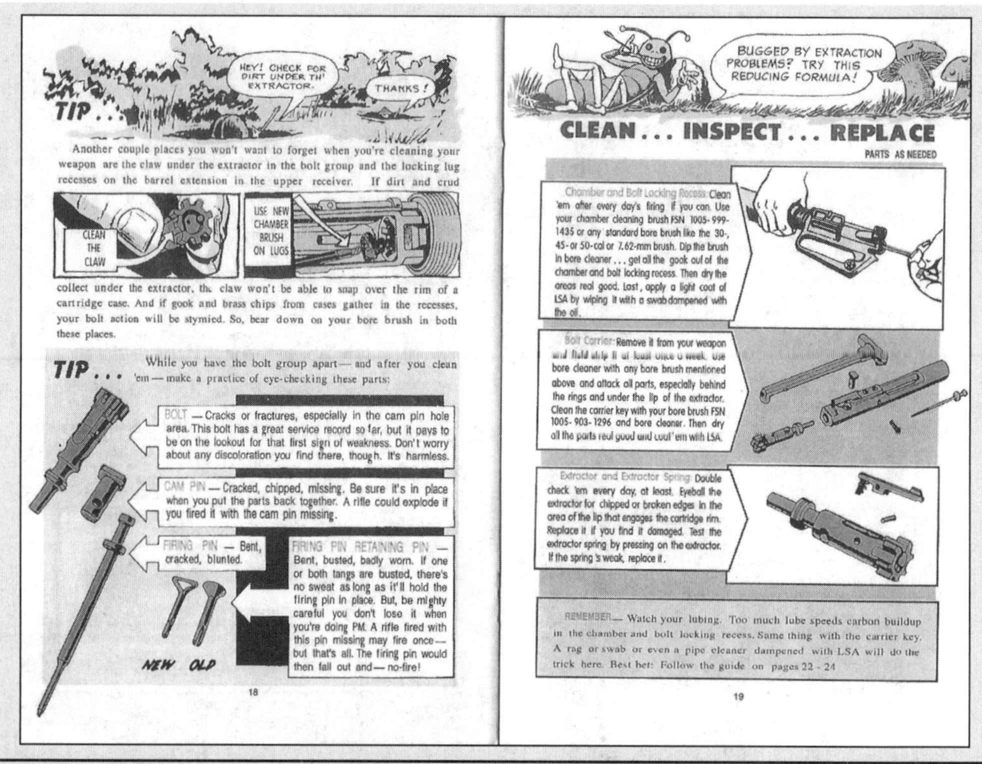

Figure 9.11 Sample pages from the U.S. Army manual for the M16A1. (Source: U.S. Army Materiel Command, 1969.)

Some readers might be tempted to view the M16 manual as a puerile attempt at a "real" manual. However, if one takes the time to find a copy on the internet and look at it, that reader may see that it effectively and concisely conveys much information regarding rifle construction, disassembly, troubleshooting, lubrication, cleaning, safety, and ammunition loading. It is easy to read—or to scan for specific information—perhaps even when under enemy fire.[30]

Recently, the U.S. Navy used this medium for its graphic novel to tell the story of four Navy corpsmen deployed to Iraq (Kraft, Peeler, and Larson 2010). Readers are encouraged to download[31] this document and study its ability to communicate information to readers quickly using both drawings and simple words. The use of drawings, rather than photographs allows the artist and writer to control the background and eliminate unnecessary distractions to the reader. The use of "word balloons" for dialogue within the drawings reduces the need for excessive verbiage to get the same point

[30] While on the topic of unusual facilitators and the military, there is one reported incident when a U.S. Marine was having problems with his rifle while under enemy fire. His solution: telephone the manufacturer's customer-service help line. It worked ("Barrett .50 Cal Won't Fire. So These Marines Called Customer Service. During a Firefight." 2017. Tribunist.Com. 2017. https://tribunist.com/military/barrett-50-cal-wont-work-so-these-marines-called-customer-service-during-a-firefight-video/?utm_source=LRD.)

[31] https://www.med.navy.mil/sites/nhrc/Site%20PDFs/The_Docs_(sm).pdf

across with the traditional owner's manual with text and photographs. This case study serves to demonstrate the versatility and efficiency of this medium.

Presently, the graphic novel is an artform as well as a written medium popular with many younger people. Thus, is it time to reconsider its use for product-safety purposes?

The following case study in an accident-prevention manual shows how a facilitator may use simple graphics as an effective medium to quickly determine potential and previous accident situations. This manual helps commercial-vehicle operators and supervisors, all "users," create a better operational environment that should result in fewer accidents and injuries. Although the manual is used in an industrial and commercial setting, the extensibility of this method to other more product-oriented situations should be evident. For example, a manual or other educational materials for a product whose users would benefit from instruction or education could be constructed and tailored to each product and yet still be versatile enough for a variety of situations.

Case Study 9.3: Instructional Manual for Accident Prevention

This is an example of an accident-prevention manual for users, in this case employee truck drivers. Also in this case, they operate in an uncontrolled setting. This manual is the result of a U.S. DOT/Federal Highway Administration (USDOT/FHWA) contract to help reduce the number of commercial-vehicle accidents. Since such trucks are so large, typically up to 80,000 pounds (36,000 kg), but sometimes heavier, any accidents involving trucks have the potential to be devastating in both damaged property and lost lives. So, the hazard is heavy trucks in motion and FHWA's approach, in this case, was to lower the *probability* portion of truck-accident risks. There was little that could be done directly about the *severity* portion of trucking-accident risks through this effort.

The manual resulting from this work (Uzgiris, Hales, and Dilich 1997) has successfully been used[32] by several large trucking companies, operating their own fleet of trucks, in reviewing truck accidents with their drivers to reduce future trucking accidents. The writing of this manual included a circulation of a draft manual to a group of reviewers followed by a meeting of the researchers, FHWA regulators, and other reviewers who included trucking companies, a trucking association, truck drivers, law-enforcement officers, insurance personnel, and a training-materials supplier. Feedback and other comments were noted and incorporated in the final manual.

Since many drivers in this case are employees, not independent truck drivers, the trucking companies are able to enforce some rules and procedures for vehicle operation. Such would not necessarily be the case with independent truck drivers. However, unlike a factory-floor setting, the open road is a highly *uncontrolled* environment with endless combinations of factors affecting safe commercial-vehicle operation. For example, weather, ambient temperature, ambient-lighting conditions, road-surface conditions, roadway hazards, driver (in)attention, driver fatigue, FHWA regulations on hours of operation, vehicle condition, and other roadway users are among the uncontrollable parameters affecting each moment of each trip made by a truck driver. Also of note was FHWA's desire to keep the manual at a reading-comprehension level of eighth grade. It was important to consider all of this when designing the manual.

---

[32] From the author's communications with members of the trucking industry.

The result was a manual which heavily relied upon pictograms to represent Accident Situations and look-up tables with Potential Causes and Countermeasures. A sample page of the manual is shown as Figure 9.12.

Some of the accident situations are easy to explain and could well be described with words; however, some accident situations are simply better identified and discussed through the use of pictures. Sometimes pictures are a much better communications medium than words. Feedback from the trucking companies confirmed that reviewing an accident with a driver after the fact was indeed assisted through the graphical nature in this manual. The truck driver and supervisor were both able to readily identify the appropriate accident scenario and then proceed to investigate how such accidents could be avoided in the future.

In the case of "right turn squeeze" accidents, the third row of Figure 9.12 shows a pictogram of a truck and trailer, during a right turn, squeezing-out an automobile in the right-hand lane of traffic. The Accident Situation is "Right turn squeeze." Remaining in the same row and moving to the right by one column indicates that Potential Causes could be for this accident situation include:

- Failure to scan space to the right
- Failure to use turn signals
- Failure to block area to right
- Failure to use mirrors

Possible countermeasures to future right-turn squeeze accidents, from the third column of the same row, include:

- **B6 Turning left and right**
- B1 Defensive Driving
- B2 Right-of-way
- B8 Using and changing lanes

The most-effective countermeasure for the right-turn squeeze accident is in bold lettering. In this case, it is "Turning left and right." Figure 9.13 shows the content of the manual's page for that countermeasure in bold letters, "TURNING LEFT AND RIGHT B6." The very-top row of this countermeasure shows the accident-situation pictograms for which this countermeasure is applicable. In this case, there are three accident situations that this countermeasure can help reduce. They are a right squeeze, a left squeeze, and a left turn across opposing traffic all of which are shown in Figure 9.12. This countermeasure block of this sheet includes entries under Objective, Description, Questions for Management, Maintenance checks, and Driving tips for both right and left turns. Pertinent trucking-regulations and references appear in a block at the bottom of the page.

There are numerous approaches to getting information and concepts across to readers of the manual. Although Figure 9.13 contains *prescriptive* and *advisory* statements, it also includes *consultative* questions. Each of these is explained below.

- A *prescriptive* statement is what is typically seen in manuals and warning signs. Such statements of the form "Thou shalt…" or "Thou shalt not…." Here, an

TABLE OF ACCIDENT SITUATIONS AND COUNTERMEASURES

ACCIDENT SITUATION	POTENTIAL CAUSES	COUNTERMEASURES
Over-the-centerline head-on 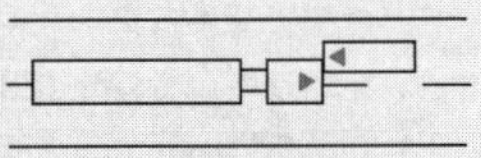	Illness or fatigue Drug impairment Adverse conditions Inattention or drowsiness Mechanical defect	**B8 Using and changing lanes** B5 Passing B11 Driving in adverse conditions A5 Driving and substance abuse A6 Illness and fatigue C1 Preventive maintenance and inspection procedures
Intersection collision 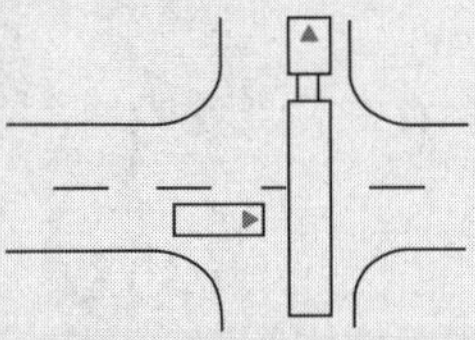	Misjudging speed and closeness of vehicles Misjudging time for vehicle to clear intersection Failure to obey traffic control device Failure to use mirrors	**B7 Crossing intersections** B1 Defensice driving B2 Right-of-way C10 Vehicle lighting and conspicuity
Right turn squeeze 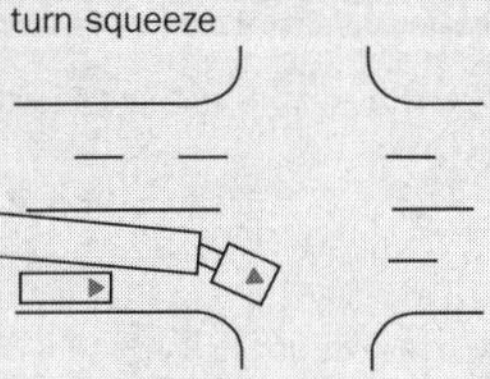	Failure to scan space to the right Failure to use turn signals Failure to block area to right Failure to use mirrors	**B6 Turning left and right** B1 Defensive driving B2 Right-of-way B8 Using and changing lanes
Left turn squeeze 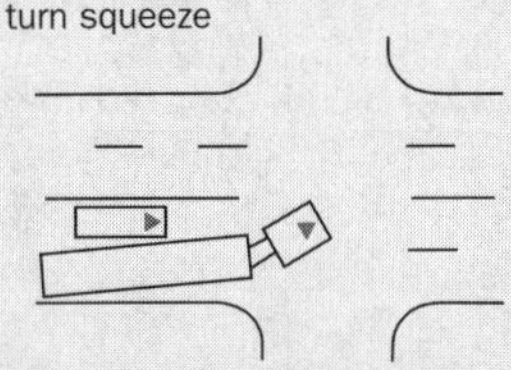	Turning from wrong lane Failure to use turn signals Failure to use mirrors	**B6 Turning left and right** B1 Defensive driving B2 Right-of-way B8 Using and changing lanes C10 Vehicle lighting and conspicuity
Left turn across opposing traffic 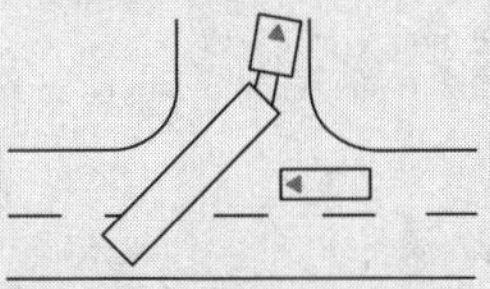	Misjudging speed of oncoming traffic Misjudging time for vehicle to clear intersection Failure to obey traffic control device	**B6 Turning left and right** B1 Defensive driving B2 Right-of-way C10 Vehicle lighting and conspicuity
Obstructing traffic flow when entering roadway 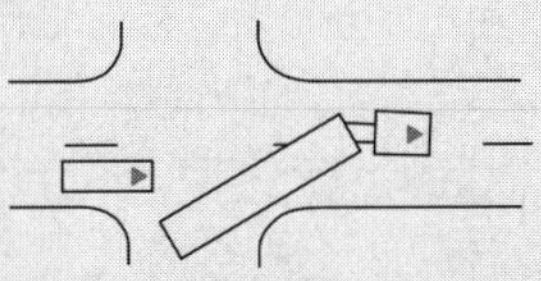	Failure to give right-of-way to passing traffic Assuming other driver will see and avoid Aggressive or reckless driving attitude Misjudging speed of oncoming traffic	**B2 Right-of-way** B1 Defensive driving C10 Vehicle lighting and conspicuity

FHWA Commercial Vehicle Preventable Accident Manual — Triodyne Inc.

4

FIGURE 9.12 Commercial Vehicle Preventable Accident Manual—Accident Situations, Potential Causes, and Countermeasures. (Used by permission of Triodyne Inc.)

TURNING LEFT AND RIGHT | B6

TYPICAL ACCIDENT SITUATIONS

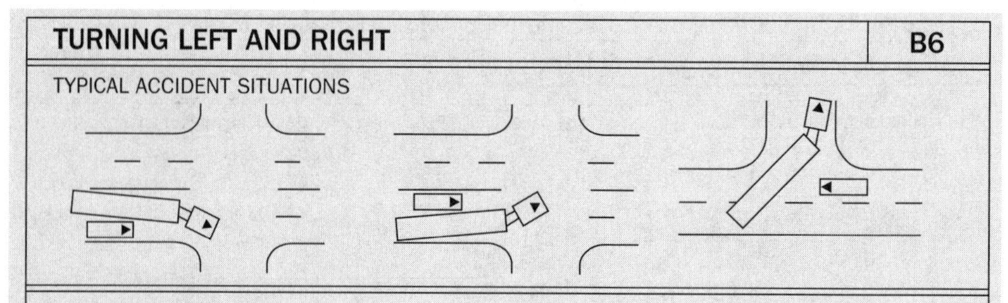

COUNTERMEASURE

Objective:
To prevent turning accidents by anticipating the hazards involved and knowing how to safely avoid them.

Description:
Making left or right turns with long vehicles created problems that automobile drivers do not have. Blind spots make it difficult to see other vehicles. Vehicle length forces drivers to make wide turns, encroaching upon adjacent lanes of traffic. Improper tracking of vehicles makes it difficult for the driver to judge position. Turning takes longer to complete, thus increasing exposure time to hazards. Drivers should recognize the hazards created while turning and follow proper procedures to minimize them.

Questions for management:
- Have your drivers been trained regarding safe turning procedures? How? When? By whom? To what standard of performance?
- Do you know if your drivers are practicing sage turning procedures?
- Do you ever have qualified personnel ride with your drivers to assess sage driving habits?
- Have you examined routes used to minimize travel and turning at difficult or hazardous intersections?
- Have you considered attaching "Wide Right Turn" decal on rear of vehicles?

Maintenance checks:
- Broken mirrors, loose mountings, and mirror adjustment.
- Tail light, brake light, and turn signal function.

Driving tips:
Right turns:
- Move to the right lane well in advance of intersection, positioned to make a safe turn.
- When turning, keep rear of vehicle to the right, blocking other vehicles from passing on the right.
- If encroaching upon other lanes, wait for other vehicles to clear and then turn slowly.
- Be careful that improper tracking does not cause the vehicle or trailer to ride up onto curb or strike stationary objects.

Left turns:
- As you approach turn with signal on, watch for drivers who may misinterpret this signal as an intention to turn somewhere before your intended turning point.
- Don't start turning until there is enough time for the rear of vehicle to clear the intersection without forcing opposing drivers to slow down or swerve.
- Don't assume opposing drivers will see you. They may be looking elsewhere.
- Be careful that improper tracking does not cause the vehicle or trailer to interfere with pedestrains and other vehicles.

REGULATIONS: FMCSR 383.111; 383.113; 383 Appendix to Subpart G.
REFERENCES: M1; M2; C2; D1.

FHWA Commercial Vehicle Preventable Accident Manual — **Triodyne Inc.**

27

Figure 9.13 Commercial Vehicle Preventable Accident Manual—Example Countermeasure: Turning Left and Right. (Used by permission of Triodyne Inc.)

example of prescription under Driving Tips, Right turns is "Move to the right lane well in advance of intersection, positioned to make a safe turn." It is an exhortation must or should always be followed.

- An *advisory* statement is an informative communication, such as "Did you know that?..." An example of this is under Description where the manual says that "Blind spots make it difficult to see other vehicles." This declaration simply passes on information to the truck driver that can be used when driving. It does not attempt to coerce anyone into always doing anything.
- A *consultative* question is the approach where open-ended questions are asked of, for example, trucking managers. In this case, management is asked if they have "examined routes used to minimize travel and turning at difficult or hazardous intersections?" in the fourth bullet point. The manual's authors are not stating that management *must* do these actions. The authors rightfully rely upon managers to consider their individual drivers, needs, and situations and then use their best judgment to see if route (re-)evaluation would be a useful exercise. Each trucking company is in a unique situation regarding trucks, types of loads, regular deliveries, urban versus rural setting, roadway restrictions, climate conditions (cold winters or hot summers), and other factors that can only be assessed by the trucking company itself. These factors may also change over time, especially with the change of seasons and the onset of inclement weather. This approach makes the manager and the truck driver *each* part of the solution to the task at hand which is reducing preventable accidents.

The approach of the above manual is simply one tool to help prevent accidents and their resulting injuries. However, in another manual, the same organization and drafting process that produced the commercial-vehicle manual created another one, this time for the safe transportation of hazardous materials (HAZMAT) (Hales, Goebelbecker, and d'Entremont 1996). This HAZMAT manual included those *incidents* considered "close calls" rather than full-blown *accidents*. The authors realized that perhaps only sheer luck—in the form of inches or seconds—may separate a close call from an actual accident with injury. The inclusion of near-accidents should, therefore, be considered for both facilitators and engineering design.

The first case study presented involved a bicycle-user's manual and its safety warnings. Its format is not a significant departure from traditional layouts used in typical product manuals. It provides hazard, risk, and accident-outcome information to product users. However, the extent to which consequence information is provided to bicycle riders is much more specific and extensive than most product manuals provided today. Rather than just saying that serious injury or death may result, the manual informs product users of the potential for paralysis and a lifetime of wheelchair use. This manual properly reinforces the warning signs on the product itself.

The second case study used the comic book as a content-delivery mechanism. This would be considered unconventional today and may not be appropriate for all products. Yet, a design-engineering team is not designing *all* products. It is designing one product or several related products. Consequently, the engineering-design team must select the best way to effectively deliver the necessary product-safety content to the product's users. Again, one size may not fit all products and users. Engineering ethics should make product-safety facilitation, by the team, a search for the best methods of reaching the consumer with important information. This is an example of one effective communications medium to convey information beyond that which would fit on a warning sign—yet this comic book remains quite compact in size. It, therefore, contains some educational and training content—information required at various points of product use, but perhaps not for each use. Although troubleshooting information is contained, there is also preventive-maintenance (PM) information for post-use product care.

The third case study is that of a heavy-truck driving manual intending to prevent traffic accidents. This operating environment is a perfect example of the PUE model from Figure 1.3. There is, of course the product, the heavy truck, but its level of safe operation is greatly affected by the user, the truck driver and company, as well as by the environment, including the roadway, other vehicle operators, and ambient conditions. Truck safety cannot be addressed through product design alone. This example is a good demonstration of limitations to engineering design. Due to the high variability in the operating environment, the manual's authors do not attempt to dictate all of what should be done. Instead, these authors ask consultative questions that should help those parties involved become part of the solution through short set of prescriptive and advisory statements included where appropriate. This is a qualitative departure from many manuals.

Although, safety information for both warning and educational/training purposes were contained in these above case studies, they are just a few examples of where non-traditional methods of safety communications have been used to better educate product users. These departures from norms should come as no surprise since the ultimate goal of product-safety engineering is *product safety*—not *compliance* with typical approaches.

9.8 Education and Training

This chapter has covered some product-safety facilitators including warning signs and manuals. Warning signs and simple manuals may be insufficient to convey the information, or skills, needed for proper and safe use of some products. In some cases, education or training[33]—rather than simple warning signs and warning messages—may be necessary in order to reach risk acceptability.

One pair of authors states that (Barnett and Brickman 1985, 4):

> Where large quantities of safety information must be communicated, warning signs cannot be used and one must resort to training. Thus, in complex situations, training is more effective than warning, which disproves the consensus safety hierarchy.

Therefore, on complex systems, such as nuclear reactors, the operators must be trained—and not necessarily only warned—about the hazards and risks of systems, or products, under their control. This chapter will not delve too deeply into the education

[33] The author prefers the word "education" over "training," although using the word "training" is often unavoidable. The author believes that training carries with it a connotation of having people—or animals—produce a *particular* response from a given stimulus. Education, on the other hand, is more open-ended than training. Education tries to equip a person with the awareness and skills needed to assess a given situation and be able to determine an *appropriate* response even if that situation is new.

or training of humans. Educational and training products can become quite involved and sometimes include areas such as psychology, education, and human factors in addition to engineering (Bailey 1996).

Well-understood safety messages may merely need to be mentioned in order to "trigger" user awareness. In these instances, warning signs are generally sufficient. However, if the awareness of such risks is lacking or is easily forgotten, then instructions in the manual or even an education/training program may be called for. Similarly, if skills are needed for safe product use and these skills are unfamiliar to users, then training may also be a necessary facilitator to be developed by the designer, manufacturer, or industry involved. A user-training manual and instructor training will be discussed in the next section.

A good user-training program may include such aspects as the determination of needs and objectives, an evaluation of sufficiency, the provision of instructors and instructor materials, measures to assure repeatability and reproducibility of the training, the use of repetition, the use of positive reinforcements, the use of quizzes and examinations to evaluate trainees/users, "hands-on" use of products in a laboratory or outdoor setting, encouragements to continue learning on one's own, and ultimately helping the user become part of the solution to her/his own problems. The details of any training program and its materials will depend upon many factors including the product, the users, and the environments—both physical and social—per the PUE model in Figure 1.3. There should be no contradiction between facilitation materials and the product-safety messages provided in them. The use of the internet, computers, and smart phones may be suitable for some products and necessary tasks depending upon the objectives and needs of the training program. Although aimed toward workers in an occupational safety and health (OSH) setting, rather than users in a product-safety setting, Jensen provides good material on establishing a sound safety program (Jensen 2020).

9.9 Innovative Facilitators

As discussed earlier in the chapter, at times, it may be necessary to go beyond printed words in warning signs and manuals or video content on a website in to increase product safety through facilitators to reinforce warnings, to provide practice or training, or both. Such traditional methods may prove perfectly suitable for many products and situations. There do exist, however, products and educational needs that could benefit from additional facilitation devices or tools. Three such case studies are provided below.

Case Study 9.4: Firearms Training

This first example is one in which the product itself, through its intrinsic characteristics, may not be suitable for use in safety instruction to its users. As ironic as that sounds, in the case of handguns and handgun-safety training, the use of an actual handgun may be unwise for several reasons. One reason for not using an actual firearm in safety-training sessions is that the presence of a handgun could, in some settings, be mistaken for an active-shooter situation by people who may not know that user instruction is taking place. This could well trigger an unnecessary law-enforcement response. Another reason for not using a real firearm in such safety-training sessions is the possibility—however improbable—for a negligent discharge[34] of the firearm. Despite the finest of intentions and efforts to assure that only unloaded firearms be used for instruction, negligent discharges of handguns in instructional settings continue to take place. This situation is further aggravated by the probability that some people taking such handgun training will have limited experience in handling firearms and could negligently fire a loaded firearm. However, even seasoned instructors have been known to make this mistake.

For these reasons, it is commonplace for handgun-safety instructors and courses to employ an obviously "fake" gun, or replica, when delivering in-person handgun instruction. One such "demonstrator" handgun is shown in Figure 9.14 alongside the handgun of which it is a replica. The actual, functioning handgun is black and is above and to the left of the relatively harmless replica which is white in color. These inert replica guns are dimensionally correct but, in this case, weigh less than an actual loaded handgun. Some replicas have the basic features of the real guns they represent in order to enhance the training experience. These features could include the size and locations of the trigger, trigger guard, safety latch, magazine release, and gunsights. However, these replica handguns do not hold ammunition and, consequently, cannot fire bullets—even through negligence and accident. The replica shown is plastic and is injection-molded using a non-traditional handgun

FIGURE 9.14 Actual and demonstrator handguns.

[34] The term "negligent discharge" is used to indicate an instance when a firearm discharges a bullet when a person pulls the trigger without intending to do so. An "accidental discharge" is the discharge of a bullet through a failure in the firearm itself.

color, white. Some training handguns are brightly colored and have other features to further visually distinguish them from real handguns even to casual observers. The replica shown contains many molded "relief" areas. Although these relief features simplify the manufacture and reduce the cost of the replica handgun, these features also help distinguish it from an actual firearm. The use of such a replica firearm makes handgun instruction a much less-risky activity for both the course participants and the instructor.

Case Study 9.5: Auto-Injector Trainer

A product which requires that users must perform an action with which they are unfamiliar is another instance where a unique facilitator, or training tool, might be beneficial or even necessary. Consider the case of people who react to the exposure of an allergen through anaphylaxis.[35] This is an emergency situation and its treatment is an immediate injection of epinephrine.[36]

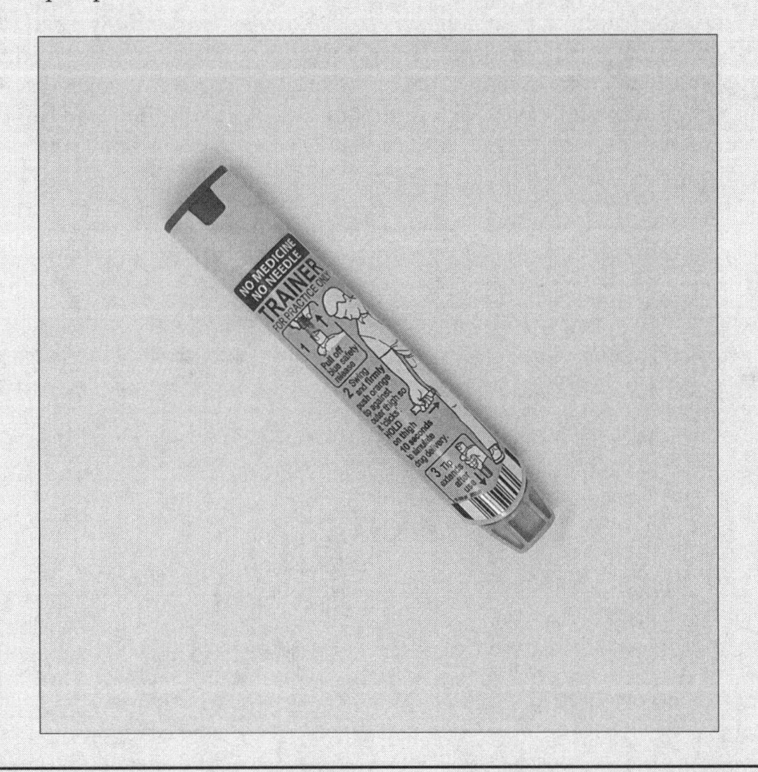

Figure 9.15 Auto-injector trainer in pre-injection state. (EpiPen® is a registered trademark of Mylan Inc.)

[35] Anaphylaxis is a severe, sometimes life-threatening, reaction to allergens such as insect bites and foods that can occur within minutes of allergen exposure. The condition can be fatal since it can affect both breathing and blood pressure.

[36] This is truly a *training* situation since there is but one proper *reaction*—a self-injection by the user—following the sole *stimulus*—an exposure to an allergen.

Many people who know that they may develop anaphylaxis in the case of a bee sting, for instance, will carry with them an "auto-injector" device. The auto-injector is a single-use hypodermic injector of epinephrine with which users inject themselves following an insect bite, for example. Since most of the population is probably unaccustomed to self-injection, the maker of at least one auto-injector provides a facilitating "training" device when the auto-injector is purchased at a pharmacy. One such auto-injector practice device is shown, before its use, in Figure 9.15. This practice device does *not* contain a hypodermic needle, but is otherwise similar to the real device. Through the use of this device, a future user of the device can practice the steps necessary to self-deliver a dose of epinephrine. Instructions for use are also included on the device. The trainer also reinforces the need for the user to remove a pin prior to injection (Figure 9.16). In addition, the future user may get a "feel" for how hard to inject oneself with the actual product in a time of need. Not only does this trainer help build confidence in the device user, but it should also prevent mis-injections of epinephrine at times of critical need.

One set of authors (Benbadis, et al. 2020) does a particularly good job of critiquing instructions for an auto-injector device. In addition, they discuss the effective presentation of safety material and discourage the use of excessive detail and color in facilitator graphics. This is what this book's author calls "color pollution" and could confuse users in a time of need.

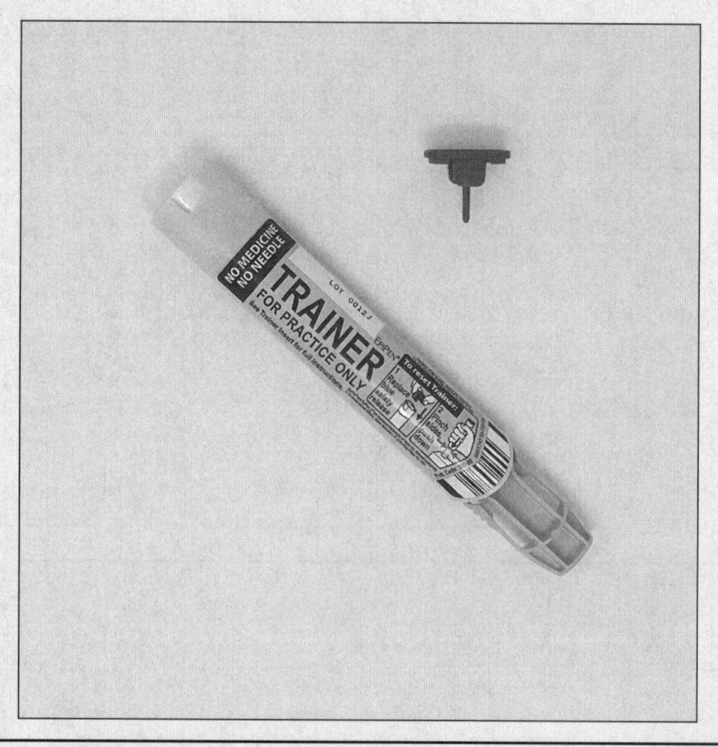

FIGURE 9.16 Auto-injector trainer shown in post-injection state with top pin removed. (EpiPen® is a registered trademark of Mylan Inc.)

Case Study 9.6: Train-the-Trainer Program

There may be instances where a product is innovative and not yet understood by the consuming public. A product may also be potentially hazardous if not used properly or could be used by an *at-risk* population. A product may also be *all* of the above. For such products, there may be a need to establish an education-and-training program for users of that new product. In some of these situations, there may be a further need to train people who will, in turn, be *training users* on the safe and proper use of the product. This is called a "train-the-trainer" program.

There are numerous examples of such instructor-training models. Such a product-and-instruction example is firearms which was just mentioned in a previous case study. Various groups offer firearm-safety training for future instructors who will go on to train employees, members, customers, and the general public. The powersports industry—motorcycles, ATVs, SSVs, snowmobiles, PWCs, and other marine products—has several examples of such programs. For example, the motorcycle industry has a program (*Basic RiderCourse RiderCoach Guide* 2001) that is recognized in many states as part of the on-road motorcycle operator-licensure program. Similarly, the ATV industry supports its own operator safety-training course (*ATV RiderCourse Handbook* 2004). Such programs have specially created materials that cover the necessary content both for product users and for product-use trainers. For motorcycles and ATVs, portions of the user-training and train-the-trainer programs are held both indoors, in a classroom setting, and outdoors, on a closed course designed for that specific purpose.

For these train-the-trainer programs, a course notebook is provided to each instructor. This notebook contains numerous aspects of the training including course requirements for interior and exterior space, the layouts of outdoor space for each hands-on exercise, the objectives of each section of training, and other materials that assist the trainer explaining explain exercises to students. These "range cards" contain exercise objectives, range setup, demonstration points, checklists, and diagrams for trainers to go through as they instruct students away from the classroom. The classroom sections permit instructors to cover things such as the pre-ride vehicle inspection and defensive product operation whether in motor-vehicle traffic or on an unimproved trail. Outdoor sections help students handle their vehicles in prescribed maneuvers to teach certain operational skills through repetitive practice.

One product that resulted in a train-the-trainer program was for a *new* type of product, a set of sit-on skis, used for a *hazardous* activity, downhill skiing, and by *at-risk* population, paraplegics and tetraplegics. The *TetraSki* ("The TetraSki Project" 2019) is a product permitting people with those or other disabilities to enjoy downhill skiing with some degree of autonomy. The TetraSki is shown in Figure 9.17. Although there is generally a "partner" with whom the TetraSki operator rides, the operator is able to steer the product through either a joystick or a straw-like device connected to a "sip-and-puff" steering controller operated through a straw (Figure 9.18) and given commands by the operator through either sucking on or blowing into the straw. This controller is programmed to respond to either joystick commands or to the positive and negative gauge-pressure straw signals from the operator to steer the TetraSki.

FIGURE 9.17 TetraSki in operation. (Photograph used by permission of Tetradapt. Photographed by Lee Cohen.)

Due to the uniqueness of this product and to the vulnerability of its user population, the proper training of users through the proper training of user instructors was a top value of the rehabilitation and adaptive-recreation engineers, physicians, and researchers involved in the TetraSki project.[37] Trainers are taught to consider the unique abilities of each user during training.

The first case study of innovative facilitators was that for handguns. The replica gun shown in the case study permits the realistic classroom training of handgun users without any risk of injury from any type of firearm discharge. Although firearms contain a well-known hazard, the severity of the injury possible from a negligent discharge as well as the reaction that a real handgun may provoke in some settings—makes the use of a replica handgun a wise choice in many cases. The mere substitution of a "prop" for a real handgun should help reinforce to students the hazard and risks of handling firearms.

Another good use of an innovative facilitator is when a user is faced with a unique task—one for which the user is unskilled—and time is limited. This was illustrated in the second innovative-facilitator case study. The demonstration device shown for the auto-injector is a wise application of a facilitator for patient-safety reasons. The auto-injector is a unique product only intended to be used in emergency situations. Indeed, it is hoped that all people carrying an auto-injection device never have to actually use it. The potential user of the auto-injector is given the opportunity to practice self-injections so that she or he can perform the task properly if the need arises which will also mean

[37] Technical papers on the TetraSki have not yet been written.

Figure 9.18 TetraSki *sip & puff* controller. (Photograph used by permission of Tetradapt. Photographed by Lee Cohen.)

that a rapid response will be essential. Although potential users of auto-injectors should have already familiarized themselves with the use of the product before any event necessitating its use, the instructions are repeated on the product and practice device for ready reference. It takes little time to re-read the simple instructions, but the user should have already developed a "feel" for the self-injection procedure.

The third case study of an innovative facilitator is that for a unique product for a hazardous activity by at-risk people. This unique skiing product places those who may already have suffered an injury to engage in a fun, but potentially risky, activity on a new type of product despite any disabilities. A train-the-trainer program helps ensure that the product users learn how to properly and safely use it from well-trained instructors. The references for this example are good examples of how a training program can be constructed and conducted to provide trainers and training, with education, for new skills needed to properly and safely use certain products.

The proliferation of affordable technology has opened the door to other innovative forms of product-safety facilitation. Many modern products have MFD panels that either indicate the state or status of the product to users, or help the user operate the product. For example, a coffeemaker may display its current state through a displayed message including "Warming up" or "Ready to brew." For some products, this may be a new avenue to reinforce[38] the safe use of a product to its users. Many automobile and automotive systems, such as GPS systems, already give their users safety guidance upon start up. Maybe more products should consider this use of the MFD. Yet, the same message is usually always provided. Since the users have already seen this safety "splash screen" many times, they may tend to ignore it and its worthwhile message. If new messages were "rotated" through the vehicle or device start-up process, perhaps users would pay more attention to them.

Such technology is no longer limited to more-expensive systems such as automobiles. More and more products are now "connected" via the internet. It is now possible for a homeowner to easily control a home's thermostat, activate its lighting, and monitor its security systems remotely through a computer or a smart-phone "app," or application. This *Internet of Things* (IoT) opens a multitude of new possibilities and conveniences for average citizens. However, the IoT also brings with it a variety of new risks, some of which involve home security and product safety. It may be possible for a consumer to remotely and accidentally activate an appliance at home that could lead to an unattended house fire. These problems, and more, will face product-safety engineers at an ever-increasing rate in the next decade. Fortunately, there is at least one standard under development for such "connected products" ("Subcommittee F15.75 on Connected Products" 2019) for consumers. At this point, this effort is still at an early stage but should provide some baseline guidance for these widely varying types of products.

Those responsible for product-safety engineering should continue to look for new ways, even if they are non-traditional, to help consumers use their products safely. It is hoped that this section has given design and product-safety engineers ideas of what can be accomplished when innovative facilitators are applied to unique products.

9.10 Conclusion

Those involved with product design and product-safety engineering should see that developing the product-safety facilitators covered in this chapter are a necessary and a vital part of integrated product design. The *design* of these materials should not be delayed until the product is complete and everyone is tired of thinking about the product. These facilitators can sometimes provide an aspect of overall product safety that the physical engineering design alone cannot deliver despite great efforts and good intentions. This construction of a comprehensive product-safety system is an important part of *designing safety into* a product through DfPS. However, the use of such product-safety supporting materials should never serve as a substitute to further risk reduction attainable through additional and feasible engineering-design effort.

Each element of product-safety facilitation discussed in this chapter should together form a cohesive *system* of product-safety education/instruction and warnings to users of that product. Therefore, these materials must be harmonious and should never contradict to one another.

---

[38] Such MFD messages should only remind the user of information contained elsewhere, such as in warning signs or manuals. They should never provide new information to the user that has not already been provided.

Facilitators must:

- Be consistent with:
    - Other facilitators
    - Product intended use
    - Instructions for product use (IFU)
    - Limitations of product use
- Demonstrate proper use/operation of product—and accessories
- Be communicative
- Be as effective as practicality and feasibility permit

Not all products are alike; therefore, it stands to reason that not all facilitators of these different products will look alike. Dare to consider applying new approaches to product-safety facilitation when a new product needs it—or when engineering ethics demands it. When designing a system of product-safety facilitators, it should be remembered that the goal of the work is not to produce *compliant* warning signs and manuals. The goal should be to produce *effective* warning signs and manuals that truly reduce product-use risks when engineering design can go no further.

The design of this product-safety facilitator system is best done concurrently with other engineering design. Should an engineering-design team wait until the end of a product-development project to write the manuals, attach warning signs, and do other tasks, simple hazards and risks may be overlooked or not addressed as well as possible.

It is easy for product-safety professionals to adopt an attitude of despair when considering the design of facilitators such as warning signs and manuals. Commerce is replete with horrible examples of both. However, it is hoped that readers instead take a fresh and active approach to the design of important product-safety facilitators to find new ways to protect product users from injury.

References

ATV RiderCourse Handbook. 2004. Irvine, CA: ATV Safety Institute.

Bailey, Robert W. 1982. *Human Performance Engineering: A Guide for System Designers*. First. Englewood Cliffs, NJ: Prentice Hall.

———. 1996. *Human Peformance Engineering: Designing High Quality, Professional User Interfaces for Computer Products, Applications, and Systems*. Third. Englewood Cliffs, NJ: Prentice Hall.

Barnett, Ralph L. 1994. "The Principle of Uniform Safety." *Triodyne Safety Brief* 10 (1). http://www.triodyne.com/SAFETY~1/SB_V10N1.PDF.

———. 2001. "Safety Potpourri." *Triodyne Safety Brief* 17 (4). http://www.triodyne.com/SAFETY~1/Sb_v17n4.pdf.

Barnett, Ralph L., and Dennis B. Brickman. 1985. "Safety Hierarchy." *Triodyne Safety Brief* 3 (2). http://www.triodyne.com/SAFETY~1/SB_V3N2.PDF.

"Barrett .50 Cal Won't Fire. So These Marines Called Customer Service. During a Firefight." 2017. Tribunist.Com. 2017. https://tribunist.com/military/

barrett-50-cal-wont-work-so-these-marines-called-customer-service-during-a-fire-fight-video/?utm_source=LRD.

Basic RiderCourse RiderCoach Guide. 2001. First. Irvine, CA: Motorcycle Safety Foundation (MSF).

Benbadis, Alexandra, Kimmy Ansems, Linda Giesselink, and Benjamin Hannon. 2020. "Trends and Leading Practices for Developing Autoinjector Instructions for Use." *AAMI Biomedical Instrumentation & Technology*, July 2020. https://lnkd.in/e8t42kf.

"Bicycle Owner's Manual." 2018. Wilton, CT: Cycling Sports Group, Inc. https://www.cannondale.com/~/media/Files/PDF/Dorel/Cannondale/Common/owners-manuals/cannondale/MY19/131264 Rev 0718 CD OM Bicycle Owners Manual.ashx.

Bridger, Robert S. 2018. *Introduction to Human Factors and Ergonomics*. Fourth. Boca Raton, FL: CRC Press. https://www.crcpress.com/Introduction-to-Human-Factors-and-Ergonomics/Bridger/p/book/9781498795944.

Dongarra, Jack J. (Argonne National Laboratory), Cleve B. (University of New Mexico) Moler, James R. (University of California-San Diego) Bunch, and G.W. (University of Maryland) Stewart. 1979. *LINKPACK User's Guide*. Philadelphia, PA: SIAM. https://doi.org/https://doi.org/10.1137/1.9781611971811.

"Final Consent Decree." 1988. Washington, DC: U.S. District Court (DC).

Green, Joey, Tony Dierckins, and Tim Nyberg. 1998. *The Warning Label Book*. New York, NY: St. Martin's Griffin.

Hales, Crispin, John G. Goebelbecker, and Kenneth L. d'Entremont. 1996. *Hazardous Materials Incident Prevention Manual: A Guide to Countermeasures*. First edition. Washington, DC: US DOT/FHWA.

Henning, Torsten. 2006. "No Title." Commons.Wikimedia.Org. 2006. https://commons.wikimedia.org/wiki/File:DIN_4844-2_Warnung_vor_gef_el_Spannung_D-W008.svg.

Hunt, Todd. 2002. *Pardon Me, But That's a Really Stupid Sign*. Chicago, IL: The Hunt Company. toddhuntspeaker.com.

In Compliance. 2016. "Languages and Your Product Safety Labels," December 2016. https://incompliancemag.com/article/languages-and-your-product-safety-labels/.

Jensen, Roger C. 2020. *Risk-Reduction Methods for Occupational Safety and Health*. Second. Hoboken, NJ: John Wiley & Sons. https://www.vitalsource.com/products/risk-reduction-methods-for-occupational-safety-and-roger-c-jensen-v9781119493976?term=roger+jensen.

Jensen, Roger C., and Erin Jenrich. 2008. "Formats for Section Safety Messages in Printed Manuals." *Proceedings of the Human Factors and Ergonomics Society2*, 2008.

Jensen, Roger C., and Andrew W. McCammack. 2003. "Severity Message from Hazards Alert Symbol on Caution Signs." In *Human Factors and Ergonomics Society*, 17671771. Santa Monica, CA: Human Factors and Ergonomics Society.

Kraft, Heidi, Russ Peeler, and Jerry Larson. 2010. *The Docs*. San Diego, CA: Naval Health Research Center. https://www.med.navy.mil/sites/nhrc/Site PDFs/The_Docs_(sm).pdf.

Mohan, Geoffrey. 2020. "You See the Warnings Everywhere. But Does Prop. 65 Really Protect You?" Los Angeles Times. 2020. https://www.latimes.com/business/story/2020-07-23/prop-65-product-warnings.

NEMA. 2017a. "ANSI Z535.4-2011: American National Standard: Product Safety Signs and Labels (R2017)." Rosslyn, VA: ANSI.

———. 2017b. "ANSI Z535.6-2011: American National Standard: Product Safety Information in Product Manuals, Instructions and Other Collateral Materials (R2017)." Rosslyn, VA: ANSI.

Newton, Isaac. 1687. *The Principia (Philosophiæ Naturalis Principia Mathematica).*

Peacock, Brian, and Lila Laux. 2005. "'Warning: Do Not Use While Sleeping'-The Role of Facilitators." *Ergonomics in Design*, 2005. https://journals.sagepub.com/doi/10.1177/106480460501300403.

Portable Generators; Final Rule; Labeling Requirements (Correction). 2007. USA: Federal Register. https://www.cpsc.gov/Regulations-Laws--Standards/Rulemaking/Final-and-Proposed-Rules/Portable-Generator-Labels.

Poulton, E. C. 1967. "Searching for Newspaper Headlines Printed in Capitals or Lower-Case Letters." *Journal of Applied Psychology* 5 (5): 417–25. https://doi.org/https://doi.org/10.1037/h0025098.

Ross, Kenneth. 2005. "More Guidance for Warnings and Instructions." *For the Defense*, December 2005.

———. 2011. "Warnings and Instructions: Updated U.S. Standards and Global Requirements." Chicago, IL: Defense Research Institute, Inc. http://www.productliabilityprevention.com/plparticles.html.

———. 2019. "The Risk of Overwarning." *In-House Defense Quarterly*, 2019.

"Subcommittee F15.75 on Connected Products." 2019. Astm.Org. 2019. https://www.astm.org/COMMIT/SUBCOMMIT/F1575.htm.

Tillman, Barry, David J. Fitts, Wesley E. Woodson, Rhonda Rose-Sundholm, and Peggy Tillman. 2016. *Human Factors and Ergonomics Design Handbook.* Third edition. New York, NY: McGraw-Hill.

The M16A1 Operation and Preventive Maintenance Manual. 1969. 1 July 196. Washington, DC: U.S. Government Printing Office.

"The TetraSki Project." 2019. Www.Tetradapt.Us. 2019. https://www.tetradapt.us/tetraski-adaptive-skiing.

Uzgiris, S. Carl, Crispin Hales, and Michael A. Dilich. 1997. *Commercial Vehicle Preventable Accident Manual: A Guide to Countermeasures.* Edited by Kenneth L. d'Entremont. Third. Washington, DC: US DOT/FHWA.

CHAPTER 10

Product-Safety Engineering Methods

You cannot expect people to stop acting like people.
—ROBERT S. ADLER[1]

10.1 Introduction

The above utterance should never be forgotten by the product-design engineer and others involved in the safety of consumer-product users—including company managers and executives. In a few simple words, the speaker, the Acting Chairman of the US CPSC, distilled one important—but sometimes forgotten—reality into a simple, common-sense statement that should be kept in mind by those involved in Designing for Product Safety (DfPS). Despite whatever guards, safeguards, and other engineered characteristics or communicative devices that are designed into a product, people will—to a large degree—continue being people and continue doing "people things." Such people things are not always recommended or wise. This should be remembered by engineers throughout the design process.

This flaw among *homo sapiens* may sometimes mean that humans do not always act in their own self-interests, even when the benefits are known. This character flaw of the species is not limited to product safety. It has been observed by others studying peoples' actions with respect to wealth, health, public policy, and education (Thaler and Sunstein 2008; Halpern 2015).

In spite of all of this—or perhaps because of it—design and product-safety engineers should continuously work towards providing reasonably safe products for the consuming public. These products should be ones that the engineers and other personnel of a design-and-manufacturing company would unreservedly use themselves—and also have their family and friends use. Humans are not always rational beings. Yet, engineered-product designs should anticipate this and reduce risks to users to the extent feasible.[2]

As the end of a design project approaches, the design engineers should ask themselves: "Would I want my loved ones using this product?" Not only is the answer itself

[1] "United for Senior Safety." Keynote presentation, ICPHSO Annual Meeting, Washington, DC, February 28, 2019.

[2] What is "feasible" is sometimes as hotly debated as what is "safe enough."

significant, but also telling is the amount of time an engineer might take to answer the question. If it took more than one second to respond "yes," then the engineer may have an inner ethical conflict about the product that might require the reevaluation of that product's design.

Previous chapters have addressed numerous concepts including product safety, engineering ethics, hazard, and risk. Also covered have been engineering-design considerations, sources for design guidance, and product-safety facilitators. An overall product-design process (PDP) along with the role and integration of the product-safety function and engineer to operate within the PDP have been already covered. What will now be presented are risk-management engineering methods for identifying, evaluating, and controlling hazards and their risks during the Engineering Design, Development, and Testing Phase (EDDTP) of the PDP. These are sometimes called hazard-analysis or safety-analysis techniques by some authors (Ericson 2016; Gullo and Dixon 2017). Most of these methods already exist in the literature; however, one tool, the Product-Safety Matrix (PSMx), is original to this book. Examples are used to illustrate the methods and some of these examples will include the hand-powered winch, or simply "winch," described in Appendix B.

10.2 Preliminary Hazards List

The preliminary hazards list (PHL) is the most basic of methods to use when approaching the task of hazards and risk assessment and reduction. It is a *primary* risk-assessment method (Ericson 2016) since it is so basic that it will likely need follow-up risk assessments using more-advanced methods. As its title indicates, the PHL is essentially a listing of either known, suspected, or possible hazards presented by a product being designed. The product-safety engineer (PSEr), or the design engineer providing the product-safety function, studies the proposed product—along with its intended use and its early design specifications—to see what hazards are, or could be, embodied by the product design elements. Many engineers know, from prior experience, the primary types of risks that may lead to accidents and injuries. This is especially true if the product is not an altogether new product, but instead a product that is derived from another of the company's products. The more an engineer knows about the construction (design) and use (hazards and risks) of the product being studied, the more accurate and useful the resulting PHL will be. It may be helpful for the engineer to consult a list of possible hazards, such as that found in Appendix A, and include those hazards that may have been overlooked or may only be considered secondary—or less prominent—in nature. It may also be beneficial to include hazards that exist, but might have also been sufficiently addressed already through preliminary engineering design or through discussions. It is frequently useful to cast a "big net" when considering initial product hazards. Unrealized hazards can always be disregarded at a later stage in engineering design. This inclusion of many hazards adds depth to the risk assessment and also demonstrates to observers that the engineering staff has addressed a multitude of hazards and is being conscientious and conservative in its design process. It may also serve as a reminder to reconsider a particular hazard, already addressed, if the design changes and product-design elements evolve or change drastically. Historical company and industry materials should also be scoured to see what types of issues have arisen during the use of this or similar products. These materials may include "post mortems" or "lessons learned" from prior marginal-to-poor engineering-design efforts, product-safety recall information from that company and its competitors, in-house survey or focus-group data, and any other sources of information, such as websites, that may provide insight into what to expect once the new product reaches consumers and begins operation.

Because the PHL is an elementary risk-assessment technique, it can be used early in the PDP. It is suitable for use in Phase II, the Concept Selection, and early in Phase III, the EDDTP of the PDP. The PHL alone will likely be insufficient to serve as the sole risk-assessment technique for a complex, new product. That point notwithstanding, the PHL can serve as an important initial document for a product-design program, especially if the program is ambitious and not a typical product for the company. The PHL can help identify potential problems areas for new products even only being considered. These potential problem areas may call for a concerted plan of action to address a particular concern that may arise from an innovative new product. The PHL is a simple and quick method of looking at a proposed product early in the PDP.

A worksheet that may be used for the preliminary hazards list is shown in Table 10.1. In addition to the title block which identifies the project, engineer, date, and page, the PHL worksheet contains columns for a number, component or sub-system, hazard, cause, effect(s), and notes.

Consider a two-slice toaster, such as the one shown in Figure 4.12. Assembling a PHL for this toaster may produce something similar to the PHL shown in Table 10.2. In this *partial* PHL example, three product design elements have been identified. These three isolated design elements are the heating elements, electrical components, and the heating controls. They are numbered as 1, 2, and 3, respectively. The hazards related to the three product-design elements are thermal, electrical, and thermal.

Preliminary Hazard List (PHL)					
	Project: ______________				Date: ___/___/___
	Engineer: ______________				Page: _____ of _____
Number	**Product Design Element**	**Hazard**	**Cause**	**Effect(s)**	**Notes**

Table 10.1 Worksheet for Preliminary Hazard List (PHL)

Preliminary Hazard List (PHL)					
	Project: _Two-Slice Toaster_____________			Date: 04 / 01 / 2020	
	Engineer: _J. Smith_____________________			Page: 1 of 1	
Number	**Product Design Element**	**Hazard**	**Cause**	**Effect(s)**	**Notes**
1	Electric-Heating Elements	Thermal	Electrical resistive heating	Injury or death from fire	High temperatures could lead to burn injuries or melting of heat-sensitive parts of product and fires
2	Electrical Components	Electrical	Unintended current flow	Injury, or death from electric shock	Components include heating elements, electrical cord, and internal wiring. Consumers could be seriously injured or killed by uncontrolled electrical-current flow. Consumer may attempt to extract toast near energized heating elements
3	Control System	Thermal	Toaster may not shut off electricity	Injury or death from fire	Stuck bread may prevent toast ejection and toaster shut-off
...	...	...	...	...	...

TABLE 10.2 Preliminary Hazard List for Toaster

Row 1 of the PHL shows that the cause of the thermal hazard is the resistive heating of the toaster elements. The first hazard is used by the toaster to perform its intended function; therefore, the electrical hazard cannot be eliminated. Because of the high temperature needed to toast bread, there is the possibility of property damage, injury, or even death in extreme circumstances. The notes for the *first* row speak of injury and potential fires. In the case of the electrical hazard, the *second* row, an unintended current flow could result in injury from electric shock or death from electrocution. The need for a high-temperature environment to perform the toasting function requires an energy source and electricity is that source of energy. It is doubtful that another, safer energy source is practical for widespread use. The *third* row shown in the table relates to the toaster's control system which ejects the toast and de-energizes the toaster's heating elements. Its hazard, in this example, is thermal in nature. If the toast-ejector mechanism fails to function for whatever reason, the continued electrical heating of the toaster and toast can lead to a toaster fire developing into a house fire resulting in potential injury or death. If, when the toast is stuck, a consumer attempts to manually extract stuck toast from the toaster without first unplugging the product, then that user may face shock or even electrocution and this points back to the second row of the table, the electrical hazard. Certainly, this short list does not contain all hazards, causes, and effects affecting the operation of a toaster. This is one simple partial example of a PHL.

As a second example, the hand-operated boat-trailer winch is considered. This product is described in detail in Appendix B. This simple, mechanical product will be used throughout this chapter in order to demonstrate the development of deeper and broader risk-assessment and risk-reduction methods. The, again, simple PHL of the winch is included as Table 10.3. The end of the appendix contains an intended-use statement as well as limitations for the use of the winch. Appendix B should be read and understood before proceeding.

Only three product design elements are considered in this elementary example. These are the spool, the strap-and-spool combination, and the transportation of a load on the trailer. With the *first* design element, the risk is receiving a cut from the sharp edge, or a burr, on the edge of one of the winch's spool halves. This would likely only be a minor injury. The *second* design element, the strap and spool, presents an in-running nip point to users of the boat winch. Since this winch is hand-powered and is also mechanically straightforward, if a user's fingers or hand become entrapped by the in-running nip, then the user will simply stop cranking the handle after sensing the condition. There is no electrical control system to power the winch since it is a manual device. In addition, the spool rotates quite slowly, so time should be available before injury occurs. The *third* and final hazard is that of impact should the boat be released by the winch, strap, or hook. Either a person or a vehicle on the road carrying people could be hit by a boat in motion. There are several scenarios that could lead to serious injury or death from a loose boat. There are, of course, more hazards to consider during the actual design of an actual real-world winch.

It is noteworthy that the results of the simple PHL may be used as a basis for additional, deeper risk assessments of a product as the EDDTP continues. Because the PHL is an elementary exercise, it might be sometimes useful to include non-technical program management into the PHL discussion to illustrate the obstacles that an ambitious engineering design and development program could face. Such a discussion could also be useful for mapping out a course of action to address any foreseen product-safety challenges for the program.

Since the PHL is such a simple task, the PSEr and others may accidentally stray from being systematic in their approach. Consistency and discipline, with even the elementary PHL, are crucial for maximum effect. Consult checklists of hazards, such as Appendix A, and consider all aspects of the product—its design, its users, and its operating environments—even if the design or concept is still being formed.

10.3 Preliminary Hazards Analysis

The preliminary hazards analysis (PHA) is the next logical step beyond the PHL but remains a *primary* risk-assessment method. Unlike the PHL, the PHA goes beyond the mere listing of hazards and begins to assess the risks of hazards that are presented by product design elements of the product undergoing design and development. The PHA can also include specific risk-mitigation measures, or actions which should reduce the risk of the product to its users. Ericson claims that the PHA is "probably the most commonly performed hazard analysis technique" (Ericson 2016, 126). This is a reasonable assertion given its utility and its simple inputs.

The PHA takes the PHL and adds to it several aspects for consideration and analysis. Therefore, the output from a PHL often serves as the starting point for a PHA. Table 10.4 shows a worksheet that may be used for the PHA. The first five columns in both the PHL and PHA are identical.

Preliminary Hazard List (PHL)					
	Project: Hand-Powered Winch				Date: 04 / 01 / 2020
	Engineer: J. Smith				Page: 1 of 1
Number	**Product Design Element**	**Hazard**	**Cause**	**Effect(s)**	**Notes**
1	Spool	Laceration	Sharp edge	Cut on finger	Minor injury
2	Strap and Spool	In-Running Nip Point	Fingers/hand caught in strap	Crushed finger(s)	Hand-powered system; user will stop cranking
3	Transported Load	Impact	Contact with moving load	Numerous	Moving load can injure in several ways
…	…	…	…	…	…

TABLE 10.3 Preliminary Hazard List (PHL) for Winch

Preliminary Hazard Analysis (PHA)													
	Project: ____________											Date: ___/___/___	
	Engineer: ____________											Page: 1 of 1	
					Initial				Revised				
Number	Product Design Element	Hazard	Cause	Effect(s)	S	P	RISK	Recommended Action(s)	S	P	RISK	Notes	Status

TABLE 10.4 Worksheet for Preliminary Hazard Analysis (PHA)

The reader will also notice the addition of several columns to the PHA when compared with the PHL. The next three PHA columns following "Effect(s)" contain the risk estimate of the product as the engineering design currently stands, or the *initial* risk estimate. These columns contain the severity, probability, and resulting risk level as assigned by the PSEr or the engineering-design team. The risk criteria that are used here are those from Tables 5.2, 5.3, and 5.4. The initial risk-estimate column is followed by recommended action(s) for the engineering-design team to reduce the initial product risk. This column is followed by a set of three more columns which will contain the *revised* risk estimate. This revised risk estimate represents the risk's new severity, probability, and level to which taking the recommended action(s) will reduce risk. Of course, this "revised" product does not exist; however, *engineering judgment* [3] is generally employed to create a mental picture of the redesigned product for conceptual assessment. Of course, this redesign must always be reevaluated, later in the EDDTP, once the recommended engineering-design changes been implemented to see if the redesign's expectations agree with its reality.

The final columns of the PHA are for the PSEr's notes as well as a status indicator for the design effort. The status is important because even though a product may still be in design, a particular aspect of that design possibly affecting product-safety may no longer be under development. It is important for the PSEr to understand that a design element may no longer see any improvement and to be part of the decision to close that design task if appropriate. This PHA layout also permits the PSEr to update this document as the product proceeds within the EDDTP, for example, from Design Review 1 to Design Review 2.

The example of the two-slice toaster from the PHL section is continued in a PHA. Consequently, the example from the PHL—numbers, product design elements, hazards, causes, and effect(s) are carried over into a PHA example. This PHA form is shown in Table 10.5. The reader will see that the initial columns of the new table are identical to those in Table 10.2. The same three hazards appear in the rows: thermal, electrical, and thermal. The cause and effect(s) also remain unchanged for each hazard.

The initial risk estimate columns come next after the "Effect(s)" column. For hazard one, thermal, the risk severity is considered to be Extreme, or 4, while its probability is considered to be Occasional, or 3. These severity and risk values are listed in the electric-heating element row as "4-X" and "3-O," respectively. The resulting risk estimate, from Table 5.4, is "Severe" which is also listed on that row of the table. The reason that these risk severity and probability values are written as "4-X" and "3-O," respectively—with the number *before* the letter—is to permit the rapid visual scanning of a large PHA in order to identify the largest risk severities and probabilities. It is easier to mentally rank order numbers than it is to do the same with letter abbreviations. The letters representing the severity and probability values are retained to help remember those categories, for example, "X" denotes extreme and "O" denotes occasional.

For the second hazard, electrical, the corresponding severity and probability levels are "3-C" and "3-O," for "Critical" and "Occasional," respectively, producing a risk

[3] "Engineering judgment" is the technical equivalent to "common sense." Authors and readers, alike, dislike using the term "engineering judgment" due to its subjectivity and its frequent use in avoiding difficult discussions. It is hoped that applied engineering ethics and rational thought play pivotal roles in making such engineering judgments.

Preliminary Hazard Analysis (PHA)													
	Project: Two-Slice Toaster											Date: 04 / 01 / 2020	
	Engineer: J. Smith											Page: 1 of 1	
					Initial				Revised				
Number	Product Design Element	Hazard	Cause	Effect(s)	S	P	RISK	Recommended Action(s)	S	P	RISK	Notes	Status
1	Electric-Heating Elements	Thermal	Electrical resistive heating	Injury or death from fire	*4-X*	*3-O*	*Severe*	Thermally insulate the heating elements; use heat-resistant materials; comply with industry standards; laboratory & field testing; competitive benchmarking	*4-X*	*1-I*	*Medium*	High temperatures could lead to burn injuries, melting of heat-sensitive parts of product leading to fires	Open
2	Electrical Components	Electrical	Unintended current flow	Injury, or death from electric shock	*3-C*	*3-O*	*High*	Use polarized electrical plug to ground elements and chassis to Neutral; instruct users to only use with GFCI-protected outlet, in dry conditions, and undamaged power cord	*3-C*	*2-R*	*Medium*	Consumers be injured, but likely not killed, by uncontrolled electrical-current flow	Open
3	Control System	Thermal	Toaster may not shut off electricity	Injury or death from fire	*4-X*	*3-O*	*Severe*	Instruct consumers on proper use including maximum bread thickness and to unplug toaster before removing stuck toast	*4-X*	*1-I*	*Medium*	Continued heating may cause toast to burn inside of toaster—toaster or house fire could result; Consumer may attempt to extract toast near energized heating elements—electrocution could result	Open
…	…	…	…	…				…				…	…

Table 10.5 Preliminary Hazard Analysis for Toaster

level of "High." The third hazard, thermal, has a risk with severity and probability of "4-X," "Extreme," "3-O," and "Occasional," respectively, resulting in a risk level of "Severe."

Due to the elevated initial risk levels, there are recommendations made to reduce the probabilities for each of the risks. There is nothing that can be done, in this example, to reduce the severity of some risks, however. The use of additional insulation is among the suggestions for the first hazard. Recommended actions also include compliance with any industry standards, both laboratory and field testing, along with the benchmarking of competitor's products to see how and how well they accomplish this task. The expected revised risk is reduced through these actions and by way of a lowered probability to "Medium" for the first thermal hazard. Similarly, risk estimates are expected to be reduced for both the second and third hazards, using the respective recommended actions, to "Medium" through lowered risk probabilities.

Engineering observations are included in the "Notes" column. The final column in the table shows that each of these risks is still being actively pursued by engineering-design teams.

Returning to the winch example already discussed, the PHA for this product is shown in Table 10.6. The same three hazards, causes, and effect(s) are reproduced from Table 10.3. Added to the PHL are the columns containing the initial risk estimates. The first winch risk, laceration, is determined have a severity of "Nominal" and a probability of "Occasional." This produces a risk estimate of "Low." The second risk, crushed fingers has a severity of "Marginal" and a probability of "Remote." This combination of severity and probability results in a risk level of "Medium." The third hazard, injury from impact, has a severity and probability pair of "Extreme" and "Remote," respectively. This produces a risk estimate of "High."

Recommended engineering-design actions to lower risk include having suppliers de-burr the winch components before shipment to the OEM (Original Equipment Manufacturer)—the design-and-manufacturing company. Effectively doing so should decrease the probability of injury to "Remote." This probability reduction results in a risk level that remains "Low." Even though the initial and revised risk levels are both "Low," the post-recommendation probability is lower than the initial risk probability. The risk is, therefore, lower after engineering-design changes although both severity-probability pairs produce a "Low" risk level.

The probability of the second winch risk has its probability lowered from "Remote" to "Improbable" through instructions and warning signs. Despite this reduction in probability, the risk estimate remains "Medium" although risk reduction was actually accomplished through the recommended design changes.

The third initial winch risk estimate, being injured by a loose boat, is reduced through similar instructions and warning signs, from "Remote" to "Improbable" to deliver a risk-level reduction from "High" to "Medium." In none of these winch cases was it possible to reduce the severity of any of the risks. This risk-severity situation is common in engineering-design practice, but design and product-safety engineers should never stop looking for ways to reduce risk severity.

By comparing the PHL to the PHA, the reader can see that the PHA produces information and product-safety engineering-design guidance that the PHL simply cannot provide. The PHA provides a risk estimation and permits the PSEr, and others on the engineering-design team, to suggest design revisions or alternative approaches to a

Preliminary Hazard Analysis (PHA)													
	Project: Hand-Powered Winch											Date: 04 / 01 / 2020	
	Engineer: J. Smith											Page: 1 of 1	
					Initial				Revised				
Number	Product Design Element	Hazard	Cause	Effect(s)	S	P	RISK	Recommended Action(s)	S	P	RISK	Notes	Status
1	Spool	Laceration	Sharp edge	Cut on finger	*1-N*	*3-O*	*Low*	Have supplier de-burr edges of spool halves	*1-N*	*2-R*	*Low*	N/A	Open
2	Strap and Spool	In-Running Nip Point	Fingers/ hand caught in strap	Crushed finger(s)	*2-M*	*2-R*	*Medium*	Instruct winch users to keep fingers and hands away from winch strap in Instruction leaflet, Owner's Manual and packaging	*2-M*	*1-I*	*Medium*	Hand-powered system; user will stop cranking	Open
3	Transported Load	Impact	Contact with moving load	Numerous	*4-X*	*2-R*	*High*	Instruct winch users to secure boat with cargo straps or ropes in Instruction leaflet, Owner's Manual, and packaging	*4-X*	*1-I*	*Medium*	Moving load can injure in several ways; winch not intended to secure load during transport	Open
...	...	...	...	...				...				...	...

Table 10.6 Preliminary Hazard Analysis for Winch

product-design problem and then estimate the risk reduction that might accompany that change. One drawback of the PHA, when contrasted to the PHL, is that the PHA requires more than simply a conceptual design in order to be useful. An engineered product must have somewhat of a detailed design in order to apply the PHA method. Therefore, each method does have its advantages depending upon the stage of the product's engineering design. The PHL and PHA may also serve to complement some of the other risk-management tools in this chapter.

In the field of system safety, the PHL and PHA *alone* do not provide the tracking of the system of the engineered safety countermeasures used to provide acceptable risk to users of a product. There do exist such methods for tracking issues within the system-safety operation, for example the Hazard-Tracking System in MIL-STD-882E (USA/DoD 2012). However, these system-safety elements may not exist outside of an organization doing military or aerospace projects of large technical and financial magnitude. In many such cases, these types of system-safety measures are routinely expected, or even contractually required, by a demanding customer such as the U.S. Department of Defense or NASA (National Aeronautics and Space Administration). However, in the *lean*[4] engineering staffs found at some successful companies in fast-paced and competitive industries, resources—both human and time—are scarce. Engineers should be happy, or at the very least content, with the results of their work—especially when product safety is involved. Design engineers and PSErs should be given the resources and management support that engineers need to deliver the products expected by consumers. Lean consultants speak of cutting the "fat" from bloated processes; however, at times, the cutting removes "muscle" and "bone." Enlightened engineering management will discern that, while lean is a *priority*, product safety is, or should be, a *value*. As stated at the beginning of the book, priorities pass with time leaving true values, and their consequences behind.

10.4 Product-Safety Matrix

An original product-safety engineering tool is presented in this section. The author has, during an industrial career, successfully tracked the development of many real-world products in widespread use today based on a worksheet, or matrix, methodology. This has since been developed into the following methodology. The initial method had been used during the engineering-product development phases of designing both incrementally improved as well as unique and innovative products. This is a simple and logical next step from the PHA method but has benefits to an engineer tracking the product safety of a real-world product within the EDDTP of a PDP.

This matrix may be used and produced at instances when a product-safety review is conducted. Yet, the matrix is more than that. It is a tool that *develops* as the product itself *develops*. The matrix permits the PSEr to design a system of product design elements as well as a system of product-safety facilitators—including manuals, warning signs, and website material—to increase the level of safety provided to the ultimate consumer of the product. This method permits as much safety to be *designed into* the product as practical and also provides a systematic and comprehensive set of facilitators to further

[4] Lean is a process of continuous improvement to accomplish more with less and produce minimal waste.

enhance the product's safety properties. For lack of a better term, this tool is simply called the product-safety matrix or PSMx.

The PSMx consists of two components. The first component is product safety; the second, product compliance. Although compliance is a separate characteristic than safety is, in reality, a design engineer or PSEr may well be responsible for both—at least as far as *safety compliance* is concerned. As will be shown, a company can produce a fully compliant product with respect to standards, regulations, and agreements, but still have a product with unacceptable levels of risk. That is why readers should not depend upon compliance documents to provide useful product-design guidance and performance criteria.[5] These two sections will be called the "safety PSMx" and the "compliance PSMx" for ease of discussion.

Before discussing the PSMx further, it is necessary to state that—as with many of the risk-management methods shown in this chapter—there is not a unique and correct way to construct a PSMx. Different people and teams will dive to different depths—and perhaps take different routes—to arrive at similar good results.

The PSMx will be presented in its two parts, each with a table showing the particular format for its use. The PSEr, or another design engineer, will be responsible for tracking the product through the PDP and EDDPT. The design of the PSMx permits that engineer to follow the progress of a product through its various design-and-development phases. The discussion will begin with the safety PSMx and be followed by the compliance PSMx.

The safety PSMx is shown in Table 10.7. The title block is similar to that seen in the earlier PHL and PHA examples. The first nine columns of the safety PSMx are almost identical to the PHA. The exception is the safety PSMx's third column, which is inserted to denote if a hazard exists. This may sound strange at first. After all, why would a hazard be included if it does not exist? The reason for the inclusion of this column is to permit the future consideration of many hazards during all stages of product design so that, if the early product design changes significantly and now presents a hazard which has not previously been considered, then the construction of the safety PSMx now can bring the new hazard to the attention of the design team. Therefore, there is a justifiable reason for inserting this column into the safety PSMx. Furthermore, if the design—several years in the future—comes under intense scrutiny, then the inclusion of such supplemental hazards for design consideration can help indicate the thoroughness of the engineering-design team's efforts.[6]

After a product design element has been identified and numbered, the same hazard, cause, and effect(s) columns are filled in as was done in the PHA. Similarly, the initial risk estimates using severity, probability, and risk level—from Tables 5.2 through 5.4—are determined. The next column shows the divergence between the PHA and safety PSMx.

---

[5] At the same time, regulators, advocates, and attorneys should avoid unnecessarily proliferating product criteria through standards, regulations, and agreements in order to *force* improvements to product safety. Consumer-product engineering teams typically do not *design to a specification* so that their product will only minimally satisfy its requirement. Instead, engineers often design a product as they believe that it should be designed and, only then, *verify* that the product indeed meets the requirements imposed upon it.

[6] Product-safety and design engineers must, of course, take full advantage of the capabilities within the PSMx by truly considering these potential hazards.

	Product:										
	Model / Description:										
	Unit Reviewed:			(ID, SN, VIN, other Identifer)							
	Product-Development Stage:										
	Completed by:										
	Date:	00 / 00 / 202X									
						Initial Estimate			Design-Engineering Countermeasure		
Number	Product Design Element	Hazard(s) Exists?	Hazard	Cause	Effect	S	P	Risk	Engineering Notes	Instructions	Owne Manu

TABLE 10.7 Product Safety Matrix (PSMx): Safety Section

It is at this point in the safety PSMx that engineering-design countermeasures are both determined and tracked through the remaining stages of engineering design and development. Under "Engineering Notes," the design team can indicate general ways in which the risk was evaluated, steps that will be taken, and also other information that can provide insight into the path taken or decision made. Some of these comments, especially if made the PSEr or PSMr, can be included in the final column titled "PSEr/ PSMr Notes."

The columns between "Engineering Notes" and the "Revised" risk estimate are for content that either should or must be contained within the product-safety facilitators supplied with the product or to be available to consumers—including pre-sale advertising content. If the particular product's industry has suitable material for including into facilitators, then that column should designate either the material or the document. Suppliers, too, can play a crucial role in the final risk level of the product delivered to consumers. If there is a critical need from the supply chain, then an appropriate note could be included here. This "Supplier" column is also helpful if a particular supplier has had problems in the past with a safety-critical design element of its parts. A note in this column can help the Purchasing Department of the design-and-manufacturing company work to correct any problems "upstream" and before the delivery of the supplier's parts for OEM production.

The last of the columns in this set is the "Retailer/Distributor" column. This area is for notes related to how retailers, dealers, and distributors can help facilitate product safety. There are sometimes point-of-sale (PoS) materials available to dealers and retailers which either contain positive safety information or advise against improper use and behavior. In addition, there are regulations against the sale of some products to those not meeting a minimum-age requirement.

ards rces:	1. N/A												
	2. N/A												
									Revised Estimate				
rning n(s)	Hang Tag	Packag'g & Tags	DVD Content	Sales & Mktg	Website Content	Industry Materials	Suppliers	Retailer/ Distrib'r	S	P	Risk	Status	P-S Engineer Notes

Such help from these third parties can be hard to secure and enforce since some parties are separate business entities and may not be contractually obligated to help the design-and-manufacturing company. However, it is often in their best interests to do so since there are sometimes penalties from agencies for retailers and dealers violating regulations. In the case of all-terrain vehicles (ATVs), dealers can be penalized by the designer/manufacturer if they knowingly sell an adult ATV for use by a person under sixteen (16) years of age ("The Consumer Product Safety Improvement Act (CPSIA)" 2008). Actions resulting from an ATV dealer continuing to sell for under-age use include the termination of the dealership agreement.

Similar to the PHA, following the detailed, itemized engineering countermeasures to risk are revised risk severity, probability, and level. This risk-level revision presumes that the prescribed engineering countermeasures are both conducted as detailed and produce the desired result. There will be instances in which someone in engineering fails to make these changes or when testing both in the laboratory and in the field will be necessary to validate the suggested engineering-design changes. Regardless, the status of the risk—as either "Open" or "Closed"—will assist the engineering team at the time of the next design review. The final column, "P-S Engineer Notes" was partially discussed along with the "Engineering Notes" column. In addition to general product-safety notes, the PSEr or PSMr can leave reminders to speak with a particular person about a topic or to review a test report that will be available prior to the next design review.

The compliance PSMx is now discussed. It is presented in Table 10.8. In this section of the PSMx, design-and-manufacturing company "safety" obligations are explicitly identified and listed. Unlike the PHA and safety PSMx, there is neither hazard identification nor risk assessment conducted in this table. The reason is simple.

	Product:							
	Model / Description:							
	Unit Reviewed:			(ID, SN, VIN, other Identifer)				
	Product-Development Stage:							
	Completed by:							
	Date:	00 / 00 / 202X						
							Compliance Measures	
Number	Standard, Regulation, or Agreement	Section	Specification Type	Compliance Specifications	Compliance Criterion	Compliance Needed?	Compliance Notes	Instructior

TABLE 10.8 Product Safety Matrix (PSMx): Compliance Section

All compulsory hazards have already been identified *for* the company by an authority. In addition, acceptable levels of risk have already been authoritatively determined for the company through performance or design criteria. There is little to no thought involved in this section. All of the thinking has already been done for the engineering team. Rather than being an *engineering* exercise, the compliance PSMx becomes an *accounting* exercise. Namely, make sure that all of the boxes are simply checked off. Once that is done, the compliance PSMx is complete—and there cannot be any argument with it. All that was required has been done. *Period!* The compliance PSMx is a completely different exercise than is the safety PSMx. The compliance PSMx must be done, but its completion—although necessary—will not automatically lead to the safe products that are the focus of this book.[7] This disparity is due to the different natures of safety and compliance as discussed in Chapter 2.

The compliance PSMx begins with the identification and numbering of standards, regulations, or other agreements, which place specific safety obligations on a company's products. After identifying a standard, for example, the section of that standard is listed in the next column. After that, the type of specification is listed in the next column. For example, types of specifications include strength, construction, geometry, along with instructions and warning signs. The column after "Specification Type" is for "Compliance Specifications." This column indicates how a designer/manufacturer demonstrates compliance with a requirement. The next column, labeled "Compliance Needed?" signifies if a company's particular product must comply with this requirement. If a particular section of a given standard for the company's product specifies that the strength of polymer product component must be demonstrated to be at least some amount, but this company uses a stainless-steel element instead and, as a result,

[7] This state is demonstrated later in this section.

Owner's Manual	Warning Sign(s)	Hang Tag	Packag'g & Tags	DVD Content	Sales & Mktg	Website Content	Industry Materials	Suppliers	Retailer/ Distrib'r	Status	P-S Engineer Notes

this section does not apply, then this column will indicate this. Again, even though this requirement for products with polymer components does not apply to this product at the moment, it may be wise to keep this criterion in front of the engineering team in the case that a future design effort attempts to reduce product weight by moving to a polymer component.

The next set of compliance PSMx columns under "Compliance Measures"[8] includes "Compliance Notes" as well as the same eleven product-safety facilitator columns as the safety PSMx. "Compliance Notes" are simply the information useful for a reviewer looking at verifying that compliance has been achieved and any details or reasoning that may be helpful at a later date. The facilitator entries are similar to those for the safety PSMx, but applying to strict compliance and not to safety. The columns "Status" and "PSEr/PSMr Notes" serve the same purposes as those columns within the safety PSMx.

An example is now presented to show the use and utility of the PSMx. This example will use the hand-powered winch used earlier in the chapter and as shown and described in Appendix B.[9] The stage at which this example takes place is Design Review 1, so the product has not reached its final-design point and engineering changes will be made prior to production. A departure in presentation will be made and that is that the compliance PSMx will be discussed before the safety PSMx for the winch is presented. The reason for this will be made clear after covering the compliance PSMx.

An example compliance PSMx for the winch is presented in Table 10.9, and uses the same columns as those shown earlier. This compliance-PSMx exercise begins with

---

[8] These actions are not considered "*counter*measures" since nothing—such as a risk—is being *countered*. These actions are merely measures used to achieve a pre-specified goal.

[9] The toaster example will not be used further in the book since the toaster's design has not been discussed to the extent needed to evaluate its details.

	Product:	Hand-Powered Winch						
	Model / Description:	W-1						
	Unit Reviewed:	DEV-01		(ID, SN, VIN, other Identifer)				
	Product-Development Stage:	Design Review 1						
	Completed by:	J. Smith						
	Date:	04 / 01 / 2020						
							Compliance Measures	
Number	**Standard, Regulation, or Agreement**	**Section**	**Specification Type**	**Compliance Specifications**	**Compliance Criterion**	**Compliance Needed?**	**Compliance Notes**	**Instr**
1	SAE J1853: JUN2014, "Hand Winches—Boat Trailer Type"	5.1	Drum diameter for wire rope	N/A	N/A	No. Strap used instead of wire rope	N/A	N/A
2	" "	5.2	Winch load 1	Force (overload) and operability	Sustained load (force) of two times (2x) the rated load at outer wound-strap radius for one minute without releasing load; winch must be operable afterward	Yes	Passed Test, Engineering Test Report 2020-0106, dated 03/09/20	N/A
3	" "	5.3	Winch load 2	Force (extreme overload)	Sustained load (force) of three times (3x) the rated load at outer wound-strap radius for one minute without releasing load	Yes	Passed Test, Engineering Test Report 2020-0107, dated 03/10/20	N/A
4	" "	6.1	Attachment of winch	Force (extreme overload)	Winch must be bolted or welded such that three times (3x) the rated load can be applied without attachment failure	Yes	No time duration specified. Passed Test, Engineering Test Report 2020-0108, dated 03/11/20	N/A

Table 10.9 Product-Safety Matrix for Winch: Compliance Section

Owner's Manual	Warning Sign(s)	Hang Tag	Packag'g & Tags	DVD Content	Sales & Mktg	Website Content	Industry Materials	Suppliers	Retailer/ Distrib'r	Status	P-S Engineer Notes
N/A	N/A	N/A	N/A	N/A	N/A	N/A	N/A	N/A	N/A	Closed	
N/A	N/A	N/A	N/A	N/A	N/A	N/A	N/A	N/A	N/A	Closed	Verify that no engineering-design changes compromise compliance with SAE J1853
N/A	N/A	N/A	N/A	N/A	N/A	N/A	N/A	N/A	N/A	Closed	Verify that no engineering-design changes compromise compliance with SAE J1853
N/A	N/A	N/A	N/A	N/A	N/A	N/A	N/A	N/A	N/A	Closed	Verify that no engineering-design changes compromise compliance with SAE J1853

							Compliance Measures	
Number	**Standard, Regulation, or Agreement**	**Section**	**Specification Type**	**Compliance Specifications**	**Compliance Criterion**	**Compliance Needed?**	**Compliance Notes**	**Instr**
5	" "	6.2	Winch line attachment to drum	Force (mild overload)	With three turns of strap around spool axle, spool must withstand 1.5x rated load	Yes	No time duration specified. Passed Test, Engineering Test Report 2020-0109, dated 03/11/20	N/A
6	" "	7.1	Breaking strength: Winch Line	Force (mild overload)	Winch strap and its attachment must withstand at least 1.5x rated load	Yes	No time duration specified. Passed Test, Engineering Test Report 2020-0121, dated 03/17/20	N/A
7	" "	7.2	Breaking strength: Winch Hook	Force (mild overload)	Winch-strap hook must withstand at least 1.5x rated load (for non-wire rope strap)	Yes	No time duration specified. Passed Test, Engineering Test Report 2020-0122, dated 03/17/20	N/A
8	Winch industry agreement with U.S. CPSC (December 2019) *	3.2	Transported load	Warn users to use straps or ropes to secure boat to trailer during transportation	Provide specific safety content in winch owners manual and on-product warning sign	Yes	In process: technical publication 20-022, rev.0 (owners manual) and engineering drawing 051-314, rev.0 (warning sign)	N/A
...	...	...	...	...	...	...	...	...

Table 10.9 Product-Safety Matrix for Winch: Compliance Section (*Continued*)

Owners Manual	Warning Sign(s)	Hang Tag	Packag'g & Tags	DVD Content	Sales & Mktg	Website Content	Industry Materials	Suppliers	Retailer/ Distrib'r	Status	P-S Engineer Notes
N/A	N/A	N/A	N/A	N/A	N/A	N/A	N/A	N/A	N/A	Closed	Verify that no engineering-design changes compromise compliance with SAE J1853
N/A	N/A	N/A	N/A	N/A	N/A	N/A	N/A	N/A	N/A	Closed	Verify that no engineering-design changes compromise compliance with SAE J1853
N/A	N/A	N/A	N/A	N/A	N/A	N/A	N/A	N/A	N/A	Closed	Verify that no engineering-design changes compromise compliance with SAE J1853
Include: "ALWAYS securely attach boat to trailer with strap or rope. Failure to do so may result in serious injury or death."	Include: "ALWAYS securely attach boat to trailer with strap or rope. Failure to do so may result in serious injury or death."	N/A	N/A	Include: "ALWAYS securely attach boat to trailer with strap or rope. Failure to do so may result in serious injury or death."	Include: "ALWAYS securely attach boat to trailer with strap or rope. Failure to do so may result in serious injury or death."	Include: "ALWAYS securely attach boat to trailer with strap or rope. Failure to do so may result in serious injury or death."	N/A	N/A	Display dealer sign 20-147: "ALWAYS securely attach boat to trailer with strap or rope. Failure to do so may result in serious injury or death."	Open	Assist with design of OM and WS; verify before release
...	...	...	...	...	...	...	...	...	...	...	...

the identification of two pertinent documents which place design-and-manufacturing obligations onto the company engineering, and ultimately selling, the winch.[10] The first document is a real-world industry standard that would apply to this product—were it an actual product. The second document is a fictitious agreement used to demonstrate the full breadth of the compliance PSMx. Assume that all of the required winch testing has been completed and that the winch passes each test requirement.

Returning to the first document, it is an industry standard for hand-operated boat-trailer winches ("SAE J1853-2014: Hand Winches—Boat Trailer Type" 2014). This is precisely what the imaginary product is intended to be. Therefore, the standard applies as do its requirements.[11] The design engineer or PSEr will begin completing the compliance PSMx by filling in the *first* row. In this standard, Section 5.1 requires a certain relationship between the winch spool and drum. However, this winch uses a polyester strap instead of a wire rope. Consequently, this section does not apply to the example winch. "N/A" appears in the "Compliance Notes" column and those before it. The *next two* rows concern standard Sections 5.2 and 5.3 pertaining to winch loading. The first of these two columns addresses an overload test in which two times the rated load is applied to the winch for one minute. The winch must be operable after this testing is completed. The remaining columns for this row indicate a passing of the test, complete with an engineering-test report number and date. No facilitator measures are needed to comply with the industry standard." This design matter is "Closed" since the winch passed the testing requirement; however, the engineers must be vigilant that no later engineering-design changes compromise the passing of this test.

The *third* row of the compliance PSMx is similar to the second with the exception that the applied load is now an extreme-overloading case where three times the rated load is applied for one minute and that the winch is not required to remain operable as a result of the testing. Similar notes for the number and date of the corresponding engineering-test report are entered.

Rows *four* and *five* pertain to the strength of winch components, namely, the chassis and the strap or wire rope, respectively. Each of these tests applies a load to either the winch or its component as specified by the listed details of the standard's sections. One noteworthy shortcoming in this standard, which is *by no means unique to this standard*, is the lack of a time duration for which the load must remain applied. In most cases, it would probably not be material whether the load is applied for one second or one day. The test results would probably not differ between load-duration periods. Yet, it would be beneficial, for the sake of consistency and the test engineer, for the committee involved in developing this standard to include a *pro forma* time duration for all applied loads.

The *sixth* and *seventh* rows of the compliance PSMx provide minimum-strength tests for the winch line and hook, respectively. Because the testing is straightforward and because the winch's components passed these tests, there is nothing noteworthy about these rows.

---

[10] The author has not performed a thorough investigation to determine how many more standards, regulations, or agreements might apply to the design and manufacture of trailer-mounted boat winches. An actual winch company would necessarily do this. This set of two documents—one of them imaginary—is merely a simple example that is manageable in magnitude for this book's purposes.

[11] As far as the author knows, this standard remains a voluntary standard without any direct penalties for noncompliant products. However, it will be presumed that the design-and-manufacturing company wishes to completely comply with this standard.

It is at row *eight* of the compliance PSMx that the fictional agreement, which was mentioned earlier, is introduced into the PSMx methodology. The author asks the reader to imagine a set of fictitious circumstances in which numerous people—including, perhaps, a beloved celebrity—have recently been injured or killed by boats which have come off of their boat trailers. The causes of these accidents have been that the boat owner had relied upon the boat winch, alone, to secure the boat to the trailer. The political and public-safety fallout has hit the boat, trailer, and winch industries heavily. The results of regulatory and public pressures have forced the U.S. CPSC and these industries to reach an agreement in December 2019—again fictitious—to have, as mandatory, specific product-safety information provided to each winch consumer regarding proper winch use. This information tells the boat owner to always use cargo straps or ropes to secure a boat to a trailer prior to transporting the boat. The winch is intended to load and unload a boat—but not to secure it during transportation. This product-safety facilitation material shall appear on the product through a warning sign and be included in other appropriate materials including any manuals, website, and advertising content. Since this is a new requirement, the winch design-and-manufacturing company has yet to complete the engineering-design changeover—although the specifications for compliance have been established.

Row eight of the compliance PSMx shows how the ongoing compliance effort is proceeding for this new product requirement. Noted in this row at issue is the transportation of the load which—although is not part of the product's intended use—becomes the winch company's problem, regardless. The row also notes that the company is required to warn winch users about the need to use straps and ropes, to secure their boats, through on-product warning signs, manuals, and other company-controlled information. It is noted that both an owner's manual and a warning sign are being drafted to meet this requirement. Assume that all "instructions" are contained within the owner's manual. The particulars of successfully meeting this requirement are listed for each of the affected facilitators. Since this compliance effort is still ongoing, the status of this item is "Open." Before release of the manual or the warning sign, the PSEr should review the deliverables against the requirements to assure compliance of the winch with the CPSC agreement.

Now that the product, the winch, has complied with all necessary standards, regulations, and agreements, the compliance-focused engineer might conclude that the product is now "safe" and ready for sale. Readers who have been paying attention will likely realize that there are safety-related aspects to the winch's design that have, thus far, been addressed in no way. This set of "holes" left by a compliance exercise demonstrates that the compliance portion of safety perhaps is a necessary condition, but is certainly not a sufficient condition.[12] This is the reason why the safety PSMx was delayed until the compliance PSMx had been completed.

The first three rows of the first nine columns of the safety PSMx worksheet for the winch are shown in Table 10.10. They are, again, essentially those of the PHA for the winch Table 10.6. However, there have been nine rows added to the safety-PSMx worksheet to include hazard-and-risk considerations not yet included in the discussion.

[12] As mentioned elsewhere in the book, *blind* compliance with a requirement may compromise product safety rather than advance it. It may also be ethically necessary for engineers to "push back" against regulators, an industry, or lawmakers in cases when compliance is contrary to safety.

Product:	Hand-Powered Winch		
Model / Description:	W-1		Hazards Sources
Unit Reviewed:	DEV-01	(ID, SN, VIN, other Identifer)	
Product-Development Stage:	Design Review 1		
Completed by:	J. Smith		
Date:	04 / 01 / 2020		

						Initial Estimate						
Number	**Product Design Element**	**Hazard(s) Exists?**	**Hazard**	**Cause**	**Effect**	**S**	**P**	**Risk**	**Engineering Notes**	**Instructions**	**Owners Manual**	**Warnin Sign(s)**
1	Spool	Yes	Laceration	Sharp edge	Cut on finger	*1-N*	*3-O*	*Low*	Conducting field testing programs; competitive benchmarking conducted	N/A	N/A	N/A
2	Strap and Spool	Yes	In-Running Nip Point	Fingers/ hand caught in strap	Crushed finger(s)	*2-M*	*2-R*	*Medium*	Slow-moving system under manual control of user; conducting field testing programs; competitive benchmarking conducted	Always keep fingers and hands away from strap and spool	Always keep fingers and hands away from strap and spool	Always keep fingers hands a from st and sp
3	Transported Load	Yes	Impact	Contact with moving load	Numerous inury scenarios	*4-X*	*2-R*	*High*	Conducting thorough engineering analysis with laboratory and field testing programs; competitive benchmarking conducted; CPSC Agreement (December 2019)	[See Item 8, Compliance PSMx]	[See Item 8, Compliance PSMx]	[See Item 8, Complia PSMx]
4	Chassis	Yes	Stability	Fastening hardware fails at winch mount	Winch separation from trailer	*3-C*	*2-R*	*Medium*	Conducting thorough engineering analysis with laboratory and field testing program for vibration and durability; requirements testing of SAE J1853:JUN2014	Always read, understand, and follow the Owners Manual	Use the supplied winch-mounting hardware to secure the winch to the trailer	Always read, unders and fol the Ow Manual

Table 10.10 Product-Safety Matrix for Winch: Safety Section

.. Appendix A: Product-Safety Hazards Checklist (d'Entremont 2021)
:. European Commission Directive 2006/42/EC

							Revised Estimate				
ackag'g & ags	DVD Content	Sales & Mktg	Website Content	Industry Materials	Suppliers	Retailer/ Distrib'r	S	P	Risk	Status	PSEr Notes
/A	N/A	N/A	N/A	N/A	De-burr spool halves prior to shipment		*1-N*	*2-R*	*Low*	Open	Confirm, through Purchasing Dept., engineering-design changes at Design Review 2; revise risk estimate
lways eep ngers and ands away om strap nd spool	Only show intended use and safe operation of winch	Only show intended use and safe operation of winch	Only show intended use and safe operation of winch	N/A	N/A		*2-M*	*1-I*	*Medium*	Open	Confirm engineering-design changes at Design Review 2; revise risk estimate
See em 8, ompliance SMx]	[See Item 8, Compliance PSMx]	[See Item 8, Compliance PSMx]	[See Item 8, Compliance PSMx]	N/A	N/A	Offer suitable cargo straps for sale	*4-X*	*1-I*	*Medium*	Open	Confirm engineering-design changes at Design Review 2; revise risk estimate
ways ead, nderstand, nd follow e Owners lanual	Always read, understand, and follow the Owners Manual	Always read, understand, and follow the Owners Manual	Always read, understand, and follow the Owners Manual	N/A	Reliable source for 1/4-20 hardware	Offer winch-mounting services to customers	*3-C*	*1-I*	*Medium*	Open	Verify BOM at Design Review 2; review test report(s) for vibration and durability tests for mounting hardware and recommended torque specifications; revise risk estimate

						Initial Estimate						
Number	**Product Design Element**	**Hazard(s) Exists?**	**Hazard**	**Cause**	**Effect**	**S**	**P**	**Risk**	**Engineering Notes**	**Instructions**	**Owners Manual**	**Warning Sign(s)**
5	Control System	Yes	Stability	Ratchet pawl in wrong position	Unanticipated movement of load and crank handle	*3-C*	*3-O*	*High*	Intuitive operation of pawl lever and can be visually verified in operation; have conducted thorough engineering analysis with laboratory and field testing program; competitive benchmarking	Always read, understand, and follow the Owners Manual	Show operation and positions of pawl lever	N/A
6	Strap hardware	Yes	Fracture	Hook breaks	Unanticipated movement of load	*3-C*	*2-R*	*Medium*	Have conducted thorough engineering analysis with laboratory and field testing program; competitive benchmarking; requirements testing of SAE J1853:JUN2014	Always read, understand, and follow the Owners Manual	Always check the condition of the winch before each use	Always r… underst… and foll… the Owr… Manual
7	Strap Material	Yes	Tensile Failure	Wear/ Damage	Unanticipated movement of load	*3-C*	*2-R*	*Medium*	Strap will withstand significant wear prior to failure; have conducted thorough engineering analysis with laboratory and field testing program; competitive benchmarking; requirements testing of SAE J1853:JUN2014	Always read, understand, and follow the Owners Manual	Always check the condition of the winch before each use	Always… underst… and foll… the Ow… Manual
8	Bolted joints	Yes	Loss of shafts	Loss of torque and nuts	Unanticipated movement of load	*3-C*	*3-O*	*High*	Use locking elements or components on bolted joints; laboratory and field testing	N/A	N/A	N/A
9	Physical Environment	Yes	Structural or Mechanical Failure	Corrosion	Unanticipated movement of load	*3-C*	*2-R*	*Medium*	Galvanize all steel components; steel components will withstand significant corrosion prior to failure; have conducted laboratory and field testing program	N/A	N/A	N/A

Table 10.10 Product-Safety Matrix for Winch: Safety Section (*Continued*)

							Revised Estimate				
Packag'g & Tags	DVD Content	Sales & Mktg	Website Content	Industry Materials	Suppliers	Retailer/ Distrib'r	S	P	Risk	Status	PSEr/PSMr Notes
N/A	N/A	N/A	N/A	N/A	N/A	Offer to demonstrate winch operation to customers	*3-C*	*2-R*	*Medium*	Open	N/A
N/A	N/A	N/A	N/A	N/A	Reliable source for hook hardware	N/A	*3-C*	*1-I*	*Medium*	Open	Verify that Test Engineering has conducted laboratory and field testing program on hook strength and durability—including latch and torsion spring; revise risk estimate
N/A	N/A	N/A	N/A	N/A	Reliable source for strap material and stitching	Carry replacement parts for winch	*3-C*	*1-I*	*Medium*	Open	Verify that Test Engineering has conducted laboratory and field testing program on web strength and durability—including stitching; revise risk estimate
N/A	N/A	N/A	N/A	N/A	Reliable source for any necessary locking elements on bolted joints	N/A	*3-C*	*1-I*	*Medium*	Open	Verify that Test Engineering has conducted laboratory and field testing program for vibration and shock on bolted joints; revise risk estimate
N/A	N/A	N/A	N/A	N/A	Galvanize all steel parts or assemblies	Inspect winches when trailers are being serviced	*3-C*	*1-I*	*Medium*	Open	Verify that Test Engineering has conducted laboratory and field testing program for corrosion (salt spray); revise risk estimate

						Initial Estimate						
Number	**Product Design Element**	**Hazard(s) Exists?**	**Hazard**	**Cause**	**Effect**	**S**	**P**	**Risk**	**Engineering Notes**	**Instructions**	**Owners Manual**	**Warning Sign(s)**
10	Electrical	No	N/A	N/A	N/A	-	-	-	N/A	N/A	N/A	N/A
11	Chemical	No	N/A	N/A	N/A	-	-	-	N/A	N/A	N/A	N/A
12	Energy	No	N/A	N/A	N/A	-	-	-	N/A	N/A	N/A	N/A
...	...	...	...			...	...	...	...	...		...

TABLE 10.10 Product-Safety Matrix for Winch: Safety Section (*Continued*)

The risk severity, probability, and level discussions for the PHA of the winch from Table 10.6 will not be repeated here in order to save time and space. Therefore, the "Design-Engineering Countermeasures" columns are the first point of discussion.

The engineering notes for the *first* risk, which is a cut from a sharp spool edge, indicate that having the supplier de-burr the spool halves prior to shipment to the OEM will alone sufficiently address the problem. As a good measure, the test engineers will field test the new parts to see if the edges have been dulled enough to prevent finger cuts. Furthermore, the engineering team will evaluate competitors' products to see how well they have dulled their spool edges. This process is known as *competitive benchmarking* and is common in many industries. The item's status remains "Open" and the PSEr has left a note to check with the Purchasing Department about the status of this by suppliers at the Design Review 2. There is no need to remind users to keep their fingers away from the edge of the winch spool *regarding the laceration hazard* because that hazard's risk level should be sufficiently reduced through the supplier's de-burring operation; however, the next item reminds users to keep their fingers and hands clear of the winch's spool for *other* reasons.

The *second* row of the safety PSMx is for the in-running nip point hazard and risk from the strap-and-spool configuration. The notes again indicate that the winch is slow moving and under manual control—there is no electric control system to elevate the hazard's risk. Despite these favorable characteristics of the winch, it is decided to remind users to keep their fingers and hands away from the nip point by using the manual provided with the winch and a warning sign prominently attached to the winch to accomplish this. This message will also be repeated on the winch packaging for good measure. In addition, other company-controlled materials, such as videos, will only show the winch being used as intended.

The PSMx's *third* row addresses hazard and risk from transporting a boat using only the winch strap for securement. This aspect the PSMx has already been largely addressed in Item 8 of the compliance PSMx through the CPSC agreement. These measures are referenced in the engineering countermeasures in that row. Even though this issue has been addressed, it is still wise for engineering to analyze and test the winch to see what can be learned or improved regarding this type of non-recommended use that a consumer might still accidentally use. Also specified are the facilitators from the compliance PSMx. Additionally, this company has chosen to encourage dealers to offer for sale appropriate cargo straps or ropes to secure a boat to a trailer.

The added *fourth* row covers a hazard and risk combination which was not contained in the original PHA but which is addressed in the industry standard for

							Revised Estimate				
Packag'g & Tags	DVD Content	Sales & Mktg	Website Content	Industry Materials	Suppliers	Retailer/ Distrib'r	S	P	Risk	Status	PSEr/PSMr Notes
N/A	N/A	N/A	N/A	N/A	N/A	N/A	-	-	-	N/A	N/A
N/A	N/A	N/A	N/A	N/A	N/A	N/A	-	-	-	N/A	N/A
N/A	N/A	N/A	N/A	N/A	N/A	N/A	-	-	-	N/A	N/A
...	...	...	...	...	...	...	...	...	...	...	...

boat-trailer winches. This particular winch-mounting strength requirement, the astute engineer will realize, only addresses the winch's response to the application of a single load. It will be necessary to design and test for more than just this. Engineers will have to consider the effects of vibration and shock on the winch-mounting hardware and the recommended fastener torques. These factors could make a winch mounting, which is able to withstand the necessary single load, fail after being exposed to driving down the road over several hours due to roadway-induced vibrations. Product-safety facilitators are specified as needed. Dealers are encouraged to offer winch-mounting services for those who may not trust themselves to do it properly. This item is closely related to item *eight* which is a more-explicit coverage of the bolted joint at the winch-mounting surface.

Row *five* covers the winch's control system or, rather, the lack of one. This is related to item 2 in that this manual control system is slow-moving and has components visible to the user. If the ratchet pawl is placed in the incorrect position by the user, no harm is likely to result due to the overall nature of the product, user, and environment. As with many of the items in this safety PSMx, the user is reminded to *read*, *understand*, and *follow* the instructions and warning signs provided by the winch.

Items *six* and *seven* of the safety PSMx cover the strap hardware and strap material, respectively. It is interesting that both of these items are covered by the industry standard listed in the compliance PSMx. These items remain in the safety PSMx to validate that, despite passing the compliance testing, these components—as implemented in the subject winch—are indeed sufficiently designed to serve their intended purposes. Compliance with a standard is no assurance that the design will be sufficient for its intended use. Item *eight* was discussed with item three.

Item *nine* is a consideration of the product's physical environment. In this case, corrosion is evaluated. In particular, how corrosion, or rust, may affect the mechanical integrity of the winch system over an extended period of time. The proposed solution to rust prevention is galvanization of all winch components or assemblies. A boat winch will be subjected to extreme environmental conditions. This includes temperature and moisture. This moisture may be rain, mud, fresh water, or salt water. The winch must be able to last a reasonable length of time under conditions consistent with its use environment. To investigate the winch's corrosion resistance, a set of salt-spray tests will be conducted as well as extensive field testing to reveal the true properties of the winch's materials and surface treatments. Rows *ten*, *eleven*, and *twelve* are merely inserted into the safety PSMx to remind the engineering team to be alert for new hazards which could be created through future engineering-design changes.

Although only a logical progression from the PHA, the PSMx worksheets permit the design of a *system* of countermeasures to hazards and risks inherent to a product and its use. Because the PSMx advances with the product as it proceeds through the EDDTP, this system of countermeasures can be reviewed at each design stage and modified as needed to suit the evolving product. The PSEr can use the PSMx, through verifying and validating its *designed-in* countermeasures, to assess the final design's readiness for manufacturing and release to retailers. Several points within the PSMx permit the PSEr or PSMr to make comments and notations that will be useful in explaining the current project status—or even decisions which may be questioned years later.

An important part of the engineering-design process is to *design-in* as much safety to products without resorting to the use of product-safety facilitators. Despite any shortcomings in the safety hierarchy, its main contribution to product safety is its reiteration of only using warning signs, manuals, and user education/training after *designing-out* as much hazard and risk as practical. Only then should facilitators be employed and PPE recommended to further product safety for users. Therefore, the author placed the chapter on product-safety facilitators *before* this chapter where the engineering-design team can construct the system of countermeasures to hazards and risks. The reason for this is that the design engineer and the PSEr must understand the utilities and limitations of facilitators before relying upon them to deliver the final measure of product safety to consumers. Design teams might expect—and management might hope—facilitators to provide more safety to their product than can realistically be delivered. However, it is also useful for the design team to know about the variety of facilitators available so that they may be part of the overall solution to any safety problems with the product.

Engineers should always select the tools and methods that are the most suitable for the task at hand. The modification of existing practices may be necessary for a given project or program. The simple tracking of effective countermeasures to hazards and their risks is included in the PSMx and appears to work well for small- and medium-sized projects and programs. Large-scale programs, such an on-highway vehicles, may indeed require the administrative engineering overhead of a complete system-safety program.

Depending upon the product being designed, developed, and tested, there may be significant amounts of information outside of the PSMx needed to show and document the risk levels being provided to users. Such information might include analyses, hand calculations, laboratory-test reports, and result of competitive benchmarking. *It is important to modify the PSMx worksheets to suit the particular purposes of the engineering-design team.* It is vital that engineers remembers that the PSMx exists to serve the product-safety function. It is not to be served. The PSMx is not sacred. The PSMx worksheets can—and should—be modified to suit the needs of the product, the engineering-design team, and the PSEr/PSMr. Columns may be freely added or deleted. It should be tailored to the product-design process in place at each company and industry, but the systematic approach to looking at all known and potential hazards, risks, and risk levels, should be retained to the extent possible. The PSMx is a tool to help in the delivery of product safety to users through an integrated PDP which includes product-safety engineering.

Up to this point in the chapter, the discussion has been limited to hazard and risk. Before moving on to the topic of faults and failures, the reader is provided with a discussion of design philosophies for addressing product failures.

Discussion: Engineering-Design Philosophies to Counter Product Failures

At this point, the topic of reliability and its relationship to product safety are revisited. The following discussion shows how reliability is one of several methods that may be used to achieve product safety through countering failures of or within products. Reliability is a generally desirable property for a *useful* product. However, with some products, reliability is not a necessary characteristic for a *safe* product. The case study in Chapter 2 illustrated the differences between reliability and safety. Also, a product that never works may be a safe product indeed, but it will not lead to "repeat customers" for the designer-and-manufacturer of that product. A product considered to be "dangerous" is, in fact, much safer if it never operates and is, therefore, never used.

There are two distinct engineering-design philosophies to countering product failures that are practiced, sometimes unwittingly, by engineers. Figure 10.1 is a diagram showing the relationships between the different elements used in this discussion. At the top or first level is the title of the discussion itself. At the second level of the figure are the two approaches to engineering-design philosophies. The first of these approaches is the *fault-averse* approach; the second, the *fault-tolerant* approach. There are two options available to the design engineer under the fault-averse approach: *reliability* and *preventive maintenance* (PM). If the product exhibits a fault-tolerant nature, then the two available operational paths are: *fail-safe design* and *manifest danger*.

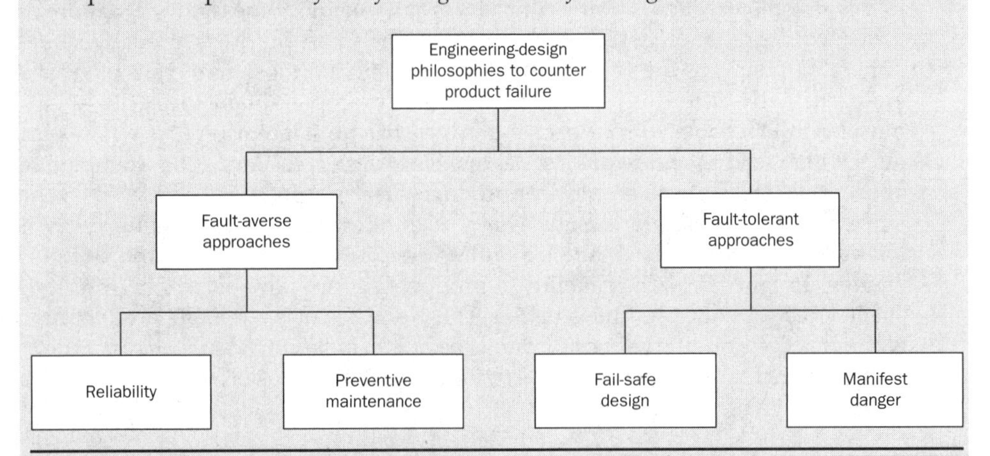

Figure 10.1 Engineering-design philosophies to counter product failure.

The fault-averse approach results in a product design that cannot tolerate a failure. The fault-tolerant approach yields products which can withstand failures—to a degree. Each of these approaches breaks down into two avenues of pursuit available to engineering designers. There are also two different approach types—*passive* and *active*—to each of these avenues. The resulting 2 × 2 matrix of options is shown as Table 10.11. Each of these options, or matrix elements, will be explored in turn.

	Passive	Active
Failure Averse	Reliability	Preventive Maintenance
Failure Tolerant	Fail-Safe Design	Manifest Danger

Table 10.11 Engineering-Design Approaches to Counter Product Failure

I. Reliability

Reliability, in lay terms, is the design approach to make a product which does not fail—perhaps for the life of the product. A most obvious example for reliability is, again, from aerospace and weaponry. A single-use weapon—such as a defensive missile and all of its systems, sub-systems, and components—which is designed, and needed, to function whenever needed and to perform flawlessly in extreme and hostile environments. Product failure must not occur; mission completion is vital. Military systems usually do require high levels of reliability due to the disastrous potential outcomes from the failures of these products.

Often, the high-reliability approach to engineering design can add considerable cost to an engineering-design project. The creation of very reliable products is frequently accompanied by large quantities of component analytic and testing data produced through significant laboratory- and field-testing programs. Cost is often considered to be secondary to the primary function in some such engineering programs. The use of high-quality, low-tolerance, and expensive components with large mean times to failure (MTTF) in such products is routine. Reliability engineering is absolutely an important and complex discipline, but is not necessary for all products.

For some products, however desirable reliability is, it simply may not be the most important design aspect to cost-conscious consumers. There are, of course, exceptions to this statement. Some consumer products can tolerate some degree of failure or performance degradation without user injury. Yet, some products used by vulnerable populations, such as the very young, the very old, and those with various disabilities, are particularly *at-risk* to product failures. For example, infant toys with small parts must be quite reliable with respect to product structural integrity. Toys whose parts can detach may lead to child-choking deaths. Each design case must be examined for the appropriate measures to employ for product-safety purposes.

A reliable product, by design, is not tolerant of failure—or its failure may not be tolerable by its user. Reliability is a fault-averse, or fault-intolerant, approach to engineering design. Through reliability, a product will not fail and continues to function as intended without user intervention. Therefore, because a reliable product functions without actions or interventions from its user, it may be considered a *passive* method to avoid product fault or failure. An *active* method requires user input or intervention as will be shown next.

II. Preventive Maintenance

Preventive maintenance, or PM, is also a failure-averse design method. However, unlike reliability, PM requires the intervention of people to assure that a product does not fail. Therefore, PM is an *active* approach to engineering design. Through PM, sub-systems and components of a system are replaced at regular intervals or as indicated by the results of diagnostic testing. Unlike pure reliability, the failure of the product is *anticipated*. Through the PM approach, such anticipated failures are countered by servicing or replacing important components of the product *before* any failure occurs.

Using the example of a commercial airliner—which is itself a highly reliable system—the designers and operators of airliners know that, however reliable a plane is, the *sustained* use of the plane will require a thorough and systematic PM program. Critical parts of the airliner will need to be serviced or replaced prior to failure. This PM program will be necessary to maintain the levels of function, availability, and safety required by both business needs and public-safety values.

III. Fail-Safe Design

Fail-Safe design is a fault-tolerant approach to engineering design where the product will fail in a *safe* manner rather than an unsafe manner. Popular culture often conjured up the concept of "fail safe" in movies of nuclear weapons during the "Cold War" of the latter half of the twentieth century. In keeping with this theme, the fusing or firing mechanism for a nuclear device, or bomb, should fail benignly whenever it does fail. For example, if the bomb's fuse were to fail, it would be vastly preferable for the bomb to be rendered "inert" rather than have the fuse inadvertently detonate the device. Importantly, Barnett states that:

> There is no such thing as a fail-safe *design*;
> there are only *modes* of failure that can be made fail-safe.[13]

In support of Barnett's disclaimer, presume that the nuclear device with a failed fuse is, in fact, inert. That inert bomb, however, still contains radioactive materials and the entire nuclear bomb still presents hazards with risks. Therefore, only one *mode* of failure—that is, fusing—has failed safely.

Another example of a fail-safe design mode is that of the common motor-vehicle traffic-control light system. If the electronic timing and controls system was to fail, it is likely that many such traffic lights would default to a relatively safe operating mode, such as flashing red lights in all directions or no lights whatsoever in any direction. This requires vehicle traffic to stop and, then, proceed only when it is safe to do so. A poorly designed traffic-light system could fail to the state of green lights in all directions, a much more dangerous failure state than the all-red state for the traffic light. Despite this relatively safe failure state, the traffic-light system remains electrically energized and, thus, still able to electrocute maintenance personnel. The lights themselves remain elevated above the roadway where structural failure could cause the lights to collapse upon vehicles and people resulting in injury. Therefore, both hazards and risks still exist after a fail-safe designed failure of a single operational mode of the nuclear weapon and a traffic-light system.

Through fail-safe design, a product is permitted to fail safely in a particular mode of operation and requires no outside intervention in order to do so. Thus, fail-safe design is a passive method.

IV. Manifest Danger

Manifest danger is the design approach taken by some engineering designers when complete product reliability is either unnecessary or unattainable and when there may be a way in which a product can itself detect imminent failure and signal this condition to the user. Manifest danger can also be applied in situations where there is no comprehensive or disciplined PM program in place. Through this approach, Barnett says that a product "communicate[s] to users that its safety has been compromised before an injury occurs" (Barnett 1992).

---

[13] Emphases by the book's author.

Although an ornate term,[14] the term "manifest danger" is true to the meaning of the word "manifest." Barnett's definition of the word being: "capable of being readily and instantly perceived by the senses; not hidden or concealed; capable of being easily understood or recognized at once; evident, obvious, apparent, plain, clear, conspicuous, or open." Therefore, a product design incorporating manifest-danger property is one in which the product clearly indicates to the user that it will soon fail unless some intervention occurs. Thus, manifest danger is an active approach to failure-tolerant design.

A manifest-danger approach to engineering design, according to Barnett, can be accomplished through using either *direct cues* or *fault indicators*. Although not explicitly discussed as such by Barnett, direct cues may be considered *raw* signals, whereas fault indicators may be considered *processed* signals. Direct cues require no further action to notify users of a problem. Fault indicators require a secondary system, or process, to notify product users or operators. Therefore, fault indicators may be considered "indirect" cues.

Direct cues include feedback to the user that could be visual, audible, force/tactile/haptic, olfactile, gustatory (taste), or other in nature. A change in appearance, a sound, an increased effort needed, a smell, or a taste are each examples of a direct cue. A frayed multi-strand winch cable would be a direct cue of the visual type. The imminent failure of the cable through strand breakage is intrinsic to its compromised and decaying state. This cable may still be quite strong, but its strength has already been compromised. This cable should be replaced before using it further.

Fault indicators send signals to users, through an additional component or subsystem, that something is wrong with the product. Those users must then perform a corrective action before ultimate failure. Such output signals include lights, odorants, control forces, buzzers/bells/sirens, utterances, analogs, and other signals or displays. An indicator light (sometimes called an "idiot light") is one type of fault indicator and signals to the user that a product has determined that its condition has been compromised with respect to safety. Such an indicator light is the low brake-fluid level indicator found on many vehicles. The product itself, the vehicle, has detected one imminent failure, brake loss, through a sensor. The brake-fluid level sensor sent an electrical signal that the vehicle then processed through either computer-hardware/software system or a simple switch. The vehicle then sends a signal—a lighted indicator on the dashboard—to the vehicle operator so that the operator can then respond properly in order to avert a loss of service brakes.

The olfactile sensation—an odor or smell—appears both as a direct cue and as a fault indicator in Barnett's scheme. This is because scent can arise in different ways to require different responses. In the case of household natural gas, an odorant is added to the substance itself. Consequently, if a natural-gas leak exists, it will present itself through the odor of sulfur which smells similar to rotten eggs. This signal should alert building occupants to immediately leave and to, then, alert emergency responders. The smell of natural gas is intrinsic to the modified natural-gas product and needs no further processing to be detected by residents.[15] In a similar fashion, Barnett tells of

[14] In deference to its originator, the legal-sounding term "manifest danger" will be used instead of a more engineering-sounding term.

[15] Building residents must know that the smell indicates a natural-gas leak. Therefore, the education of consumers is necessary to maximize the efficacy of the odorant.

railroads using an odor-producing substance in its older railcar axle-bearing lubrication systems. As oil levels dropped, the oil heated up and triggered a foul smell from the odorant. This odor could be detected by railroad personnel and passengers toward the rear of the train. Although *always present* in the oil, the odorant would not signal a compromised-safety condition unless, and until, the oil became sufficiently hot to cause the necessary chemical reaction to produce the odor. The odor was not intrinsic to the normal state of the axle-bearing oil. However, even when the oil odorant was activated, the people on the train had to then notify the train engineers of the impending problem. Hence the odorant became a fault indicator.

Let us consider two additional examples of where the manifest-danger approach is used. The first is the automotive disc-brake pad found on many motor automobiles today. Many such brake pads have on them a brake-wear indicator component that contacts the brake rotor when the brake pads are reaching the end of their useful lives. Therefore, the noise generated is a signal to the vehicle operator to have the brakes serviced. This sound is an audible squeaking sound and is a *direct cue*. The second example is the "chirping" sound emitted by fire or smoke detectors when their battery voltages run down. This is an indicator to homeowners to replace the battery in the fire or smoke alarm. This is an example of an audible *fault indicator* since the battery level is sensed by the fire or smoke detector which then produces the selected sound at selected time intervals until the situation is resolved.

What is interesting in the above two examples is that both products have a necessary and, sometimes prescribed, PM schedule. Yet, in some cases, these PM programs are simply not followed by the product users. Therefore, a designed PM approach may *degenerate* into a *de facto* manifest-danger approach where manifest-danger design elements of a product becomes a "safety net." In the case of motor-vehicle brakes, many vehicle owner's manuals prescribe a PM schedule to replace or inspect vehicle brake-system components. Many people, however, do not follow the recommended PM schedule set forth by the vehicle manufacturer and simply rely upon the brake-pad sound to tell them to have the brakes serviced.[16] Similarly, the fire/smoke detector example is one where society has an informal PM schedule whereby citizens, through media public-service announcements, are often reminded to replace the batteries in their detectors whenever there is a time change from Standard Time to Daylight Saving Time or vice versa. As with the brake-pad example, many citizens routinely rely upon the "chirp" to remind them to change batteries which are running low.

If manifest-danger characteristics are not embodied in the engineering design of a particular product, that product may "fall through the cracks" of the four engineering-countermeasure options. In such cases, safe product use may become the *practical*[17] responsibility of the user. The compromised states of some products are evident to prudent product users; other products, however, may not exhibit readily apparent signs of distress to their users before failing. Engineers must remember this during the PDP. Also remember that direct cues are simpler and more reliable than fault indicators since fault indicators rely upon another system that itself could be prone to failure.

---

[16] This is also a user dependency upon the brake pads and a misuse in kind as covered in Chapter 4.

[17] "Practical" is used to indicate that since safeguards have failed, the responsibility for product safety has now been thrust upon the user. There is no intent to indicate that this current state of user responsibility is necessarily correct, ethical, or legal.

This discussion of the four above options to address potential failures in and of products should help the reader better understand that not all failures disable a product and harm its user. This discussion also indicates that, at times, a user may be able to play a large part in their her or his own safety—proper and ethical engineering-design responsibilities notwithstanding. Each of the four fault-and-failure approaches has its own set of strengths and weaknesses. It is up to the design engineer to understand these approaches and consciously select the most-appropriate method to design and control product failure.

10.5 Failure Modes, Effects, and Criticality Analysis

The methods covered in this chapter up to this point have involved the simple hazard identification and risk assessment for characteristics of a product, or system, that are already known or are readily identifiable. These were *primary* methods of risk assessment. The next method is a *secondary* risk-assessment method. If hazards and risks of a new product are latent, or hidden from plain sight and intuition, then it is somewhat probable that these risks might be overlooked at the early engineering-design stage. It is hoped that these latent risks are uncovered as product characteristics emerge through further engineering analysis, laboratory testing, and field testing within the EDDTP. Although it should go without saying, product problems are much better addressed during the EDDTP before the product gets into consumer hands. Product-safety recalls are an inefficient and expensive way to correct design and manufacturing defects. The discovery of post-sale product-safety issues also puts the public at risk of injury. Therefore, it is important to identify product risks and to mitigate them as needed. Such mitigation measures can either reduce the severity of a potential injury and/or reduce the likelihood of that injury. The *Failure Modes, Effects, and Criticality Analysis* (FMECA) is one such method available to the engineering-design team and the PSEr.

The Failure Modes and Effects Analysis (FMEA) is a more widely used acronym within the engineering and reliability fields than FMECA. The difference between the FMECA and the FMEA is that the FMECA includes a criticality analysis. This criticality analysis has a risk assessment and often includes recommended actions to further reduce the risk level. Criticality analyses may be either qualitative or quantitative in nature. Some FMEAs actually include a criticality analysis without calling the work an FMECA.

It is important to remember that only *single* failures are considered in FMECAs, in general. If multiple failures are required to result in an injury accident, the FMECA will not necessarily show an accident and its outcomes that as product effects. Furthermore, an FMECA cannot show the consequence of the failure of a part or component that does not exist. For example, strictly speaking, an FMECA will not assess the occupant-ejection risk of an SSV which does not have any doors on it.[18] Of course, any intelligent engineer would recognize such an SSV hazard, but solely relying upon an FMECA to detect and address this—and some other—hazard is an incorrect reliance upon the method.

Furthermore, the FMECA is an *inductive* analysis method. The differences between induction and deduction were discussed in Chapter 4. Another method to be covered

[18] Many, if not most, SSVs will have seat belts which often help prevent occupant ejections, but these restraints may or may not be used by vehicle occupants.

after the FMEA is Fault-Tree Analysis (FTA). FTA is a *deductive* analysis method. To paraphrase Ericson, the FMECA is a bottom-up approach to analysis where the "what if this part failed" question is asked; the FTA is a top-down approach where the "how can this event happen" question is asked (Ericson 2016, 64). As a result, the FMECA starts off with the individual failures of each component of a product to determine their effects with the goal of preventing a failure or an accident with unwanted outcomes. The FTA starts off with an undesirable event—a fault or a failure—and identifies the ways in which that could happen. The FMECA and FTA go in opposite directions as will be shown later.

Returning the discussion to the FMECA, there are several *types* of FMECA. They can be used to analyze an engineering design and become a Design FMECA, or a *DFMECA*. An FMECA can also be used to evaluate a manufacturing process. This becomes a Manufacturing FMECA or *MFMECA*. Similarly, if a design-and-manufacturing company wants to evaluate product difficulties when that product is being used by consumers, the company can conduct a Use FMECA or *UFMECA*. In systems engineering, the failure modes of a function may be of interest so that a functional FMECA or *FFMECA* is conducted. These are just four types of FMECAs that can be constructed and completed. The DFMECA will be of sole interest in this section.

The FMECA is one of several system-safety engineering methods that was applied within the defense and aerospace industries. It is now widely used within the automotive industry. The rapid advances in aerospace and electronic technologies that started in the middle of the twentieth century led to ambitious, complex, ground-breaking, and sometimes questionable ideas that were dependent upon immature technologies for success. The Convair B-58 Hustler example used in Chapter 8 is one such instance of a technologically ambitious and complex system. In such systems, thousands of part failures could lead to a multitude of effects. To make such a problem manageable, it was necessary to create a discipline of systematic methods to understand large, complex systems. The resulting discipline became known as *system safety*.[19]

The field of system safety grew and matured from the 1940s through the 1970s (Ericson 2006). The U.S. military and NASA experienced several destructive and deadly accidents, including the Apollo I accident mentioned in Chapter 5, during this time period. The FMEA/FMECA started to be used regularly by the aerospace and automotive industries in the 1960s and 1970s.

One early FMECA reference source is the old military standard ("Procedures for Performing a Failure Mode, Effects, and Criticality Analysis (MIL-STD-1629)" 1980). Although it has been discontinued ("Procedures for Performing a Failure Mode, Effects, and Criticality Analysis (MIL-STD-1629) Notice 3" 1998), it remains a readily accessible, simple, and free document for people learning FMEA. The FMECA can easily become an intensive and deep process. Many people have spent considerable time diving into the FMECA discipline. In fact, NASA has assembled a thorough bibliography of FMEA resources and examples (NASA 2000) broken down into subject divisions. There are also several good system-safety textbooks and standards including these (Ericson 2016; Gullo and Dixon 2017; "Potential Failure Mode and Effects Analysis in Design (Design FMEA; Potential Failure Mode and Effects Analysis in Manufacturing and Assembly Processes (Process FMEA)—SAE J1739_200901" 2009). The treatment of

---

[19] System safety is sometimes called *systems safety*.

FMECA in this book can be considered to be only a superficial treatment of this particular analysis which is frequently conducted by FMECA specialists. Perhaps due to this deep and exhaustive nature of the FMECA, many people do not understand its necessary *resources* and its potential *results*. It is hoped that this section helps the reader to better understand these issues.

The practice of system safety exists at the intersection of engineering, science, mathematics, and statistics. Ability in each of these fields is helpful in order to fully apply system-safety methods to real-world problems. (The PHL and PHA, shown earlier in this chapter, were also products of the development of system safety as a discipline.)

The primary use of FMECA—and many other system-safety methods—is to determine and improve the *reliability* of a system or product. Other possible goals of the FMECA include high levels of quality, dependability, availability, mission completion, security, and safety. As has been discussed elsewhere in this book, high levels of these other product characteristics *may* lead to a high level of safety, but they do *not* necessarily do so. This *conflation* of reliability and safety leads many people to assume that FMECA is a safety tool. In fact, the output from many FMECAs is the *RPN* (Risk Priority Number). The RPN is, in essence, a quality or reliability metric—not a safety metric.

Much FMECA work in industry is done at the interface between OEMs and suppliers. When a supplier delivers a system for a larger product, engineers from both sides often assemble to conduct an FMECA to determine the effects of potential faults at the interface of the OEM's larger product and the supplier's system. For example, the OEM's and the supplier's engineering teams may meet to go over an electrical subsystem. They may go to the extreme of looking at individual electrical-plug connectors. From there, they may construct a table of possible short circuits due to bent pins or worn electrical insulation and then determine the effects of each possible short-circuit path. This takes much time and much engineering effort, but poor electrical-plug design practices could be exposed through such an exercise. This exercise will be conducted as much for reliability issues as for product-safety ones.

For the purposes of this chapter, only an elementary discussion of FMECA can be undertaken. However, a sample safety-based FMECA worksheet is provided along with an example using the winch discussed earlier in the chapter and book.

The FMECA, in this case a DFMECA, exhaustively looks at the failure of each component in a product. For each component, each way in which it may fail, or failure mode, is examined. The DFMECA worksheet appears in Table 10.12. This FMECA worksheet examine causes, immediate effects, means of detection of failure, current controls to prevent failure, hazard, and risk for each part and failure mode. Risk is composed of severity and probability to produce a risk level. The values of risk severity, probability, and level used in this FMECA are those from Tables 5.2, 5.3, and 5.4. Also included in the FMECA are actions that the design-engineering team should undertake as listed by the PSEr. After the listing for the particular component—an item, part, or assembly—there appears a column for each of these pieces of information as shown in Table 10.12. The banner atop the worksheet is similar to those seen in the PHL, PHA, and PSMx.

The hand-powered winch described in Appendix B is used for this Design FMECA example which will simply be called an "FMECA" hereafter. This appendix contains a parts listing and a bill of materials (BOM) as Tables B.1 and B.2, respectively. Since the winch's parts list contains some assemblies, this parts list alone does not provide sufficient component detail for the FMECA. The BOM is needed and the first four columns of Table B.2 are used as the items evaluated in the example FMECA. The first column of

	Product:											
	Model / Description:											
	Unit Reviewed:		(ID, SN, VIN, other Identifer)									
	Product-Development Stage:											
	Completed by:											
	Date:	____/____/____										
	Item/Part/Assembly	**Failure Mode**	**Causal Factors**	**Immediate Effect**	**Product Effect**	**Means of Detection**	**Current Controls**	**Hazard**	**S**	**P**	**Risk**	**Recommended Action**

TABLE 10.12 Design FMECA (DFMECA) Worksheet

the FMECA worksheet, Table 10.13, is expanded to accommodate the multi-tiered numbering from the winch BOM. Several parts of the FMECA for the winch will be explored without going into full detail for every part or assembly.

The first item, part number 1, is the spool assembly. Since this is an assembly, the FMECA will first break down this part into its components and operations before looking at failure modes and effects. This is because it is important to realize that components, such as bolts, may fail and that operations, such as welding, may be performed incorrectly. The first component to the spool, part number 1.1, is the free, non-driven, side of the spool which does not have a *probable* failure mode and effect. Part numbers 1.1.1 and 1.1.2 are operations performed on the free side of the spool before assembly. These two operations are related to part numbers 5 and 6 which apply to the anchoring of a wire rope or cable for winching the load. Since this winch employs a polyester strap to move the load, the cable-related parts and operations are not applicable ("N/A").

Part number 1.2 is the driven side of the spool and is an assembly having operations numbered as 1.2.1. These holes are low-precision operations, just as operation 1.1.1, and should not have a significant safety effect if not executed properly. Therefore, they, too, are not applicable. As the reader can see, all of the work put into the FMECA so far has produced nothing of product-safety interest. However, the FMECA effort begins to produce results with part number 1.3, the driven ring gear.

The driven ring gear is the large gear on the spool that transfers the cranking effort by the user into tension on the strap used to move the boat on the trailer. This is a hearty part and the most visible failure mode is tooth failure. In true FMECA fashion, the focus will be on the failure of a *single* tooth of the ring gear. Such a tooth failure could be due to an overloading condition during winch use. The immediate effect of the single-tooth failure will be the engagement of the gear teeth adjacent to the failed gear tooth. Due to the particular design of the gear train, the spool will not freewheel; instead, the gear train will jam and render the winch inoperative.[20] The user will notice the inability to turn the cranking handle but will also be able to see the jammed gears. This condition can be controlled, or prevented, through engineering analysis and both laboratory and field testing of the winch as well as through quality-control efforts. Since there is no injury hazard to users due to the design of the gear train, that column and the next three risk columns to not apply.[21] Because this driven ring gear interfaces with the drive sprocket, part number 8.2, the FMECA row for that gear is identical to this gear's row.

The next part number is for a spot-welding sequence of operations, 1.3.1—six spot-welding operations to be precise—that attach the driven ring gear to the driven side of the winch spool. The most likely failure mode for this weld is in shear. This shear failure could be caused by winch overloading or by poor welding. Since the FMECA considers single-item failures, the failure of any single weld will not cause winch failure or even decreased performance—there are five more welds attaching

[20] At least, this is what the engineering team believes will happen upon the failure of a single tooth on the driven ring gear. Since the design has not yet been tested, this effect of this failure mode must be verified. Hence, the FMECA truly is an inductive method.

[21] This treatment of the FMECA may be different from those of other authors since this presentation and treatment is not intended to be exhaustive and is focused solely on product-safety hazards and risks.

Design FMECA (DFMECA)														
#			Item/Part/ Assembly	Failure Mode	Causal Factors	Immediate Effect	Product Effect	Means of Detection	Current Controls	Hazard	S	P	Risk	Recommended Action
1			Spool assembly	N/A										
	1.1		Free side of spool	N/A										
		1.1.1	Stamped holes (11)	N/A										
		1.1.2	Tapped holes (2)	N/A										
	1.2		Driven side of spool	N/A										
		1.2.1	Stamped holes (3)	N/A										
	1.3		Driven ring gear	Broken tooth	Overload	Other teeth engage	Jammed spool	High effort, visual	Engineering Analysis, Laboratory and Field testing; QC	N/A				N/A; no safety-related FMs due to gear design and failure of a single tooth
		1.3.1	Spot Welds (6)	Shear	Overload; poor weld	None (for single weld failure)	None	Visual	Engineering Analysis, Laboratory and Field testing; QC	N/A				N/A; Probability of injury approaches zero
	1.4		Spool shaft	Bending	Overload; Poor material	Spool halves misalign	Winch remains operable	Visual (possibly)	Engineering Analysis, Laboratory and Field testing; QC	N/A				N/A; Probability of injury approaches zero
				Torsion	Overload; Poor material	One half of spool rotates	operable winch; misaligned strap	Visual	Engineering Analysis, Laboratory and Field testing; QC	N/A				N/A; Probability of injury approaches zero

Table 10.13 DFMECA Worksheet for Winch

#			Item/Part/ Assembly	Failure Mode	Causal Factors	Immediate Effect	Product Effect	Means of Detection	Current Controls	Hazard	S	P	Risk	Recommended Action
Design FMECA (DFMECA)														
		1.4.1	Fillet welds (2, one each side)	Shear	Poor weld	Partial rotation of loose half of spool	Small movement of load; inoperable winch	Visual	Engineering Analys s, Laboratory and Field testing; QC	Crush	3-C	1-I	Medium	None; load might only move inches before strap-retainer bolt arrests spool/strap motion
2			Spool-shaft bolt	Tension	Poor material	Jammed, non-rotating spool	Inoperable winch; load stuck	Visual	Engineering Analys s, Laboratory and Field testing; QC	N/A				N/A; Probability of injury approaches zero
				Bending	Poor material	Jammed, non-rotating spool	Inoperable winch; load stuck	Visual	Engineering Analys s, Laboratory and Field testing; QC	N/A				N/A; Probability of injury approaches zero
3			Spool-shaft nut	Separation; lateral movement of shaft	Low torque; poor process control	Jammed, non-rotating spool	Inoperable winch; load stuck	Visual	Engineering Analysis, Laboratory and Field testing; QC	N/A				N/A; Probability of injury approaches zero
	3.1		Thread-locking element	Loss of locking function	Poor process control	None	None	N/A	Engineering Analysis, Laboratory and Field testing; QC	N/A				Mechanical torque of bolted joint will maintain clamp load
4			Spool-shaft bushing	Wear	No lubrication; end-of-life	Load hard to move	None	High effort, audible	Lubrication during assembly	N/A				N/A; Probability of injury approaches zero

5			Strap-retainer bolt	Tension	Overload; improper winch use	Loss of complete strap retention	Movement of load	Visible	Engineering Analysis, Laboratory and Field testing; QC	Impact	3-C	1-I	Medium	None; testing has shown suitable strap-retention qualities
				Bending	Overload; improper winch use	Loss of complete strap retention	Movement of load	Visible	Engineering Analysis, Laboratory and Field testing; QC	Impact	3-C	1-I	Medium	None; testing has shown suitable strap-retention qualities
6			Strap-retainer nut	Separation	Low torque; poor process control	Loss of complete strap retention	Movement of load	Visible	Engineering Analysis, Laboratory and Field testing; QC	Impact	3-C	1-I	Medium	None; testing has shown suitable strap-retention qualities
	6.1		Thread-locking element	Loss of locking function	Poor process control	None	None	N/A	QC	N/A				Mechanical torque of bolted joint will maintain clamp load
7			Strap-retainer bushing	N/A										N/A; no safety-related FMs probable
8			Drive-sprocket shaft	N/A										An assembly; look at components below
	8.1		Shaft	Torsion	Overload	Loss of connection between handle and spool	Movement of load	Visual	Engineering Analysis, Laboratory and Field testing; QC	Impact	3-C	1-I	Medium	None; testing has shown suitable shaft qualities
				Bending	Overload	Plastic deformation of shaft	Jammed spool	High effort, visual	Engineering Analysis, Laboratory and Field testing; QC	N/A				N/A; no safety-related FMs due to gear design

Table 10.13 DFMECA Worksheet for Winch (*Continued*)

Design FMECA (DFMECA)														
#			Item/Part/ Assembly	Failure Mode	Causal Factors	Immediate Effect	Product Effect	Means of Detection	Current Controls	Hazard	S	P	Risk	Recommended Action
		8.1.1	Threaded end	Stripped threads	Improper use	Loose handle	Winch harder to use	Visual	Engineering Analysis, Laboratory and Field testing; QC	N/A				None; suitable threads
		8.1.2	Flats (2)	Wear	Overload; end-of-life	Cannot turn drive shaft	Cannot move spool with the handle	Visual	Engineering Analysis, Laboratory and Field testing; QC	N/A				None; the two flats will fail in parallel; flats are suitable for use
		8.1.3	Groove	N/A										N/A; no safety-related FMs probable
	8.2		Drive sprocket	Broken tooth	Overload	Other teeth engage	Jammed spool	High effort, visual	QC	N/A				N/A; no safety-related FMs due to gear design and failure of a single tooth
		8.2.1	Fillet weld (one side)	Shear	Poor weld	Spool disconnects from drive shaft	Spool can "free wheel"	Visual	QC	Impact	3-C	1-I	Medium	None; testing has shown suitable shaft qualities

9			Drive-sprocket shaft bushing (2)	Wear	Overload; end-of-life	Drive shaft hard to turn	Makes noise	Audible	QC	N/A				N/A; no safety-related FMs probable
10			Drive-sprocket shaft C-clip	Fracture	Poor material	Drive shaft can shift laterally	Noise	Audible	Engineering Analysis, Laboratory and Field testing; QC	N/A				None; other design features retain keep shaft in position
11			Strap assembly	N/A										An assembly; look at components below
	11.1		Webbing, black, 50 mm x 5 meters	Tension	Overload; Poor material	Broken strap	Movement of load	Visual	Engineering Analysis, Laboratory and Field testing; QC	Impact	3-C	1-I	Medium	None; testing has shown suitable material properties
		11.1.1	Loop End	Tension	Overload; Poor material	Broken strap	Movement of load	Visual	Engineering Analysis, Laboratory and Field testing; QC	Impact	3-C	1-I	Medium	None; testing has shown suitable material properties
		11.1.2	Hook End	Tension	Overload; Poor material	Broken strap	Movement of load	Visual	Engineering Analysis, Laboratory and Field testing; QC	Impact	3-C	1-I	Medium	None; testing has shown suitable material properties
	11.2		Hook Lead	N/A										N/A

Table 10.13 DFMECA Worksheet for Winch (*Continued*)

Design FMECA (DFMECA)														
#			Item/Part/ Assembly	Failure Mode	Causal Factors	Immediate Effect	Product Effect	Means of Detection	Current Controls	Hazard	S	P	Risk	Recommended Action
		11.2.1	Webbing, black, 25 mm x 600 mm	Tension	Overload; Poor material	Broken strap	Loss of hook control during attachment to load	Visual	Engineering Analysis, Laboratory and Field testing; QC	N/A				None; hook lead is not used when winch is under load
		11.2.2	Stitching	Unstitching	Poor process control	Unfastenting of webbing loop	Movement of load	Visual	Engineering Analysis, Laboratory and Field testing; QC	Impact	3-C	1-I	Medium	None; stitching should fail slowly over time while maintaining sufficient strength for safe use
		11.2.3	Warning label	N/A										No FMs for the DFMECA
	11.3		Hook assembly	N/A										An assembly; look at components below
		11.3.1	Hook	Fracture	Overload; Poor material	Loss of load attachment	Movement of load	Visual	Engineering Analysis, Laboratory and Field testing; QC	Impact	3-C	1-I	Medium	None; testing has shown suitable component properties
		11.3.2	Latch plate	Open hook attachment	Misuse	Open hook	Movement of load—possible only if strap is extremely loose	Visual	Engineering Analysis, Laboratory and Field testing; QC	N/A				None; latch plate is not involved in securing load once strap is taut

		11.3.3	Latch torsion spring	Open hook attachment	Misuse	Open hook	Movement of load—possible only if strap is extremely loose	Visual	Engineering Analysis, Laboratory and Field testing; QC	N/A				None; latch plate is not involved in securing load once strap is taut
		11.3.4	Latch pin	Open hook attachment	Misuse	Open hook	Movement of load—possible only if strap is extremely loose	Visual	Engineering Analysis, Laboratory and Field testing; QC	N/A				None; latch plate is not involved in securing load once strap is taut
		11.3.5	Latch-pin rivet	Open hook attachment	Misuse	Open hook	Movement of load—possible only if strap is extremely loose	Visual	Engineering Analysis, Laboratory and Field testing; QC	N/A				None; latch plate is not involved in securing load once strap is taut
12			Cable-retaining clamp	N/A										N/A; strap is used instead of wire rope/cable
	12.1		Stamping and forming, body	N/A										N/A; strap is used instead of wire rope/cable
	12.2		Stamping, holes (2)	N/A										N/A; strap is used instead of wire rope/cable
13			Cable-retaining machine screws (2)	N/A										N/A; strap is used instead of wire rope/cable
14			Handle assembly	N/A										An assembly; look at components below

Table 10.13 DFMECA Worksheet for Winch (*Continued*)

Design FMECA (DFMECA)														
#			Item/Part/ Assembly	Failure Mode	Causal Factors	Immediate Effect	Product Effect	Means of Detection	Current Controls	Hazard	S	P	Risk	Recommended Action
	14.1		Handle, rubber	N/A										N/A; no safety-related FMs probable
	14.2		Handle shaft	Bending	Misuse	Difficult to crank winch	None	High effort, visual	Engineering Analysis, Laboratory and Field testing; QC	N/A				None; testing has shown suitable material properties
		14.2.1	Handle-shaft rivet (1)	Pull through	Misuse; poor process control	Not possible to crank	Inoperable winch	Visual	Engineering Analysis, Laboratory and Field testing; QC	N/A				None; testing has shown suitable material properties
	14.3		Handle lever	Bending	Misuse	Difficult to crank winch	None	High effort, visual	Engineering Analysis, Laboratory and Field testing; QC	N/A				None; testing has shown suitable material properties
		14.3.1	Stamping	N/A										N/A; no safety-related FMs probable
		14.3.2	Forming	N/A										N/A; no safety-related FMs probable
		14.3.3	Hole with 2 flats	Rounding of hole	Misuse; poor process control	Not possible to crank	Inoperable winch		Engineering Analysis, Laboratory and Field testing; QC	N/A				None; testing has shown suitable material properties

15			Handle retainer	N/A										N/A
	15.1		Stamping	N/A										N/A
	15.2		Forming	N/A										N/A
	15.3		Hole with 2 flats	Rounding of hole	Misuse; poor process control	None	None		QC	N/A				None; flats on handle lever will continue to function
16			Drive-sprocket shaft nut	Separation; lateral movement of shaft	Poor process control	Handle detaches from shaft	Inoperable winch	Visual	Engineering Analysis, Laboratory and Field testing; QC	N/A				None; testing has shown suitable material properties
	16.1		Thread-locking element	Loss of locking function	Poor process control	None	None	N/A	QC	N/A				Mechanical torque of bolted joint will maintain clamp load
17			Ratchet-shaft bolt	Tension	Overload; Poor material	Loss of ratchet shaft	Unlocking of spool; movement of load	Visual	Engineering Analysis, Laboratory and Field testing; QC	Impact	3-C	1-I	Medium	None; testing has shown suitable material properties
				Bending	Overload; improper winch use	Loss of ratchet shaft	Unlocking of spool; movement of load	Visible	Engineering Analysis, Laboratory and Field testing; QC	Impact	3-C	1-I	Medium	None; testing has shown suitable material properties
18			Ratchet-shaft nut	Stripped threads	Poor process control	Loss of ratchet shaft	Unlocking of spool; movement of load	Visual	Engineering Analysis, Laboratory and Field testing; QC	Impact	3-C	1-I	Medium	None; testing has shown suitable material properties

Table 10.13 DFMECA Worksheet for Winch (*Continued*)

Design FMECA (DFMECA)												
#	Item/Part/ Assembly	Failure Mode	Causal Factors	Immediate Effect	Product Effect	Means of Detection	Current Controls	Hazard	S	P	Risk	Recommended Action
19	Ratchet-shaft spacer bushing	N/A										N/A
20	Ratchet-shaft flat washer	N/A										N/A
21	Ratchet pawl	Tooth bending	Overload; misuse	Unlocking of spool	Movement of load	Visual	Engineering Analysis, Laboratory and Field testing; QC	Impact	3-C	1-I	Medium	None; testing has shown suitable material properties
22	Ratchet tensioner	Bending	Misuse	Unlocking of spool	Movement of load	Visual	Engineering Analysis, Laboratory and Field testing; QC	N/A				N/A; Probability of injury approaches zero
23	Ratchet tension spring	Breakage	Poor material	Release of ratchet pawl	Unlocking of spool; movement of load	Visual	Engineering Analysis, Laboratory and Field testing; QC	Impact	3-C	1-I	Medium	None; testing has shown suitable material properties
24	Winch chassis	N/A										An assembly; look at operations below
24.1	Stamping, body	Misalignment of sides	Poor process control	Winch cannot be assembled	Rejected winch			N/A				N/A; winch will never be sold
24.2	Stamping, holes (10)	Misalignment of holes	Poor process control	Winch cannot be assembled	Rejected winch			N/A				N/A; winch will never be sold

	24.3		Stamping, holes with keyways (2)	Misalignment of holes	Poor process control	Winch cannot be assembled	Rejected winch			N/A				N/A; winch will never be sold
	24.4		Forming	Misalignment of holes	Poor process control	Winch cannot be assembled	Rejected winch			N/A				N/A; winch will never be sold
25			Winch-mounting bolts (3)	Loss of bolt	Poor user attachment	None; two bolts remaining	None			N/A				N/A; redundant mounting bolts
26			Winch-mounting flat washers (6)	N/A										N/A; no safety-related FMs probable
27			Winch-mounting locking washers (3)	Loss of clamp load	Poor user attachment	None; two fasteners remaining	None			N/A				N/A; redundant mounting fasteners
28			Winch-mounting nuts (3)	Loss of bolt	Poor user attachment	None; two fasteners remaining	None			N/A				N/A; redundant mounting fasteners

Table 10.13 DFMECA Worksheet for Winch (*Continued*)

the ring gear to the winch spool. Upon close inspection, a user might be able to detect a weld failure—but probably not. The chances for such a weld failure can be lowered through analysis, laboratory and field testing, as well as through QC measures. Again, there is no injury hazard from the failure of a single spot weld and "N/A" is entered into that column.

It is at part number 1.4, the spool shaft which attaches together the two spool flanges, that some of the breadth of the FMECA method reveals itself. There are two possible failure modes for this single component. Therefore, the line for the spool shaft contains two sub-lines, or failure modes—bending and torsion. It is possible, however unlikely, that the spool shaft could fail in bending since the strap transfers a load at right angles to the center of the shaft which is supported at each end as a "simple beam." A relatively much more likely, but still unlikely, failure mode is that of torsional failure due to the input load on the spool only being directed to one side of the spool—the driven side. The free side of the winch spool is driven through the torque provided through the spool shaft and fillet weld. If the spool shaft was to fail in bending it could be due to overloading the winch or poor materials. Upon bending failure, the shaft would plastically deform and misalign the two halves resulting in the spool appearing abnormal to the user when the winch was in use. This would be visible to the user, but the winch should continue to perform. Prevention of this failure would be accomplished in the same way as the earlier failure modes. There would be no injury hazard from this condition so hazard and risk would not apply. With regard to the torsional failure of the spool shaft, it could result from overloading or poor material. Spool-shaft failure in torsion under winch-strap loading would result in plastic deformation which could result in some rotation of free half of the spool—but very little. The winch would remain functional after this, but the strap may not want to "track" properly due to rotational misalignment of the two spool halves. Engineering analysis, laboratory and field testing, and QC measures could prevent such a failure mode. As with bending failure, torsional failure would result in no injury hazard.

The two halves of the winch spool are connected to one another through the spool shaft. However, the spool shaft is connected to each spool half through a fillet-weld operation, part number 1.4.1. The failure mode for such a weld in such an application is shear failure. In this product, the welds are essentially in "series" since the failure of either of the two welds will result in the disconnection of one spool half from the other.

It is at this point that the inductive nature of the FMECA appears. If either of the fillet welds was to fail it might lead to an injury or it might not. The two spool halves would still remain connected through the strap-retainer hardware mentioned earlier. At this point, the PSEr does not know if the load would be released with the potential of injuring a bystander or not. If the product is already prototyped, laboratory testing could easily determine through testing the effects of weld failure and the strength of the strap-retaining hardware. If the product is not yet prototyped, then the PSEr will need to conduct analysis, consult with design engineers, and exercise engineering judgment to determine any hazard and risk. As with any form of inductive reasoning, one often cannot be certain of an answer from analysis and consultation. For this section's purposes, it will be assumed that the strap-retention hardware will hold the winch's load and only permit the load to move a few inches (50 mm) at most. This failure and effect will render the winch inoperable and its condition will be visually evident to the user. Of course, analysis and laboratory and field testing, plus QC efforts will help minimize the probability of this failure mode.

Due to the unexpected motion potential of this failure mode, there is a crush hazard. For example, although a boat may not move many feet (one or more meters), there is the chance that someone's hand or fingers could be crushed by the unintended motion of the boat. Therefore, risk severity is considered to be 3, or Critical, and risk probability is rated as 1, or Improbable. This combination of severity and probability produces a risk level of Medium. Again, Tables 5.2 through 5.4 are used for this risk estimate.

The description of the winch example will go no further. The DFMECA is included in its entirety. The reader is asked to examine the remainder of Table 10.13. This FMECA exposes only one hazard. The crush and impact hazard arise from the same unexpected motion of the load due to the failure of winch parts or operations. On the one hand, this hazard is not something that an intelligent person could not have figured out without the assistance of the FMECA. On the other hand, this FMECA reveals *some* of the many ways in which the design can fail leading to an accident.[22]

No two people or groups will likely produce identical FMECA worksheets. There will probably be variances in terminology and, perhaps, immediate and product effects. As stated earlier in the book, it is not so important what something is called, but it is important that product-safety is provided regardless of nomenclature. Once some level of FMECA expertise is reached, it will become important to be consistent with terminology. However, engineering students and readers should not be afraid of potentially making mistakes; making mistakes is an integral part of learning.

As with any method, the FMECA has both benefits and drawbacks. With large- and even medium-sized products or systems, the expansiveness of a complete FMECA can be overwhelming. The FMECA is not a particularly useful, efficient, or effective way to assess the *safety* of a newly designed product. System-safety author, Leveson, agrees: "FMEA (Failure Modes and Effects Analysis) is sometimes used as a hazard analysis technique, but it is a bottom-up reliability analysis technique and has very limited applicability for safety analysis" (Leveson 2011, 211). It is important to temper expectations of the product-safety results from an FMECA—versus the effort put into it—by remembering that many or most of the FMECA's results will not directly point toward an immediate and actionable product-safety risk. It is easy to see, however, how useful the FMECA is for reliability analysis. It may take several part failures, or a particular combination of failures, within the Product/User/Environment (PUE) model in order to manifest an injury risk. The FMECA remains limited to, but still available for, the evaluation of the effects of single failures in a product.

One serious limitation of the FMEA for product-safety applications is that it strictly does not identify hazards unless a *failure* is the cause. It is quite possible to be injured by a product that does not fail. The instance of someone cutting their hand with a kitchen knife involves no failure of the knife; the knife is simply functioning as designed. In another example, no product failure is involved, generally speaking, when an automobile operator hits a tree alongside the road. Unless there was a form of mechanical failure—such as one affecting steering, braking, or throttle control—the vehicle was functioning as intended. It remains unclear if this will still be the case in the near future as advanced driver-assistance systems (ADAS) countermeasures, including lane-departure warnings and interventions, collision-avoidance interventions,

[22] Remember that the FMECA only investigates the effects of single failures. There may exist *combinations* of failures that could result in events which will not be uncovered by the FMECA.

and semi-autonomous and fully autonomous vehicles, become more commonplace in vehicles. One engineering professional organization has started addressing aspects of such autonomous and intelligent systems (A/IS) (*Ethically Aligned Design: A Vision for Prioritizing Human Well-Being with Autonomous and Intelligent Systems* 2019).

If the new product is based upon similar existing products, then much is already known about its hazards and risks. For many products, the primary risks are well known and a product-safety analysis would be better served by a risk assessment method that can leverage such knowledge from prior product experiences. Even if a product is new and innovative, it is often possible to imagine some of the hazards that its users will face. One such method for using such knowledge, the FTA, is covered in the next section of this chapter. However, if a completely new feature or novel way of performing a function is incorporated into the new product, then an FMECA might still be useful and productive in identifying some of the hazards and potential risks of the new product. This is especially if that product is large, highly complex, and perhaps otherwise unmanageable.

10.6 Fault-Tree Analysis

As was briefly mentioned in the prior section on FMECA, the FTA is risk-assessment method that substantially differs from the FMECA presented in the last section. Although the FTA is also a *secondary* risk-assessment method, it is a top-down deductive methodology unlike the FMECA's bottom-up inductive approach.

The FTA falls under the umbrella of system safety and shares some of the origins of such methods. The FTA originated at Bell Telephone Laboratories in 1962 (Hammer 1972, 238). More-detailed descriptions of FTA can be found in the following references, among others (Ericson 2016; NASA 2002; NASA/STI 2000).

In the FTA, the PSEr starts off by identifying an event that she or he does not want to happen. This could be a particular accident with injury or death as an outcome. Such undesirable events could include fires, explosions, ORV rollover accidents, unintended start-ups, inabilities to shut-down, and many more. For numerous products, the accidents which lead to injury outcomes are often well known. Therefore, the FTA method permits the PSEr to focus efforts in a more-efficient manner than the FMECA method. The FMECA is an exhaustive and sometimes exhausting method to implement for real-world projects. In addition, while the FMECA output is quite useful for those pursuing reliability and quality problems, the method often falls short when it comes to providing actionable insight into product-safety problems. The FMECA can be helpful to product-safety engineers, but, in many cases, the FTA is time and effort better spent on improving consumer-product safety.

This chapter will give but a brief introduction to FTA. The FTA, as with FMECA, is a field where an interested engineer could spend much time in study. It is hoped that some readers will be so motivated after the rudimentary treatment of FTA provided here. FTA is a fascinating area for both understanding how the use of a product may lead to an accident and user injury as well as how the probability of such an event may be reduced. This is true in both qualitative and quantitative senses. Both approaches will be investigated here.

This FTA exploration will be limited to the use of only four symbols. These symbols are the event, the basic event, the AND gate, and the OR gate. Each of these symbols is shown in Figure 10.2.

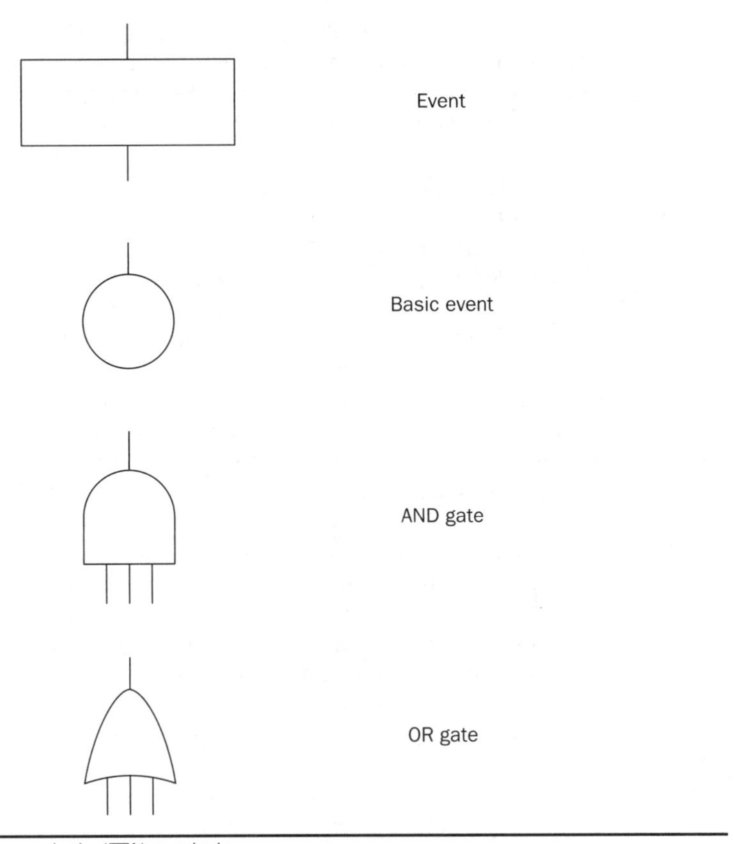

Figure 10.2 Fault-tree analysis (FTA) symbols.

The first of these FTA symbols is the event which is represented by a rectangle. A generic event will have connections both at the top and at the bottom. There is one particular event, the top-level event (TLE), which only has a connection on the bottom since the TLE sits atop the FTA. The TLE is the one event that the design engineer and PSEr do not want to occur. Such an event is the result of other events and gates.

Another type of event which is separate from the above TLE is the basic event. The basic event is an independent event. These basic events are represented circles with only one connection at the top of each. A single basic event may appear multiple times as inputs to gates leading to other events, but will appear at the bottom of an FTA branch. The graphic representation of these events is discussed soon.

There are two types of "gates" used in the FTA presented here. The two gate types are *AND* and *OR*. These gates represent conditions that must be met before the gate lets pass whatever events are trying to get past it in order to permit another event to take place. For example, with the AND gate, *all* of the events immediate below that gate and connected to it must have taken place in order to lead to the event above that gate. In the case of the OR gate, *any*[23] of the events taking place below and connected to the gate will lead to the event above the gate.

[23] One or more.

An example of a non-specific FTA is included as Figure 10.3. In this example fault tree, which is qualitative in nature, at the very top of the figure is the TLE. It is not important what the TLE is but, in any case, it can be seen from Gate 1, an AND gate, that the TLE cannot happen unless both Event A and Event B take place.

Event A can only take place if Event C OR Event D take place because of Gate 2. Event C can only occur if both Basic Event 1, B1, AND Basic Event 2, B2, take place due to Gate 4. Event D can only occur if either B3 OR B4 take place because of Gate 5.

Event B is prevented from happening unless Event E, Event F, Event G, and Basic Event, B1 all take place due to AND Gate 3. Event E is prevented from happening by AND Gate 6 unless both Basic Events B1 AND B5 take place. Event F only happens because of OR Gate 7 which requires that either B6 or B7 occurs. Event G cannot take place unless OR Gate 8 sees that either B2, B3, OR Event H happen. Finally, Event H will not take place unless AND Gate 9 sees that both B4 AND B6 have happened.

From looking at the fault tree in Figure 10.3, the reader will notice that gates are not connected to gates; events separate gates. Similarly, events are never connected to events; gates separate events. Gate-to-gate and event-to-event connections make no sense in the FTA methodology and are avoided.

When looking at a qualitative (QL) FTA, take notice of the AND gates and the OR gates and how they are feeding into one another—of course, through events. Since an AND gate is harder to get past than an OR gates, branches of the fault tree with many AND gates will be less likely to lead to the TLE. The opposite is also true. A branch having many—or only—OR gates can quickly lead to the undesirable TLE. If a branch only has OR gates leading directly to the TLE, then any of the basic events at the bottom that branch will produce a single-point failure (SPF) for the system. Basic events B3 and B4

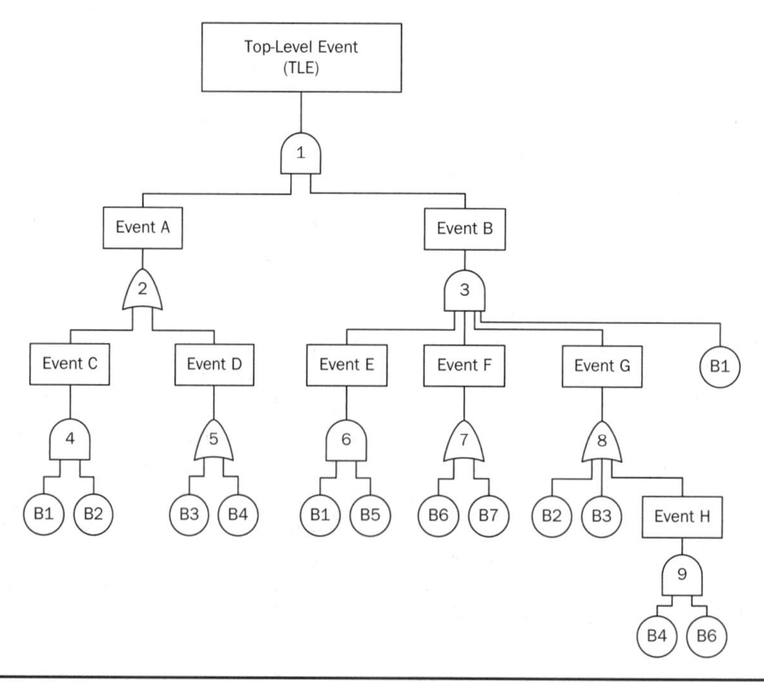

FIGURE 10.3 Fault-tree analysis (FTA) example—qualitative.

proceed upward through OR Gate 5 to Event D and then through OR Gate 2 onto Event A. It is only at AND Gate 1 that this failure path is stopped from being a SPF leading to the TLE.

The FTA can also be used for quantitative estimates. If the probabilities of the basic events are known, then an estimate of the TLE can be calculated. It is important to remember, however, that the quality of any such TLE estimates depends upon the quality of the input values. Additionally, the probability calculated for the TLE represents the occurrence of only that event, or failure, and not all undesirable events or failures.

Assume that the probabilities of the basic events in Figure 10.3 are available and are as given below:

$$P(B1) = 0.01$$
$$P(B2) = 0.02$$
$$P(B3) = 0.03$$
$$P(B4) = 0.04$$
$$P(B5) = 0.05$$
$$P(B6) = 0.06$$
$$P(B7) = 0.07$$

It will now be possible to calculate the probabilities for each event leading to the TLE using mathematical properties for the AND and OR gates.

The probability of each basic event is shown beneath each basic event in Figure 10.4. The first Event for which the probability must be calculated—working from left to right and from the bottom up—is Event C which depends upon Basic Events B1 and B2. These basic events are connected to Event C through an AND gate. The probability for an AND gate is simply the product of the probabilities of its input events, in this case B1 and B2. Therefore, the probability of Event C is calculated as:

$$P(C) = P(B1) \times P(B2) = 0.01 \times 0.02 = 0.0002$$

This calculation also works for three or more events.

In the case of an OR gate, such as the gate connecting Event D to B3 and B4, the event probability is *approximated* by the sum of the basic-event probabilities. Therefore, the probability of Event D is approximately the probability of event B3 plus the probability of event B4, or:

$$P(D) = P(B3) + P(B4) = 0.03 + 0.04 = 0.070$$

This calculation also works with three or more events—as will be needed to calculate the probability of Event G, but it must be remembered that this OR-gate approximation only works for *small* values of probability. As the individual input probabilities increase, the calculated output probability will be an overestimate of the true values. Ericson provides an excellent discussion of this effect and greater accuracy will require a mathematical "expansion" of the relevant terms (Ericson 2016, 257, 258). As a quick illustration of the failure of the OR gate using this simplified calculation, consider an OR gate that depends upon tossing a coin three times and getting at least one "heads."

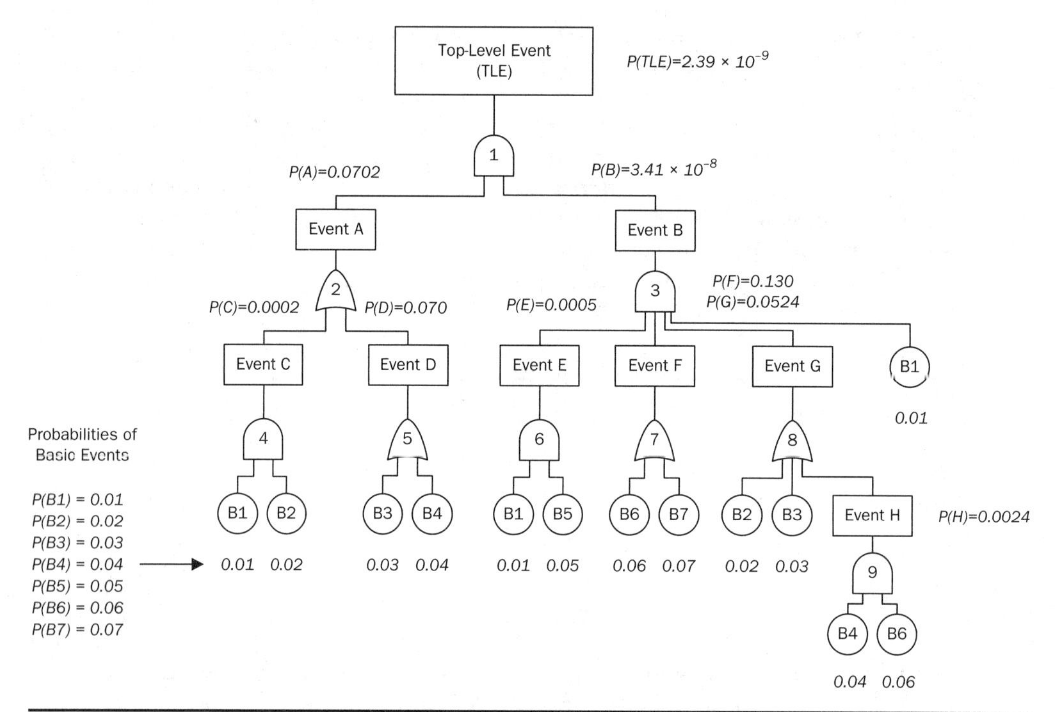

FIGURE 10.4 Fault-tree analysis (FTA) example—quantitative.

Since this fair coin will produce "heads" fifty percent (50%) of the time, the probability of getting at least one "heads" is calculated by the following:

$$
\begin{aligned}
P(\text{"At least One Heads"}) &= P(\text{Heads: Toss 1}) + P(\text{Heads: Toss 2}) + P(\text{Heads: Toss 3}) \\
&= 0.5 + 0.5 + 0.5 \\
&= 1.5
\end{aligned}
$$

This misapplied calculation produces an incorrect probability for tossing a coin three times and getting at least one "heads" of 150% which is greater than the possible upper value of 100%.This contrived example shows how large probability values must be avoided unless care is taken to consider including the higher-order terms of the OR-gate probability calculation.

Since the probability values used in this example are relatively small, it is safe to continue completing the probabilities for the events, including the TLE, in Figure 10.4. The results of these event-probability values are listed in the figure near the events. The calculated probability of the TLE is determined to be 2.39×10^{-9} by using the basic-event probabilities provided above.

As a confirmation of the SPF discussion above for the qualitative FTA in Figure 10.3, comparing the probabilities of events A and B, P(A) and P(B), respectively, in Figure 10.4, shows that the presence of the multiple OR gates does lead to P(A) being much larger than P(B).

The hand-operated winch example will now be revisited through an FTA example. The faults, or failures, leading to the unintended movement of the boat, or load, are

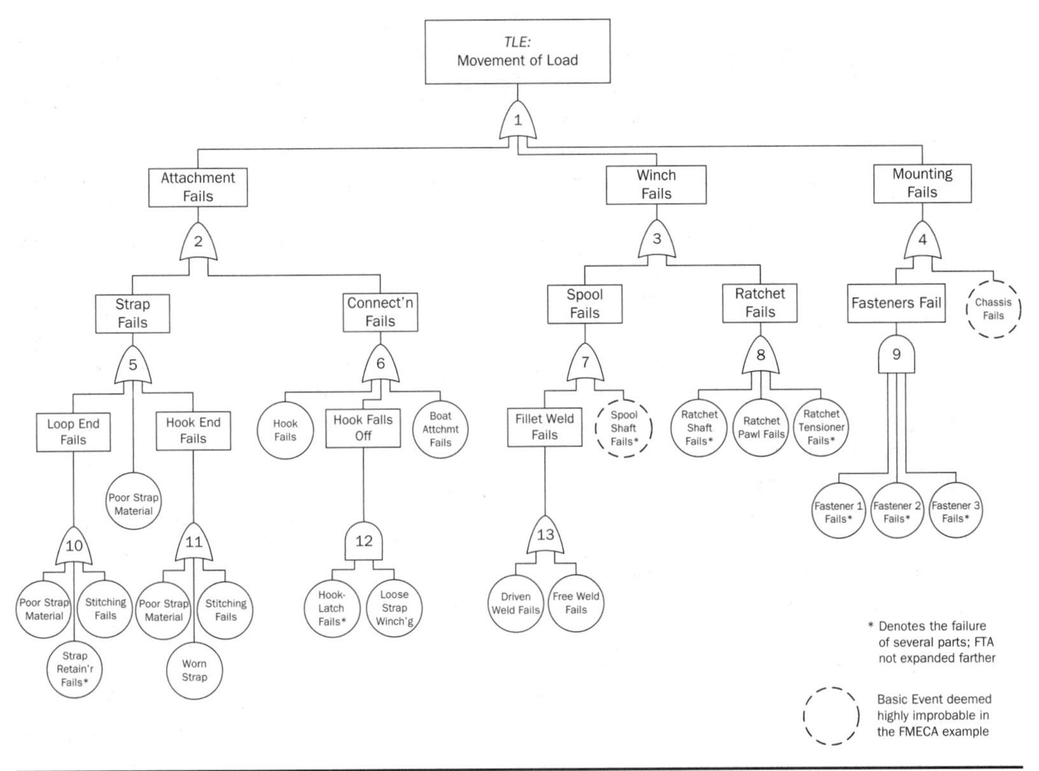

FIGURE 10.5 Fault-tree analysis (FTA) for winch.

shown as Figure 10.5. The unintended motion of the boat on, or from, the boat trailer is the TLE. This TLE is shown at the top of the fault tree in the figure. Gate 1 shows that there are three types of ways that the hand-powered winch can fail leading to the movement of the load: the attachment to the boat through the strap fails, the winch itself fails, and the winch-to-boat mounting joint fails. Since any one of these failures will lead to the TLE, Gate 1 is an OR gate.

Considering the *first* failure under Gate 1, the attachment to the boat is compromised or lost, through a generalized strap failure. There are two possibilities, each leading to attachment failure. Consequently, Gate 2 is an OR gate. The two failures immediately beneath Gate 2 are Strap Fails and Connection Fails.

Neither event just under Gate 2 is a basic event. Strap Fails and Connection Fails are the results of other events, so they must each be expanded farther.

Gate 5 is under Strap Fails and is an OR gate since any of the three events below it will lead to failure. These three events are Loop End Fails, Poor Strap Material, and Hook End Fails. Loop End Fails takes place if any of the events below OR Gate 10 occurs. These three events are all basic events and are Poor Strap Material, Strap Retainer Fails,[24] and Stitching Fails.

[24] The strap-retention system consists of at several components or operations: bolt, nut, bushing, and locking feature. Failures of these will not be investigated independently. The asterisk, "*", in Figure 10.5 will be used to designate similar aggregations of components or operations to simplify the FTA example.

Poor Strap Material, directly under Gate 5, is a basic event and needs no further expansion. Hook End Fails is not a basic event and will occur if any of the three basic events below it takes place. Consequently, Gate 11 is an OR gate. These basic events below Gate 11 are Poor Strap Material, Worn Strap, and Stitching Fails. Although Worn Strap could happen at either end of the strap, as well as the middle, Worn Strap is included at the hook end of the strap since that is where the wear on the strap will be more severe.[25]

It can be seen that Poor Strap Material appears three times and Stitching Fails appears two times just under Strap Fails. This confirms that a basic event may appear multiple times within a fault tree.

The other event under Gate 2, Connection Fails, is a failure in the connection between the strap and the boat. The strap-boat connection can be lost if any of the three events below Gate 6 takes place. These events are Hook Fails, Hook Falls Off, and Boat Attachment Fails. Both Hook Fails and Boat Attachment Fails[26] are basic events that need not be further expanded. The failure of the hook would be a fracture or other failure releasing the boat from the strap while the failure of the boat attachment would be either a fracture of the boat's attaching loop or the loop pulling out of the boat due to poor structural integrity of the boat. The non-basic event under Gate 6, Hook Falls Off of boat, is the result of a combination of the two events below Gate 12. In order for the hook to fall off of the boat, the Hook Latch must fail and leave the hook "open" and there must also be slack in the winch strap, Loose Winch Strap, to permit the hook to fall out of the boat's attachment loop. This makes Gate 12 an AND gate. Since the hook latch is composed of several parts or operations, there is an asterisk beside its name.

The *second* failure under Gate 1 is Winch Fails. There are two ways in which the winch fails under OR Gate 3—Spool Fails and Ratchet Fails. The spool can fail in two ways: Fillet Weld Fails and Spool Shaft Fails. Since only one of these failures is necessary to lead to Winch Fails, Gate 7 is an OR gate. For Fillet Weld Fail, only one of the two fillet welds—driven side or free side—must fail, so Gate 13 is an OR gate. Under Gate 7 also appears Spool Shaft Fails. Since it would be the failure of several components, an asterisk is used in its name. This failure was not elevated to an injury-causing hazard in the FMECA, but was added at the author's discretion to show that it is worth continued consideration by a design engineer at a conceptual level.[27] As a result, this basic event is represented by a dashed line instead of a solid line.

Also appearing under Gate 3 is Ratchet Fails. Ratchet failure results from any of three failures below Gate 8. Gate 8 must be an OR gate, as a result and has Ratchet Shaft Fails, Ratchet Pawl Fails, and Ratchet Tensioner Fails. The first and third of these events are failures of several components and are denoted by an asterisk after their basic-event names.

The third and final failure under Gate 1 is Mounting Fails. In this case, a structural failure takes place that either tears away significant componentry from the winch's

[25] This is an application of "engineering judgment."

[26] It is interesting to note that Boat Attachment Fails points to the contribution of non-winch design elements to an accident situation. This external component, beyond the control of the designer/manufacturer, would fall under the Physical Environment in the PUE model.

[27] The contribution of Spool Shaft Fail to the probability of the TLE could be assessed through a quantitative FTA, but it is not conducted here.

chassis or results in the complete separation of the entire winch from the boat trailer's mount. This can occur in two ways: all three of the fastener assemblies must fail or the winch chassis itself fails. Since either of these alone is sufficient, Gate 4 is an OR gate. Since all three fasters assemblies—including bolts or nuts—must fail, Gate 9 is an AND gate. Chassis Fails is a basic event which was not included in the FMECA. Therefore, it is within a dashed line.

There will be no attempt made to quantitatively evaluate the probability of the TLE: Movement of Load. This is because this winch is a fictional product that may still be in design at this stage of the PDP. It is also because that the lack of hard numbers does not significantly reduce the utility of the FTA. The FTA's purely qualitative representation of the failure paths provides insight into the occurrence of the TLE. These paths can guide both the design engineer in product design and development, but it can also provide insight for the PSEr when designing a system of product-safety facilitators. This is described below.

A quick look at Figure 10.5 shows that there are numerous paths to the TLE resulting from just one failure. These are the SPFs. In fact, any failure beneath any gates other than the AND gates—Gate 9 and Gate12—immediately lead to the TLE. The existence of AND gates help prevent SPFs, but in many products, such AND conditionals may not exist. Depending upon the product, this may be a problem. However, for many products, a user will see and recognize an imminent failure—such as a badly frayed winch strap—and discontinue use of the product. If it is possible to readily identify such conditions, then a set of facilitators may be required. Examples include an on-product warning sign asking users to inspect a product before each use, an owner's manual exhorting users to read, understand, and follow all on-product warning signs, and a PM schedule so that worn product components may be replaced before failure. Through such facilitators, a set of "barriers" to SPFs can be constructed. This barrier identification, and creation, is considered within *bow-tie analysis* (Jensen 2020), but is not presented here due to space considerations.

Another virtue of the FTA is that it can expose needs for new countermeasures during the design phase of a product. In the earlier case of the ORV occupant-ejection example, the FMECA may not show the consequences of not having doors or nets on the vehicle from a bottom-up approach focusing on failures of existing parts. However, the top-down approach of the FTA could show a need for doors or nets on an SSV to design engineers if the TLE under consideration was occupant ejection.

There are additional aspects of FTA that should be investigated by readers with an interest in system safety. These include other types of gates as well as cut sets and success trees. Information on these, and other, FTA topics can be found in the references.

Leveson states, "Bottom-up reliability engineering analysis techniques, such as failure modes and effects analysis (FMEA), are not appropriate for *safety* analysis. Even top-down techniques, such as fault trees, if they focus on *component* failure, are not adequate"[28] (Leveson 2011, 14). This insightful observation notwithstanding, the FTA is a powerful tool for examining the safety of existing and future products. It is often a more productive exercise that and FMECA and the FTA can still lead to valuable insights in a fast-paced product-development environment. Its deductive nature and its ability to focus on safety and specific known or suspected TLEs—rather than on reliability—make the FTA a valuable tool to the PSEr.

[28] Emphases by this book's author.

10.7 Conclusion

This chapter provides some systematic tools to use for the task of product-safety engineering. These methods, coupled with product-safety concepts from Chapter 4, along with hazard and risk identification and estimation from Chapter 5, will help a design-and-manufacturing company design, manufacture, and deliver safe products to its consumers through an integrated and systematic *Design for Product Safety* effort.

Some of these tools are quite elementary, such as the PHL and PHA. Another method, the PSMx is a logical extension of the PHL and PHA. The PSMx has useful and practical additions that can be helpful, and even tailored, to design or product-safety engineers in real-world product-development environments.

There is also, in this chapter, an interesting discussion of fault-averse and fault-tolerant engineering-design approaches. These include reliability, PM, fail-safe design, and manifest danger. Each of these has its own strengths and weaknesses. Each approach may be better suited to particular uses or products within the engineering-design spectrum.

Two different and more advanced risk-analysis methods are presented for additional risk control. These are the FMECA and the FTA. These methods differ in significant ways. The FMECA is an inductive, bottom-up method, while the FTA is a deductive, top-down method. There can also be significant differences in their input requirements and their output results. Just as it is important that an engineer always makes ethical decisions, so too must the engineer select the best tool for a particular task and understand its abilities and its limitations.

Many engineering students are eager to use numbers—any numbers—for analyses. Remember *GIGO—Garbage In, Garbage Out*. The simple use of precise numbers does not give an analysis method trustworthiness unless those numbers are truly representative and meaningful. It can be quite difficult to get such input for analysis. It is more important to be *accurate* than to be *precise* when evaluating risk.

The tools presented in this chapter must be thoughtfully integrated into the particular engineering-design effort facing the engineer. Not all methods are useful for all problems all of the time. By carefully matching the product-safety analysis methods available to the engineering-design problem being faced, the design engineer or PSEr can design safety into the final product for its consumers.

References

Barnett, Ralph L. 1992. "The Doctrine of Manifest Danger." *Triodyne Safety Brief* 8 (1). http://www.triodyne.com/SAFETY~1/SB_V8N1.PDF.

Ericson, Clifton A. 2006. "A Short History of System Safety." *Journal of System Safety*. 2006. https://system-safety.org/ejss/past/novdec2006ejss/clifs.php.

———. 2016. *Hazard Analysis Techniques for System Safety*. Second edition. Hoboken, NJ: John Wiley & Sons.

Ethically Aligned Design: A Vision for Prioritizing Human Well-Being with Autonomous and Intelligent Systems. 2019. First edition. Piscataway, NJ: IEEE. https://ethicsinaction.ieee.org/#read.

Gullo, Louis J., and Jack Dixon. 2017. *Design for Safety*. First edition. Hoboken, NJ: John Wiley & Sons. https://doi.org/10.1002/9781118974339.

Halpern, David. 2015. *Inside the Nudge Unit*. London: Penguin Random House UK.

Hammer, Willie. 1972. *Handbook of System and Product Safety*. First edition. Englewood Cliffs, NJ: Prentice Hall.

Jensen, Roger C. 2020. *Risk-Reduction Methods for Occupational Safety and Health*. Second. Hoboken, NJ: John Wiley & Sons. https://www.vitalsource.com/products/risk-reduction-methods-for-occupational-safety-and-roger-c-jensen-v9781119493976?term=roger+jensen.

Leveson, Nancy G. 2011. *Engineering a Safer World: Systems Thinking Applied to Safety*. Cambridge, MA: The MIT Press. https://mitpress.mit.edu/books/engineering-safer-world.

NASA/STI. 2000. "Fault Tree Analysis: A Bibliography (NASA/SP-2000-6111)." Hanover, MD: NASA/STI. http://ntrs.nasa.gov/archive/nasa/casi.ntrs.nasa.gov/20000070463.pdf.

NASA. 2000. "Failure Modes and Effects Analysis (FMEA): A Bibiography." Hanover, MD: National Aeronautics and Space Administration Scientific and Technical Information Program. https://ntrs.nasa.gov/search.jsp?R=20000070720.

———. 2002. "Fault Tree Handbook with Aerospace Applications." Washington, DC.

"Potential Failure Mode and Effects Analysis in Design (Design FMEA), Potential Failure Mode and Effects Analysis in Manufacturing and Assembly Processes (Process FMEA) —SAE J1739_200901." 2009. Warrendale, PA: SAE International. https://www.sae.org/standards/content/j1739_200901/.

"Procedures for Performing a Failure Mode, Effects, and Criticality Analysis (MIL-STD-1629)." 1980. Washington, DC.

"Procedures for Performing a Failure Mode, Effects, and Criticality Analysis (MIL-STD-1629) Notice 3." 1998. Washington, DC.

"SAE J1853-2014: Hand Winches—Boat Trailer Type." 2014. Warrendale, PA: SAE International.

Thaler, Richard H., and Cass R. Sunstein. 2008. *Nudge: Improving Decisions About Health, Wealth, and Happiness*. New Haven, CT: Yale University Press.

"The Consumer Product Safety Improvement Act (CPSIA)." 2008. Cpsc.Gov. 2008. https://www.cpsc.gov/Regulations-Laws—Standards/Statutes/The-Consumer-Product-Safety-Improvement-Act.

USA/DoD. 2012. "MIL-STD-882E—System Safety." Washington, DC.

CHAPTER 11

Product-Safety Defects and Recalls

If you have to eat crow, eat it while it is young and tender.

—THOMAS JEFFERSON[1]

11.1 Introduction

Although America's third president was likely speaking about politics at the time, his above words remain particularly applicable to the topic of product-safety defects and recalls. No one ever wants to err; however, whenever a mistake is made, it generally goes better for the culpable when that error is promptly acknowledged and quickly corrected than if it is simply ignored. Large egos, vested interests, poor leadership, and even prejudices tend to make the digging in of heels *de rigeur* behavior by some people and parties on certain issues—including product-safety recalls. Such a reflexive response discourages the flow of information and slows constructive discourse, and its resulting actions, at a critical time—a time during which product users could sustain injury. Product-safety problems must be identified and corrected quickly. Although there truly exist instances where information beyond an initial report is necessary to fully understand an issue, delays only worsen the outcomes for those involved in these product-safety problems.

In the field of consumer-product safety, no side is immune from credible claims of obstinance or narrow-mindedness whether engineer, manager, executive, regulator, advocate, attorney, or consumer. Several parties may each be guilty of derailing either the identification of and solution to an existing problem or the free and open dialogue needed to properly respond to a future predicament that has yet to present itself. Engineers and their managers may make poor design decisions having effects which will not be discovered due, perhaps, to a sub-optimal engineering-test program; regulators and advocates may have a "default position" or unrealistic product-performance expectations of products; attorneys and the injured party may nonsensically pursue a manufacturer through litigation if that manufacturer is the only "deep pocket" involved.[2] It is

[1] https://quotestats.com/topic/jeffersons-quotes/

[2] Readers should not conclude that the author favors the interests of design-and-manufacturing companies. The author opposes limiting exposure from the design or manufacture of defective products. Any truly egregious behavior should not be protected by a legal system.

important to remember that no side is either always right or always wrong. Therefore, it is counterproductive to categorize parties into the classic American cowboy-movie "white hats" and "black hats," the good guys and the bad guys, respectively.

Objectivity and critical thinking are crucial at such an important potential inflection point in the safety of product users. It is unfortunate that the adversarial legal process does nothing to support such reflection and cooperation for the greater good of society. Although each problem situation must be looked at on its own merits, its own facts, and its own evidence, it is unreasonable to expect good behavior from another party when one's own party has been guilty of a pattern of dishonorable or untenable behavior. This is why unethical conduct by one or more parties involved can compromise the consumer-product safety of many.

Not all accidents happened as stated by an injured user or by an eyewitness. This is not to say that any party is necessarily lying. There have been studies indicating that individuals may not recall events accurately even without any incentive to waver from the truth. Loftus first wrote on this phenomenon in the late 1970s (Loftus 1979) and updated this work with a new edition two decades later (Loftus 1996). Inaccurate recall may especially so for those witnesses viewing events that happened in the periphery, happened quickly, or were unexpected in nature (Simons and Chabris 1999) Laney and Loftus 2010).

Also, not all products are as safe and easy to use as believed by a design engineer and the manufacturing corporation. It is hoped that all parties involved can remain ethical, look at facts and evidence, and avoid rushing to either an ill-informed judgment or a convenient fallback position. No party's values necessarily need to be violated in order to do this so long as the process is not delayed unnecessarily.

This chapter is dedicated to investigating potential product-safety defects or post-sale product-performance issues that may rise to a level posing an unacceptable safety risk to its users. Obviously, to do this task well, true issues must be quickly identified as the problems that they are. As will be seen, doing this requires effort and good character from both individuals and companies.

Failure to properly identify product-safety problems—and fixing them completely—may result in a design-and-manufacturing firm recalling the same product for the same safety problem multiple times. Although doing any product-safety recall can be humbling, if not humiliating, for a company, recalling a product *once* is not nearly as bad as recalling that same product *repeatedly* (Silvestro 2020). Needless to say, during all of this time, consumers may be injured or even killed by the product that was *recalled* but not actually *repaired*.

11.2 Defects

By definition, a defect is "an imperfection that impairs worth or utility."[3] When this definition is extended to a product defect, the result is a product that is of lesser worth or utility than was either *planned* by the designer or *expected* by the consumer.

Not all defects are safety defects or, in this book, product-safety defects. Many defects have no impact upon safety whatsoever. Cosmetic defects are simply that, those related to aesthetics or appearance. For example, the exterior paint on an automobile's

[3] https://www.merriam-webster.com/dictionary/defect

doors, hood, and other body panels not matching one another identically. Likewise, many product-quality defects have no effect on user safety. An example here would be a toothpaste tube that was not securely sealed at the fused or crimped end due to poor process controls by the packaging company. When the user of this toothpaste tube squeezes the tube, the toothpaste might come out of the back, sealed end rather than the front, dispensing end. No one will be hurt by either the defective paint job or the defective toothpaste tube. Thus, these defects would not be product-safety defects.

In addition, the dictionary definition above does not address the possible safety characteristics of defects. The sale of a particular children's toy, the lawn dart, was prohibited in 1988 by the U.S. CPSC ("CPSC Safety Alert: Lawn Darts Are Banned" 2012).[4] This is the result of injuries and deaths to children playing with lawn darts. Although the toys were deemed to be unsafe, they functioned properly and were not of lesser utility. Therefore, it is important to grasp the multiple aspects of product-safety defects.

At some point in time, all product-safety engineers—and many design and manufacturing engineers—will likely be asked to participate in evaluating a product's performance after it has been sold to and used by consumers. Prior to performing a product-safety recall evaluation, the ethical engineer and other evaluators will want to know what the recall criteria are. This is a perfectly reasonable request—unless products are performing so poorly in the field that any reasonable person would immediately conclude that a recall is needed. These product-safety recall criteria and other useful considerations are provided in the following sections of this chapter.

11.3 Product-Safety Defect and Reporting Criteria

The concept of a safety defect has been discussed. Yet, design engineers and those reviewing post-sale product-performance data must know more about what is required of them by various regulatory agencies in order to properly perform their task. Engineers do not want to design defective products. Simply tell them what a defect is and they would be more than willing to *not* design such products—on both professional and personal levels. Likewise, those tasked with reviewing data on the field performance—or failure—of their company's products must be given guidance to properly assess the risks of products.

Two sets of criteria for either safety-defect recalls or issue-reporting are provided below. These criteria are not exhaustive. These criteria cover the United States alone and not Canada or Europe. Some cover consumer products, others cover motor vehicles; none covers medical devices, foods, pharmaceuticals, or medicines. These are but two sets of the criteria a design-and-manufacturing engineer or corporation may need to meet by a company. These are from:

- U.S. Consumer Product Safety Commission (CPSC)[5]
- U.S. Department of Transportation (DOT)/National Highway Traffic Safety Administration (NHTSA)[6]

[4] Interesting to note is that the sale of lawn-dart components still takes place today. However, the heads and the tails of lawn darts cannot be sold in their assembled state. The author was able to find at least one retailer of unassembled lawn-dart components in a recent internet search. A reference to that website will not be provided here.

[5] www.cpsc.gov

[6] www.nhtsa.gov

These agencies serve the United States as governmental regulators for different product types. The CPSC regulates consumer products and the NHTSA regulates motor vehicles.

11.3.1 U.S. CPSC

The product falling under the regulatory authority of the CPSC in the United States include: all-terrain vehicles (ATVs), appliances, bicycle helmets, bicycles, (some) chemicals, children and infant clothing, cigarette lighters, cribs, furniture, garage-door openers, lawnmowers, portable generators, and toys.[7] There are, of course, other products not listed here.

According to the U.S. CPSC (U.S. CPSC n.d.), the following events require that the CPSC be *contacted* regarding an issue with a product falling under its regulatory scope:

> If you are a manufacturer, importer, distributor, and/or retailer of consumer products, you have a legal obligation to immediately report the following types of information to the CPSC:
>
> A defective product that could create a substantial risk of injury to consumers;
>
> A product that creates an unreasonable risk of serious injury or death;
>
> A product that fails to comply with an applicable consumer product safety rule or with any other rule, regulation, standard, or ban under the CPSA or any other statute enforced by the CPSC;
>
> An incident in which a child (regardless of age) chokes on a marble, small ball, latex balloon, or other small part contained in a toy or game and that, as a result of the incident, the child dies, suffers serious injury, ceases breathing for any length of time, or is treated by a medical professional; and
>
> Certain types of lawsuits. (This applies to manufacturers and importers only and is subject to the time periods detailed in Sec. 37 of the CPSA.)
>
> Failure to fully and immediately report this information may lead to substantial civil or criminal penalties. CPSC staff's advice is "when in doubt, report."

The above CPSC guidance to businesses contains the words *report*, *defective*, *substantial*, *unreasonable*, and *risk* in addition to specific trigger events and warnings of fines and penalties. CPSC provides no definitions for these terms.

Furthermore, the CPSC document goes on to provide the following rules regarding the timing for reporting:

> A company must report to the Commission within 24 hours of obtaining reportable information. The Commission encourages companies to report potential substantial product hazards even while their own investigations are continuing. However, if a company is truly uncertain whether information is reportable, the firm may spend a reasonable time investigating the matter…
>
> The company's investigation to determine whether to report to the CPSC should not exceed 10 working days, unless the firm can demonstrate that a longer time is reasonable under the circumstances. Absent such circumstances, the Commission will presume that, at the end of 10 working days, the company has received and considered all information that would have been available to it had it undertaken a reasonable, expeditious, and diligent investigation.

---

[7] https://cpsc.gov/Regulations-Laws—Standards/Regulations-Mandatory-Standards-Bans/

Although the manufacturer of a potentially defective product is given explicit guidance on when it must report an incident to the CPSC, there is little specific guidance on what constitutes a product-safety condition which must be reported. The practice of engineering ethics becomes crucial at this point. In an actual evaluation of a product-performance problem experienced by consumers, it is likely that legal counsel will become involved quickly and would provide guidance to company personnel reviewing the product and post-sale data on how to proceed based on how such matters have been viewed in case law.[8] Regardless of how another company might have handled such matters in the past, ethical engineers must rely upon what is right to them, regardless of history and any resistance put up by others—if any.

No definitions are provided for the first two of CPSC's five reporting criteria. These two conditions are:

- Substantial risk of injury and
- Unreasonable risk of serious injury

Engineers must be true to themselves when determining whether or not *substantial* or *unreasonable* risk exists in their company's products. It must be reiterated that, just because the use of a product *could* lead to injury or death, that product is not necessarily defective and should not necessarily be reported to regulators. This is why it is so vital to both read and understand post-sale product-safety data as well as determine how the product was being used. Hazardous products with large user-activity levels, from Figure 4.12, can be quite difficult to assess. Differing value systems between evaluators can lead to even ethical people reaching different conclusions in good faith.

11.3.2 U.S. DOT/NHTSA

The *second* regulatory body, NHTSA, regulates motor vehicles in the United States. Motor vehicles include automobiles, pickup trucks, SUVs, vans, commercial trucks and trailers, and motorcycles that operate on public roads. NHTSA does not regulate OHVs. Many OHVs fall under CPSC supervision. A small sample of the components and aspects of motor-vehicle design and performance regulated by NHTSA is contained in the Haddon-Matrix example shown in Table 4.11.

NHTSA regulates only the design, construction, and performance of motor vehicles. Highway infrastructure falls under the U.S. DOT/FHWA (Federal Highway Administration).[9] This includes highway design, the design and placement of roadside appurtenances—or hardware such as street-light posts, guardrails, and signage—and long-term planning for roadways.

NHTSA operates under the *Motor Vehicle Safety Act* (*Motor Vehicle Safety Act* 2005) initially put forward in 1966. This Act, and another NHTSA document, contain several important definitions relevant to product-safety defects and recalls, including the following:

- *Motor Vehicle*: "[A] vehicle driven or drawn by mechanical power and manufactured primarily for use on public streets, roads, and highways, but does not include a vehicle operated only on a rail line"

[8] Case law is the historic record of court rulings and opinions from similar cases.

[9] www.fhwa.dot.gov

- *Motor Vehicle Safety*: "[T]he performance of a motor vehicle or motor vehicle equipment in a way that protects the public against unreasonable risk of accidents occurring because of the design, construction, or performance of a motor vehicle, and against unreasonable risk of death or injury in an accident, and includes nonoperational safety of a motor vehicle"
- *Defect*: "[A]ny defect in performance, construction, a component, or material of a motor vehicle or motor vehicle equipment"
- *Safety Defect*: "[A] problem that exists in a motor vehicle or item of motor vehicle equipment that:
    - Poses a risk to motor vehicle safety, and
    - May exist in a group of vehicles of the same design or manufacture, or items of equipment of the same type and manufacture" ("NHTSA's Process for Issuing a Recall" n.d.)

Once an issue has been determined to be a safety defect by the motor-vehicle manufacturer, NHTSA requires that the manufacturer report the defect "[n]ot later than 5 working days" later. Although this is a manufacturer-initiated action in many cases, NHTSA also requires that motor-vehicle manufacturers also periodically report to NHTSA certain instances that could be indicative of an undiscovered product-safety defect. The *Transportation Recall Enhancement, Accountability, and Documentation* (TREAD) Act, which was incorporated into the Motor Vehicle Safety Act in 2002, mandates the following:

> The manufacturer of a motor vehicle or motor vehicle equipment shall report to the Secretary, in such manner as the Secretary establishes by regulation, all incidents of which the manufacturer receives actual notice which involve fatalities or serious injuries which are alleged or proven to have been caused by a possible defect in such manufacturer's motor vehicle or motor vehicle equipment in the United States, or in a foreign country when the possible defect is in a motor vehicle or motor vehicle equipment that is identical or substantially similar to a motor vehicle or motor vehicle equipment offered for sale in the United States.

Much of the impetus to quickly create and pass the TREAD Act came from the perception, by consumers and lawmakers alike, that the automotive and pneumatic-tire industries failed the American motoring public during events leading to the Ford Explorer and the Firestone-tire recalls in the late 1990s (Greenwald 2001). These events will likely serve for decades as a bad example of how to identify and fix a product-safety problem.

11.3.3 U.S. Safety-Defect Criteria

The preceding two sub-sections contain the criteria for reporting potential product-safety problems and identified product-safety defects to the appropriate regulatory agency. A summary of these criteria is contained in Table 11.1.

This book only looks at two regulatory agencies in the United States. Each of these agencies regulates different products and, although the defect and reporting criteria differ, the reader can see that the intents of these two agencies is similar if not identical.

#	Agency	Defect Criteria	Defective-Product Reporting Criteria	Reporting Period	Reporting Notes
1	U.S. CPSC	N/A	Can create a substantial risk of injury to consumers. Creates an unreasonable risk of serious injury or death	Within 24 hours	Company investigation not to exceed 10 working days; "when in doubt, report"
2	U.S. DOT/ NHTSA	Any defect in performance, construction, a component, or material of a motor vehicle or motor vehicle equipment	Poses a risk to motor-vehicle safety	Within 5 days	Periodic TREAD Act reporting on on-going basis

TABLE 11.1 Defect or Reporting Criteria by Regulatory Agency

They each wish to protect users of these products from unreasonable injury and death risks. However, clear guidance on exactly that is, or is not, a defective product is lacking. What poses either a "risk" or a "substantial risk" to a product user may vary from one person to another especially when field reports of incidents, accidents, or injury are few or nonexistent. This, the author hopefully says, should not be the case when field reports of accident and injury are numerous.

If engineers find themselves working for a company involved with products regulated by both agencies,[10] then what might have been a mere academic perplexity becomes a practical quandary. Instead of having only one set of unclear instructions, the engineer now has two. Rather than worrying about the differences between these two sets of guidelines, it is better to focus upon the similarities in the "spirits" of these requirements. With the possible exception of hard reporting deadlines, an ethical engineer should be able to effectively evaluate product-safety defect criteria and reach an appropriate conclusion. Again, this engineer—or other product-safety reviewer—should be able to present that *conclusion* to executive management for an objective—and ethical—reception. A process for looking at field data, evaluating the same, and making decisions which may lead to product-safety recalls is provided later in this chapter.

11.4 Types of Product-Safety Defects

There are several *types* of product-safety defects. In contrast, *quality* defects may be classified according to the *extent* of the defect (Knack 2019). For example, *minor* defects

[10] A reality in the powersports industry.

do not affect the function of the product, *major* defects affect product function or appearance, and *critical* defects affect the operability—ultimately even the safety—of the product.

On the other hand, product-safety defect *categories* are qualitative, not quantitative, in nature. These product-safety defects differ in *nature*, but each could prove fatal to product users if their effects are severe enough. Product-safety defects generally fall into one of three categories. Each category has potentially serious safety implications for consumers:

1. Manufacturing Defect
2. Design Defect
3. Marketing Defect

There is general agreement on these three categories of product-safety defects (Ross 2019; Pine 2012). The first two types of defects, Manufacturing and Design, are generally direct results of the engineering process. The third category, Marketing, is less affected by the engineering process than by the sales and marketing departments and their personnel, still the prior chapter included a method to help engineers prevent these defects through a comprehensive approach to designing product safety into a product *system*, which includes facilitators, during the PDP.

11.4.1 Manufacturing Defect

The first category of product-safety defects is the manufacturing defect. A manufacturing defect is purely one in which the product was not manufactured as intended by its design engineers. Some manufacturing defects are simple and obvious to detect; others, much more subtle and difficult to detect. Such a defect may be the fault of the manufacturing firm, its supplier, or their supplier's supplier (sub-supplier) to follow engineering drawings, manufacturing processes, materials specifications, or assembly procedures.

Examples of manufacturing defects include:

- Use of the incorrect material for a part
- Using a bad batch of parts from a (sub-)supplier
- Using highly variable manufacturing processes (such as blow molding and welding)
- The improper treatment or finishing of a part made from the correct material
- The improper torque on fasteners (such nuts and bolts) due to poorly calibrated assembly tools
- The improper routing of wires, cables, or hoses
- Damage to a product from shipping and handling

It is hoped that many of manufacturing defects are detected and corrected prior to the shipment of a product. These follow-up actions prior to product release are categorized as "rework" and often occur at the end of the assembly line. There are, of course, other reasons that a product is reworked. For example, if one or two parts of a product were not available during the product production, there may be a decision to proceed with assembly and hold—or "embargo"—affected products until the missing parts arrive and are installed.

Many manufacturing defects may be exacerbated or even caused by poor engineering-design decisions. If a part is quite difficult to install on a product during its assembly, then it will be more likely to be misassembled. If there is no access for a worker to install and then verify proper assembly, then improper assembly may go unnoticed. Furthermore, if parts are difficult to position and attach to other parts, then solid connections may not be established in the case of cables, wiring harnesses, and fuel lines. Such a condition may also lead to the damage of other components if excessive force or awkward worker ergonomics are required for assembly. This, too, can go unnoticed. This negative effect on product safety is recognized in Figure 5.11 as the "as delivered" increases in product risks, S_0 and P_0.

The upside to a manufacturing defect, if there exists one, is that not all of the product units may be affected. For example, if a manufacturing defect is discovered, it may be that the mis-torqueing of a bolted joint only took place during a given time period—perhaps *one* shift of *one* day by *one* assembly-line worker. Good production record keeping could indicate that the normal assembly worker missed one day of work due to illness and was replaced on one shift by another worker who was unfamiliar with, or untrained for, a specific task that resulted in the under-torqueing of a critical fastener nut. Not only are such employee records quite helpful in sorting out such issues, but the use and recording of serial numbers will reduce the number of product units to be recalled in this case. Similarly, an inventory-control system, such as FIFO (First In/First Out), may help isolate the affected product-unit population for recall if a (sub-) supplier provided a single shipment of bad components for production. This problem type is shown in Figure 6.5 as the arrow going from Tier 1 to Phase IV of the PDP.

11.4.2 Design Defect

The second category of product-safety defects is the design defect. Unlike the manufacturing defect which arises because engineering drawings and other specifications were *not* followed, design defects arise when those drawings and specifications *are* indeed followed. In these cases, the product *as designed* was defective. Therefore, a design defect affects *all* of the product units manufactured and released to the public. A disciplined EDDTP during the PDP will reduce the number of design defects. Yet, despite the best and most-ethical of efforts, design defects can and do occur. As stated earlier in this book, innovative products may be prone to a higher frequency of design defects since precise product use may not be known and because there is no history or experience in designing such a new product. Even the finest of design or manufacturing companies do not have perfect "batting averages." What is important is not engineering perfection, but the active recognition and rapid remediation of product-safety defects.

There is a generally accepted meaning to the word "defect" in casual conversation. However, that word is strictly a *legal conclusion* in the world of product safety and means that an unacceptable product-safety risk exists with a product. The CPSC requires that defective consumer products be remediated—repaired, replaced, or bought back from customers.

Some authors (Garner 1999) use these same categories for product-safety recalls but may include different elements—such as instructions and warnings—within Marketing defects, but there is generally agreement on these product-safety defect categories. For this book, warning signs and manuals will be included under design defects since they are included in the PSMx.

11.4.3 Marketing Defect

The third and final category of product-safety defect is the marketing defect. This is different than the manufacturing defect or the design defect. Engineers may rarely be faced with a marketing-defect evaluation. This is because many claims of marketing defects arise from lawsuits and are, consequently, handled by attorneys without involving engineers.

As stated in the previous sub-section, some people may include warning signs and manuals under the marketing-defect category. This may not be improper in general; however, since this book emphasizes an integrated approach to engineering design, including product-safety facilitators, failures in warning signs and manuals have been included under design defects.

Examples of marketing defects include:

- Not warning of hidden dangers of a product
- Not instructing users on how to properly and safely use a product
- Showing the incorrect or unsafe use of a product in marketing materials[11]

The first two examples above could be classified as failure-to-warn allegations. All of these examples are often legal matters rather than engineering ones. Even if a product did not contain either a manufacturing or a design flaw, a litigant or attorney might argue that a marketing defect existed. Some of the larger failure-to-warn actions have involved widely used products sold for decades by certain companies and industries such as pesticides, tobacco, and asbestos. In addition, many large alleged marketing defects pertain to health rather than to safety. Smoking a single cigarette is unlikely to harm person immediately; however, a lifetime of smoking will adversely affect a person's long-term health outlook.

Unlike manufacturing defects, but like design defects, marketing defects do affect all product units. Perhaps if a unique marketing campaign was used for a "special edition" of a product and was accused of being defective, then not all normal-production units could be affected. Sources of possible marketing defects include advertisements, catalogues, websites, promotions, sales representations, videos, packaging, and accessories (Ross 2019). The example of lawn darts discussed earlier might fall under the category of marketing defect.

11.4.4 Summary of Types of Product-Safety Defects

Although there are three categories of product-safety defects, this book will only look at the first two: manufacturing defects and design defects. The category of marketing defect will not be covered for two reasons: (1) allegations of such defects typically arise from legal proceedings and are, therefore, addressed by people other than design or manufacturing engineers and (2) the construction of product-safety facilitators, such as warning signs and manuals, is treated as an engineering-design task within the EDDTP.

Depending upon the category of product-safety defect, not all units produced will be affected and need to be recalled if such a decision is finally made by the company or by regulators. On what basis and by what method product-safety recall decisions

[11] Some advertisements have small print at the bottom of the screen or page telling viewers and readers to not attempt the demonstrated actions themselves because a professional driver is shown or because the action was performed under controlled conditions.

are made are discussed later in the chapter. However, in the author's experience, many product-safety recalls do not involve scientific or technical problems only solvable by rocket scientists; instead, many recalls result from simple and fundamental failures in elementary engineering design and manufacturing.[12] This is both good and bad: good because they should be readily identifiable—given sufficient information—and bad because they might also have been avoidable. It is good to study prior recalls of both your own company and others to see what mistakes have been made and then work to avoid repeating them.

11.5 Product-Safety Defects and the EDDTP

The only way that an organization can avoid ever making a defective product is to never make a product. So long as decisions are being made, poor decisions will be among them. A modern consumer product involves thousands of engineering-design decisions ranging in size from the miniscule to the massive. A truly innovative product will require more decisions than a simple revision to an existing product. Some of the decisions for a ground-breaking product may never have been made before and, thus, history will be unable to guide the design engineer. The massive decisions are usually recognized and pondered accordingly. The miniscule decisions, on the other hand, can often go unnoticed. Some of these small decisions may never even be explicitly contemplated let alone thoroughly analyzed and tested by engineering. For instance, many plastic parts of some products are black in color by default. Black is a visually neutral and inconspicuous color, so being black in color usually does not affect the aesthetics of a product. Yet, making that part black in color is a design decision with its own potential effects on product safety. Such effects are usually benign in nature, but are not always so. There is one case, with which the author was involved, where the standard color of a part was black. This color was later changed to red, as an option, and this design decision eventually led to a product-safety recall (US CPSC 2004). Table 11.2 shows some material from that recall as published on the U.S. CPSC website. This was an instance in which the color of a standard black part was changed to red in order to offer consumers an accessory ski to customize the appearance of their snowmobiles. With neither malice nor intent, the engineering-design change was approved and the ski manufactured without going through the normal, rigorous testing program which would have included shock, vibration, heat, cold, and light-exposure testing. It is reasonable to assume that many such parts have made it into the hands of consumers from many manufacturers and have done so without incident. It is likely that the design engineer simply did not appreciate that the use of the untested red colorant to the plastic, instead of the tested black colorant, would compromise the ultra-violet (UV) light stability of the resulting skis. Although, many snowmobile skis are not exposed to much direct sunlight during use due to the short daylight period during the winter period of use, when snowmobiles are put away at the end of season, they often sit parked directly in the sun during the sunlight-intense summer. It was during these summer times that UV-light exposure could do the most damage to the red-plastic material and lead to the

[12] The author has seen many product-safety recalls related to topics including improper nut-and-bolt torques, misrouting of cables and wire harnesses, improper materials selection, insufficient laboratory and field testing, and supplier and sub-supplier mistakes.

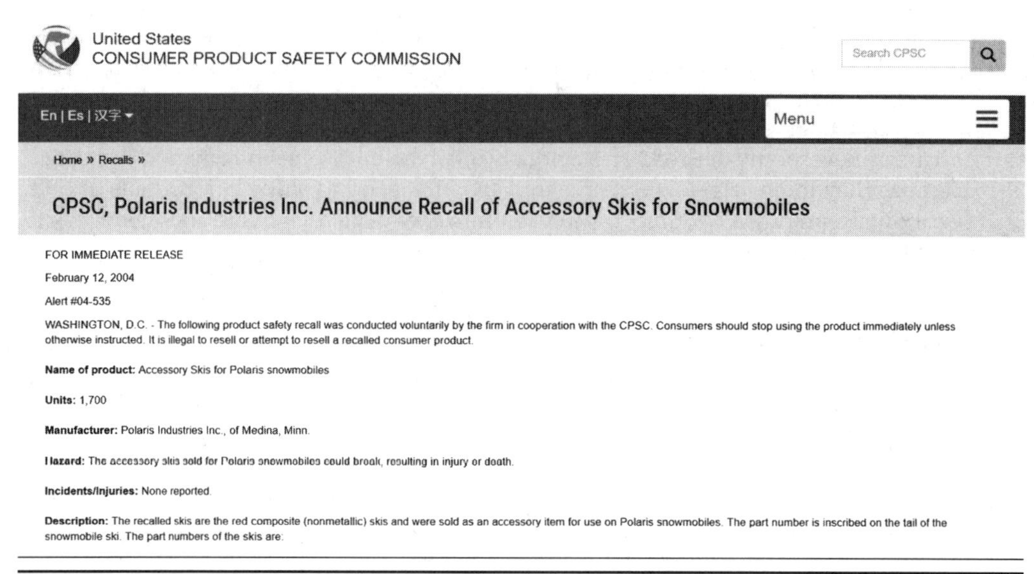
United States
CONSUMER PRODUCT SAFETY COMMISSION

Search CPSC

En | Es | 汉字

Menu

Home » Recalls »

CPSC, Polaris Industries Inc. Announce Recall of Accessory Skis for Snowmobiles

FOR IMMEDIATE RELEASE

February 12, 2004

Alert #04-535

WASHINGTON, D.C. - The following product safety recall was conducted voluntarily by the firm in cooperation with the CPSC. Consumers should stop using the product immediately unless otherwise instructed. It is illegal to resell or attempt to resell a recalled consumer product.

Name of product: Accessory Skis for Polaris snowmobiles

Units: 1,700

Manufacturer: Polaris Industries Inc., of Medina, Minn.

Hazard: The accessory skis sold for Polaris snowmobiles could break, resulting in injury or death.

Incidents/Injuries: None reported.

Description: The recalled skis are the red composite (nonmetallic) skis and were sold as an accessory item for use on Polaris snowmobiles. The part number is inscribed on the tail of the snowmobile ski. The part numbers of the skis are:

Table 11.2 U.S. CPSC Recall of Snowmobile Skis

breakage of these skis during use the following winter. These events led to this recall. Testing by the manufacturer did show that UV-light exposure could finally lead to ski breakage, but this was only discovered after the release of the accessory skis. A broken ski on a snowmobile can lead to the loss of control (LoC) of a snowmobile which can result in an accident with injury. Fortunately, and to its credit, the manufacturer was able to respond and offer replacement of skis before any consumer was injured. This example illustrates that minor, even seemingly insignificant, design decisions can sometimes lead to product-safety recalls.

From the above example, it should be evident that all products, even insignificant ones with minor changes, should be developed with care. The disciplined application of the EDDTP to all products—whether tried and true or new and innovative—can help reduce the safety risks that these products present to consumers.

11.6 The Inevitability of Product-Safety Recalls

Earlier in this chapter, it was stated that the only way to avoid product-safety recalls is to never make a product. Even if the probability of a recall is low, not making any product equals no possibility whatsoever for product-safety defects and the ensuing recalls. From this one might ask the excellent pair of questions below:

1. What if no design changes are made to the product so that the same product is made over and over again each batch or model year?
2. Can recalls now be avoided?

The surprising answer is *no*. Today, for the competitive mass producer of products, the realities of efficient production operations and international business make "no design changes" a practice that may still lead to defects and recalls—even if the *initial* product was *not* defective. Even when the design engineer makes no design changes, over time, a *conscientious* purchasing department *will* strive to reduce cost—by virtually any means necessary. This means that purchasing agents will be pressuring their suppliers to reduce the costs of parts. How does the supplier respond? By moving production to a lower-cost state or country, by modifying the manufacturing processes used, by changing the materials used, by selling your parts to your competitors, and more. These are ways that suppliers can either decrease cost or increase revenue. Of course, the same is true of the manufacturer, too. By the way, the purchasing agents for the supplier will be doing the same thing to their suppliers. This cost-reduction work goes on all of the way down the supply chain.

These changes are manufacturing issues and not design issues, in general. So, any product-safety defects may become manufacturing defects for the original-equipment manufacturer (OEM). However, for the supplier, this could well be a design defect if changes to specifications, processes, or materials were made to reduce cost and resulted in the product-safety defect.

Consider the case where a supplier has had their part submitted to the OEM's Quality Assurance (QA) department for the Production Part Approval Process (PPAP) assessment. The sample part submitted by the supplier to the manufacturer passes the PPAP with flying colors. Thus, this part is approved for use in production. Assume that this part is a grommet to be used to seal in a vehicle fuel system. Since it is used in a fuel system, that part must be designed and manufactured to specifications particular to its application. When vehicle production begins, the supplier provides their part on schedule to meet the manufacturer's assembly demands. Vehicles are produced. As the vehicle is used by consumers, the manufacturer receives reports of leaking fuel systems—a very serious matter. Upon investigation, the vehicle manufacturer learns from the supplier that that supplier had changed its own supplier. That sub-supplier had, for production purposes after the PPAP, manufactured the grommet from a different material that could tear as the part was being inserted into a fuel tank during the vehicle-assembly process and was also unstable when exposed to gasoline resulting in the grommet swelling when used. Thus, the part passing the PPAP was *not* the part validated by test engineering, approved by QA, verified during assembly prototyping, and then ultimately used in production.

The manufacturer did not change the design of the part, the sub-supplier did. It was not the designer/manufacturer's *fault*, but it was truly their *problem*. Although another party was the transgressor for the resulting product-safety recall, the OEM's name is on the product and on the resulting U.S. CPSC recall press release (US CPSC 2003). It is a rare industry in which one company is the sole source of all product-safety recalls. In the case of off-road vehicles (ORVs), such as the one just mentioned, virtually all ORV manufacturers have had to recall vehicles as a simple search of the U.S. CPSC's website[13] will show. The more innovative the products, the more recalls there seem to be. Although this may help explain a quantity of product-safety recalls, this by no means serves as an excuse for insufficient or slow responses by a manufacturer to its product-safety defects.

Even if an OEM manufactures its own products in-house without external suppliers, tooling eventually wears out and must be replaced by new tooling, suppliers

[13] www.cpsc.gov.

go out of business or no longer make the original materials or parts so replacements must be found, and significant manufacturing-department employees possessing valuable institutional knowledge retire or quit the company without leaving behind their experience. Regardless of whether or not a company is changing a product's design, that design *is* changing. Those in manufacturing are well aware of this reality; those in engineering design and management must come to acknowledge it. This is but one illustration of the importance of competent supply-chain management to producing safe products. A strong relationship between the purchasing and the engineering departments will go far in reducing the likelihood of product-safety defects.

Every successful and innovative company will have recalls, but one does not necessarily imply the other. A non-innovative company producing lackluster products may also have numerous product-safety recalls. Innovation is the nature of modern business and retaining the contemporary consumer. Competition is fierce and today's consumers demand the most from each and every purchase.

An example of extreme *design stability* is the iconic—and formerly ubiquitous—New York City taxicab made famous in movies and on television from the 1960s through the 1980s. The Marathon was a vehicle model manufactured by Marathon Motors between the years of 1962 and 1982 ("Checker Models A-11 and A-12" 2019). A Marathon taxi cab is shown in Figure 11.1. Only a Checker Marathon *aficionado* has any hope of telling a Marathon of one model year (MY) from a Marathon of another MY since the vehicle never substantially changed in its design. The vehicle was available to the general public, but Checker Motors' main clients were taxicab-fleet operators. Mechanics for the taxicab operators knew how to fix the Marathon and only needed one set of parts for routine maintenance and repair. There truly was benefit to Checker Motors in not changing their design. Yet society, along with the automobile and taxicab industries, was changing. In the end, Checker Motors was forced to shutter its factory as it went out of business. Today's industrial environment is even more dynamic and its consumers even more demanding than those of decades ago.

Given that consumers have a demand for better and better products, designs for these products must be continually updated. Few businesses can survive, let alone thrive, in the modern business climate without continuous change in, and improvement to, its products. Engineers will be making many decisions to improve on the designs

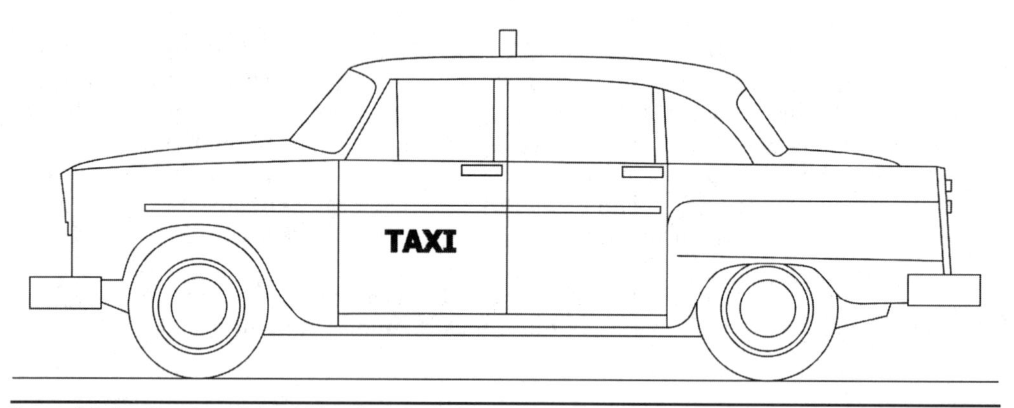

Figure 11.1 Checker Motors' Marathon taxicab.

of their products. With so many decisions being made, not all of these decisions will be the proper ones. Even the most well-intentioned of engineers will not have a perfect record. It is likely, and hoped, that many of these design errors will be inconsequential in nature. Occasionally, one may lead to a product-safety defect. What is most important are the proper identification and complete remediation of such product-safety problems quickly whenever the need arises.

11.7 Post-Sale Product-Safety Surveillance

Once a product has been successfully designed, re-designed, analyzed, tested, manufactured, and delivered to retailers, *some* responsibilities of the design-and-manufacturing organization are only beginning. Even when numerous engineering design, analysis, and test engineering teams involved in delivering a product to customers have made their best-faith efforts, it is possible for product-safety design defects, of all types, to exist in the products and potentially harm consumers. It is, therefore, crucial that organizations not only be vigilant for signs of such defects but also actively seek out evidence of them.

There are many sources for information about problems that consumers are having with products. One simple list is included as Table 11.3. Organizations should make use of any such valuable source of consumer difficulties, not only for product-safety reasons, but also for customer-service grounds.

The column-wise sources of potential product-safety information in Table 11.3 are the design-and-manufacturing organization's employees. It is suspected that many readers would expect that consumers should be the first source of information. The order of these sources is intentional, however. Even if an organization's countermeasures to defect prevention, such as engineering analysis and testing as well as design reviews, do not work as intended for whatever reason, it is hoped that conscientious employees in various tasks can raise a warning to the company about potentially hazardous products. Such employees include assembly-line workers, design engineers, and manufacturing engineers. Such employees are listed early because they may be able to stop a possibly defective product from ever being released in the first place to the public. In addition, even after product release and retail, these employees generally have a great deal of understanding of the product, its properties, and its manufacturing. Furthermore, claims made by knowledgeable employees are perhaps more likely to be

Employees	Distributors	Customer Service Dept.	Legal Dept.
Engineering Dept.	Sales Dept.	Emergency Responders	Social Media
Manufacturing Dept.	Purchasing Dept.	Fire Departments	Friends/Neighbors
Warranty Dept.	Suppliers	Law Enforcement	Outside Attorneys
Technical Service Dept.	Sub-Suppliers	Media	Regulators
Retailers/Dealers	Consumers	Healthcare Providers	Any Other Sources

Table 11.3 Sources of Post-Sale Product-Performance Information

taken seriously within the company than those made by others outside of the organization. Thus, assembly workers who experienced problems with product assembly, design engineers refining the product for the next MY or production run, and manufacturing engineers noticing the miscalibration or malfunctioning of production equipment may be the very-first people to realize that a product-safety defect may exist. The proximity of these personnel to the product-safety function of the organization gives them the unique ability to quickly raise legitimate issues to those needing to know about potential problems.

Similarly, the Warranty and Technical Service Departments of an organization are also important in identifying problem areas with products. Not all problem areas will, of course, involve consumer safety. The Warranty and Technical Service personnel will often work with Retailers, authorized Dealers, and Distributors of the product. These parties will also be working with representatives of the Sales Department. It is through working with these people that some problems may be encountered with a product and given attention even before a consumer takes custody of that product. Such problems may involve product damage during shipment, incorrect parts and instructions, or even observations about fitness of the product for its use. It is important to realize that warranty data and dealer information may no longer be available for products still in use once the product's warranty period has expired. For some products, a consumer has no choice but to contact the manufacturer or dealer for service or replacement parts, but for other products there exist third-party repair and maintenance options for consumers. The design-and-manufacturing company will not generally see the information regarding third-party service work. Therefore, if an organization offers a six-month warranty for a product, then they will likely see a decrease in field information for that particular unit (serial number, VIN, or other identifier) after six months of use. It is important for the Warranty, Technical Service, and Sales personnel to notify the appropriate engineering groups about real or potential problems that they have seen or heard of in the field promptly. Some products will also exhibit seasonality in field data. For example, reports of field problems on snowmobiles disappear with the snow. Likewise, field reports for products used in outdoor water activities may decrease in the winter. Any decrease in field reports from seasonal products should not be confused with either the non-existence or the elimination of a product problem.

It will not be uncommon for suppliers and sub-suppliers to notify Purchasing personnel of problems with parts that they have produced for, and delivered to, the manufacturing unit. This information must be quickly conveyed to the design, testing, and manufacturing engineers for review. Again, the delivery of this information may well be before any hazard in the product has resulted in an injury to a user. Therefore, such time must be valued and not wasted if remediation will ultimately be needed.

The direct voice of the product consumer is addressed next. At this stage, the consumer and the organization's Customer Service Department interact. It is hoped that information from consumers will be non-safety related complaints rather than reports of injury. Customer Service personnel should be skilled in listening to consumers and then gathering as much information as possible at the time. They should also arrange for follow-up communications with internal personnel skilled on the issue at hand when this is necessary. All information and complaints from consumers, as with all other incoming information, should be documented, preserved for market-surveillance purposes, and reviewed regularly by product-safety personnel looking for problem areas.

Computerized, electronic databases are best for this purpose since they are searchable and relatively permanent.

If an accident does take place with the product, sometimes records or reports from emergency responders such as fire departments, emergency medical services (EMS) workers, and law-enforcement officers such as police departments, sheriff's offices, and highway patrols may provide useful information about what these responders observed and experienced when arriving at the accident scene. These reports may also contain accounts of the accident from the injured or from witnesses. Additionally, photographic and video evidence may also be available to help investigate the accident. Some of this material may come from the local news media although some media reports are either imprecise or inaccurate. It is also possible that physical evidence was collected at the scene by some people.

Health care workers can also be a valuable source of information. Strict patient-privacy policies and laws, e.g., Health Insurance Portability and Accountability Act (HIPAA), prevent the unauthorized disclosure of personal information. Yet, reports are released through appropriate channels and may provide information helpful to an accident investigation. Furthermore, healthcare providers may also be able to signal to the public an increase in a particular accident scenario (Kass 2019) or single type of product involved in numerous accidents (Hill 2019).

Reports of many injuries from a single product, or type of product, will likely become public through media reports, whether in print, through over-the-air broadcast, or on internet websites. Even though initial media reports frequently lack the detail needed by product-safety and accident investigators, these reports are usually timely and can serve as notification to designers and manufacturers of the products involved. Because of the need to publish the news in whatever medium, the incomplete information sometimes found in initial reports can later prove misleading. All parties should be wary of committing to a course of action before sufficient information becomes available. Conclusions should be supported by the available evidence. However, once reliable information is received through any source, such as those others in the table, any necessary decisions and actions should swiftly follow.

11.8 Reviewing Post-Sale Product-Safety Data

The prior section covered potential sources of post-sale product-performance data and other information available to a company designing, manufacturing, and retailing a product. Precisely what these data might look like and how they may be used is covered in this section.

Example data for a fictitious ORV will be provided to the reader and then discussed. These data are not indicative of any particular product on the market. The ORV is used in this example since they are of historic and current interest to regulators at the U.S. CPSC and to consumer advocacy groups regarding "voluntary standards and fire hazards" ("Caroleen[e] Paul and Other CPSC Staff Meeting with Members of the Speciality Vehicle Institute of America (SVIA), Recreational Off-Highway Vehicle Association (ROHVA), and Outdoor Power Equipment Institute (OPEI)...Topic: ATV/ROV [June 27, 2019]" 2019) due to past and recent accidents, injuries, and deaths (*The Argus Observer* 2019). Additional illustrative value is added to this example by the ORV being a highly user-active product with potentially high levels of risk as illustrated in Figure 4.12.

With the ORV, there are many hazards inherent to the product, its environment, and its user. These include: speed, mass, kinetic energy, potential energy, linear momentum, combustion heat, gasoline, flammable vehicle materials, electricity, rotating parts, collision, rollover, occupant ejection, bad weather, extreme ambient temperatures, immovable objects,[14] operating in remote riding areas, uneven and sloped terrain, uncontrolled other operators and bystanders, animals, thrill seeking, stunt driving, alcohol and drug consumption, bystander enablement, and competition with other operators (even when not in an organized race). At first glance, the only hazards absent from ORV operation appear to be firearms and ionizing radiation.

The overall performance of a product, in this case the ORV, is a combination of the three factors shown in Figure 1.3, the PUE model. These factors must be kept in mind as post-sale field data are collected and analyzed. The hazards just mentioned represent all three of these factors and can affect the overall safety of the PUE system.

As just discussed, post-sale data for a design, manufacturing, and distribution company can come from a variety of sources such as retailers, regulators, and company staff. These data also come in a variety of forms as well. In whatever manner the data come, they should be aggregated and stored for both immediate and long-term access in an electronic searchable database. The data should be monitored regularly for potential product-safety related problems in order to quickly identify those issues that are, or might be, true problems posing unacceptable risks to the safety of product users. It is through such rapid identification of product risk by scrutinizing post-sale reports that user safety can best be maintained by the product's designer and manufacturer.

Some product-safety problems are quite readily apparent; some others are more a subtle shade of gray than purely black-and-white in nature. Many problems of the latter kind are only conclusively visible after looking over incidents, accidents, or other information over a time span greater than that which may be permitted by regulators working with perfect hindsight. This can place even the ethical manufacturer of products in a difficult spot: to wait for more data (which means potentially more injuries including death) or to respond immediately, reacting without full information and understanding of the cause of the issue. Reacting before fully comprehending the problem can lead to *unintended consequences* due to a rushed or incomplete product "fix" which may carry with it additional hazards and risks for a product user. Some regulators, attorneys, and advocacy groups are prone to engage in "Monday-morning quarterbacking" when they see manufacturer data. In hindsight, it may truly be possible to see an "obvious" pattern that should have been recognized by the manufacturer much earlier than reported. However, the manufacturer cannot move into the future to get the retrospective viewpoint available only to these reviewers at a later date. Manufacturers are reminded that not having a remedy to an acknowledged or potential problem is neither a reason nor an excuse for not reporting a product-safety defect to the proper regulatory authority.

Recall decisions are among the hardest decisions made by a designer and manufacturer of a product. It is hoped that the data upon which a recall decision is made are timely, complete, and accurate. Table 11.4 is provided to offer the reader some

[14] Which are hard, sharp, protruding, or flammable (Popovich 2018)

#	Date	Product	PIN	Source	Location	Description
1	04/01/19	MY2020 X-5 ORV	000123	Dealer	US	Was smoking. replaced plastic fender (DSM said goodwill it)
2	05/14/19	MY2020 X-5 ORV	000314	Dealer	CA	la machine ,elle fumait; elle est allé «boum»; grand feu; brûlé mon garage
3	07/08/19	MY2020 X-5 ORV	000511	Consumer	US	Was just riding along, hit tree, wheel was off vehicle, broken arm and leg, contacting lawyer
4	10/31/19	MY2020 X-5 ORV	001009	Regulator	US	Operator's friend fell out of vehicle while moving
5	02/14/20	MY2020 ORV Winch, 2-Ton, Electric	EW2-0163	Dealer	US	the wench is dead
6	03/16/20	MY2020 X-5 ORV	001387	Dealer	US	vehicul esploded fule leekd Billy sez he dont' know why evryone wanst killed they shuld'uve bin this can kill thosands Figurd out why vehicul leked fule . we fixd all 7 vehiculs on fllor will bill co for 5 hrs labor & hardware store parts have seen hunderds like this

Table 11.4 Example of Post-Sale Product-Performance Data Received by Designer/Manufacturer of ORV

perspective of just how difficult it may be to objectively evaluate field data for products once they have been sold and delivered to consumers. This table contains six fictional reports of field problems with an equally fictional ORV—in this case a Side-by-Side vehicle[15] (SSV) and an OEM accessory. Table 11.4 shows the data as they might appear in a spreadsheet or from a searchable electronic database maintained by as SSV design- and manufacturing company and reviewed by product-safety engineering personnel

[15]A Side-by-Side (SSV) vehicle differs from an all-terrain vehicle (ATV) in that, among other differentiators, the SxS has bucket or bench seating with a steering wheel, while the ATV has straddle seating with handlebars. SxS vehicles may also be known as UTVs, ROVs, and MOHUVs. "ORV" includes all of these. A separate ANSI standard exists for the ATV, the ROV, and the MORUV. The media frequently use "ATV" to represent any ORV.

to scan for potential product problems. Assume that the design-and-manufacturing organization in question makes an SSV called the "X-5" which is manufactured according to an MY and numbered serially. The reports in the table have been limited to one model year and one product model with one accessory exception. During the course of business, this imaginary design-and-manufacturing company would receive, again from a variety of sources, information that it would accumulate continuously and review regularly. These table entries were created for this *ad hoc* example review, but are typical[16] of what is often received by a manufacturer and reviewed to look for potential product-safety problems. Each entry will first be reviewed individually and then reviewed collectively.

As the data in Table 11.4 are read, it is important for objectivity's sake that the reader realize and remember that, although some ORV manufacturers may have more product-safety recalls than some others (Weintraub 2019), most if not all ORV manufacturers do or will have product-safety recalls, ("Arctic Cat Recalls Textron Recreational Off-Highway Vehicles Due to Crash Hazard (Recall Alert) [May 22, 2019]" 2019) and ("American Honda Recalls Recreational Off-Highway Vehicles Due to Crash and Injury Hazards (Recall Alert) [June 13, 2019]" 2019). Even when product-safety recalls are limited to fire hazards, numerous ORV manufacturers have had such recalls, ("Kawasaki Recalls All-Terrain Vehicles Due to Fire Hazard [August 10, 2017]" 2017) and ("American Honda Recalls Recreational Off-Highway Vehicles Due to Fire and Burn Hazards (Recall Alert) [November 8, 2018]" 2018). Nor is the ORV industry the only industry to have struggled with reports of fire. The emergence of electric automobiles with lithium batteries have brought new hazards to the motoring public and problems for several automobile manufacturers (Klein 2019). It is not uncommon for appliance manufacturers to also face problems with fire ("GE Recalls Dishwahers Due to Fire Hazard" 2012). What is important as a design engineer is not necessarily focusing on any *one* manufacturer or industry for a particular type of fault with a product, but instead learning from history and then working diligently—and ethically—to improve safety for *all* users of regardless of the product at hand.

The *first* table date entry, in Table 11.4, is for the MY 2020 X-5. In it, a U.S. dealer has reported that something was smoking and that a plastic fender was replaced. Furthermore, a manufacturer's district sales manager (DSM) said to replace it on a "goodwill"[17] basis. From this, one might assume that the fender on the customer's X-5 was smoking, melting, or even on fire and that the "it" referenced by the DSM was, in fact, that fender. This is not exactly what this incident report says. The report only says that something was smoking, that a fender was replaced, and that the DSM said to goodwill "it."

This report describes an incident which is troubling because there was evidently an unidentified source of heat that resulted in the melting of something remaining unspecified. Those in charge of studying field reports such as these should follow up with both

---

[16] At least in the author's experience.

[17] "Goodwill" is a term used by some OEMs to denote covering a repair as a gesture of *goodwill* to a consumer whose product is no longer under warranty coverage. Ironically, one pitfall for an OEM having a generous goodwill policy is that, if the product is ever recalled for a related problem, such well-intentioned acts by the company to satisfy a customer could be seen, at a later date, by regulators as an implicit admission that the OEM knew that it had a safety problem at that early date.

the reporting dealer and the DSM. More information is needed to fully comprehend the events described in this field report. For example:

- What was smoking?
- Why was the fender melted?
- Why did the DSM agree to goodwill "it?"

This information, and more, will likely be necessary to fully analyze these events, its causes, and its potential outcomes including injury or death.

The *second* table entry is for the same product as the first, but for a different "unit" as shown by a different set of last six characters of the PIN (Product Identification Number) for the second X-5. This information came from a Canadian dealer and was entered into the database using the French language. Translated, the dealer report says that the consumer said that "the machine was smoking; went 'boom'; there was a large fire; and the consumer had her or his garage burn down."

Ignoring the content of this dealer report for the moment, this dealer report was not difficult to translate since French is a language known by many in the United States, however rudimentarily. There are also on-line translation websites to assist someone encountering such a report. Furthermore, the French language generally uses the same Latin alphabet as does the English language. Therefore, in this case, translation was simple and little appears to have been lost in the translation. However, it would be prudent to have a native speaker of Canadian French follow up with this dealer to obtain the most information possible.

Returning to the content of the report indicates that this was indeed a major event. Given the event description and resulting property damage, the loss of life was certainly possible. Fortunately, it appears that no one was injured. This is certainly an accident requiring attention to understand and address since insurance companies and attorneys will likely be contacting the manufacturer for reimbursement and compensation for the property damage as a minimum. It would serve the manufacturer well to have a competent person visit the site, inspect the vehicle, review the vehicle's service, repair, and manufacturing histories, and then speak with the owner and witnesses. In order to eliminate causes other than a safety defect, the manufacturer will need to eliminate some other potential causes for the accident. These might include:

- Had the ORV been modified?
- Had the ORV been involved in any accidents that could have damaged the vehicle?
- Had the ORV been properly maintained and by whom?

Although a cynic may see such questioning as a manufacturer's attempt to shift blame, an ethical engineer must truly know the answers to such questions in order to objectively determine the cause, or eliminate other causes, of this accident. Doing so may prevent future accidents and injuries. There may also be analysis and testing required to fully understand the events. There is legitimacy to asking such questions and conducting such analysis and testing, but they should never delay a prompt response by the manufacturer to regulators and consumers.

The *third* table entry is the record for an accident reported to the manufacturer directly by a consumer. In it, the consumer said that he was simply operating the X-5 when the ORV struck a tree. After the impact, one wheel of the ORV was no longer attached to the vehicle. The consumer, unfortunately, suffered a broken arm and a broken leg. An attorney is being contacted. Implicit in this report is that the consumer believes that the wheel came off of the ORV before the impact with the tree and caused that accident. An engineer with the designer/manufacturer tasked with reviewing this accident will want more information including photographs of the damaged ORV and the accident scene. The company should also want to inspect the ORV.[18] Since an attorney is now involved to protect the consumer's interests and the manufacturer's Legal department will no doubt have become involved as well, it is improbable that a *timely* inspection of the vehicle will now take place. Such resulting legal "stand-offs" are detrimental to consumer and product-safety interests.

The *fourth* table entry is a report from a regulatory body, such as the U.S. CPSC, indicating that it had received news of an accident in which the operator's friend fell out of the X-5 ORV while it was in operation. There is no indication of injury to the friend. The manufacturer will be glad to hear of this. But there is also not indication of any extenuating circumstances such as if seat belts being used, if the friend fell out of the passenger compartment or out of the cargo box, and other information which may shed more light on this unfortunate-but-lucky event. The manufacturer needs to know if there was any equipment failure or other product shortcoming in order to make the greatest use of this report. The responsible engineer should follow up with internal people at the manufacturer who received the notification from regulators for more information as well as the regulators themselves who should be happy to share their information with the product manufacturer. Only after taking such steps can many incident reports be fully understood and, thereby, prove useful in assessing the safety risks posed by a particular product.

The *fifth* table entry is not for an X-5, but for an accessory sold by the ORV designer and manufacturer for the X-5. It is important to review incident reports for a product's accessories as well as for the product itself. An accessory may present risks that the product alone does not possess. This is especially true for product accessories powered by the product's electricity, hydraulic, pneumatic, or mechanical systems. Aftermarket accessories can also adversely affect the weight and balance of a product significantly. Each such accessory puts a greater load on the "host" product than that product would see without that accessory. This aspect should be fully evaluated during the EDDTP. In this instance, the incident description appears to be a benign accessory-winch failure and probably contains a typographical error made by a U.S. dealer's employee.[19]

The *sixth* and final table entry is a report of a serious accident of some kind, but it is unclear if this report can initially be taken at face value. While the report states that a vehicle possibly "exploded," the remainder of the report assaults the credibility and judgment of the dealer staff reporting the accident.

---

[18] For some types of accidents, the physical evidence provides clear indication of the true failure mode of a product. In this case, the physical damage to the ORV may reveal whether or not the wheel was attached to the vehicle at the moment of impact with the tree. For some products and failure modes, it can be quite straightforward to determine if failure caused the accident or was a consequence of the accident. This shows the value of experienced accident investigators.

[19] These are the actual words seen by the author for one warranty claim filed by a dealership.

The *first* aspect of the sixth dealer report tells the reader that the dealer's staff does not have a firm command of the written English word and, consequently, may not be the finest reporter of events. This may not be the case, yet it is easy to disregard the veracity of this dealer's comment and the quality of the dealer's opinion based upon the grammar and spelling used. The same could be said of the fifth report. Regardless, it is apparent that the manufacturer must follow up with this dealer in order to ascertain the truths behind this report—and do so quickly.

Second, the report indicates that the dealer's staff may have a penchant for exaggeration. There are two instances where dealer staff likely engaged in hyperbole. These two scrawlings—deciphered as "can kill thousands" and "have seen hundreds like this"—would be extremely embarrassing to the designer/manufacturer should it have to provide this dealer report to a regulator during the course of a product-safety recall or to the attorney of an injured party during litigation. However embarrassing the production of this record may be for the manufacturer, it may also simply be *untrue*. That is the largest potential fault with the report. It is potentially both inflammatory and false. It is unlikely that thousands of people truly would die from a repeat of this event. It may also be equally untrue that the dealer has literally seen hundreds such as this unless that dealer is a large-volume retailer. It is unclear to someone keeping an eye out for potential problems exactly what was meant by the inferred term "explosion." Follow-up questions might include those below:

- Did the vehicle literally explode and send shrapnel, or parts, flying in all directions?
- Was there *only* a fire—and *not* an explosion? (This is still not good.)
- Was there any warning that something was going to happen?
- Where on the product did it initiate?
- How did the fire or explosion initiate?
- Are there any photographs of the vehicle or the property damage?

These questions are all unanswered by the dealer's report. This sixth record reinforces the need to include as *much* fact—and as *little* opinion and speculation—as possible during the collection and entry of post-sale product-performance information.

Third, the dealer and staff may have just created even more problems for the manufacturer with their improvised repair of vehicles which are in dealer possession. As illustrated earlier in this chapter, even simple and seemingly straightforward engineering-design changes should be tested thoroughly before the release of a product. Although the precise nature of the vehicle modifications performed by the dealer's staff is as of yet unknown, the manufacturer must keep these affected vehicles from reaching the hands of consumers. It would be difficult to know just what design considerations might have been ignored by dealer staff in attempting to fix the one single problem. Even if that single problem has been addressed through these *impromptu* repairs, the dealer staff may have unwittingly created several more hazards and risks that could manifest themselves at a later date under conditions not imagined by the dealer's employees using the hardware-store parts and their mechanical skills.

Looking back upon all of the entries in Table 11.4, the events of particular interest involve fire. Entries 3, 4, and 5 can be ignored in this discussion although they may prove important for different discussions. The three remaining events—events 1, 2, and 6—taking place over a period of eleven months do indicate that there may be a fire-related problem with the X-5 ORV needing the attention of the designer/manufacturer.[20] It may be that some action or set of actions has already been undertaken by the manufacturer, with the cooperation of regulators, well before the sixth event. Regardless, each of these three events would likely deserve its own follow-up investigation to determine why excessive heat arose and why fires or smoking occurred. It would be indefensible for a manufacturer to wait until some arbitrary number, say *N*, fires have taken place before taking a serious look at a fire problem. Each potential problem should be thoroughly investigated, evaluated, and rapidly responded to in such a manner as demanded by the hazard and its risks.

The evaluation of accidents—and *potential* accidents—is discussed next. When looking at a large collection of field-accident reports, it is both important and fair to recognize that hindsight is 20/20. What may be easy to see looking backwards over much historic data may not have been easy for even the most ethical of reviewers to see at those earlier points in time.

11.9 Product-Safety Recall-Process Model

This section introduces one conceivable model for an administrative process for evaluating potential product-safety defects and recalls. This model should be considered to be neither unique nor authoritative. This is but one possible path to take in evaluating post-sale data, the safety risks of products to users, and the potential need to recall products. Each individual company, industry, regulatory body, and body of accident report data should be considered in its own light. Remember that mandated regulatory criteria, such as the timing for reporting incidents, accidents, and issues, should always be observed by during any product-safety recall process. However, the onus for meeting these requirements should be upon the company's legal personnel providing that they are promptly notified of potential problems. Consequently, engineering personnel must always keep legal staff apprised on incidents, issues, and their statuses.

A flow chart of the overall product-safety recall process presented here is shown in Figure 11.2. The product-safety recall process shown is divided into a total of six stages of analysis. However, only five stages are given Roman numerals. These six stages are:

Nulla Data Collection and Mining
I. Initial Analysis
II. Intermediate Analysis
III. Advanced Analysis
IV. Final Analysis
V. Product-Safety Recall

[20] Depending upon what was discovered by further investigating events 1 and 2, the need for remedial action could have been clear well before the receipt of event 6. Events 1 and 2 happened within a month and a half of each other, while event 6 took place almost a year after event 1.

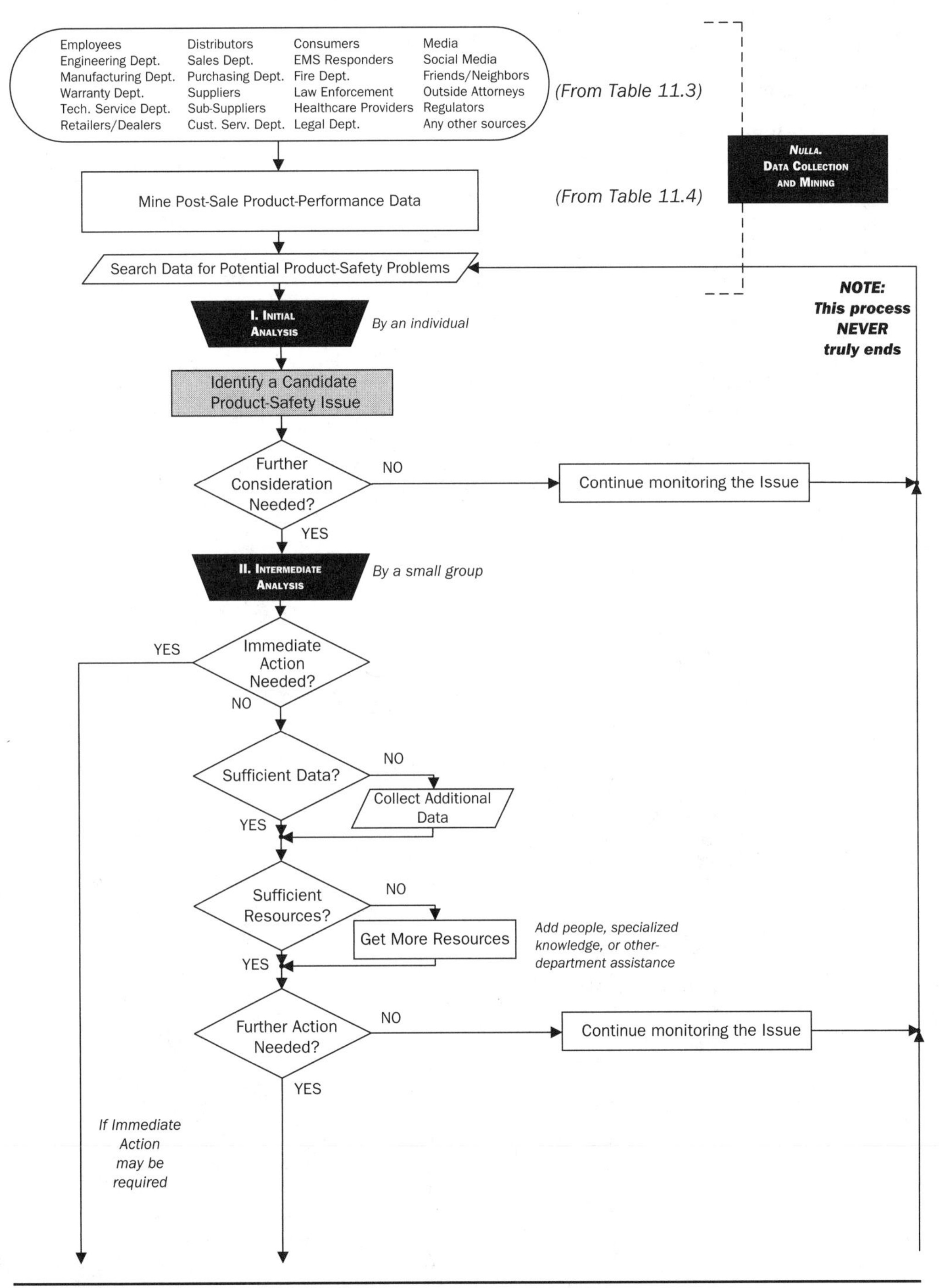

Figure 11.2 Flowchart of the product-safety recall process.

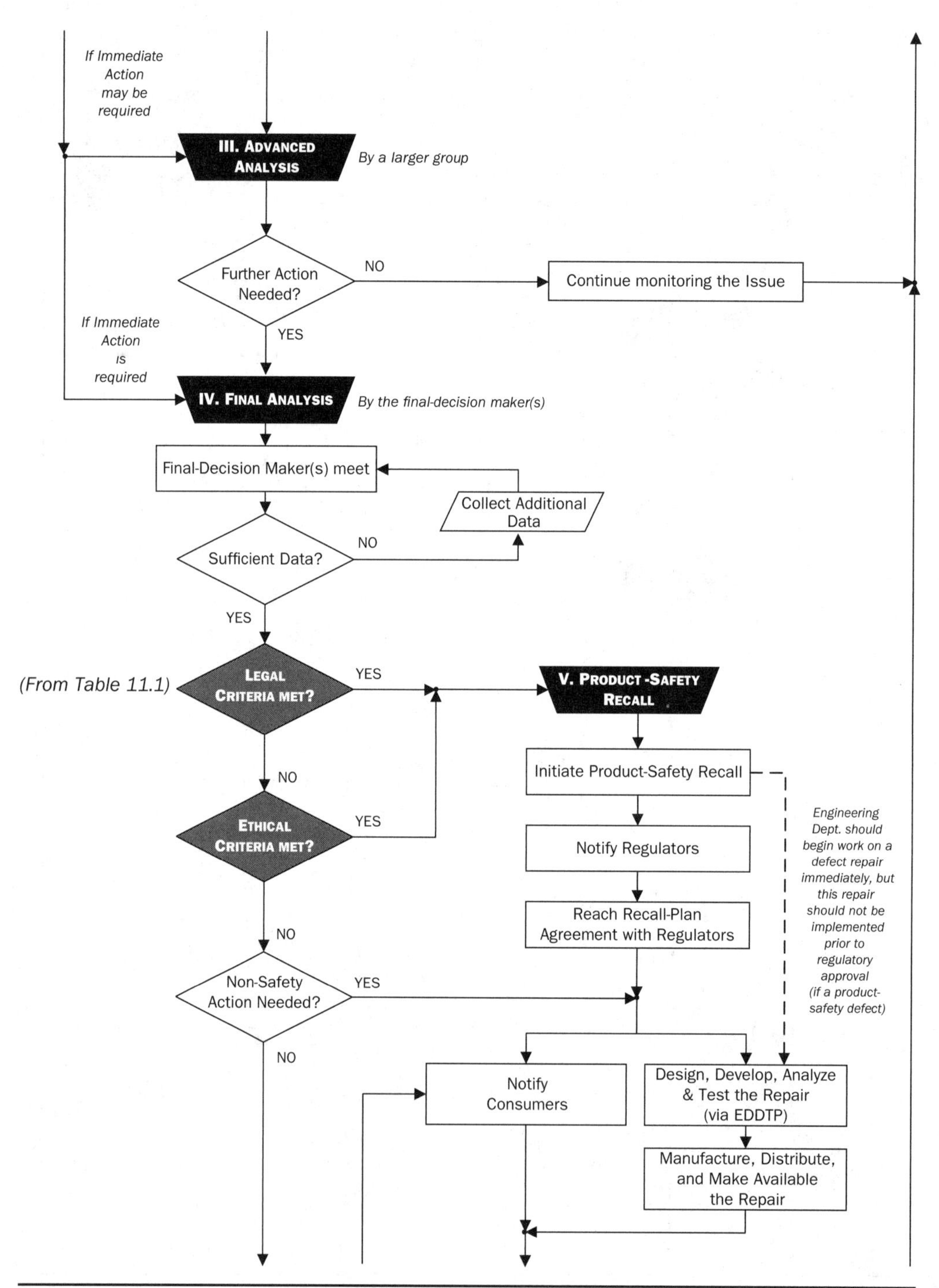

Figure 11.2 *(Continued)*

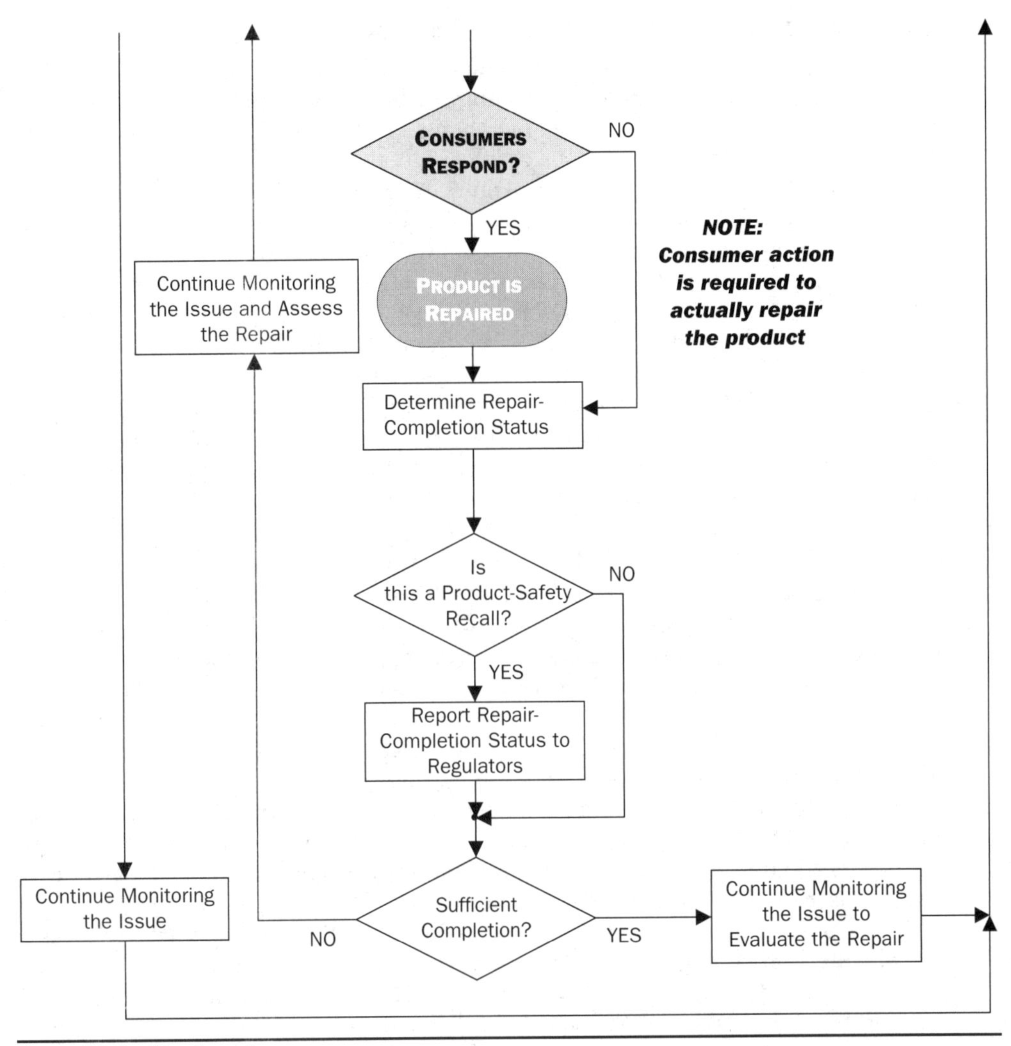

FIGURE 11.2 *(Continued)*

Each stage will be discussed in turn. Not every stage must take place, however. There may arise urgencies issues when skipping steps will be necessary to facilitate a rapid company response to protect the safety of consumers. Also, if no problem is concluded, then subsequent steps become unnecessary. This discussion will outline an internal organizational structure that a corporation may consider for purposes of reviewing post-sale product-performance data which could lead to product-safety recalls. Individuals and groups are involved in this process who consider and evaluate such data at regular intervals, either weekly or as needed, and make decisions on product safety. It goes without saying that these people should be ethical, competent, and readily available for their important duties. These people should also be free from business pressures to keep the company profitable while considering the safety of consumers.

11.9.0 Stage Nulla: Data Collection and Mining

The constantly on-going collection, aggregation, and searching of post-sale product-performance data and other information for possible problems is what is being called *Stage Nulla*[21] in this section of the chapter. Some of the activities taking place in this stage have been described in previous chapter sections.

The dotted lines at the top right of Figure 11.2 show the reaches of Phase Nulla. Post-sale data are continually collected from the sources listed in Table 11.3 and, once searched, may look similar to the data shown in Table 11.4. That search of the database may look for keywords such as fire, smoke, hot, accident, injury, perhaps a similar euphemism, and more. Databases may permit those entering post-sale reports to mark, or "flag," a report that may require follow-up from others within the company.

This stage is reproduced here because it is both a necessary and a crucial portion of an (pro-) active and successful product-performance monitoring process. It is numbered as "zero" since it is always on-going and may not lead to any further action. Such an organizational infrastructure should be in place in order to successfully perform and manage a post-sale product-performance monitoring program.

11.9.1 Stage I: Initial Analysis

Of course, the initial-analysis stage, Stage I, will be the formal start of the process for any identified potential product-safety recall. This stage starts with the review and screening of post-sale product-safety related data collected from the field from such sources as those listed in Table 11.3. These data could look something similar to the examples shown in Table 11.4. For a large company, there may be much data to review for numerous different products. For very-large corporations, it may be wise to have separate post-sale product-performance monitoring processes and staffs. This is true not only due to possible data volume, but each product group may require unique skills and abilities for data evaluation. For example, evaluating toasters may be quite different from evaluating gasoline lawn mowers due to different hazards (such as power sources), user actions, and usage environments.

Those post-sale product-safety data showing potential product-safety problems may initially be reviewed by a single person. There are advantages to having an individual "screen" the data before having a group of people look at them. This individual must, of course, be both competent and ethical. A lone person responsible for the initial data review will be able to better accommodate conflicting meetings and out-of-office days that typically arise when using a regularly scheduled meeting of even a small group of people to review data. This person can retrieve and look at the data whenever her or his schedule permits and not be at the mercy of attempting to reschedule a regular group meeting. This aspect of the initial analysis ensures that no needless delays arise when responding to new data. In addition, this person will be able to initiate simple follow-up actions on issues that have caught her or his attention. As a result, this person will be able to provide answers to many questions that the small group will undoubtedly ask in the next stage of a potential product-safety recall. The efficiency of a single person within this recall-process model at Stage I, as already mentioned, may help prevent the unnecessary lapsing of time.

---

[21] *Nulla* is the closest descriptor the author found for "zero" in Roman numbering.

The person conducting the initial data review may be a design or test engineer, a product-safety engineer, or simply a skilled technician. The level of expertise needed by an initial reviewer will depend upon the complexity of the products being reviewed. For simple products, no particular expertise may be required at this stage. Conversely, some products and their issues will take some expertise to thoroughly review, hence, the need for a more-qualified initial reviewer and need for Stage II of the product-safety recall process.

The tasks within Phase I are shown in Figure 11.2 and are relatively simple and intuitive. From the searched database provided by Stage Nulla, data collection and mining, the Stage-I analyst or reviewer looks at the database-search results and identifies any potential product-safety problems that should be reviewed by the small group in Stage II of the process. In exigent cases, this Stage-I reviewer will notify others of the need to respond immediately rather than wait for the regularly scheduled Stage-II meeting. If the database search shows nothing unusual or suspicious for a given issue, then that issue is not forwarded to the Stage-II group. This data reviewer should make a note of this issue—either mentally or in writing[22]—to keep an eye out for it, or similar issues, as time progresses and new data arrive.

An efficient post-sale product-tracking process would have reviewed the data such as those shown in Table 11.4 well before the more-than-eleven-month period shown by the table's data had elapsed. Those involved in Phases I and II should not be seeing old data for the first time if the data are considered to be indicative of a potentially serious post-sale issue. At times, however, it might be necessary to generate all recorded reports of a particular issue with a given product in order to illuminate a possible product-safety problem. Such a report may contain older data such as those shown in Table 11.4.

Those performing this stage are at the *vanguard* of a company's post-sale product-safety effort. If a potential product-safety problem is never identified, it can never be addressed by the company responsible for its product performance with consumers. Accordingly, it is hard to overemphasize the importance of the job function that is Phase I. The company's legal personnel should be notified of reportable incidents as required by regulators.

11.9.2 Stage II: Intermediate Analysis

If the initial reviewer of post-sale product-safety data determines that more attention should be paid to a particular issue seen in the data, the potential problem proceeds from Stage I of the recall process to Stage II. During Stage I, an individual person looked at data; at Stage II, a small group of people look at the data on the product and any prospective issues. Up to this point, there has not yet been an official determination of whether or not a product-safety defect exists. Instead, the competent initial reviewer in Stage I has merely decided that a closer look at one or more reports by a group of people is needed—unless an obviously urgent product condition has arisen. This Stage-II group may consist of the initial reviewer, product-safety engineers, reliability or quality engineers, and design, test, and manufacturing engineers for that product. The group should also include the product-safety manager or director. (If not directly involved in these meetings, the Engineering Manager or Director for the product should be kept up

[22] The mechanism for remembering a note will depend upon the volume of data reviewed, a company policy, and the memory ability of the Stage-I reviewer.

to date by their engineers in this meeting.) The Stage-I initial reviewer should lead the meeting since she or he has the greatest familiarity with the issue(s) being discussed. There may be several products and issues being offered for discussion at a single Stage-II meeting.

While the person performing Stage I sees all of potential product-safety problems, those participating in Stage II only see those data forwarded from Stage I. Therefore, the ability, discretion, and ethics of those performing Stage I is of paramount importance in identifying and correcting product-safety defects.

The group in Stage II may make several determinations while executing their responsibilities. First, the group may decide that immediate action is required.[23] The resulting actions from this decision may depend upon whether a high risk to consumers is somewhat elevated or whether that risk is almost certain. This is shown at the start of Stage II in Figure 11.2.

Second, the small group must decide if there exists sufficient data to accurately assess the potential risk to consumers. If not, additional data or other information are generated or collected as rapidly as possible in order to resume the intermediate analysis.

Third, the small Stage-II group must then determine if the group possesses the necessary human capital, specialized knowledge, and access to department-specific answers needed to properly proceed in Stage II. If not, this group must ask for additional resources such as more people, specialists (e.g., the product's design and test engineers or materials-science experts), and different departments (e.g., the Purchasing Department to get answers on how a supplier manufactured a particular component).

Once the Stage-II data and resource needs have been met, the small group can make its determination on whether there is a potential safety risk to product users. If so, the recall process moves on to Stage III. This and the above decisions are shown in Figure 11.2. If not viewed as a possible product-safety issue, the issue is no longer discussed by the small group at this time.[24] However, the initial, Stage-I data reviewer must remain vigilant for future recurrences of the same issue. If future post-sale data on the same issue arise, they may contain new information, shed greater light on the issue, and result in a different Stage-II group decision later on. These new data may contain details that were not available when the issue was initially reviewed in Stage II.

---

[23] In extreme cases of product-safety risks to consumers, the initial Stage-I data reviewer may correctly reason during Stage I that immediate action is necessary. However, it seems unlikely that the initial review will proceed further without contacting one or more people—some of whom would be in the Stage-II small group. The Stage-II group member(s) should assist or lead the Stage-I reviewer as needed.

[24] Because of the seriousness of many decisions being made in Stage II of the recall process, it may be wise for Stage-II personnel to *elevate* a possible product-safety related matter to Stage III—even if Stage-II personnel do *not* necessarily consider the particular issue to be a product-safety defect. Even if the Stage-II group *properly* decides that no defect exists at the time, that decision could haunt the company if regulators later disagree with that decision or if that issue later presents itself as a true product-safety problem. Therefore, in some matters, it may be wise to present the Stage-II evidence and resulting decision to the larger Stage-III group for manager and executive "buy-in." This is an example of Rule 3 of Table 3.1's *Rules of applied engineering ethics* of making managers and executives earn their salaries by having them *explicitly* agree with the Stage-II decision. They cannot, then, claim ignorance or assign blame in the important matter, or disagree with the Stage-II decision, at a later date. Furthermore, in Stage III, that group may *disagree* with the Stage-II decision after looking at the evidence and, in fact, *proceed* with a product-safety recall.

New information in such a matter should be brought to the Stage-II small group for reconsideration whenever available and necessary. These new data may, indeed, prompt a product recall at this later time since more information is now available. The Stage-II group may again properly decide that there is no need for action at this time. As with Stage I, notify the company's Legal Department regarding reportable issues and incidents to keep within regulatory-reporting deadlines.

11.9.3 Stage III: Advanced Analysis

The third stage of the product-safety recall process is similar to Stage II except that a larger group of people will now be involved. It is envisioned that the Stage-II group be included within the Stage-III group due to the Stage-II group's familiarity with the possible product-safety issue and its importance to the company's overall product-safety efforts. Not only will people be added in Stage III, but the Stage-III group members will generally have more-senior positions within the organization than group members from Stage II. The appropriate Engineering Directors or Managers should attend these meetings as well as the Chief Technical Officer (CTO). Other company officers, such as General Managers, could also be invited depending upon the company's structure and operating procedures. If the need for a product repair could arise, it might be wise to include the technical-service department manager or other personnel as well. It is probable that the corporation's legal counsel will be involved by this point. They should be notified of any meetings so that they, themselves, can monitor activities and report to regulators as required.

Although they are not shown in Figure 11.2 as being repeated in Stage III, the larger Stage-III group may have the same initial requests as the Stage-II group—namely for more data and resources. These requests should be accommodated while remaining vigilant of regulatory-reporting deadlines. As with any stage, the larger Stage-III group may rightfully decide that no product-safety recall is necessary.

Stage IV is entered if the Stage-III group decides that a product-safety recall may be needed. Depending upon an organizational's structure and administrative procedures, the Stage-III group could have the authority to make a binding corporate decision regarding a product-safety recall, in which case Stage V is entered and Stage IV bypassed. As always, there do exist legitimate instances in which there is no need to conduct a product-safety recall, even after entering Stage III of the process. In these cases, the company continues to seek and monitor post-sale data for re-occurrences of the issue at hand.

11.9.4 Stage IV: Final Analysis

If a product-safety issue enters Stage IV of the recall process, the company is obligated to make a binding decision on the need to report and recall a product for a safety-related defect. Decisions made in the prior stages of the recall process have been in more of an "advisory" function and, usually, not binding for the company.[25] There may, indeed, be more information available to Stage-IV group members than in earlier-stage meetings. Therefore, this final decision may be made with more-complete information than previous decisions.

[25] A company's legal counsel will provide *legal* guidance on these matters.

The events to be described for Stage IV are shown in Figure 11.2. As with Stage III, not all decisions and requests are shown. Not shown is the potential need for additional resources. This decision is omitted because the likelihood of that need has been reduced through the completion of the prior stages. That need may still arise, however, in Stage IV, as could the need for additional product analysis or testing.

Depending upon corporate policies, the recall decision made during the Final Analysis could be the result of the Chief Executive Officer (CEO), President, General Manager of the product's division, a product-safety committee, or some other entity. In many instances, the Stage-IV group will include members with profit-and-loss (P&L) interests for the corporation. These P&L interests could involve the overall profitability of a corporation or just of one of its divisions or product lines.

At this point in the discussion, it is worth noting that Blasius makes interesting observations about corporate-management styles and chances of product-safety recalls (Blasius 2020). Although he addresses quality control, in particular, in his article, substituting "product safety" and "safety" for "quality control" and "quality," respectively, does not change the thrust of his message.[26] Blasius speaks about his experience as a regulator of consumer products and his investigations to remove "dangerous products from the marketplace" below:

> A weak corporate culture that unwittingly tolerates shortcuts and lapse quality control may be the real culprit...Undue pressure may fall on your quality control, compliance, and legal staffs to "find a way" to approve product changes, and the results can be disastrous. If the very people charged with holding the line on quality are reluctant to sound the alarm when problems appear, leadership is failing. It takes strong leaders to ensure that the organization doesn't lose track of the need to keep the safety of their customer's front and center, even at the expense of slower growth or reduced profit margin.

Blasius also recognizes the dichotomy between *management* and *leadership*. His article's title mentions "management style," but its critical evaluation of weak corporate culture speaks of failed "leadership." Leadership takes more courage than simple management does.

For a corporation, conducting a product-safety recall involves spending money on products it has already sold. There is no revenue *upside*, only cost *downside*. There is no short-term business impetus for doing a recall. There are, however, long-term business incentives for doing so—and for doing so sooner rather than later as adduced via references throughout this book. And, not to forget, there are regulatory requirements for reporting and repairing defective products.[27]

Unlike fine wines, product-safety recalls do not age well. Putting off a recall may result in the accumulation of even more accident reports—and possibly more injuries to users. These additional later data points may make the earlier data points stand out as clear early indicators of a problem to regulators who, upon seeing the data for the first time, will justifiably ask why there was no response by the company.[28] Long-term

---

[26] This was confirmed through personal communication with Dennis Blasius on January 21, 2020.

[27] It is hoped that ethical responsibilities are also considered in such times.

[28] Some will argue that "hindsight is 20/20" as mentioned earlier. Not all problems with products in a complex Product/User/Environment setting are visible from a small dataset early on; however, some problems truly are.

business incentives for prompt ethical behavior also include less adverse publicity and the potential avoidance of regulatory fines. Criminal charges are also a possibility—however remote—from unethical conduct.

It is hoped that ethical and social considerations triumph over meeting business goals and the resulting personal rewards during times of product-safety recall deliberations. The final recall decision may be up to a simple majority of group members, the general *consensus*[29] of the group, or an executive decision. This, again, is up to the organization's administrative policies.

When approaching a final decision on conducting a recall in Stage IV, as in other stages, the final-decision maker(s) may need additional information regarding the recall. This may be perfectly appropriate. There are times, however, when one or more members of the final-decision group—or any other group—may, for whatever reasons, want to derail the process by demanding answers to questions that are unlikely to truly advance the discussion. Facetious and fictitious—yet illustrative—examples include the following questions:

- How many of these accidents took place during a full moon?
- Was the stock market up or down that day?
- Is the product-safety failure severe enough to *(fill in the blank)*?

These are examples of what the author calls product-safety "off ramps." Although the person asking such inane questions wants to appear as though the questions being asked actually advance the discussion toward the final decision, some questions serve no real purpose except to delay a decision which one might avoid—or *escape* from—if that decision is delayed long enough. People might become distracted and lose a sense of urgency in the interim. Off ramps may appear at any stage of the product-safety recall process, but may be more prone to appear as a final product-safety recall decision approaches.

Ultimately, there must be a decision made on whether or not the possible product-safety issue rises to the level of a *legal* product-safety defect. It is probable that this decision will be made in conjunction with legal counsel. That decision may be either "Yes" or "No." However, there could exist a higher standard than the legal[30] one—namely, the ethical standard. If business pressures make the determination of a product-safety defect "unpalatable," then it may be the ethical obligation of, one or more, Stage-IV participants to proceed with the product-safety recall, regardless. The ethical requirement may be *higher* than the legal one.

It may be necessary for a manufacturer to report a *potential* safety defect on its product to regulators before making a final decision on whether or not the problem being experienced by users truly is a product-safety defect. There are instances in which both the company and the regulatory body have agreed that there is, in fact, *no* product-safety defect. Of course, the company's legal counsel could be involved throughout the product-safety recall process.

It is also possible that the decision in Stage IV is to conduct a non-safety related repair program. This may be a completely justifiable and ethical decision. Of course, the

[29] It is the author's experience that a good working definition of *consensus* remains elusive.

[30] The author reiterates that he knows attorneys who are among the finest of people by any standard.

particular issue and evidence are important. It may be that the Stage-IV group decides that, although there is no safety implication, there may be reasons why a consumer may be otherwise displeased with the product. The company may choose to fix that product, perhaps only if the consumer complains, without expense to the consumer. Regulatory obligations may not exist in such cases thereby permitting the company to proceed as it wishes.

Again remember, that the final, Stage-IV decision may rightfully be that no product-safety recall is needed. However, the group responsible for the Stage-I and Stage-II product-safety and data reviews must continue to look through the data for new occurrences of this same issue in the event that new data could provide information or greater insight that could change a previous decision of "no recall needed."

11.9.5 Stage V: Product-Safety Recall

The fifth and final stage of the product-recall process is the product-safety recall itself. This phase is shown along the right-hand side of Figure 11.2. What is shown is a generic product-safety recall procedure. Depending upon the regulator, there may be particularities not shown in this figure which must be followed. For example, for consumer products, there are specific requirements for the design-and-manufacturing company from the U.S. CPSC that are outlined in sources including their *Recall Handbook* ("Recall Handbook" 2012). Although there exist two international standards providing guidance on consumer-product safety and recalls, ("ISO 10377:2013 Consumer Product Safety—Guidelines for Suppliers" 2013) and ("ISO 10393:2013 Consumer Product Recall—Guidelines for Suppliers" 2013), respectively, these standards will not be covered here because of the regulatory requirements of the CPSC.[31] Likewise, motor-vehicle safety recalls are covered by specific requirements including the U.S. Department of Transportation's *Motor Vehicle Safety Act* (*Motor Vehicle Safety Act* 2005).

Once a company has made the decision that the product possesses a safety defect, then the pertinent regulatory body must be notified. Before initiating the product-safety recall, however, the company should reach an agreement with regulators regarding several issues. For example, the regulatory agency should agree that the condition is a product-safety defect. There is rarely, if ever, disagreement from regulators on this point. However, there may be regulatory dissention regarding the repair to and the affected population of the product. The design-and-manufacturing company may propose a repair of the admittedly defective product while the regulators might want a complete replacement of the product. Similarly, the company may want to only recall units manufactured between date 1 and date 2. It is not uncommon for motor-vehicle recalls to be limited to regions that are either hot and humid or cold with roadways regularly treated with salt depending upon the defect. Regulators may, instead, want the recall of *all* units produced. Prior to contacting consumers and repairing affected product units, the company should reach agreement with regulators on the remedial measure, the fix, and the units to be corrected. Despite these potential hold-ups to executing a recall, the company may not want to waste time by waiting for regulatory approval before turning engineers loose on designing, developing, analyzing, and testing a repair to the defect. Again, before touching and fixing any products, the regulators

[31] In these standards, "suppliers" are those parties supplying a product or a service to the public. Thus, a design-and-manufacturer would be considered to be a supplier.

should agree with the company's plans. Once the fix is working properly and available for shipment, the company should make the remedy available to consumer, retailers, dealers, and distributors. In some cases, the repair kit, if in "kit" form, will automatically be sent to retailers and consumers; in other cases, the repair will be made available when requested by an affected consumer.

At the time of the recall decision or after regulator notification, the company may announce to the public to stop using the product without certain countermeasures or to stop using the product altogether. During this time, the company should issue a "stop-sale" notice to retailers and consumers of the product to expose no further people to the defect and its risks. Companies may also issue a "stop-shipment" order to warehouses to no longer ship products to retailers.[32] Such stop-sale and stop-shipment notifications, of course, have adverse effects on company and retailer revenues.

Once regulatory agreement for remediation has been secured, the company is free to manufacture and release the remedial measure to the public. At the same time, the company will announce to the public the availability of the recall repair measure. Depending upon the product, some fixes may be accomplished by consumers through simple actions, or product repair may require retailer or dealer work. These product-safety repairs should never be at an expense[33] to consumers and should remain available to users for a considerable length of time.

As mentioned for non-safety related actions in Stage IV, the company may announce the availability of no-expense repairs for certain product conditions. These announcements may go to all consumers, or only to retailers or dealers.

Once the no-expense repair has been announced, it is vital that the company conduct an effective publicity campaign to announce to affected consumers both the existence of the product-safety defect and the availability of a repair. Even assuming that this publicity effort is effective, it remains imperative that consumers respond. Without consumer response, it is *impossible* for a company to conduct an effective product-safety recall. Consumers must act in their own self-interests to affect an effective product-safety recall. However, it also remains imperative that the company reach as many consumers as possible. With email, the internet, and social media, new avenues through which to reach consumers are now available to companies.

Periodically, perhaps monthly or quarterly, the company people tasked with conducting the product recall—who may be a different group than those in the recall-phase decision making—will need to determine the number of affected recalled product units that have been repaired. In the case of regulated recalls, periodic completion-status reports are filed with regulators until a certain time has elapsed or a certain level of completion is attained. Regulators may insist that the company re-announce the recall and the availability of the repair. In the case on non-safety related actions, the company may end the availability of the repair when it alone is satisfied with the results.

It is necessary that the company continue to monitor the field reports, such as those seen in Stage I, even after the product-safety recall has started. It is through such

[32] In some instances, it may be reasonable to permit shipments of the affected product to dealers with the understanding that none of these units is to be sold before making the recommended product-safety recall repair(s). This is especially true when the repair is simple and easy to execute by a dealer.

[33] "Expense" is different than cost. While the consumer will not need to pay for the repair as an expense, the consumer will likely incur some "cost" due to loss of use or lost time in order to obtain the "free" product-defect repair.

information that the company can see if the defective state of the product has actually been fixed. If reports of product problems continue to come in on "repaired" product units, then it may be that the defective state of the product was not adequately addressed. This could, unfortunately, lead to successive recalls of the same problem with the same product. This is truly an ugly situation for all parties involved. There are no "winners" in this case. Consequently, it is in the best interests of the company to quickly acknowledge a problem, *thoroughly* address that problem, and work toward high completion rates as rapidly as possible.

It is here that the importance of the Stage-I persons is reiterated. The need to continually "monitor" the in-coming field data is stated in numerous places in the recall process. This vigilance is necessary to maintain the integrity of the post-sale product-safety data-analysis process. Admittedly, the word "monitor" is vague and could be believed by some to be forget about the issue since "that bullet has been dodged." However, the character and ethical behavior of those involved in Stage I, and Stage II, of the process should not permit any issue to be forgotten. These people should never be reluctant to raise such issues when the field data suggest a change in status of such any prior issues.

11.10 Evaluating Post-Sale Product-Safety Data

The flow chart shown in Figure 11.2 and discussed in the prior section provide an overview of one possible set of mechanics[34] for conducting a product-safety recall without going into the details that necessarily arise when the product-safety recall process is followed in real-life situations. The first stage involving screened data collection aggregated from various sources into a single database are reasonably straightforward. However, once post-sale product-performance data have been collected from the various sources, these data and any other information must be evaluated by skilled and ethical people. This, too, has been outlined in the previous section, albeit only generally.

The reader has an advantage that company personnel may not have when reviewing data such as in Table 11.4. Routine pressures of employment and business goals can rush or cloud the judgment of data reviewers. This is a potential *reason*—but not *excuse*—for poor decision making within a company during a product-safety recall decision-making process. Companies must foster an environment conducive to ethical decision making regardless of the financial costs. This must be a continuing *value* of the company, not just a transient *priority*. If engineering ethics is merely a passing whim within the company, some personnel may legitimately reason that doing the right thing now might be punished later on in various ways once an urgency has passed. Doing the right thing now may end up being reflected in poorer performance metrics such as budgeting, scheduling, or other types of productivity. A *values*-driven company will ensure that people do not suffer for doing the right thing. A *priorities*-driven company may indeed punish the virtuous in indirect ways such as through lesser raises, bonuses, and promotions. It is hoped that people in the company find the fortitude to take the correct actions regardless of the repercussions of doing so. In extreme cases, leaving the company, or even whistleblowing, may be warranted to protect oneself from being perceived as complicit with poor company decisions. There is limited whistleblower

[34]Practicing engineers should consult with their company's legal counsel about the specific requirements for their product(s).

protection for employees. Some attorneys are eager to take on whistleblowers as clients, but many of these attorneys only want to represent clients where substantial monetary claims will result from corporate malfeasance—such as with fraudulent Department of Defense contracting. Engineers faced with such prospects are urged to proceed carefully, but follow their ethics and consciences on this matter as well as obtain legal counsel.

There are also other ways in which judgment may become confused, even for the ethical. This is true on both the professional and the personal level. Some types of products may be easy to evaluate when consumer problems arise in the field. The simple two-slice toaster shown in Figure 4.10 is a product low in both risk and user-interactivity according to the User Activity-Risk Plot in Figure 4.12. Once plugged in, the user only inserts bread and pushes down the spring-loaded operating slider. The toast soon appears to the user after sufficient browning time. Aside from selecting the desired darkness level of the toast, the user exercises little decision making when operating the product. The user's behavior should have little-to-no effect on the safe operation of the toaster. If properly grounded electrically, the toaster is even a relatively safe product with respect to electrocution risk when not in operation. Moreover, many toasters have thermal insulation such that the toaster housing remains cool to the touch even during use. From this, many people will make the *prima facie* assumption, due to the design-safety countermeasures taken by engineering designers along with the undemanding operator interaction with the toaster, that any toaster accident is a sign of a product-safety defect.[35] This is not always the case, even with such a simple product.

If one is dealing with a product of higher risk and user activity than a toaster, then it becomes increasingly more difficult to properly attribute the primary event or chain of events leading to an accident. This may be true even for the most competent and ethical of engineers. If evaluating a product which has high levels of both risk and user-activity, then finding the cause[36] of the accident can be much more complicated than for simple low-risk, low-user-activity products. Therefore, properly evaluating ORV accidents may be much more involved than evaluating toaster accidents.

Looking at Table 11.4 and focusing on only the third row shows just one particular ORV accident report. Imagine a product-safety engineer looking through accident reports. If properly chosen, this engineer will be familiar with the design, manufacturing, and use of ORVs. Confronted with this accident report, several questions arise such as:

- Does this engineer have enough information to make a determination of product-safety defect?
- How should this presumptively responsible engineer proceed?

The prior section strongly suggests that an inspection of the subject ORV is a must to completely evaluate this accident. However, as also stated earlier, the probable delay

[35] The author does not make this presumption having learned that simple products may have hidden intricacies due to engineering-design details that complicate such simple product evaluations. Each product and situation should always be carefully evaluated on its own merits.

[36] "Cause" should never be confused with "blame."

due to legal considerations in getting this evidence cannot be used as an excuse to do nothing in the interim. The ethical engineer should ask the following questions:

- What is objectively *known* about the accident—the accident scene, photographs, witness statements, ...?
- Have there been similar issues in the past with similar products made by the company?
- Have there been similar issues in the past with similar products made by other companies?

Until more information comes in about this accident and maybe even thereafter, the ORV's design-and-manufacturing company will have to consider numerous factors to evaluate the true risk of the product to consumers including:

- The "facts" of the report
- The effects of engineering-design characteristics
- The effects of user inputs to the product
- The effects of the environment—both physical and social
- The prior field-performance history of this product
- The prior field-performance history of similar products

The ORV is a product that is used by many people for many uses and in many places. It is important to understand these aspects of ORV use and how they might be affected by engineering designers, consumers, and environments. In an attempt to better understand the hazards of operating an ORV, the Haddon matrix is used once more. This time, however, instead of listing *safeguards* to injury from an accident involving a product, the Haddon Matrix is used to indicate potential *hazards and risks* of ORV operation to its occupants—the operator and any passenger(s). Table 11.5 is one such Haddon matrix. As already mentioned, ORV operation has many hazards and risks. Some of hazards bring with them high risks and serious potential injury for ORV occupants[37] when factors such as the operator behavior, the environmental conditions, or the product itself are unfavorable to safe ORV operation.

There exist a myriad of hazards facing the occupants of ORVs[38] as attested to by the elements in Table 11.5. Hazards and risks may arise from factors such as ORV design and manufacture, occupant behavior, and environmental factors. In addition, the Haddon matrix indicates the phases of an accident where such hazards could indeed lead to injury or death to ORV occupants.

The behavior of occupants can influence their own ORV-riding safety as shown in the "Occupants" column of Table 11.5. Reckless operation combined with poor or alcohol/drug impaired judgment can be a common cause of ORV accidents and injuries. Occupants should use appropriate PPE (Personal Protective Equipment) such

[37] Occupants include the operator plus any passenger(s).

[38] The set of hazards and risks listed in Table 11.5 is by no means exhaustive, unique, or beyond challenge. There also exist hazards to other factors such as harm to the physical environment, but these hazards are not covered since product safety is being discussed.

as helmets and eye protection, as a minimum for most ORVs and some conditions. Furthermore, ORV safety equipment such as doors/nets should not be removed, speed-limiting devices should not be overridden, and any safety-related interlocks should not be tampered with. Seat belts, doors/nets, and handholds should be used by all occupants during ORV operation. In addition, occupant PPE can help prevent or reduce injury in some situations. One association recommends operating ORVs only when wearing a helmet, gloves, long-sleeved shirt, long pants, and boots ("Tips Guide for the Recreational Off-Highway Vehicle Driver" 2017). Riding in an ORV—either as an operator or a passenger—should not be done when impaired or if thrill seeking. As the "captain of the ship," the operator has the safety of passengers to also consider when driving, inspecting, maintaining, and modifying the ORV. The recommendations of the ORV designer and manufacturer should be kept in mind when operating ORVs.

Similarly, there are several *potential* ORV design or manufacturing properties that could adversely affect occupant safety. Some of these are listed in the "ORV" column of Table 11.5. Certainly, vehicle-stability characteristics[39] as well as the durability and integrity of safety-critical systems, such as fuel, steering, brakes, and electrical, will affect the safety of occupants. Occupants and vehicle components should be adequately shielded from engine and exhaust-system heat. Appropriate seat belts, doors/nets, seating, and Rollover Protective Structure (ROPS) should be provided (and used by occupants). Operators and any passengers should be provided with information and warning signs sufficient to permit them to safety operate the ORV under reasonable conditions. Occupants should read, understand, and follow the advice provided to them including operator-age restrictions.

The physical environment may also negatively impact the safety of ORV occupants. Such factors are listed in the corresponding Table 11.5 column. Unlike automobiles which operate upon paved and maintained public roadways a majority of the time, ORVs generally operate on unprepared and unmaintained unpaved surfaces for the majority of the time. Also, unlike automobiles which operate on roads often limited in grade, or incline, ORVs frequently climb substantial grades and navigate cross-slopes. These unpaved surfaces may be of grass, mud, sand, rock, gravel, snow, or ice—or even combinations of these. The terrain itself may lead to a simple event, such a veering off of a marked trail, into a serious accident if a large vertical drop-off is present just off of that trail. Having an otherwise benign mechanical failure in a remote area can lead to life-threatening results if assistance cannot be obtained. ORVs are generally not totally enclosed to the elements and, thus, their occupants are not well sheltered from weather and temperature extremes. Reliable mobile-telephone service is frequently non-existent in ORV-riding areas. Even when emergency services can be contacted to respond to an accident, long distances and inaccessibility may hinder a prompt response for those injured. Although automobiles also strike large animals such as deer, the construction of ORVs and the rural and isolated areas in which many ORVs operate exposes ORV occupants to even greater probability of injurious animal encounters than automobile occupants.

In addition to the physical environment being able to compromise the safety of ORV occupants, the social environment also can lead to injury in generally indirect ways.

---

[39] ORV-stability characteristics should be evaluated in conjunction with the environment, for example "operating on a side slope," and the operator activity, for example "doing doughnuts."

Phase		Factors			
		ORV Occupants	ORV	Physical Environment	Socio-Economic Environment
1	Pre-Accident	• Unskilled operator • Impaired operator • Improper use of ORV accessories • Operator driving aggressively or irresponsibly • Operator driving on hazardous terrain • Operation in hazardous weather • Ignoring trail-use rules and laws (if applicable) • Operating on public roads • Operator trying to scare or impress passenger(s) or bystanders • Operator fails to inspect or maintain critical systems such as: tires, suspension, steering, brakes, seat belts, and doors/nets • Improperly filling and capping the fuel tank • Spilling oil onto the engine and exhaust system • Consumer removal or override of safety-related equipment • Occupants not using seat belts, doors/nets, and/or hand holds • ORV load is improperly located or secured • Carrying unsecured fuel containers • Operator did not provide passenger(s) with PPE • Occupants(s) not following manufacturer's warning signs • Operating the ORV in poor combinations of environmental conditions (such as deep water plus below-freezing temperatures) • Making modifications to the ORV compromising pre-accident safety (such as additional passenger seating or lift kit) • No experience or poor judgment when loading/unloading and securing vehicle	• Finite directional-stability properties • Finite lateral-stability properties • Gasoline in fuel tank and fuel lines • Unpredictable power-delivery through ORV drivetrain • Insufficient durability of steering, braking, fuel, suspension, and electrical systems • Occupant compartment is insufficiently isolated from engine and exhaust-system heat • Occupant compartment is insufficiently isolated from engine-exhaust emissions • Vehicle components are insufficiently isolated from engine and exhaust heat • Insufficient ORV lighting • Insufficient information to operator and passenger(s) regarding proper ORV use through warnings, manuals, videos, and websites • Manufacturer marketing messages inconsistent with the Intended Use and Limitations of Use for the ORV • Authorized dealer fails to adequately prepare the ORV for customer delivery • Lack of, or poor, ORV loading/unloading or securing guidance	• Uneven terrain • Steep terrain • Loose terrain • Deep water crossings • Open water on partially frozen bodies of water (with "open water") or on thin ice • Inclement weather • Poor operator visibility due to terrain, nighttime, or weather conditions • Recent inclement weather degrading terrain and waterways • Other ORVs • Automobiles and trucks (when operating on public roads) • Bystanders on trail or in riding area • Domestic animals • Wild animals • Human-made malicious hazards ("booby traps") • Improper ORV transportation equipment • Improper ORV unloading/loading and securing equipment	• Irresponsible operators of other ORVs • Social enabling of risk-taking behavior (such as consuming alcohol and drugs, performing stunts, not using PPE, and operating on public roadways) • Lax enforcement of DUI laws and trail-use rules including sand-dune flags (if applicable) • Poorly marked trail sections (if applicable) • Not preventing ORVs from licensure for public-road use

TABLE 11.5 Example Haddon Potential Hazard-and-Risk Matrix for ORV Occupants

Phase		Factors			
		ORV Occupants	ORV	Physical Environment	Socio-Economic Environment
2	Accident	• Occupant(s) not wearing PPE • Occupant(s) limb extends outside of the ORV • Occupant(s) riding in ORV cargo box • ORV load is improperly secured • Making modifications to the ORV compromising during-accident safety (such as additional seating, ROPS, or seat belts)	• Insufficient ORS (Occupant-Retention System including ROPS) properties • Insufficient fuel-system integrity • Insufficient electrical-system integrity • Insufficient heat-resistance of ORV components • Hard or sharp surfaces within occupant compartment	• Terrain may worsen some accidents through slopes, drop-offs, and sharp, hard, protruding features (such as rocks and tree branches) • Environment may pose a drowning threat	• N/A
3	Post-Accident	• Not carrying a fire extinguisher • Not carrying a first-aid kit • Not having an operable mobile phone • Operator does not turn off the ORV ignition system • Making modifications to the ORV compromising post-accident safety • Not letting others know of your riding plans	• Fuel system may leak • Engine and fuel pump may continue to operate • ORV constructed from flammable materials • ORV battery may leak electrolyte (acid) • Occupant compartment is insufficiently isolated from fuel tank and battery • Accident damage to vehicle may make occupant removal or self-extraction difficult • Not having a fire extinguisher • Not having a first-aid kit	• Location may not have cellular-phone service • Location may not be reachable by on-road EMS vehicles such as ambulances, fire engines, or other rescue vehicles • Emergency responders may take much time to arrive • Hospital may be far away • Poor weather may hinder timely emergency response • Poor weather may preclude helicopter evacuation of the severely injured • Extreme heat or cold (may lead to exposure injuries for isolated and stranded occupants)	• Bystanders may not initially recognize the severity of an accident • Bystanders may not have the ability or supplies to render first aid to the injured • Bystanders may worsen an injury by moving injured person • Bystanders may not be able to provide adequate location information to EMS responders • Bystanders may flee scene rather than assist the injured

Table 11.5 Example Haddon Potential Hazard-and-Risk Matrix for ORV Occupants (*Continued*)

Such hazards and risks are listed in the final column of Table 11.5. Social enabling can be a significant factor in ORV accidents. Simply watch internet videos of ORV operation to see how. There are sometimes intoxicated crowds of bystanders urging an operator to perform a feat, or stunt, more daring than the ORV prior operator. There is also sometimes an atmosphere which discourages the use of helmets and other protective gear and safety equipment. It would appear that many young operators and passengers are vulnerable to such exhortations from throngs.

The elements of the Haddon matrix in Table 11.5 must be kept in mind while studying accident reports such as those from Table 11.4. This is often an extremely difficult task for a conscientious and ethical engineer to perform. Numerous parties are involved. Information received is often lacking the specificity needed to pinpoint causation. Unfortunately, some of parties may have ulterior motives which could affect the truth of a report or the thoroughness of the information contained therein.[40] An ORV operator may not want to admit that an accident truly was operator error. Law-enforcement officers may not have the proof necessary to actually write down on an accident report that an operator was intoxicated. Attorneys are advocates and pursue the interests of their clients. Regulators have a mandate to protect the public's safety. Despite all of this, remember that these parties may not be the only group with less-than-sterling motivations at this point in time. Some companies simply will not want to conduct yet another product-safety recall and may quash any such discussions.

The numerous ways in which an ORV occupant can be injured can be overwhelming to a design or product-safety engineer. It is easy for someone to give in to extreme impulse and say either that the ORV cannot be made reasonably safe, or all ORV accidents are caused by operator error. Neither of these two perspectives advances the goal of better product safety for consumers. It remains clear that there is a strong desire, if not need, for ORVs by some portion of the population. The elimination of ORVs by regulators would certainly reduce the number of new ORVs, each of which has the potential to injury its occupants. It is likely that, if ORV sales were forbidden, "hand built" ORVs constructed by off-road enthusiasts would appear on the trails to fill the continued need. It is frightening to think of the perils that homemade ORVs might pose to their users. Similarly, the blaming ORV operators for all accidents by design-and-manufacturing companies ignores the positive effect that conscientious engineering design, through the EDDTP, can play in reducing the number of accidents and injuries occurring or in decreasing the severities of injuries resulting from the ORV accidents that will still take place.

11.11 The Making of a Product-Safety Recall Decision

In many companies, the decision to commit to a product-safety recall is made by a group or groups and, perhaps, in the stages shown in Figure 11.2. One example of this is that the small group in charge of reviewing the initial post-sale data decides in Stage II that a product-safety recall might be, or is, required. It is highly unlikely that this small, ethical, and well-meaning group is authorized to bind the corporation to a product-safety recall. Therefore, there can be no recall yet.

[40] Although not contained in the data in Table 11.4, a not-uncommon request heard by some companies is that the consumer wants the company to buy back the product. Some of these instances might appear to be cases of "buyer's remorse" once the consumer has started paying for a significant purchase.

From this level, the potential product-safety recall should advance to Stage II where more people with greater corporate positions and authorities become involved. Again, at this point, some corporate models may still not yet permit a binding decision on a recall. Therefore, another meeting, or several more meetings, may be required before the corporation reaches an official decision on recalling a product for safety reasons.[41] The product-safety recall "off ramp" discussed earlier is an example of a stall tactic which someone might choose to employ once a recall decision is necessary. In extreme situations, a group of people may instead work to prevent—even unconsciously—such an issue from ever reaching a point requiring a decision.

A previous section of this chapter covered one possible recall-decision model in detail. However, assume that at this point in this discussion, a final decision to recall—or not—has yet to be made by the corporation. Decisions made along the way will be made by groups of people. Because individual people are imperfect, it is logical to conclude that—or at least entertain the possibility that—groups of people will also be imperfect. Whether an individual is more perfect than a group or a group more perfect than an individual depends upon both the individual and the group. A sole person's good judgment on product-safety recall matters could be outweighed by many people with poor judgment. Conversely, one person's poor judgment may also be overwhelmed by the good judgment of the larger group. Suffice it to say that, whenever people are involved, there exists the chance to make poor decisions. If a sequence of human decisions must be made, then the probability of reaching a final, proper decision will never be greater than the least of these individual probabilities. Ultimately, although many companies have formal groups and processes in place for product-safety purposes, the ethical constitutions of the individuals involved are key to the success or failure of the overall product-safety effort and the protection of consumer wellbeing.

Time does not permit the inclusion of a separate section to cover, in great detail, one particular group decision-making model showing the pitfalls that groups might make. This model, of course, would not be limited to product-safety related decisions. The literature on this model, known as *groupthink*, evaluates important decisions made by groups involving military campaigns, economic policies, the space program, and even a political cover-up (Janis 1972; 1982). The original work appears to be by W.H. Whyte (1952; 2012). Although the product-safety issues being considered in this book by readers may be, *in toto*, less dramatic than an ineptly planned and executed military campaign, the fact that lives may be seriously affected by similar thinking and decision making legitimizes the mention of groupthink.

In essence, Janis (1972, pp. 35, 36) says that "members of any small cohesive group tend to maintain *esprit de corps*[42] by unconsciously developing a number of shared illusions and related norms that interfere with critical thinking and reality testing." Members can reject thinking critically to remain in the "in-group" and to avoid relegation to the "out-group" (Rose 2011, 241). One group, for example the company, may think that it is morally superior to another group, for example the users. A group may believe itself to be omniscient as well. In addition, a group which has recently experienced failure, for example through a prior recall, may let fear cloud its judgment.

---

[41] Again, if the product truly poses no unacceptable safety risk to users, then it is perfectly reasonable and acceptable that a product-safety recall not be conducted.

[42] Italics by this book's author.

One historic incident revisited by Griffin in the context of the groupthink phenomenon was the Challenger space-shuttle accident mentioned earlier in the book (Griffin 1991). One author, who was himself involved in the incident later co-authored a book on events around the accident (McDonald and Hansen 2012).

Although groupthink does indeed have its pitfalls, "it doesn't always happen [and] it's not always bad" (Griffin, 240). Other references, for those interested in studying this particular group decision-making phenomenon include ('t Hart 1991; Esser 1998; Kramer 1998; Turner and Pratkanis 1998; Barron 2005).

Great care should be taken when making group decisions, especially when decision makers may, or will, never have the data and information that they truly want to have. Those making important decisions, such as for a product-safety recall, may need to rely upon *heuristics*, described in the following discussion, in order to reach the best possible decision.

Discussion: Heuristics and the Problem with Post-Sale Product-Safety Data

The theory of groupthink was just mentioned. It has been discussed, in some form, for over half a century. Yet, groupthink has not been validated by research and groupthink research remains an empirical rather than theoretical matter. This notwithstanding, one researcher concludes that "groupthink research has had and continue of have considerable *heuristic* value." (Esser 1998) What are heuristics and how may they be used? *Heuristics* has been defined as:

> "Cognitive shortcuts or rules of thumb that simplify decisions, especially under conditions of uncertainty. They represent a process of substituting a difficult question with an easier one (Kahneman 2003). Heuristics can also lead to cognitive biases[43]. There are disagreements regarding heuristics with respect to bias and rationality...." ("Heuristic" n.d.)[44]

Although the above quote goes on to say that even more disagreement exists about heuristics, the role of heuristics cannot remain unaddressed when discussing and making product-safety recall decisions. In many endeavors, especially within engineering, when engineers recommend taking a particular action, executives and managers alike will make the following demand: *Show me the data!* Engineering-management expectations are that they will be provided with many pages of finite-element stress-analysis results, laboratory-testing and field-testing reports, and field-incident reports. Some data may also have been statistically analyzed through appropriate hypothesis-testing models with statistical-confidence limits.

Management logically reasons that such information is needed in order to properly make important decisions. They are not necessarily wrong. And, product-safety recall decisions are indeed important decisions, after all. However, the ever-ethical and vigilant company will likely not have the number of field-incident and accident reports to generate statistically significant output for in-depth analysis and will not want to wait for mountains of data to come before responding to a potential problem. At least this is society's hope. A company may be forced to respond—either through regulatory deadlines or through ethical responsibility—well before the availability of more information

---

[43] One particular cognitive bias is discussed in the case study that follows.

[44] In engineering and mathematics, *heuristics* often refer to analysis methods based on intuition instead of on provable theory.

or before more analysis and testing results become available. If a company receives one or two reports of serious injury from the use of its product under similar circumstances, that company simply cannot afford to wait on deciding a product-safety recall until more data become available just in order to prove "statistical significance." Therefore, the traditional thorough—if not exhaustive—deep-analysis-leading-to-decision model that can be used by a company in many strategic-planning instances often cannot be used when looking at potential product-safety problems. Exigencies sometimes require decision making from partial data and information when serious injury may result from continued product use arising from inaction by the designer and manufacturer.

From the above legal and ethical restrictions, a decision on product defect must be made now, or quite soon![45] Because a deep-dive may not be possible due to lack of either information or analysis, an accelerated decision-making process must sometimes be used in order to reach a decision sooner rather than later.[46] Thus, logical "short cuts" become inevitable. One of these is the reliance upon heuristics when looking at the limited information available. This should not be viewed as unequivocally *improper*; it should instead be viewed as *necessary* for a timely response to a particular urgent situation. Although such short cuts may be permissible under these conditions, great care should be taken so that the heuristics employed are neither biased nor self-serving. This chapter's references to groupthink should be kept in mind.

Within the information constraints outlined above, a company following the law may not be able to determine, with certainty, if a safety defect truly exists within the allotted time. With the U.S. CPSC, for example, such a condition may need to be reported even if there has been no determination of defect by the manufacturer or any reported accidents or injuries. Companies do not like giving over to regulators their executive powers to decide whether or not a product contains a safety defect. There may exist distrust between the regulator and regulated. Some regulators have seen questionable behavior by the regulated; some of the regulated fear hearing the words "I'm from the government and I'm here to help you." Throughout it all, each party should remember that the safety of the public may be at risk and that poor behavior by a party has repercussions both with the immediate event and with future events.

Some important, real-world decisions may need to be made with incomplete data. This is only another reason why any proposed remedy to a post-sale product-safety problem be properly analyzed and tested through an EDDTP-level program to avoid introducing unintended consequences.

---

[45] There are instances when the cause of a problem cannot be narrowed down enough to respond immediately with a remedy. However, legal requirements to report the problem to regulators should be followed as with ethical obligations to society.

[46] During accelerated decision making, even before a problem is a recognized and actionable product-safety recall, one important decision that can be made early on in the recall-evaluation process is to create a plan to find the source, give the assigned team the necessary resources, and work as rapidly as possible to solve the problem. Legal counsel should be aware of this and report to regulators as necessary.

Just as the use of heuristics may be needed to make a decision when full information for decisions makers is lacking, it may be necessary to avoid taking other logical "short cuts" in order to reach a quick decision. Groupthink was one potential trap into which a decision-making group might fall, but there are other failures in critical thought that might be made by the same people. Some of these are known as cognitive biases which are traps to critical thought that might lead even the well-meaning into wrong decisions. A case study of one such cognitive bias follows.

Case Study 11.1: Cognitive Bias and Armor Plating

Cognitive bias is a term for a collection of numerous deviations to rational logic that lead to incorrect, and perhaps even counterproductive, actions. A cognitive bias is a term from psychology and is an instance where "human cognition reliably produces representations that are systematically distorted compared to some aspect of objective reality" (Haselton, Nettle, and Murray 2016, 968). Perhaps, to the average person, having a cognitive bias means that one is simply not seeing things as they really are. Although there are several types of cognitive biases and causes thereof, one particular type of bias pertinent to this book on product-safety engineering and this chapter on product-safety recalls is discussed. This potential error in reasoning is called *survivorship bias*.

An interesting historic example of survivorship bias is provided. The purpose of this is to make product-safety professionals aware of lapses in critical thinking and to paying disproportionate attention to distractions when evaluating a potential product-safety matter. Such distractions might include a focus on available data simply because they are available.

A number, or any other morsel of data, is not important simply because it exists. At times, a small set of data points may become "red herrings" and lead investigators away from the truth at a time when critical thinking is of paramount importance.[47] Vital product-safety decisions sometimes may, at times, depend upon information that is hard to objectively measure or even impossible to collect whatsoever.[48]

The cognitive-bias story at hand begins as follows: during World War II, all air powers were experiencing heavy losses in both aircrew and aircraft as a result of aerial combat. The U.S. Army Air Force (USAAF) was no exception. Losses from bombing raids over continental Europe, especially in the earlier years of the war, were heavy. According to The National World War II Museum, "Throughout the summer of 1943, American bomber crews sustained heavy casualties. Losses of 30 or more aircraft—300 men—were not uncommon throughout the summer" ("The Eighth Air Force vs. The Luftwaffe" 2017). USAAF bomber aircraft were being lost to flak, enemy-aircraft fire, and sometimes friendly fire. One possible solution to helping lowering aircraft losses was to armor plate these bombers against these hazards. Since armor plating is usually of steel and, therefore, quite heavy, armor plating would have to added to airplanes judiciously in order to limit the increase in aircraft weight to a reasonable level.

In their attempt to identify the portions of a bomber to protect with armor plating, the USAAF collected data from bombers returning from continental bombing raids. The USAAF had its ground personnel mark on an aircraft outline where battle damage was located on each returning bomber aircraft. These marked-up figures were aggregated into a single figure to denote the collective damage on returning aircraft.

[47] One set of authors asks the rhetorical question about large numbers of "how can something big not be important?" (Rosling, Rosling, and Rosling Ronnlund 2018, 130). Cognitive biases may also function as red herrings by diverting attention and resources from true causation when conducting product-safety defect investigations.

[48] Thus, refer to the earlier discussion of heuristics.

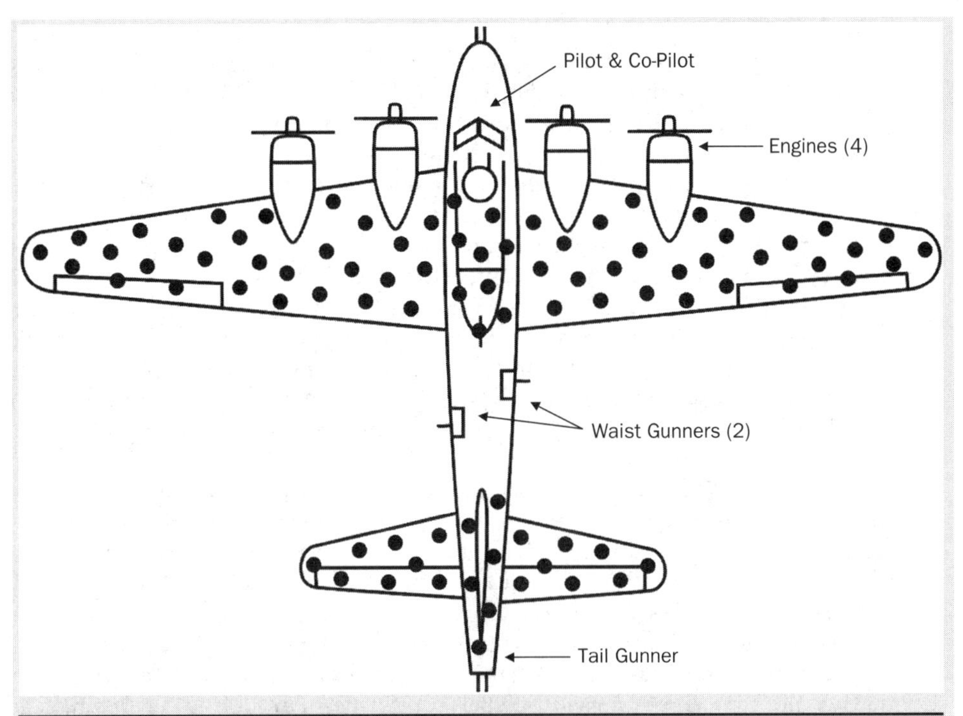

Figure 11.3 Representative battle damage to returning Allied bombers.

Figure 11.3 shows a possible example of aggregated battle damage to bomber aircraft after returning from a mission. This battle damage is displayed as dots and is consistent with descriptions of reported and collected damage reports as well with other authors' figures showing such reported damage (Wainer 1999) (McRaney 2013)[49].

When the information, such as that shown in the figure, were first seen by USAAF personnel, their first instinct was to advocate the placement of armor plating where all of the battle damage was located. These areas of high battle damage were the wings, central fuselage, and tail-gunner position. After all, that was where these aircraft were being hit most frequently by flak and enemy aircraft-mounted guns. These must surely be the most-important parts of the bombers to protect.

During this wartime period, the Statistical Research Group (SRG) was at work solving unique mathematical and statistical problems for America's war effort (Mangel and Samaniego 1984). Among the members of the SRG was a mathematician named Abraham Wald. Fortunately for the allied forces, Dr. Wald saw the folly in the USAAF's initial reaction.

Wald reasoned that, from the ground, anti-aircraft artillery (AAA) rounds and, in the heat of battle, aircraft bullets were being dispersed at bomber bombers and bomber formations randomly despite the foe's anti-aircraft and air force's best efforts to aim

[49] These returning-bomber battle-damage data appear either to have never been made public or to no longer be available.

ground-fired shells and air-launched bullet and cannon rounds. It would not be possible to aim ground artillery at one particular part of one particular bomber from 30,000 feet (9100 meters) below a bomber formation traveling at over 200 miles per hour (320 kmph), for example. Nor would it be particularly easy for a pilot flying an enemy fighter to target one part of a bomber within a dense bomber formation under typical circumstances since enemy fighters had to themselves avoid being shot down by the gunners on the bomber and by friendly escort fighters. As a result, enemy fighters in motion had to approach an allied bomber, also in motion, from strange angles, in three dimensions, with a wide range of approaching speeds.[50]

What Wald had correctly observed is that the USAAF had identified all parts of the bomber that the enemy *could* damage *without* crippling the bomber.[51] The USAAF's answer was the solution to the problem *opposite* of the *true* problem of interest. The USAAF was measuring the *surviving* population when, in reality, they needed to measure the *perishing* population. In this case, the non-survivors were the population of true interest, but they, of course, remained un-observable data points because they probably were shot down behind enemy lines or had crashed into the English Channel or the North Sea. Although the non-surviving population could not be *directly* measured, Wald correctly reasoned that putting the armor plating in the areas *not* highlighted by aggregated battle-damage figures on future planes would be the proper path to decrease allied-bomber and aircrew losses to enemy fire.

The results of this armor-plating exercise are not known to the public. Yet, what is evident is that Dr. Wald identified an error in reasoning that could have led to actions that would have compromised bomber effectiveness through lower bombloads due to extra and unnecessary bomber-aircraft weight, without significant added benefit toward bomber and aircrew survivability. The data readily available from one population served to bias the observer's reasoning. This led the observers to mistakenly apply conclusions for one population, the *survivors*, to another population, the *non-survivors*.

The reason that this bias, or phenomenon, could be important in a product-safety recall situation today is the following. Suppose that a designer and manufacturer of consumer electrical appliances has had reports of fires with its toasters. Also suppose that many parts of this product are made of plastic. Unless the fire of a product largely made of plastic is extinguished quickly, much of the evidence helpful in leading investigators to a precise identification of the cause may be consumed by that fire. Even an ethical designer and manufacturer is sometimes at a loss to explain why a product is catching on fire because the evidence is consumed by the fire.

Extend the above supposition to include that, during this time of product fires, the design-and-manufacturing company is receiving a few reports of products that are

[50] In some situations, such as a mechanically crippled bomber aircraft flying alone on two or three of its four engines and with its crew in severe physical distress from enemy actions, a skilled opposition fighter pilot could precisely aim his gunfire, *non-randomly*, at the remaining bomber engines or at the pilot's cockpit to bring the bomber down.

[51] In fairness to the United Kingdom's (U.K.) war effort, it must be pointed that, at the same time Wald was conducting his work in the United States, Patrick Blackett was performing the same analysis in the U.K. for the Royal Air Force (RAF) Bomber Command's Operations Research Section (BC-ORS) regarding the armor plating of its aircraft (Nyor et al. 2014). Initial observers in the U.K. also wanted to protect the heavily battle-damaged areas of their aircraft. Placket and the ORS properly reasoned, as did Wald, that armor plating should be located where the battle damage on returning aircraft was light.

smoking or are melting without actually combusting into product-consuming flames. Being a conscientious group, the company asks the consumers to send to the company these smoking and melting product units. The consumers happily comply with the company's request. When the company receives these units from consumers, their diligent team is able to identify a cause for these events. It does not really matter whether, in this case, the cause identified is a design flaw or a manufacturing flaw. What does matter, however, is whether the company has identified the cause of the consuming fires or, alternately, a cause for smoking and melting which can be survived by product without creating a fire. Without additional information, analysis, or testing, it could be a leap of faith to presume that the conditions leading to smoking and melting *will* lead to the large fires experienced. Just as the USAAF used surviving-bomber data to look at non-surviving bombers, the company making electrical appliances might wind up using survivor information to make decisions regarding the non-surviving—and non-observable—products that are of the greater interest.

It is at such points in the product-safety engineer's life that knowledge, critical thinking, and engineering ethics rise to the forefront. Whether or not to recommend a product-safety recall to the company is arguably the most-difficult task facing ethical people involved in product safety. Small sample sizes and difficult statistics can neither wait for nor tolerate more accidents in order to constitute a larger sample size for the statistical significance of an analysis. At times, decisions must be made using the best that mathematics and science can offer evaluated with critical thinking and driven by applied engineering ethics.

11.12 Conclusion

Product-safety recalls are a fact of life for a company making new and innovative products. This is not to say that product-safety defects are not to be taken seriously. On the contrary, they must always be. However, what is important is that the designer and manufacturer of products—engineers, managers, and executives—do their very best to prevent product-safety recalls from happening through practices such as an integrated PDP and an effective EDDTP. Constructing a competent and properly resourced post-sale product-performance surveillance program as well a work environment rewarding ethical behavior go a long way in protecting the reputation of the design and manufacturing company with both consumers and regulators as well as protecting the public. The ethical employees of that company will also sleep better at night.

Product-safety recalls will happen and innovative companies will probably have more recalls than status-quo companies. Companies must do the best job possible to prevent the avoidable recalls and also work honestly and rapidly to fix safety-defective products as soon as possible. Through a good post-sale product-surveillance program, some safety defects may be discovered before anyone is ever injured by a product.

Furthermore, the number of product-safety recalls conducted by a company should never be used as a metric for performance evaluation for company personnel because this number provides a *dis*incentive for company engineers and managers to perform recalls and because cases where a product-safety recall was considered—but ultimately not deemed necessary (even if properly so)—do not represent engineering-design triumphs. These "close calls" were engineering-design failures but, fortunately, did not rise to the level of a product-safety defect. Although the number of product-safety recalls is indeed easily *measurable* as a *priority* for a company at annual-review time, that number is not a *meaningful* metric for an ethical company pursuing product safety as a *value*.

There *should* necessarily be no shame for a design-and-manufacturing organization in admitting to a product-safety recall. Innovative and fast-moving companies have been susceptible to having more recalls than stagnant companies if for no other reason that so many unprecedented decisions are being made on new types products. No one has a perfect batting average. The only shame *could* come either from not avoiding a known or predictable problem—or from not recognizing a true problem quickly and responding immediately as needed.

Any "pop" psychology concepts briefly raised in this book are not for *dispositive* purposes, but only for *illustrative* ones. Even if an individual or a group behavior *may* take place, there is no certainty that such behavior *will* take place. However, illustrations in the references remain useful when seeking the truth and avoiding "ditches" in the logical-reasoning or critical-thinking processes. A clear analytic path—free from external influences—is necessary when pursuing the safety of the public as a corporate value.

Despite the myriad of potential product-safety investigation and recall processes that can be constructed and even followed meticulously, the necessary inclusion of imperfect beings in any such process may prove to be an obstacle even when intentions are pure. Interests vested in a specific outcome from any such remediation process—by *any* party—can turn the imprecise process into an inaccurate one. Poor behavior by any single party on one particular product-safety recall matter may lead to consequences with other, future problems. It is important to avoid the bifurcation of parties in to the *good* and the *bad*. Events are not always what they may initially appear to be. Therefore, it is also important to avoid jumping to conclusions, mandating *pro forma* reactions, and possibly creating unintended consequences when the exact phenomenon is not yet understood. Again, objective, ethical, and timely responses by all parties should not lead to the violation of any party's core values. It is hoped that all parties involved with product-safety recall process act in manners that engender the mutual respect that such an important process needs and that product users deserve. The court of public opinion moves much faster than a court of law—but the engineer's and the company's motivations must remain the public safety instead of corporate revenue and good publicity.

This chapter closes with a case study of a product-safety recall that, however embarrassing, was performed well and with sincerity for those affected by it. Website-reader comments indicated that, although some of them had never heard of the company, the reading of the open letter gave readers a positive opinion of the company of which they are now aware. The company went well beyond what might have been required of them and they exhibited regret and humility. Such sincerity goes a long way with the public.

Case Study 11.2: The Anatomy of a Good Recall

Many readers may have seen recent product-safety recalls that did not go as smoothly as the design-and-manufacturing company would have liked them to have gone. Perhaps a particular recall "fix" did not solve the problem. Perhaps this recall was a company's second or third attempt to resolve an earlier problem. Perhaps not all of the defective products were properly identified the first time. There are multiple factors and reasons for poor recall performances by companies. This is apparently true for industries which include a wide range of consumer products and motor vehicles. Regulatory agencies are not immune to contributing to this phenomenon either since such agencies themselves consist of groups of imperfect people with divergent perspectives. Regulatory staff and their commissioners may disagree on the safety of a product.

It is rare to see a company truly "come clean" on what happened and led to a product-safety recall, express true contrition to its consumers, and offer a *variety* of options for product remediation. Therefore, when the author came upon such an example as he was completing this book, he was compelled to include it in this chapter. The product being recalled was the fork for a mountain bicycle or MTB. The fork is the part of the bicycle chassis which holds the front wheel in position, absorbs shock, and transfers rider-steering inputs to the front wheel. The failure of a bicycle fork is one of the most feared failures by bicyclists due to its sudden nature and its serious injury risks. The company and author of this letter to customers graciously agreed to let this author reproduce it here. The text of this letter (Lana 2020) is found below:

> The statement we didn't want to write.
>
> 09 April 2020
>
> By Adam Lana, Curve Cycling, Victoria, Australia
>
> ---
>
> The statement we didn't want to write.
>
> Curve is currently going through its first product recall. The most recent batch of GXR forks (SKU: CVFK049-GXR SN HJ2019080XXXX) made late in 2019 is being recalled due to a manufacturing defect.
>
> Firstly, sorry.
>
> We are extremely sorry and embarrassed to have put you or any customer in this position, especially now during this current climate of hardship. Curve has directed all its energy into correcting this issue.
>
> We've managed to capture the problem quickly, but unfortunately it has delayed current Kevin of Steel orders and upcoming Kevin titanium builds. Most embarrassingly, a number of customers had actually received them. Thankfully no incidents have occurred and we've reached out to each customer for return and to resolve the issue.
>
> Will my fork be affected?
>
> No, not likely. Problem forks have only been in circulation since December 2019 and can be identified by new artwork, as shown below and by serials number as stated previously. If you are uncertain, contact us immediately for assessment.
>
> If you want all the gory details please feel free to read on, or feel free to reach out for a chat.
>
> We uncovered the problem mid batch, when QC found a few misaligned forks. This immediately prompted independent testing on the batch. The GXR forks are tested under the ISO 4210 benchmarks, and while it passed all the usual strength and bending tests, it failed the ISO 4210-6 5.6.3 brake fatigue test, which replicates 20,000 repetitions of extremely heavy braking.

One hundred forks from this batch were allocated to our Kevin of Steel project which is now on hold, and the other hundred black painted forks were intended to be sold individually or with GXR Titanium bikes. Unfortunately, 54 of the 200 forks have already been dispatched to customers or stores.

We invested heavily in developing this fork, and despite this recall, the GXR fork will continue to be a success; it has done thousands of kilometres of extreme adventures all across the globe.

So what happened here?

Our investigations have revealed a manufacturing fault in this batch, where the "over grinding" of an area created a weak spot near the brake mounts. Subsequent testing revealed that the manufacturer's internal testing equipment was not calibrated to the higher standards that we require from this fork.

What processes were already in place to protect customers?

Curve has global insurance policies in place to cover personal injury. Thankfully no customer forks have failed out in the real world and no incidents have arisen. It is Curve's main priority to ensure our customers are safe and able to confidently enjoy Curve products.

What actions have we taken?

We've traced the serial number of every recalled fork and made contact with impacted customers and shops, offering three options:

A refund for the fork at retail value.

A replacement with an equivalent third party (non-Curve branded) fork.

A replacement with a Curve GXR Fork, due in May.

We are committed to making this up to you and to ourselves. While we managed to catch this problem early, it wasn't early enough; these forks should not have been ridden. While it's "only 54 forks" and many cycling brands have experienced this on a much larger scale, it's a club that we didn't want to be part of.

We're a brand built on making products that last, but unfortunately, in this case, we have failed. We are a young brand and people have only just learned to put their trust in us. Now we must rebuild this trust and improve our processes.

In light of this we have set up new independent testing facilities to test future batches of forks: one here in Australia and an independent testing facility in Taiwan. We have also imposed more stringent and comprehensive QC testing standards at the fork manufacturing facility. The same rigorous testing standards are being applied across Curve's entire product range, from frames to thru-axles.

Once again, we extended our sincere and deepest apologies for our failure. We are working night and day to resolve the issue as quickly and possible, to deliver solutions at the highest possible standard while avoiding inconveniencing our valued customers as much as we can. If there is anything we can do to help, or if you have any questions or feedback please reach out to our friendly team via sales@curvecycling.com.

Thank you and stay safe.

Curve Crew

It should be apparent to readers that, not only did Curve Cycling admit to the problem, they went further to humbly and sincerely apologize to their customers for failing them. They admit to embarrassment and state that they will quickly move to fix the problem. The condition was also reported by Curve Cycling to the consumer-product regulatory agency in the country of recalled product sales, in this case, the Australian Competition and Consumer Commission (ACCC)[52].

Curve Cycling openly discusses their relief that there have been no incidents rather than simply stating that there has been none. The company has already reached out to customers—in typical product-safety recall protocol—to arrange for problem resolution. They also reassure purchasers of other products that their products are unaffected by this recall. This is also routine.

However, Curve Cycling goes farther than the typical recall. They provide the admittedly "gory details" and encourage people to "reach out" to them for a "chat." This indicates their commitment to customer engagement and satisfaction. They do provide details on what *precisely* happened to result in the problem. Some other companies conducting recalls only state something such as "variations in the manufacturing process" resulted in the recall. Although both approaches are correct and apparently acceptable, the willingness of Curve Cycle to not shy away from details appears to show their ability to be open and honest with their consumers. This is a refreshing change from the artfully crafted "legalese" from corporate attorneys usually found in product-safety recall press announcements.

The company is able to readily identify the affected population: all units of a particular model of fork. They also convey a clear remedy: replacement—not repair. In this case, what is both noteworthy and refreshing is that Curve Cycling is giving consumers *three* options for fixing the problem: a full refund, a replacement from a *competitor*, or a replacement—albeit delayed—from Curve Cycling. The first and the last options are not unheard of, but the competitor-replacement option is almost so. Curve Cycling understands, and is sensitive to, the needs of their cycling-passionate consumers and appears to want to help them in any way that they can—even if it is through a competitor's product.

They close their letter to customers by apologizing for the problem once again. They go on to admit that faith in their brand might have been tarnished, but that they wish to "rebuild this trust and improve our processes." They continue by providing some details of new investments to avoid a repeat of the problem. They close the letter by reiterating their "failure" and their willingness to help anyone affected.

Some readers working for large, global corporations may reason that this small bicycle company has little to lose by dispensing with the typical corporate attorney-drafted communication to regulators and customers. However, the same readers are

---

[52] https://www.accc.gov.au/

reminded to consider that Curve Cycling is not large and that the financial impact of this *relatively* small recall may indeed be quite large to them. Additionally, they may be in a growth phase for their marque, or brand, and they truly operate in an intensely competitive industry. Missteps made during such a crucial business phase can easily become serious setbacks to establishing their enterprise as a first-tier company to discriminating consumers at home or abroad.

As mentioned throughout this book, there is necessarily no shame in having to recall a product for product-safety reasons—especially with innovative products in highly competitive markets. It only becomes a problem if it is not realized quickly, acted on rapidly, and fixed to the consumer's satisfaction that disgrace may become warranted. Curve Cycling appears to have acted ethically and quickly while also demonstrating contrition and the need to earn back consumer trust. It is the author's hope that such good product-safety recall announcements no longer remain remarkable in the future.

References

"American Honda Recalls Recreational Off-Highway Vehicles Due to Crash and Injury Hazards (Recall Alert) [June 13, 2019]." 2019. Cpsc.Gov. 2019. https://cpsc.gov/Recalls/2019/American-Honda-Recalls-Recreational-Off-Highway-Vehicles-Due-to-Crash-and-Injury-Hazards-Recall-Alert.

"American Honda Recalls Recreational Off-Highway Vehicles Due to Fire and Burn Hazards (Recall Alert) [November 8, 2018]." 2018. Cpsc.Gov. 2018. https://www.cpsc.gov/Recalls/2018/american-honda-recalls-recreational-off-highway-vehicles-due-to-fire-and-burn-hazards.

"Arctic Cat Recalls Textron Recreational Off-Highway Vehicles Due to Crash Hazard (Recall Alert) [May 22, 2019]." 2019. Cpsc.Gov. 2019. https://cpsc.gov/Recalls/2019/Arctic-Cat-Recalls-Textron-Recreational-Off-Highway-Vehicles-Due-to-Crash-Hazard-Recall-Alert.

Blasius, Dennis. 2020. "Why Your Management Style May Increase Your Chances of Conducting a Consumer Product Recall." Linkedin.Com. 2020. https://www.linkedin.com/pulse/why-your-management-style-may-increase-chances-recall-dennis/.

"Caroleen[e] Paul and Other CPSC Staff Meeting with Members of the Speciality Vehicle Institute of America (SVIA), Recreational Off-Highway Vehicle Association (ROHVA), and Outdoor Power Equipment Institute (OPEI)... Topic: ATV/ROV [June 27, 2019]." 2019. Www.Cpsc.Gov. 2019. https://www.cpsc.gov/Newsroom/Public-Calendar/2019-06-27-130000/caroleen-paul-and-other-cpsc-staff-meeting-with-members.

"Checker Models A-11 and A-12." 2019. Checker World. 2019. http://www.checkerworld.org/plant---model-a11-a12.

"CPSC Safety Alert: Lawn Darts Are Banned." 2012. Bethesda, MD: U.S. CPSC. https://www.cpsc.gov/s3fs-public/5053.pdf.

Esser, James K. 1998. "Alive and Well after 25 Years: A Review of Groupthink Research." *Organizational Behavior and Human Decision Processes* 73 (2/3): 116–41. https://doi.org/https://doi.org/10.1006/obhd.1998.2758.

Garner, Bryan A. 1999. *Black's Law Dictionary*. Seventh. St. Paul, MN: West Group.

"GE Recalls Dishwahers Due to Fire Hazard." 2012. Cpsc.Gov. 2012. https://www.cpsc.gov/Recalls/2012/ge-recalls-dishwashers-due-to-fire-hazard.

Greenwald, John. 2001. "Inside the Ford/Firestone Fight." *Time*, May 2001. http://content.time.com/time/business/article/0,8599,128198,00.html.

Haselton, Martie B., David Nettle, and Damian R. Murray. 2016. "The Evolution of Cognitive Bias." In *Handbook of Evolutionary Psychology*, edited by David M. Buss, 968–87. Hoboken, NJ: John Wiley & Sons.

"Heuristic." n.d. Behavioraleconomics.Com. Accessed June 28, 2019. https://www.behavioraleconomics.com/resources/mini-encyclopedia-of-be/heuristic/.

Hill, Jessica. 2019. "'Sand up for Safety' Focuses on Reducing Beach Umbrella Injuries." Taunton Daily Gazette. 2019. https://www.tauntongazette.com/news/20190617/sand-up-for-safety-focuses-on-reducing-beach-umbrella-injuries.

"ISO 10377:2013 Consumer Product Safety—Guidelines for Suppliers." 2013. Geneva: ISO.

"ISO 10393:2013 Consumer Product Recall—Guidelines for Suppliers." 2013. Geneva: ISO.

Kahneman, Daniel. 2003. "Maps of Bounded Rationality: Psychology for Behavioral Economics." *The American Economic Review* 93 (5): 1449–75. https://www.jstor.org/stable/3132137.

Kass, Perry. 2019. "Children's Safety on Wheels." *New York Times*, June 10, 2019. https://www.nytimes.com/2019/06/10/well/family/children-bike-scooter-safety.html?smid=nytcore-ios-share.

"Kawasaki Recalls All-Terrain Vehicles Due to Fire Hazard [August 10, 2017]." 2017. Cpsc.Gov. 2017. https://www.cpsc.gov/Recalls/2017/kawasaki-recalls-all-terrain-vehicles.

Klein, Jonathon. 2019. "Parked Teslas Keep Catching on Fire Randomly, And There's No Recall In Sight." Thedrive.Com. 2019. https://www.thedrive.com/news/28420/parked-teslas-keep-catching-on-fire-randomly-and-theres-no-recall-in-sight.

Knack, Oliver. 2019. "3 Types of Quality Defects for Defect Classification." InTouch. 2019. https://www.intouch-quality.com/blog/3-types-quality-defects-different-products.

Lana, Adam. 2020. "The Statement We Didn't Want to Write." Curvecycling.Com.Au. 2020. https://www.curvecycling.com.au/blogs/news/the-statement-we-didnt-want-to-write.

Laney, Cara, and Elizabeth F. Loftus. 2010. "Change Blindness and Eyewitness Testimony." In *Current Issues in Applied Memory Research*, edited by Graham M. Davies and Daniel B. Wright, First, 142–59. New York, NY: CRC Press. https://www.crcpress.com/Current-Issues-in-Applied-Memory-Research/Davies-Wright/p/book/9780415647137#googlePreviewContainer.

Loftus, Elizabeth F. 1979. *Eyewitness Testimony*. Cambridge, MA: Harvard University Press.

———. 1996. *Eyewitness Testimony with a New Preface*. Cambridge, MA: Harvard University Press. https://www.hup.harvard.edu/catalog.php?isbn=9780674287778.

Mangel, M., and F. J. Samaniego. 1984. "Abraham Wald's Work on Aircraft Survivability." *Journal of the American Statistical Association* 79 (386): 10.

McRaney, David. 2013. "Survivorship Bias." YouAreNotSoSmart.Com. 2013. https://youarenotsosmart.com/2013/05/23/survivorship-bias/.

Motor Vehicle Safety Act. 2005. United States. https://www.law.cornell.edu/uscode/text/49/subtitle-VI/part-A/chapter-301.

"NHTSA's Process for Issuing a Recall." n.d. NHTSA. Accessed January 19, 2020. https://www-odi.nhtsa.dot.gov/owners/RecallProcess.

Nyor, Ngutor, Adamu Idama, Joseph O. Omolehin, and Kamilu Rauf. 2014. "Operations Research—What It Is All About." *Universal Journal of Applied Science* 2 (3): 57–63. https://doi.org/10.13189/ujas.2014.020301.

Pine, Timothy A. 2012. *Product Safety Excellence: The Seven Elements Essential for Product Liability Prevention*. Milwaukee, WI: Quality Press.

Popovich, Jeff. 2018. "4 Bodies Recovered from ATV Crash near Payson, Sheriff's Officials Say [September 17, 2018]." Www.Abc15.Com. 2018. https://www.abc15.com/news/state/4-missing-after-atv-crash-near-payson-sheriffs-officials-say.

"Recall Handbook." 2012. Washington, DC: U.S. CPSC. https://www.cpsc.gov/s3fs-public/8002.pdf.

Rosling, Hans, Ola Rosling, and Anna Rosling Ronnlund. 2018. *Factfulness: Ten Reasons We're Wrong About the World and Why Things Are Better Than You Think*. New York, NY: Flatiron Books.

Ross, Kenneth. 2019. "Product Liability, Regulatory Compliance, and Product Safety." Salt Lake City, UT.

Silvestro, Brian. 2020. "Harbor Freight Recalls the Jack Stands Meant to Replace the Recalled Jack Stands." Roadandtrack.Com. New York, NY. 2020. https://www.roadandtrack.com/car-culture/buying-maintenance/a33235534/harbor-freight-replacement-jack-stands-recalled/.

Simons, Daniel J., and Christopher F. Chabris. 1999. "Gorillas in Our Midst: Sustained Inattentional Blindness for Dynamic Events." *Perception* 28: 1059–74.

The Argus Observer. 2019. "UTV Fire Leads to 23-Year-Old Idaho Man's Death," May 14, 2019. https://www.argusobserver.com/news/utv-fire-leads-to—year-old-idaho-man-s/article_b25d3e32-7669-11e9-82cb-dbc4d361a03a.html.

"The Eighth Air Force vs. The Luftwaffe." 2017. Nationalww2museum.Org. 2017. https://www.nationalww2museum.org/war/articles/eighth-air-force-vs-luftwaffe.

"Tips Guide for the Recreational Off-Highway Vehicle Driver." 2017. Irvine, CA: ROHVA. https://rohva.org/driving-tips/.

US CPSC. n.d. "Duty to Report to CPSC: Rights and Responsibilities of Businesses." Cpsc.Gov. Accessed June 22, 2019. https://www.cpsc.gov/Business—Manufacturing/Recall-Guidance/Duty-to-Report-to-the-CPSC-Your-Rights-and-Responsibilities.

———. 2003. "CPSC, Polaris Industries Inc. Announce Recall of ATVs." U.S. CPSC. 2003. https://www.cpsc.gov/th/Recalls/2003/cpsc-polaris-industries-inc-announce-recall-of-atvs.

———. 2004. "CPSC, Polaris Industries Inc. Announce Recall Accessory Skis for Snowmobiles." U.S. CPSC. 2004. https://www.cpsc.gov/Recalls/2004/cpsc-polaris-industries-inc-announce-recall-of-accessory-skis-for-snowmobiles.

Wainer, Howard. 1999. "The Most Dangerous Profession: A Note on Nonsampling Error." *Psychological Methods* 4 (3): 250–56. https://doi.org/10.1037/1082-989X.4.3.250.

Weintraub, Rachel. 2019. "An Analysis of OHV Recalls: Increasing Number of OHVs Pulled from Market Due to Safety Concerns." Consumerfed.Org. 2019. https://consumerfed.org/analysis-ohv-recalls-increasing-number-ohvs-pulled-market-due-safety-concerns/.

CHAPTER 12
Conclusions

People do not like to think. If one thinks, one must reach conclusions. Conclusions are not always pleasant.
—HELEN KELLER

The alert reader will have noticed that the title of this chapter is in the *plural*. After all, this is not the end of anything. It is but the start of applying the understanding and skills gained from reading and reflection. These are complemented by the engineering ethics of the individual and should be supported by an ethically conducive work environment provided by enlightened corporate *leadership*—not management.

Also true is that product safety is often addressed through *opinion*. An opinion is like a nose—everybody has one. What is crucial in the field of product-safety engineering is to be able to reach *conclusions*. There is no opinion or conclusion that cannot, and will not, be disagreed with by someone somewhere. Since *all* people can never, ever be pleased, conclusions must please the reasoner's skills, available information, critical thought, and ethics.

Hope in magic and baseless opinions are poor strategies for meeting the product-safety engineering needs of the public. No side involved in public safety should have unreasonable demands or expectations of the other sides. Technology has yet to devise a magic knife that will readily cut a thick steak, but not the flesh attached to a living person. Advocates and regulators should beware of demanding unrealistic safety-performance levels from products lest they lose credibility and consequently be ignored by designers and manufacturers—and sometimes even by the very consumers they work to help. However, *nothing* in this book absolves any engineers working for a design-and-manufacturing company from doing anything less than their utmost best on product-safety matters. Nor does it absolve design-and-manufacturing companies which do not provide a work environment promoting—and even demanding—ethical behavior from their employees.

Companies that do not discipline those who behaved unethically might have confused their *values* with their *priorities*.[1] Employees are not stupid and will quickly—and properly—reason that their employer tolerates unethical behavior so long as the misbehaving employee otherwise performs well—or at least is never publicly exposed. Those who have worked with Ralph Barnett know that he is famous for saying, "Practicing ethics is great as long as you are not the only one doing it." Engineers should not have to put their careers on the line to do what is ethical. Companies and their executives,

[1] Refer to the opening of Chapter 1.

managers, board members, and legal counsel should never let the engineers be the only ones practicing ethics and looking out for the welfare of the public on the design-and-manufacturing side of product safety.

This chapter is "Conclusions" and is meant as the end of the book as well as the start of a journey. It also reaffirms the need for those whose actions affect product safety to *reach well-reasoned verdicts based on evidence, arrived at through critical thinking and logic, and consistent with the professional ethics of the engineer.*

Bad situations must first be acknowledged before they can be fixed. This is the first step in many twelve-step addiction-recovery programs ("Twelve Steps and Twelve Traditions" 2020).[2] Large egos and business interests may get in the way of delivering safe products to the public. Engineers designing consumer products are reminded to stay focused on engineering ethics, in general, and perhaps applied engineering ethics, in particular, when mathematics and physics can take one no further. Although not everyone will agree on what the "right thing" to do is, it is important—both as an engineer and as a human being—to act ethically. This primarily means putting the welfare of the public first in engineering design. It is, of course, a delicate balancing act, but if reasoned conclusions based on science, technology, mathematics, experience, data, and, finally, engineering ethics are used, then the task becomes simpler and clearer.[3] Doing so is still often not easy, however. If design and product-safety engineers have done their best—and *truly* their best—then what more can anyone ask of them? Period.

Engineers should always remember this.

Throughout this book it has never been the author's intention to criticize any single party except, perhaps, the corporate management who may be making it difficult for design and product-safety engineers and other professionals to do their ethical duties, or when another party has not done its best job with objectivity and integrity—even if that job was not the engineering of a product. Nor has the author's point been to preach a radical new message of engineering activism. Instead, engineers should participate in *active engineering* throughout their careers and both think and act ethically. They should also hold those around them—and above them—accountable to the same standards when product safety is at issue.

Noted physicist Freeman Dyson said, "We are scientists *second*, and human beings *first*"[4] (Dyson 1979, 6). As a scientist, Dyson's quandaries—especially the ethical ones—would no doubt have been different from a design engineer's quandaries. After all, scientists are often confronted with the ethical considerations of acts which include the release of the energy within the atom and the creation of new forms of life. Engineers struggle with the hot surfaces and sharp corners of their designs. When viewed up close, the scientist's perplexities greatly outweigh those of the engineer. If a step back is taken, however, it can be seen that the individual scientist may only have an indirect effect on a slow, albeit potentially huge, process. The engineer's actions may well result in effects that are both direct and immediate, although fewer people may ultimately be affected than by a cosmic process. Yet, if you are one of those people injured by poor design or manufacturing, the effects of an engineer's actions are immense.

---

[2] Perhaps the pursuit of profit can become an addiction to companies and their management teams.
[3] Refer to the quote opening Chapter 7.
[4] Emphasis by the present author.

Design and product-safety engineers should never shy away from taking on the ethical ramifications of their jobs. Ethicist Sabine Roeser states (Roeser n.d.) that:

> Rather than delegating moral reflection to "moral experts," engineers should cultivate their own moral expertise. They have a key moral responsibility in the design process of risky technologies, as they have the technical expertise and are at the cradle of new developments.

Such a position may come as quite a shock to engineers. Engineers are often "worker bees" who dutifully perform the work placed before them by their employers. Many of those engineers truly enjoy their jobs. The higher-education system in the United States sometimes does a poor job of equipping engineering students for anything beyond technical analysis. There simply is little room for non-technical courses in the modern university engineering curriculum. Consequently, many engineering students simply have never had the opportunity to study such issues. There has simply never been any time to do so.

Once employed, engineers usually receive a salary sufficient to keep them focused on their employers' tasks. With work to perform and families to raise, in many cases, there is still little time for engineers to think about the philosophical, broader aspects of their work and its potential consequences. Engineers are not robots. They think, feel, and even bleed. Some do ultimately conclude that some product of theirs is more dangerous than it should or could be. Yet, the work environments in which some engineers find themselves do not permit the expression of such views, let alone the actions needed to remedy such problems. Therefore, speaking out may feel decidedly foreign to an engineer and should, therefore, be encouraged by insightful corporate leadership.

There are many concepts and methods covered in this book. It may take some time for the reader to process the material in this book and in its references. Yet, what should never be forgotten is the engineer's duty to look out for the welfare of the public—even in small ways. These "small" ways may make large differences to people who were never injured because a product's risk was further reduced through additional engineering-design work. Those consumers may also never realize that they were spared injury because some engineers stood up for what they concluded were ethical engineering decisions—and were then supported by their corporate leadership.

References

Dyson, Freeman J. 1979. *Disturbing the Universe*. New York, NY: Basic Books.

Roeser, Sabine. n.d. "Engineering Ethics 2.0." Tudelft.Nl. Accessed November 5, 2019. https://www.tudelft.nl/ethics/ethics/for-educators/ethics-20/.

"Twelve Steps and Twelve Traditions." 2020. aa.org. 2020. https://aa.org/pages/en_US/twelve-steps-and-twelve-traditions.

APPENDIX A

Hazards Checklist

Design Element	Hazard
Mechanical	Sharp edge
	Nip point
	Pinch point
	Impact
	Stability
	Falling objects
	Fatigue
	Fall
	Slip/trip
	Suspended load
	Shifting load
	Translation
	Rotation
	Pinch
	Crush
	Choking
	Wear
	Vibration
	Shock
	Ejected parts
	Lubrication
Electrical	Shock
	Sparking
	Arcing
Energy	Thermal
	Noise
	Radiation (ionizing)
	Radiation (non-ionizing)
	Interference (EMI/RFI)
	Static build-up

Product-Safety Hazards Checklist

Design Element	Hazard
Chemical	Lack of oxygen (O_2)
	Corrosive
	Poisoning
	Explosion
	Toxicity
Control Systems	Inadvertent start up
	Will not shut down
	Cannot start up
	Hardware malfunction
	Hardware-logic flaw
	Software-logic flaw
	Disconnection of power
	Calibration
User	User error
	Maintenance error
	Inadvertent action
	Careless use
	Unqualified user
	Fatigue
	Poor ergonomics
	Substance abuse
Storage/Transport	Leakage of liquid
	Leakage of gas
	Movement of liquid
	Rupture
	Explosion
	Implosion
	Reversal of flow
	Contamination
Physical Environment	High temperatures
	Low temperatures
	High humidity
	Low humidity
	Earthquake
	Dust
	Lightning
	Flooding
S-E Environment	Sabotage
	Workplace violence
	Social facilitation

Product-Safety Hazards Checklist (*Continued*)

APPENDIX B
Hand-Powered Winch

A hand-powered winch is shown and described in this appendix. This winch[1] is used for examples in the book. Such a winch could be used for pulling a small boat onto a boat trailer after it has been securely mounted to a suitable part of the boat trailer. Figure B.1 is a photograph of such a winch. This winch uses a polyester strap, rather than a steel cable or wire rope, to connect the winch spool to a hook end.

This winch is manually powered, not electrically powered, and has no electric/electronic control system as a result. The winch user simply starts cranking the handle on the long arm which is connected through a reduction gear train to the strap spool to provide a mechanical leverage, or advantage, to pull a boat onto the trailer or to permit a boat to roll off of the trailer in a controlled manner. Because of the reduction gear train, the speed at which the boat is pulled onto the trailer is quite slow.

The winch has a simple control mechanism. The user simply moves the ratchet mechanism to cause strap motion in the desired direction—either toward the winch or away from the winch. The spring-loaded pawl mechanism shown in Figure B.2 is moved to either extreme position to permit ratcheted cranking in or cranking out of the strap. The pawl is the lever-shaped component which engages the teeth in the gear of the drive shaft. There is also a middle, or neutral, position of the pawl mechanism that permits the handle and strap spool to "freewheel" or move in either direction without ratcheting. The neutral position should be used with caution and the handle not released with a load on the winch strap since the handle may spin freely at a high speed (angular velocity). Not only will the load be released, but the user may be unable to regain control of the load through the winch handle without moderate-to-severe injury to her/himself.

A view of the drive shaft is shown in Figure B.3. The drive shaft is also visible in Figure B.2. but is removed for greater clarity. On the right of the drive shaft are two flat features that provide a point of power transmission from the handle which has a mounting hole with two similar flats. The threads on the shaft permit a nut to be screwed onto the shaft and hold the handle lever in place. (This nut and the stamped handle retainer, also with flats, are visible in Figure B.1.) The drive sprocket is visible in Figure B.3 next to the flats and threads. The circumferential groove on the left side of the shaft is for a "C" clip that prevents the entire drive shaft from shifting laterally and farther to the left in the drive-shaft bushings. The shaft is prevented from moving

[1] The product shown and described in this appendix accurately represents no product currently on the market—nor is it intended to. It is merely an example for one product-safety engineering effort with one artificial product.

Figure B.1 Hand-powered winch.

Figure B.2 Ratchet mechanism of winch.

farther to the right by the drive sprocket itself. The drive sprocket is permanently attached to the drive shaft by a circumferential fillet weld on one side of the sprocket, also visible.

The spool that holds the strap is shown in Figures B.4 and B.5. Figure B.4 shows the driven side of the spool and is the side hosting the ring gear that is driven by the drive shaft's drive sprocket. The free side of the spool is shown in Figure B.5. The ring gear is permanently attached to the driven side of the spool by six (6) rivets

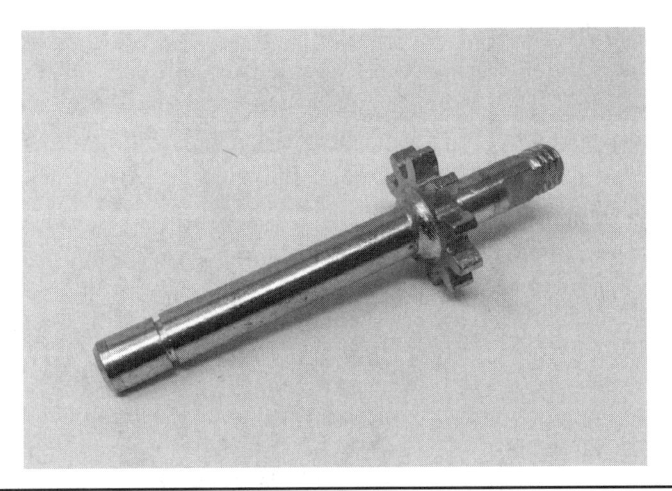

Figure B.3 Drive-sprocket shaft of winch.

Figure B.4 Driven side of winch spool.

visible in Figure B.5. This free side of the spool is ungeared and free of other significant features apart from the one cable-end retainer bracket and two machine screws. The cable option is not utilized in this winch which uses a polyester strap instead. The smaller shaft parallel to the larger shaft is the strap-anchoring shaft beneath the exposed shaft bushing. The strap-anchoring shaft fits through an end loop manufactured into the strap assembly and prevents the strap from rotating on the spool when load is applied to the strap through hook assembly. The larger shaft is the main

Figure B.5 Free side of winch spool.

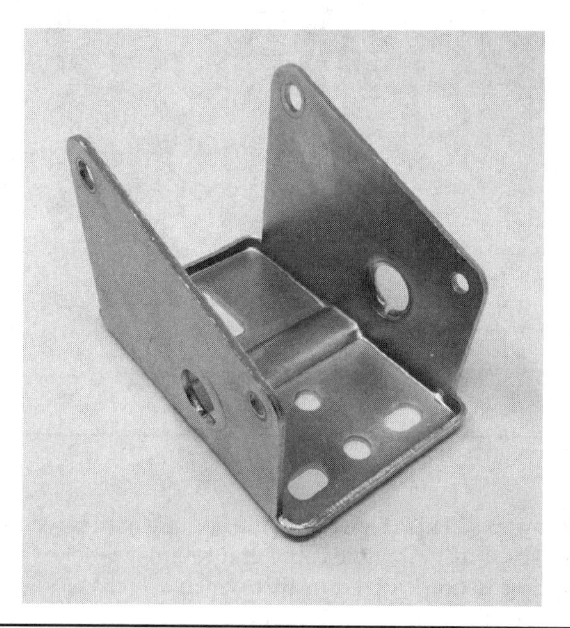

Figure B.6 Winch chassis.

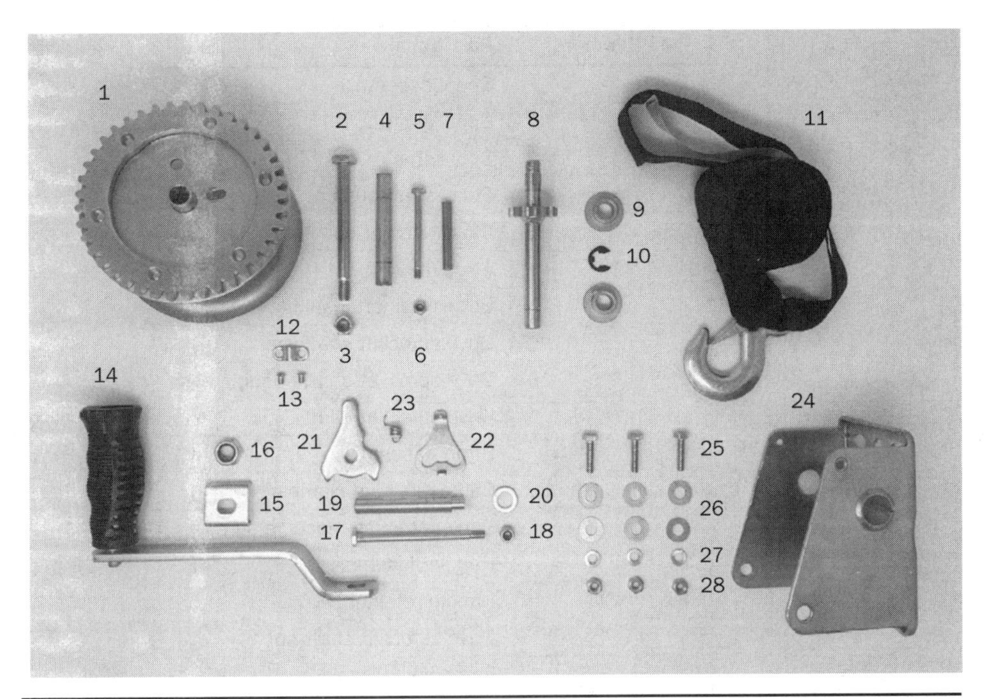

FIGURE B.7 Parts of winch.

spool shaft. Each side of the spool is permanently mounted to the spool shaft by a circumferential fillet weld on the shaft. This spool pulls on the hook at the boat-end of the strap.

A better view of the winch's main housing, or chassis, is provided in Figure B.6. This chassis locates the three shafts—pawl-mechanism, drive, and spool—relative to one another. The chassis also provides the mounting holes for securely attaching the winch to a boat trailer.

Most of the winch components are manufactured from steel due to its relative high strength and low cost. Steel is also a good material choice since many engineers and manufacturers have significant experience with steel fabrication. However, steel is susceptible to corrosion, or rust. Due to the harsh environments in which boat-trailer winches are operated and stored—such as rain, snow, ice, fresh water, and salt water, many components of the winch are galvanized, or coated with zinc, to resist corrosion. This coating enhances both the operation and the longevity of winches while still permitting the use of affordable steel. The surface treatment also preserves the appearance of the product.

Some components of the winch have already been named and shown during the discussions of the figures above. However, all parts contained within the hand-powered winch are shown and numbered in Figure B.7. Even in this simple mechanical example, there are twenty-eight (28) different types of parts used. These parts are listed in Table B.1. Some parts are used multiple times as seen in the table and the last figure.

Some of these parts are actually "assemblies" of another set of parts. This is true for the spool, the handle, the drive shaft, and the strap assemblies. An *assembly* is a

Part Number	Part Name
1	Spool assembly
2	Spool-shaft bolt
3	Spool-shaft nut
4	Spool-shaft bushing
5	Strap-retainer bolt
6	Strap-retainer nut
7	Strap-retainer bushing
8	Drive-sprocket shaft
9	Drive-sprocket shaft bushing (2)
10	Drive-sprocket shaft C-clip
11	Strap assembly
12	Cable-retaining clamp
13	Cable-retaining machine screws (2)
14	Handle assembly
15	Handle retainer
16	Drive-sprocket shaft nut
17	Ratchet-shaft bolt
18	Ratchet-shaft nut
19	Ratchet-shaft spacer bushing
20	Ratchet-shaft flat washer
21	Ratchet pawl
22	Ratchet tensioner
23	Ratchet tension spring
24	Winch chassis
25	Winch-mounting bolts (3)
26	Winch-mounting flat washers (6)
27	Winch-mounting locking washers (3)
28	Winch-mounting nuts (3)

Table B.1 Parts Listing of Winch

grouping or set of parts that cannot be further disassembled or is not meant for disassembly. Several of the nuts have a locking feature, or component, which prevents the nut from loosening significantly on its own when subjected to shock and vibration. A fictional, simple Bill of Materials (BOM) is provided as Table B.2. Unlike an actual BOM, some operations and labor required to make the winch's component parts are not included in this winch's BOM.

Part Number	Sub-Parts and Operations		Part Name	OTS or MTS	Notes
1			Spool assembly	MTS	Assembly specifications
	1.1		Free side of spool	MTS	
		1.1.1	Stamped holes (11)	MTS	
		1.1.2	Tapped holes (2)	OTS	Standard thread
	1.2		Driven side of spool	MTS	
		1.2.1	Stamped holes (3)	MTS	
	1.3		Driven ring gear	MTS	
		1.3.1	Spot Welds (6)	OTS	Standard welding
	1.4		Spool shaft	MTS	
		1.4.1	Fillet welds (2, one each side)	OTS	Standard welding
2			Spool-shaft bolt	OTS	Standard bolt
3			Spool-shaft nut	OTS	Standard nut
	3.1		Thread-locking element	OTS	
4			Spool-shaft bushing	MTS	
5			Strap-retainer bolt	OTS	Standard bolt
6			Strap-retainer nut	OTS	Standard nut
	6.1		Thread-locking element	OTS	
7			Strap-retainer bushing	MTS	
8			Drive-sprocket shaft	MTS	Manufacturing specifications
	8.1		Shaft	MTS	
		8.1.1	Threaded end	OTS	Standard thread
		8.1.2	Flats (2)	MTS	
		8.1.3	Groove	MTS	For C-clip
	8.2		Drive sprocket	MTS	
		8.2.1	Fillet weld (one side)	OTS	Standard welding
9			Drive-sprocket shaft bushing (2)	MTS	
10			Drive-sprocket shaft C-clip	OTS	
11			Strap assembly	MTS	Assembly specifications
	11.1		Webbing, black, 50 mm x 5 meters	OTS	
		11.1.1	Loop end	OTS	Standard web stitching
		11.1.2	Hook end	OTS	Standard web stitching
	11.2		Hook lead	MTS	
		11.2.1	Webbing, black, 25 mm x 600 mm	OTS	
		11.2.2	Stitching	OTS	Standard web stitching

TABLE B.2 Bill of Materials of Winch

Part Number	Sub-Parts and Operations		Part Name	OTS or MTS	Notes
		11.2.3	Warning label	MTS	
	11.3		Hook assembly	OTS	Catalog part
		11.3.1	Hook	OTS	
		11.3.2	Latch plate	OTS	
		11.3.3	Latch torsion spring	OTS	
		11.3.4	Latch pin	OTS	
		11.3.5	Latch-pin rivet	OTS	
12			Cable-retaining clamp	MTS	Assembly specifications
	12.1		Stamping and forming, body	MTS	
	12.2		Stamping, holes (2)	OTS	Standard size
13			Cable-retaining machine screws (2)	OTS	Standard size
14			Handle assembly	MTS	Assembly specifications
	14.1		Handle, rubber	OTS	Catalog part
	14.2		Handle shaft	OTS	Catalog part
		14.2.1	Handle-shaft rivet (1)	OTS	Standard riveting
	14.3		Handle lever	MTS	
		14.3.1	Stamping	MTS	
		14.3.2	Forming	MTS	
		14.3.3	Hole with 2 flats	OTS	Standard size
15			Handle retainer	MTS	
	15.1		Stamping	MTS	
	15.2		Forming	MTS	
	15.3		Hole with 2 flats	OTS	Standard size
16			Drive-sprocket shaft nut		Standard size
	16.1		Thread-locking element		
17			Ratchet-shaft bolt	OTS	Standard size
18			Ratchet-shaft nut	OTS	Standard size
19			Ratchet-shaft spacer bushing	MTS	
20			Ratchet-shaft flat washer	OTS	Standard size
21			Ratchet pawl	MTS	
22			Ratchet tensioner	MTS	
23			Ratchet tension spring	OTS	Catalog part
24			Winch chassis	MTS	Manufacturing specifications
	24.1		Stamping, body	MTS	
	24.2		Stamping, holes (10)	OTS	Standard sizes

Table B.2 Bill of Materials of Winch (*Continued*)

Part Number	Sub-Parts and Operations		Part Name	OTS or MTS	Notes
	24.3		Stamping, holes with keyways (2)	OTS	Standard size
	24.4		Forming	MTS	
25			Winch-mounting bolts (3)	OTS	Standard size
26			Winch-mounting flat washers (6)	OTS	Standard size
27			Winch-mounting locking washers (3)	OTS	Standard size
28			Winch-mounting nuts (3)	OTS	Standard size

Table B.2 Bill of Materials of Winch (*Continued*)

Also not included in this BOM are engineering part numbers, the part's supplier, the original-equipment manufacturer's (OEM's) cost, extended costs,[2] and some other data fields needed for a true BOM. However, this BOM indicates whether a part is made to specification (MTS) or is off the shelf (OTS) and includes nondescript notes.

In order to help the engineers design the winch, two statements containing winch-use information are included below. Parts of this information may also be included with the winch in some form, such as in a manual, when the winch is purchased by consumers.

Statement: Intended Use

This hand-powered mechanical winch is intended to be semi-permanently attached to an appropriate mounting platform on a boat trailer. Only use the included hardware (nuts, bolts, lock washers, and washers) to secure the winch to the trailer mount. The winch is only designed for horizontal-load applications through tension in its strap when loading and unloading a boat from a trailer.

The neutral position on the gear-drive mechanism is only to be used for feeding out winch strap without a load on the winch strap.

Statement: Limitations for Use

This winch is limited to 2000 lbf (8.9 kN, 910 kg) of tension on its strap. It is designed to only load and unload boats from a trailer when securely bolted to that trailer. It is not designed to secure a boat to a trailer during transportation.

The winch is not designed to either raise or support vertical loads.

[2]Extended cost is single part cost multiplied by the number of times that part is used.

Acronyms and Abbreviations

ABS	Anti-lock braking system
ADAS	Advanced driver-assistance system
A/IS	Autonomous and intelligent systems
App	Computer or smart-phone application or program
AR	Augmented reality
ATC	All-terrain cycle with three wheels; a type of ORV no longer sold in the United States
ATV	All-terrain vehicle with four wheels; a type of ORV
AWD	All-wheel drive
BOM	Bill of materials
CE	Conformité Européenne
CFD	Computational fluid dynamics
CO	Carbon monoxide
CPS	Comprehensive product safety
CSR	Corporate social responsibility
CTO	Chief technical officer
DFMECA	Design FMECA
DPFS/DfPS	Design for product safety
DFS/DfS	Design for safety
DOD	U.S. Department of Defense
DOT	U.S. Department of Transportation
DRL	Daytime-running light
DSM	District sales manager
EC	European Commission (of the European Union)
EDDTP	Engineering design, development, and testing phase
EDP	Engineering-design phase
EEA	European Economic Area
EFTA	European Free Trade Association
EHSR	Essential health and safety requirements

EMS	Emergency medical services
ETA	Event-tree analysis
EU	European Union
FAA	U.S. Federal Aviation Administration
FDA	U.S. Food and Drug Administration
FE	Fundamentals of engineering
FEA	Finite-element analysis
FMEA	Failure modes and effects analysis
FMECA	Failure modes, effects, and criticality analysis
FRI	Final risk index
FTA	Fault-tree analysis
GFCI	Ground-fault circuit interrupter
HAZOP	Hazards and operability (analysis)
HC	Hydrocarbon
HIPAA	Health Insurance Portability and Accountability Act
ICBM	Intercontinental ballistic missile
IFU	Instructions for use
IoT	Internet of Things
IRI	Initial risk index
JRA	Just riding along
lb_f	Pound force; *United States Customary System* (USCS) unit of force
LoC	Loss of control
LO/TO	Lockout/tagout
LSV	Low-speed vehicle
MBD	Multi-body dynamics
ME	Mechanical engineering
MFD	Multi-function display
MFMECA	Manufacturing FMECA
MOHUV	Multi-purpose off-highway utility vehicle; a type of ORV
N	Newton; *Système International* (SI) unit of force (4.4482 N = 1 lb_f)
NASA	U.S. National Aeronautics and Space Administration
NHTSA	National Highway Traffic Safety Administration of U.S. DOT
OEM	Original equipment manufacturer
OHV	Off-highway vehicle; synonym for ORV
OJT	On-the-job training
ORS	Occupant-retention system
ORV	Off-road vehicle
OSH	Occupational safety and health
PDO	Property damage-only (accident/mishap)

PDP	Product-design process
P.E.	(Registered) Professional engineer
PED	Proper engineering decision
PHA	Preliminary hazard analysis
PHL	Preliminary hazard list
PIN	Product Identification Number
PM	Preventive maintenance
PMD	Proper management decision
PoS	Point-of-sale
PPAP	Production part approval process
PPE	Personal protective equipment
PSEg	Product-safety engineering
PSEr	Product-safety engineer
PSMr	Product-safety manager
PSMt	Product-safety management
PSMx	Product-safety matrix
PTW	Powered two wheelers
PUE	Product-user-environment (model)
QA	Quality assurance
QC	Quality control
RA	Risk assessment
RAPEX	Rapid alert system of the EU
RI	Risk index
RM	Risk management
ROV	Recreational off-highway vehicle
RA	Risk analysis
RAC	Risk assessment code
RoHS	Restrictions on hazardous substances
ROPS	Rollover protective structure
RPN	Risk priority number for FMEA/FMECA
SAS	Safety-alert symbol
SDO	Standards Development Organization
SPF	Single-point failure
SRS	Supplemental-restraint system
SS	System safety
SSF	Static-stability factor
SSV	Side-by-side vehicle; a type of ORV
SxS	Another abbreviation for side-by-side vehicle
S/N	Serial number

TCO	Thermal cutoff
TCS	Traction-control system
TLE	Top-level event
TLX	(NASA) Task load index
UFMECA	Use FMECA
UPI	User-product interface
UTV	Utility-task vehicle; a type of ORV
VIN	Vehicle Identification Number
VRU	Vulnerable road user
VVT	Verification, validation, and testing
ZES	Zero-energy state
ZMS	Zero-mechanical state

Glossary

Acceptable risk A risk level that is deemed to be satisfactory; risk that the appropriate "acceptance authority" (as defined in DoDI 5000.02) is willing to accept without additional mitigation (MIL-STD-882E)

Accident An event in which someone is physically injured

Administrative controls Procedural safeguards against harm

Advanced driver-assistance system An automated, intelligent system which helps an automobile driver avoid accidents

Analysis Tearing apart a system of components or characteristics to understand the system's functions and limitations

Applied engineering ethics A subset of engineering ethics focused on the practice of product-safety engineering at a design-and-manufacturing company

Assembly A grouping or set of parts that cannot be further disassembled or is not meant to be disassembled

At-risk population A population of product users with specific needs due to a vulnerability when compared to the population in general

Augmented reality A technology which superimposes computer-generated information on top of actual information that could include images, sounds, haptics, and olfaction

Autonomous and intelligent systems Self-contained systems able to make decisions and execute them

Availability The state of being committable to perform a function

Barricade A device which presents an obstacle to passage

Barrier A device which prevents passage

Bill of materials The final collection of resources needed to manufacture a product that may include labor and processes in addition to the parts, the part numbers, and the quantities of each part

Connected product A product with a wireless connection for use or control

Caution A warning-sign signal word to indicate a hazardous situation which—if not avoided—could result in minor or moderate injury

CE mark The mark certifying, by the appropriate party, that a product has met or exceeded requirements set forth by the CE-mark directives of the EC; appears as CE

CEO Chief Executive Officer

Cognitive bias A systematic (non-random) error in thinking, in the sense that a judgment deviates from what would be considered desirable from the perspective of accepted norms or correct in terms of formal logic ("Cognitive Bias" n.d.)

Color pollution The use of color in facilitators when unnecessary sometimes confusing the reader to the detriment of effective communication

Compliance Meeting or exceeding legal or regulatory requirements; a property often confused with safety

Consensus General agreement within a group

Conspicuity The ability to be seen by others; in contrast to "Visibility"

Critical thinking The objective analysis and evaluation of a topic to reach a well-reasoned conclusion

Danger A warning-sign signal word to indicate a hazardous situation which—if not avoided—will result in death or serious injury

Declaration of conformity Certification that a product has successfully met the criteria for the CE mark in Europe

Defect A legal conclusion about a product or product condition which poses an unreasonable safety risk to a consumer

Dependability The combination of availability, reliability, safety, and security

Design development The act of refining a product's detail design from initial design to intermediate design(s) and on to final design

Design element Either a product component or an emergent property of a product that manifests itself in a product characteristic

Design for product safety (DFPS/DfPS) An integrated approach to actively implementing product-safety engineering goals through participating in the product-development process (PDP) including the engineering design, development, and testing phase (EDDTP) in order to "design in" safe product characteristics and to include the design a system of facilitators prior to product release and consumer use

Design for safety (DfS) An industrial-engineering approach to (re-)design manufacturing processes and workplaces to improve occupational-worker safety and health

Design guidance Information, recommendations, advice, and requirements for engineers synthesizing a new product

Detailed design The current design embodiment whether it is preliminary, intermediate, or final design

Emergent property A property of systems where a particular characteristic of a system arises (emerges) which is not a characteristic of that system's individual components

Engineering analysis The process of disassembling a product design in order to further evaluate its properties with respect to its specified function; in contrast to "Engineering synthesis"

Engineering controls Physical safeguards against harm

Engineering design phase A generalized version of Phase III in a PDP

Engineering design, development, and testing phase (EDDTP) A detailed version of the Phase III of PDP explicitly incorporating engineering design, analysis, prototyping, laboratory testing, and field testing in design-iteration loops and should include PSEr participation; the portion of the PDP where the majority of Design for Product Safety (DfPS) should take place

Engineering ethics The subset of ethics used by engineers

Engineering judgment The reasoning sometimes needed by educated and experienced engineers to solve problems or to assess the sufficiency of a solution to a problem; it must often be employed when information and data are unavailable or when subjective criteria must be met; akin to an engineer's "common sense"

Engineering synthesis The process of assembling a product design in order to meet a specified function subject to constraints; in contrast to "Engineering analysis"

Essential health and safety requirements The necessary health- and safety-related criteria that products must meet or exceed in an EC directive.

Ethics The branch of philosophy addressing correct and incorrect conduct

European Commission The civil-service branch of the EU government

European Economic Area Norway, Iceland, and Liechtenstein

European Free Trade Association The combination of the EU, EEA, and Switzerland

European Union The economic and political alliance of twenty-seven (27) European-area countries (as of the time of writing)

Executive A person within an organization responsible for its decisions including the determination and exercise of priorities and values

Facilitators Countermeasures, beyond the physical characteristics and design elements of the engineered product, that further enhance product safety when these countermeasures are known, understood, and followed by the user; include product/user interface, warning signs, manuals, hangtags, instructions, and other materials

Factor of safety An expression for the actual strength of an item with respect to its minimum, or necessary, strength to perform its intended function

Fail-safe (design) A design characteristic in which a failure *mode* of a product manifests itself in a benign manner when it occurs

Failure mode A manner in which a product may fail; mechanical examples include tension, compression, torsion, bending, rupture, seizure, and puncture

Failure modes and effects analysis An inductive hazard-analysis technique evaluating the effects of potential failure modes in systems and products

Failure modes, effects, and criticality analysis An FMEA which includes a criticality analysis

Fault/fault state An abnormal condition or state of a product, system, or subsystem, but not necessarily a failure

Fault-tree analysis A deductive hazard-analysis technique to identify the root causes of undesirable events

Final risk index The measurement of risk after an engineering-design change or set of engineering-design changes is made on a product

Finite-element analysis A structural analysis performed, usually through computer software, to analyze stresses in product components

Functional safety The layer of product or system safety provided by the proper functioning of a set of controls or a control system

Ground-fault circuit interrupter A fast-acting electrical circuit breaker to disconnect the supply of alternating-current (AC) electricity to an outlet

Groupthink "[A] rationalized conformity—an open, articulate philosophy which holds that group values are not only expedient but right and good as well" (W. H. Whyte 1952); a pattern of beliefs and resulting behavior into which groups making decisions *may* fall

Guard A device or barrier separating in space the hazard and the user; also see "Safeguard"

Hazard A real or potential condition that could lead to an unplanned event or series of events (i.e., mishap) resulting in death, injury, occupational illness, damage to or loss of equipment or property, or damage to the environment (MIL-STD-882E)

Hazard identification The detection of the hazards presented by a product to its users

Health The addressing hazards that can injure or harm over time; in contrast to "Safety"

Heuristics Cognitive shortcuts or rules of thumb that simplify decisions, especially under conditions of uncertainty ("Heuristic" n.d.)

Incident A "close-call" or a mishap where no injury or death resulted; a minor accident; or a mishap whose details or results are not fully known

Inherently safe design The use of protective measures which either eliminate hazards or reduce the risks associated with hazards by changing the design/operating characteristics without adding guards or safeguards

Initial risk index The measurement of risk prior to an engineering-design change or set of engineering-design changes is made on a product

Integrated product development An engineering-design process in which product design and development are considered as a whole and operated as a continuum—including the product-safety engineering function

Intended use The service or application for which a product was designed

Internet of Things The trend in consumer products permitting users to remotely control internet-connected products

Just riding along An accident situation where the operator of a vehicle describes the accident using the following phrase: "I was *just riding along*, minding my own business, when all of a sudden and without warning, [fill in accident description]"; a variant of a "just driving along" accident

Leader A person in any position within an organization who makes, encourages, and supports ethical decisions and actions—sometimes against the prevailing organizational culture

Legacy product A product of a type that a company has been producing for an extended period of time

Legality A property sometimes confused with safety

Lockout/tagout The practice and procedure required by OSHA to control hazardous energy during machinery service and maintenance as prescribed in 29 CFR 1910.147

Low-speed vehicle An on-highway vehicle with a top speed of not more than 25 mi/h (40 km/h) falling under FMVSS 500

Manager A person within an organization responsible for a group's day-to-day operations; not to be confused with a *leader*

Manifest danger A design characteristic in which a failure mode of a product manifests itself to users before it produces a failure

Mishap An event or series of events resulting in unintentional death, injury, occupational illness, damage to or loss of equipment or property, or damage to the environment (MIL-STD-882E)

Morals Values concerning what is right and what is wrong; a topic not covered in this book and only mentioned in reference to the work of other authors

Multi-function display An electronic display screen on a product that conveys information to the user including that related to product status, control, and safety

Notice A warning-sign signal word to indicate important—but not hazard-related—information for property-damage only (PDO) mishaps

Occupant kinematics The study of the motion of motor-vehicle occupants—drivers and passengers—within that vehicle during an accident and how those occupants interact with vehicle-interior surfaces

Occupant-retention system The system of components including seat belts, doors, nets, hand holds, seats, head rests, and other design features that assist an occupant in staying within the compartment along with the occupant's limbs; abbreviated as "ORS"

Occupational safety and health The discipline focused on the safety and health of people in the workplace

Off-highway vehicle See "Off-road vehicle"

Off ramp A tactic for delaying a safety-recall decision by asking numerous unnecessary questions which often have meaningless answers

Off-road vehicle A vehicle not intended or designed for on-road/on-highway use; includes dune buggies, all-terrain vehicles (ATVs) and vehicles known as utility-task vehicles (UTVs), recreational off-highway vehicles (ROVs), multi-purpose off-highway utility vehicles (MOHUVs), and side-by-side (SSV or SxS) vehicles; an administrative definition that often overlaps with OHV; some jurisdictions include road-legal four-wheel-drive vehicles in this category

Original equipment manufacturer The design and manufacturing company responsible for the final product delivered to consumers

Overwarning The phenomenon of providing a product user with more information than the user can reasonably read, comprehend, remember, and act upon

Personal protective equipment Equipment worn by a person to protect that person from safety hazards such as impact, debris, abrasion, heat, and others; this includes helmets, safety glasses, gloves, and other items; abbreviated as "PPE"

Point-of-sale materials Those materials available to consumers at, or before, the time and location of product sale

Probability An expression of the likelihood of occurrence of a mishap (MIL-STD-882E)

Product A consumer product; a product/system of lesser size and lower complexity than an aerospace or weapons system

Product design element A characteristic, feature, or component of an engineering design of a product that permits the performance of its intended function; it is a design element of the product, not an element of product design; therefore, it is a "product design element" rather than a "product-design element"

Product-design process A multi-stage procedure used to engineer new products

Product-designer matrix The model showing product characteristics versus designer characteristics in matrix form when discussing engineering-design guidance needs and availabilities

Product-safety defect A legal conclusion that a post-sale product issue rises to the level of having an unacceptable safety risk to its users

Product-safety engineer A person tasked, either fully or partially, with implementing product-safety engineering concepts within a design and/or manufacturing organization

Product-safety engineering A subset of system safety focused on small- and medium-scale systems—without the mandatory DoD and military systems-engineering requirements and deliverables—that relies upon the application of engineering ethics to help make decisions regarding the safety of product users

Product-safety recall A remediation of a product for defects negatively affecting the safe use of that product

Quality A property often confused with safety

Recall Remediation of a defective product through means such as removal of the product from the marketplace (with consumer compensation) or repair of the product (at no expense to the consumer)

Reliability A property often confused with safety

Repeatable Similar test results can be duplicated by the same group at the same test facility or laboratory

Reproducible Similar test results can be duplicated by a different group at a different test facility or laboratory

Requirements engineering That branch of engineering where a set of system requirements are discovered, documented, and maintained using systematic and repeatable techniques to ensure that requirements are complete, consistent, and relevant (Sommerville and Sawyer 1997)

Reasonably foreseeable misuse The use of a product in a way not intended in the information provided in the instructions for use, but which may result from readily predictable human behavior (based on Directive 2006/42/EC; an *administrative* definition, not a *legal* one)

Residual risk The level of risk remaining with product use after product design in the EDDTP by engineers and after product facilitators are known, understood, and followed by users

Restrictions on Hazardous Substances European Commission (EC) Directive 2011/65/EU

Risk A combination of the severity of the mishap and the probability that the mishap will occur (MIL-STD-882E)

Risk index The measure of current risk of a product as a combination of severity category and probability level; also see "Initial risk index" and "Final risk index"

Risk management A process in which hazards are identified and their risks are analyzed, assessed, estimated, evaluated, and reduced to acceptable levels; consists of risk assessment and risk reduction

Rollover protective structure A structure of a vehicle that is protective to occupant(s) in the event of a rollover accident; sometimes called a roll cage or roll bar; abbreviated as "ROPS"; a ROPS should provide occupant-*protective* characteristics but may not offer complete *protection* to occupants

Safeguard A device, technology, or scheme to protect a user from a product hazard; *guards* are one type of *safeguard*

Safety Freedom from conditions that can cause death, injury, occupational illness, damage to or loss of equipment or property, or damage to the environment (MIL-STD-882E); protection against accidental or inadvertent events (in contrast to "Security"); the addressing hazards that can injure or harm immediately (in contrast to "Health")

Safety-alert symbol A warning-sign symbol to indicate a hazard with the potential for personal injury; appears as ⚠ (or similar)

Safety compliance Those aspects of product compliance related to product-safety

Safety defect A legal determination based on product-safety risk

Safety hierarchy One of several possible rank orderings of design countermeasures to the hazards inherent in a product design

Security Protection against malicious events

Severity The magnitude of potential consequences of a mishap to include: death, injury, occupational illness, damage to or loss of equipment or property, damage to the environment, or monetary loss (MIL-STD-882E)

Single-point failure One independent failure leading to an unwanted event such as an accident

Sloganeering The use of superficial mantras to elicit desired behaviors and results

Standards Development Organization An organization accredited by ANSI to develop an American National Standard

Static-stability factor A calculated simplified metric of the lateral stability of a vehicle under static conditions; formulated as T/(2H) where T is the average vehicle track width and H is the center-of-gravity height above the ground

Supplemental-restraint system Automotive airbag

Survivorship bias A cognitive bias, or logical error, made when undue influence is given to a population passing a selection process while ignoring those that did not

Supplier The party responsible for providing components or systems to an OEM

Synthesis Combining discrete components or characteristics to create a purpose-driven, functional system; the opposite of analysis (although analysis is necessary during synthesis)

System An integrated group of components joined together to perform a specified function or set of functions; includes a "product" since a product is an integration of components

System(s) safety The application of engineering and management principles, criteria, and techniques to achieve acceptable risk within the constraints of operational effectiveness and suitability, time, and cost throughout all phases of the system life-cycle (MIL-STD-882E)

Task loading A metric of the difficulty one experiences while attempting to perform a given task

Technical file The required collection of documentation, including engineering and test information, maintained by a manufacturer for CE-marking purposes

Thermal cutoff A electrical component which interrupts the flow of current when it is heated to a specific temperature

Tier A level of the supply chain used by OEMs for purchasing systems, subsystems, components, and materials for their final products

Top-level event The primary event in an FTA that is undesirable and to be avoided

Trike A three-wheeled on-highway vehicle

Unacceptable risk A level of risk which is considered too high to permit product use as is

Unintended consequences Unanticipated side effects arising from solving a given problem without full understanding of its causation

User-product interface The manner in which a user controls interacts with a product and, through which, that product signals to or communicates with the user

User qualification Criteria which establish whether or not a potential product user is qualified to use that product; examples include requirements for minimum age and valid driver's licensure

Validation The process of showing that a product is suitable for its intended use(s)

Vehicle Identification Number A seventeen (17) character code assigned by a vehicle manufacturer to uniquely identifying a specific vehicle

Verification The process of showing that a product meets its design or performance specifications or other requirements

Visibility The ability to see; in contrast to "Conspicuity"

Vulnerable road user A road user who is not well protected from road hazards and includes pedestrians, bicyclists, and motorcyclists

Warning A warning-sign signal word to indicate a hazardous situation which—if not avoided—could result in death or serious injury

Zero-energy state The state of a product or system where there exists no energy that can be released to harm to a nearby person

References

"Cognitive Bias." n.d. Behavioraleconomics.Com. Accessed June 28, 2019. https://www.behavioraleconomics.com/resources/mini-encyclopedia-of-be/cognitive-bias/.

"Heuristic." n.d. Behavioraleconomics.Com. Accessed June 28, 2019. https://www.behavioraleconomics.com/resources/mini-encyclopedia-of-be/heuristic/.

Sommerville, Ian, and Pete Sawyer. 1997. *Requirements Engineering: A Good Practical Guide*. New York, NY: John Wiley & Sons.

USA/DoD. 2012. "MIL-STD-882E—System Safety." Washington, DC.

Whyte, William H., Jr. 1952. "Groupthink." *Fortune Mazazine*, March 1952.

Index

Note: Page numbers followed by *f*, *t*, and *n* indicate figures, tables, and footnotes, respectively.

B

C

E

F

N

S

T

U

Z